THE INDUSTRIAL LASER HANDBOOK

David Belforte Morris Levitt
Editors

THE INDUSTRIAL LASER HANDBOOK

1992–1993 EDITION

With 106 illustrations

Springer-Verlag
New York Berlin Heidelberg London Paris
Tokyo Hong Kong Barcelona Budapest

David Belforte
Belforte Associates
Sturbridge, MA 01566
USA

Morris Levitt
PennWell Publishing
Westford, MA 01886
USA

ISSN 0941-4185

Printed on acid-free paper.

Softcover reprint of the hardcover 1st edition 1992

Production managed by Howard Ratner; manufacturing supervised by Robert Paella.
Photocomposed copy prepared from the editors' XyWrite file using Ventura Publisher; camera-ready sections provided by the editors.

9 8 7 6 5 4 3 2 1

ISBN-13:978-1-4612-7707-1 e-ISBN-13:978-1-4612-2882-0
DOI: 10.1007/978-1-4612-2882-0

Preface & acknowledgments

In 1990, the editors of the *Industrial Laser Annual Handbook* celebrated 25 years of growth in this segment of laser technology. After completing the fifth edition of the handbook, the editors decided that the industry and technology had matured to the point where biennial publication would be more appropriate.

Recognizing the spread of industrial technology beyond the major producing nations, the editors also decided to associate with a publisher with international standing in the fields of engineering and science. This edition of the renamed *Industrial Laser Handbook*, then, is the first in a continuing series of such handbooks to be published by Springer-Verlag on an every-other-year basis.

Readers of the first five editions of the handbook will note that this new edition retains the same structure. Changes, while many, are mainly in production of the book and improvements in presentation.

The economic review and technical forecast has been expanded to cover 1991 and the period 1992–1993. Because of the length of this period, the editors have reduced the detailed analysis of the preceding year because this information is better placed in other, more frequently published, media.

Section 1 has been overhauled to present new processing data, much of it in updated formats. It remains our intention that data presented is general enough to cover the most popular applications and materials for the widest readership. Wherever possible we have asked contributors to provide us with practical, rather than laboratory, process rate data.

Section 2 reflects a change in editorial philosophy. After presenting the basics of industrial lasers, systems, and applications in the preceding five editions, the editors asked the editorial panel to submit articles for this section focused on the practical applications served by industrial lasers. We are indebted to our outstanding panel of experts, who have presented original articles for the continuing education of our readers.

Section 3, the supplier directory, contains the industry's most comprehensive listing of international companies supplying laser goods and services. Readers of previous editions of the handbook tell us that this section is their most-used reference source.

Section 4, related products, narrows the field of companies that support industrial lasers by identifying only those vendors that concentrate on this sector.

Thus, as with past editions, and with editions to come, the handbook provides a constantly changing review of industrial laser technology. Intended to be a growing set of references, the handbook will continually add new reference information in a timely manner.

The editors enjoy the invaluable assistance of a number of people who are necessary in compiling an international publication. We especially thank our associates at the Advanced Technology Group of PennWell Publishing Co.: Linda Wright, publisher of *Industrial Laser Review*, who served as production liaison; Kathy Kincade, managing editor; Mary Williams, directory coordinator; and the graphic arts staff who provided us with professional support.

We also thank the staff at our new publisher, Springer-Verlag, for their patience and understanding during the transition.

As always, we are indebted to our corporate sponsors in the industrial laser industry, whose support is vital to the preparation and publication of this handbook.

David A. Belforte
Morris Levitt

Contents

1992 International Editorial Board

Technical Advisory Panel

Industry Advisory Panel

Economic review & technology trends

The availability of reliable, reasonably priced laser systems in 1991 generated an industrial laser market of more than $1.3 billion. Product technology developed to meet the requirements of rapidly expanding applications has reassured the supplier industry as it adjusts and copes with changing world markets. Reflection on comments heard in face-to-face contacts and offered in response to our annual survey leaves the editors convinced that the 1990s will be *the* decade for industrial laser processing.

For this edition of the handbook, our annual survey of the world market has been expanded to cover a two-year period, consistent with the new publication schedule. Forecasting one year is a difficult task, and our challenge to suppliers to predict two years' worth of developments produced the expected apologia.

During the last quarter of calendar 1991, most suppliers began to experience a return to normalcy after many months of market lethargy. During the year certain industry segments, such as sheet-metal cutting, managed to show performance at 1990 levels. Most other segments, however, experienced the adverse effects of world financial and political actions.

On balance, the world market for industrial lasers was expected to show a small increase in 1991, thanks to continuing strength in the largest sector—Japan. North America and Europe struggled to stay even with 1990 numbers as these markets felt the combined weight of recession in the manufacturing sector and reallocation of government funding to meet changing defense and political positions.

Tables 1 and 2 represent the views of leading industry suppliers as interpreted by the editors. While the final numbers for 1991 will not be available until after publication of this handbook, one projection is clear: The world market for 1991 did well to reach an estimated 6.5% growth in unit sales over what had been a very good year in 1990.

Essentially flat sales in North America, low growth in Europe, and continued double-digit growth in Japan, coupled with a high growth rate (but low volume) in developing market areas, resulted in positive market results. Suppliers to international markets probably had a reasonable year, while those focusing on regions undergoing financial turmoil quite likely had an off year.

The addresses of all companies and organizations mentioned in this section can be found in the blue-page directory section of this handbook.

Normally optimistic suppliers showed untypical conservatism in projections made to us for unit sales in 1992–93. Many respondents to our survey, including those in Japan, predicted a small improvement over 1991 numbers for 1992, with continuing improvement in 1993.

The underlying message seems to be that the glory days of 10+% annual growth may be a thing of the past. The editors have long held the theory that once unit sales passed the magic 5000-per-year level, annual growth rates of 8%–9% would become common. Our reasoning was that machine tool sales, over time, follow a single-digit growth rate curve and that consequently laser processing, a maturing "machine tool" process, should be expected to do likewise.

Unit sales in dollars for 1991, as shown in Table 2, are expected to be about $325 million, a number readjusted to reflect a previous underestimation for the value of 1–3-kW CO_2 lasers sold in 1991. Likewise, readjusting 1990 figures produces a growth rate for total laser sales of 5.8%.

Table 1. World industrial laser sales—units (estimates)

Type	1991	1992	1993
CO_2	3225	3425	3750
YAG	2100	2250	2450
Excimer	85	100	120
Total	5410	5775	6370

Source: ILH Annual Survey

Table 2. World industrial laser sales—$ million

Type	1991	1992	1993
CO_2	241*	255	275
YAG	75	80	85
Excimer	9	12	16
Total	325	347	376

*Adjusted over 1990 sales to reflect average price underestimate.
Source: ILH Annual Survey

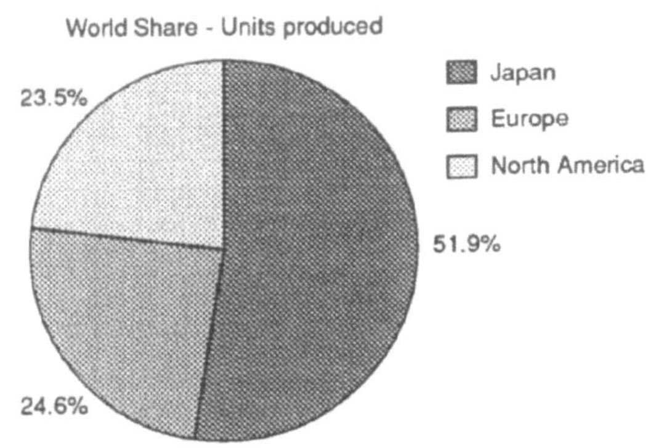

Figure 1. Segmentation of world production of industrial lasers. (Industrial Laser Review)

As has been the case for several years, CO_2 sales dominated in 1991, with 60% of the units sold representing 75% of the revenues.

Figure 1 segments the world by market share. Not surprisingly, Japan continues to be the largest producer of industrial lasers, with 51.3% of the units built and 49.4% of laser sales revenues. Europe follows at 25.2% of units built and 29.4% of laser sales revenues, while North America slipped slightly in 1991 to 23.5% produced and 21.3% of the sales revenues. Suppliers headquartered in other world markets remain too small to estimate market share.

Extrapolating laser sales into systems sales is always a numbers game. Actual sales totals for the systems built by more than 230 suppliers are not available. So, as in past years, we have employed a sliding scale of multipliers for various system categories which, when combined, yield a conservative estimate of the systems business of about $1.3 billion.

This relatively small growth in system sales, about 4%, results from extensive selling-price discounts offered to customers in a slow systems market. Also, a reduction in laser cost has finally translated into a slightly lower system-selling price. Moreover, in 1991 certain new systems such as sheet-metal cutters and marking systems were offered in certain markets at dramatically low unit prices.

The major industrial applications for CO_2 lasers continue to be cutting for 1–3-kW units and marking for units under 100 W. Welding managed some growth in 1991, but the largest increases are yet to come as automotive demands grow over the next five years.

YAG lasers gained some ground in cutting applications as the attractive fiberoptic-beam-delivery option gained popularity in auto-body cutting applications. Drilling, marking, and circuit trimming remain as the largest application segments requiring YAG lasers.

Contrary to predictions made by the editors in the past, the number of systems suppliers continues to expand. In 1991, 234 companies (listed in the blue-page directory) offered industrial lasers and processing systems. This represents a 5% increase over the number reporting in 1990. Laser suppliers have not increased; in fact, this segment of the industry has been most subject to change. This year's handbook lists more than 1000 company names in Section 3, the directory section.

The job shop industry, a major user of industrial lasers, expanded dramatically in 1991. This year, 731 companies responded to our survey, an increase of 26% over 1990. The ILH database contains the names of more than 1300 job shops in North America and Europe. Combined with another 1500 shops in Asia, this market segment represents a significant contribution to annual industrial-laser goods and services revenues.

Using a conservative multiplier of $150,000/machine/shift, we estimate 1991 sales of the job shop segment at $750 million. Details on this industry were extrapolated from data reported to us by the hundreds of companies listed in the handbook directory.

Nd:YAG lasers

At the end of 1991, industrial laser technology was moving rapidly toward a new generation of power sources that are more compact and have better beam-quality characteristics, higher average output power, and, more importantly, reduced $/watt selling prices.

The long-awaited market push for industrial Nd:YAG slab lasers developed in 1991 should occur in 1992. During 1991 the sole supplier, mls Munich Laser Systems, was forced by default to be the only supplier spokesman for this product. At Laser '91 in Munich, mls showed a 1-kW slab that, in effect, is two of their standard 500-W units running in series.

Also at the Munich show, Lasag AG made a modest introduction of a new 150-W slab that was targeted for drilling applications. Festerkörper Laser Institut Berlin GmbH showed a static model of a 1-kW slab design, which officials say should be in the market by the end of 1992.

And, confirming a long-running rumor, Fanuc Ltd. announced in October 1991 that it had obtained marketing rights to the GE slab. Until now, this has been the only such device with several years of industrial experience gained from installations at General Electric's Aircraft Engine facility in Evendale, OH.

Meanwhile, Laser Technology Associates, which had announced its intention to produce a 500-W prototype slab by mid-year, missed that mark but in September delivered 500-W output in a near-diffraction-limited beam from a lab unit.

Hoya Corp. has the only other known commercial slab unit, in both glass and YAG.

High-average-power YAG lasers of the rod type continued to move up the power scale. Industry leaders Martek Laser Inc. and Electrox Ltd. placed kilowatt-level units in the field, Martek with several 1.8-kW CW units and Electrox with a number of 1.4-kW pulsed units. Both companies announced customers for 2.5-kW units, with Electrox identifying theirs as a Spanish contract processing shop, Laser Navarro, and the Martek unit scheduled for the Applied Research Lab at State College, PA.

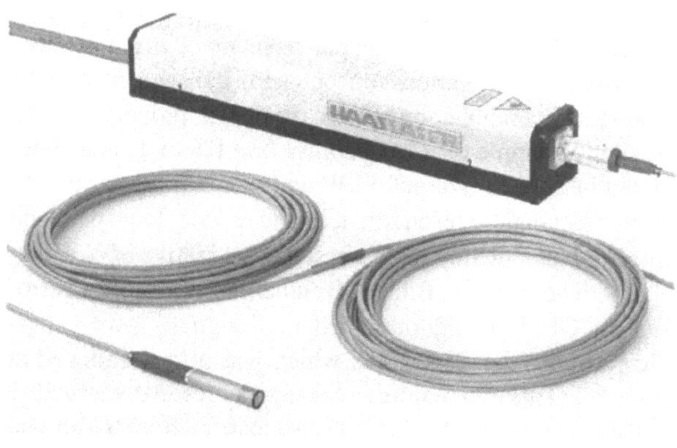

Photo 1. Medium-power Nd:YAG coupled to fiberoptic delivery cable. (Haas Laser)

The leading supplier of kilowatt CW YAG lasers, NEC Corp., did not make any significant product news at these power levels, but did introduce several promising systems, including a robot cutter powered by lower-average-power units.

Haas Laser GmbH, a major European YAG supplier, used Laser '91 in Munich to introduce its 2-kW YAG, which was featured with a fiberoptic delivery unit for automobile body processing. Photo 1 shows a lower power unit coupled to a fiber.

Conventional pulsed YAG lasers continued their return to popularity through the advantages of fiber optic beam delivery. Lumonics Ltd., Haas, Lasag, NEC and Rofin-Sinar GmbH increased shipments of fiber optic beam delivery units ranging in power from 150-1000 W. Some of these suppliers speculated that as much as 80% of the units sold with powers in excess of 150 W were equipped for fiberoptic beam delivery.

Two French companies made moves to expand markets beyond their domestic roots. Cheval Freres SA, a long-time supplier of YAG processing systems, introduced a 900 W average power multi-mode unit for metal cutting applications. Quantel, until last year not well known in the industrial market, made a major move into this segment with the 1991 introduction of industrial YAG units ranging in average power from 300–1200 W.

Industry leader Lumonics Ltd. continued to innovate by introducing an 800 W average power unit, with conservatively rated peak power to 50 kW, specifically focused on the high drilling speed or deep drilled hole market. At the upper end of their power scale the company had installed more than 15 1 kW units in the field by the end of 1991.

More attention is being given to the replacement of flashlamps by diode arrays. Diode pumped YAG lasers are more energy efficient, since the low energy demands of the diodes and the improved coupling efficiency of the diode laser wavelength to the YAG crystal enhance laser operation. Currently, diode-pumped YAG lasers are commercially available up to about 10 W. At this power only applications such as using frequency doubled lasers in circuit adjustment are practical.

Among the leading suppliers of diode pumped YAG devices are Electro Scientific Industries, Amoco Laser, and Adlas GmbH. The number of total units shipped is small, but as output power increases, opening new application areas, sales are expected to increase dramatically.

CO_2 lasers

Just when the industry seemed to have settled down to acceptance of a well established CO_2 laser technology, customer demands for compact, high beam quality units adaptable to system integration, at a reasonable selling price, arose.

Caught on the horns of a dilemma between growing demands for increased delivered power and reductions in unit size and $/watt, suppliers began to shift gears to look for alterations to the current slow flow, fast axial flow and crossflow laser technologies. The direction being taken by a number of suppliers is to diffusion cooled slab type units. Development activity was expedited in 1991 while engineering and design changes of the current products were instituted to satisfy a growing market with OEM customers. For example, Rofin-Sinar obtained the rights to a microwave excited CO_2 design from the University of Alberta, and Coherent General exhibited a 300-W sealed off device at Laser '91 in Munich.

New products that attracted buyers seemed to focus on compactness and lower cost/unit. After a relatively quiet first half year the industry made a major impact at Laser '91 with the introduction of four new resonator designs from one new and three established suppliers.

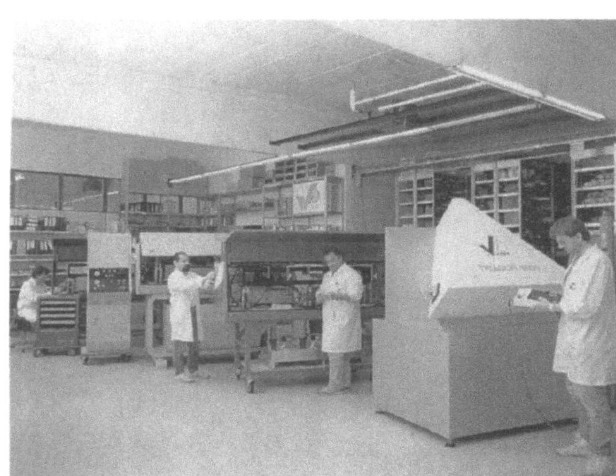

Photo 2. Triangular resonator design compared to standard fast-axial-flow unit. (WB Laser)

Multi-kilowatt units from Wegmann Baasel Laser, Trumpf GmbH, and Bystronic Laser Ltd. and a 1-kW unit from Electrox Ltd. shook traditionalists, who continued to view laser-resonator design in linear terms. The most dramatic departure was made by WB Laser, whose unique Triagon laser (see Photo 2) incorporates a triangle resonator configuration tilted 45° to the horizontal. Valid reasons for such a design include improved resonator stability and reduction in optical components. A new multistage radial compressor from Becker (Wuppertal, FRG) adds to performance improvement and helps reduce unit cost. Projected single-unit selling price for the 5-kW Triagon is $300,000.

For these reasons, and because we see the triangular design as a dramatic industry innovation, we have singled this unit out as product of the year. Performance figures from industrial installations have yet to be reviewed, so the honor goes to WB more for its "tradition be damned" approach, which we expect will set the tone for substantial product and cost improvement designs for the early 1990s.

Another free spirit among laser designers resides at Electrox where, counter to industry practice, engineers stood a resonator on its head in the first vertical orientation of a commercial high-power resonator (see Photo 3). Carrying a very low selling price ($70/W), the Electrox Pegasus, like the Triagon, treads where others fear to travel. Naysayers debunk the approach, preaching concern for contamination of the resonator mirrors. We also have reservations with the new design, but only until sufficient field results are accumulated to support performance and reliability projections. As with WB, we applaud the creative thinking at Electrox that is so refreshing in an industry segment noted for copy-cat designs.

Trumpf GmbH also nourishes a staff of engineers with an innovative bent. The company has an admirable president who does not hesitate to give young talent a chance to succeed. Trumpf's laser people have opted for the more expensive RF-excitation concept and, consequently, have experienced tough competition from DC proponents. An excellent example of RF capability (see Photo 4) is demonstrated in the new Trumpf 5 kW unit that employs a stacked double-rectangular resonator design to reduce floor space and improve beam quality, with a resultant attractive $/W value.

Not to be outdone, Bihler Machinenfabrik GmbH, heretofore an OEM integrator, opted to design its own multi-kilowatt unit, the BSR 3000, which was also introduced at Laser '91. This unit features an octagonal resonator structure that, while not radical, is certainly a departure from traditional design.

Meanwhile, the traditionalists continued to improve their standard products. Industry leaders Rofin-Sinar and Fanuc introduced new versions of the RF-excited multi-kilowatt units for markets in welding and heavier-section cutting. PRC, committed to DC excitation, continued to supply reliable, high beam-quality units and ended the year with the introduction of a series of compact units providing power up to 2 kW, specifically designed for system integration by OEM customers. PRC is a major supplier and leading exporter of OEM lasers. A 1990 joint venture with Oerlikon will improve the European market for this company's products.

Matsushita, selling outside of Japan as Panasonic, quietly gained acceptance by a leading OEM integrator through the introduction of a new series of fast axial flow devices in the 1–2.5-kW range.

Photo 3. Unique vertical resonator orientation to reduce shop-floor space requirements. (Electrox Ltd.)

Photo 4. Double-stacked square-resonator 5-kW CO_2 laser. (Trumpf GmbH)

Photo 5. Sealed-off RF-excited 100-W CO_2 unit. (Synrad)

Coherent General pursued a different path, emphasizing the well-received slow-axial-flow Arrow laser. Field performance data shows that the high beam-quality of this unit produces performance numbers equal to or better than 1.5-kW fast-axial-flow units.

Periodically, an Israeli company introduces a product with a different design concept. In 1991, Optomic Technologies Ltd. brought its ICCL 750 700-W, compact portable unit to a market identified as the moving beam segment.

Moving up the power scale to territory long the fiefdom of Rofin-Sinar and United Technology Industrial Lasers (UTIL), other established suppliers introduced higher power levels. Laser Ecosse brought an 8-kW fast-axial-flow unit to the auto market. Trumpf, already a supplier of 5-kW units, made the aforementioned change to a more integratable unit. MLI had success at 8–10 kW, becoming the only current serious challenger to UTIL. In fact, the two companies went head to head on a camshaft melting application, which had not been resolved when this review was being written.

Mitsubishi Electric Corp., dubbed the quiet company by competitors because a majority of its products are used within this giant corporation, is supplying 5-kW units to automotive manufacturers.

Meanwhile, UTIL received significant corporate support as it shifted its operation to a new 35,000 ft^2 facility dedicated to very-high-power laser manufacturing and marketing. Rumor has it that UTIL is working on a 45-kW unit said to be targeted for the structural steel industry.

In a move that startled and concerned many in the industry, the Welding Institute formed a consortium of UK companies to bring to market the very-high-power (30-kW) unit developed under EUREKA funding. Some questions regarding the proprietary nature of market information shared with TWI that could now be used competitively will need to be answered.

On the other end of the power scale, Synrad moved merrily along as the leading supplier of low-power CO_2 lasers (see Photo 5) for OEM applications in marking and insulation stripping. Melles Griot upped the market limits to 250 W with a compact, sealed-off unit compatible with interesting non-metals processing applications.

Another participant in this market is Carl Baasel, which is also planning to move up the power curve to compete in the kilowatt-level market. Baasel will join Adron Sources as a self-consumer, searching out new OEM markets.

Excimer lasers

Excimer lasers continued to puzzle analysts. This year survey results were not encouraging in terms of market growth, a situation easy to understand as the main market segment, semiconductors, had a second consecutive off year.

Another factor about excimer sales is confusion over the term "industrial." Suppliers consider any excimer sold to a manufacturing company as industrial, even though the laser may be used in prototype process R&D studies. If we were to count these units, our numbers would easily double.

We prefer the definition of industrial as a laser operated in a manufacturing environment, on a shift basis, at least five days/week. Under this definition, many excimers are excluded from our count.

In Germany, excimer laser processing is a significant part of many institute programs. This is likely the precursor to significant industrial sales in the next few years, so we, along with suppliers, remain optimistic that the corner will finally be turned and sales will skyrocket.

Systems

Changes in system technology seem to be more evolutionary than revolutionary. This is partly because the end-use applications served by these systems is also typically evolutionary as new products are adapted from older designs. A case in point is the auto industry, in which body-style changes are barely perceptible over a three-year period.

It is quite likely that most of the system concepts seen today will still be viable two years hence. This is a comforting factor for capital equipment buyers, who are always concerned about purchasing a soon-to-be-obsolete unit.

Three industry events scheduled for the next two years have, in the past, been showcases for dramatic new laser products. Witness the 1989 introduction, at EMO-Hannover, FRG, of the Maho Lasercav machine.

The two 1992 events, Chicago's International Machine Tool Show and Euro Blech in Hannover, are world-class shows that likely will produce at least one innovative system concept. The European Machine Tool exhibit, EMO-Hannover, in 1993 is far enough out to also produce some unique system solution to a common processing problem.

Several system innovations of the past two years seem to be setting standards for products to be introduced in the next two years. First, and perhaps foremost, is a dramatic reduction in system cost, brought about by innovative

Photo 6. Economical two-axis sheet-metal cutter. (Bystronic)

product designs introduced by several of the leading suppliers.

In the field of two-axis sheet-metal cutting: Mazak, Trumpf, Cincinnati, Bystronic, Amada, and Laser Lab all introduced systems capable of cutting thin-gauge sheet metal, with processing at rates consistent with output power in the 750–1200 W range. Typical of these units is the Bystronic Bysmall (see Photo 6). Certain of these units carried a base price of less than $200,000, opening up new markets for laser cutters among the smaller, less-well-financed job shops. Contrary to past efforts to produce low-cost cutters, these systems offer the buyer reasonable capability in a quality product.

The most dramatic system price decreases became evident in the marking industry, where intense competition caused suppliers to quickly adapt to a buyer's market. Led by General Scanning Inc. and Control Laser Corp. in the YAG marking sector and Applied Laser Technology in CO_2 markers, companies that recently introduced economical marking systems, the cost of laser markers has dropped dramatically. Contributing to the cost reduction in CO_2 machines was the availability of low-cost CO_2 lasers from suppliers such as Synrad. Today it is not uncommon to find galvo-type YAG markers selling for less than $50,000 and mask and galvo-type CO_2 markers for under $40,000.

In these two examples, integrators completely redefined the systems as laser-based rather than a conventional system with a laser added. This design philosophy looks to be a major change in system design, and could spread throughout the industry in the next few years.

Examples in other areas are already becoming apparent. At EMO '91 in Paris, Esab-Held, a leading five-axis laser-system supplier, introduced a new five-axis design in conjunction with its French subsidiary, Precacier. This unit, the Mobilas, has the three main axes of motion (x, y, and z) oriented vertically, rather than the conventional horizontal approach of other flying optic systems. In the future, two axes of rotary motion will provide three-dimensional processing capability.

The Esab system is very similar to one introduced earlier in 1991, and also shown at EMO, by Prima Industrie. The Prima Rapido, however, is a cantilevered arm design for x, y, and z that is arranged to place the focusing optics over large worktables.

Both the Held and Prima systems were new designs taking into account the rapid traverse capabilities of moving laser beams. Another five-axis unit, of the more conventional design but again designed specifically for moving beam applications, is the Lumonics/Laserdyne 890 BeamDirector shown at IMTS '90 and now installed in several US prototype shops.

A flat-sheet cutter that gained immediate and substantial customer interest is the Lasercat, manufactured by Trumpf Inc. for sale in North America and Europe. This unit (see Photo 7) employs 2 axes of beam motion over a shuttle table positioned between the gantry supports. Work can be brought to the cutter on a train of shuttle cars, improving sheet flow and cut part removal.

An outstanding example of a system designed specifically for laser processing is Raycon Corp.'s drilling-on-the-fly unit, which synchronizes part rotation and hole location motion with laser pulse repetition rates to enable shallow hole drilling at rates of 40 holes/s.

Many of today's multiaxis and high-speed processing systems could not have been developed were it not for the development of 32-bit CNC controls and, subsequently, dual 32-bit CNC controls that provide the computing power necessary to manipulate motion system information at rates consistent with laser processing. In fact, improvements in software and control features such as graphic representation, simultaneous monitoring of laser and process status, and data collection and storage are available features that enhance laser system capability and flexibility.

A case in point is a blank-welding system developed by Nothfelter and used by Thyssen Steel Technical Center in the US. This system has an acoustic monitoring device that generates real-time weld quality data, which can be fed back to control welding parameters. Further, the control system logs and stores information for up to 10 years to meet customer warranty and liability claims.

One trend for the future in system control is the increasing demand for nonproprietary process software. Already a limited number of programs can be purchased from software retailers, but these are not the powerful control programs needed. It is thought that standard cutting programs, complete with IBM personal computers, will become available, thereby enabling small shops to gain processing capability at low cost. By this action, the market for low-cost laser cutters is expected to grow dramatically.

Not too distant in the future are a host of new control technologies that offer systems integrators new power in system programming. Among these are neural networking, automatic intelligence, and fuzzy logic. In fact, Toshiba Corp. became the first laser system integrator to offer a line of laser welders featuring fuzzy logic control functions.

Photo 7. Gantry-style two-axis laser cutter. (Trumpf)

Applications

Cutting metals and nonmetals was the largest application for industrial lasers in 1991. As seen in Figure 2, cutting applications are 50% higher than the next largest application, marking. Taken on a region-by-region basis, the cutting numbers become even more disproportionate. For example, in Japan cutting applications account for 80% of the CO_2 lasers sold each year, whereas in North America this number is 30%.

Looking ahead to 1992–93, we see no reason for any significant decrease in the number of units installed for metal-cutting applications. We believe that applications classified generally as nonmetal-cutting will likely increase significantly as a result of the availability of lower cost, more powerful sealed-off CO_2 lasers.

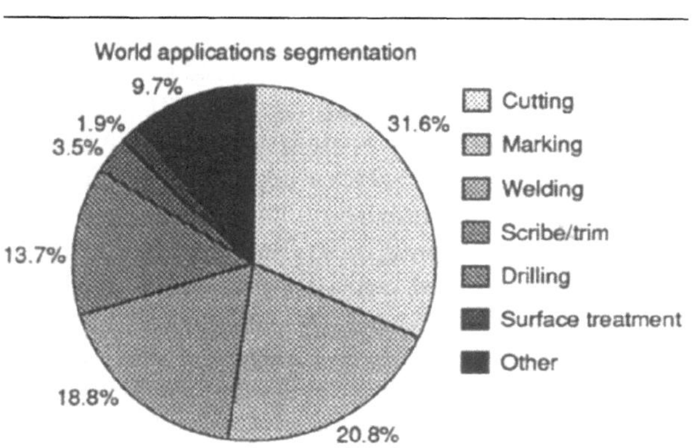

Figure 2. World applications segmentation for industrial lasers. (ILR)

Significant gains are anticipated in YAG laser cutting, although with the small number of existing units, large percentage increases will be deceiving. However, the attractiveness of fiberoptic beam delivery has already caused YAG lasers to be selected over CO_2 for applications such as on-line body-in-white cutting in the auto industry. At least two dozen YAG lasers coupled to robots through fiberoptic systems are expected to be in place in various world auto assembly plants by mid-1992.

And, as more commercial suppliers of slab YAG and high-brightness rod YAG lasers enter the market, we expect that laser cutting of thick metals, such as aerospace alloys, will become a popular application. This market does not have the size of the sheet-metal market, but nevertheless is a market of consequence, since high-precision, low heat-affected-zone cutting is a standard requirement that can be met with YAG lasers.

Applications in sheet-metal cutting, flat and three-dimensional, have increased because of continuing process improvements that now allow operators to produce tight-tolerance, clean-edge cuts with no clinging dross and imperceptible heat affected zones (see Photo 8). The technology of metal cutting has advanced rapidly in the last few years as a result of widespread and extensive applications development by the supplier industry and a host of R&D laboratories.

Assisting in the expansion of laser cutting are a number of industries such as aerospace. Through partnering relationships with suppliers, these industries are working to refine process specifications that until now restricted the amount of allowable recast metal. A number of job shops serving the

Photo 8. High-quality cuts with water-assist cutting. (Amada)

aircraft engine industry have taken a leadership position in working with their customers to define and measure heat affect on aerospace alloys such as titanium.

New process development from companies such as Held and Fanuc are extending metal-cutting applications into the thicker plate range. Already, bevel cuts on 9–12-mm-thick steel have been produced without clinging dross and a narrow heat-affected zone that disappears in subsequent welding operations. Finish cutting of plate could well be the next large application area for lasers in the coming years, especially as the cost per watt of high-power CO_2 and YAG lasers drops.

Welding

Although currently ranked as the third-largest application, laser welding is expected to undergo a dramatic increase in usage during the 90s. A number of new applications in automobile assembly, where the laser replaces conventional resistance spot welding, have shown substantial process cost reduction. For this reason, increases in auto applications are expected to drive laser welding to as much as 25% of the total applications installed annually.

One example of laser-welding advantages, detailed in this handbook's article by Mombo-Caristan of Thyssen, is expected to produce annual sales in excess of $250 million. Today most models of Toyota automobiles have their door rings, of varying metals and thicknesses, laser welded. In the US, Cadillac employs laser-welded blanks to form the center door pillar in a new body style. Developments at Armco Steel (Middletown, OH) and Utilase are also well along the experience curve, pointing toward widespread opportunities in auto, truck, and bus body-welding applications.

In the field of hermetic sealing, long a major user of pulsed YAG lasers, further growth is anticipated as the medical devices industry continues fast-paced growth. Meanwhile, integrated circuit package sealing, currently undergoing development to downsize packages, will rely more and more on precise, low total heat input YAG welding.

Multiplexed laser beams, through fiberoptic delivery for multistation or multicomponent welding, are becoming increasingly popular. In some plants a central laser location, remote from the assembly area, houses lasers served by one technician. Fibers deliver energy to remote workstations on demand, maximizing the welding efficiency of YAG systems.

These and other applications of fiberoptic beam delivery for remote welding applications promise excellent growth opportunities for high-average-power YAG units.

Marking

As a result of better reporting, the editors have moved marking to the second-largest laser application. We have already commented on the effect lower unit prices have had on marking equipment sales. Applications developments have also contributed to the spread of the technology.

High-speed galvos to manipulate YAG and CO_2 beams in scanned marking are now common. Controlled by user-friendly software, these beams are marking products and parts at increasingly higher rates, thereby lowering the cost/part and making laser marking more competitive with other processes such as ink jet.

Most suppliers were forced to review their systems' writing speeds when General Scanning introduced what was claimed to be the world's fastest laser marker in 1990. As a result, writing speed incorrectly became the criterion for assessing a system vendor. Legibility and permanence are functions of writing speed and, as a consequence, are equally important factors in selecting a system vendor.

Deep-engrave mode marking (see Photo 9), precipitated by the process needs of heavy industry such as automotive, became a reality as several system suppliers (such as Lumonics and Control Laser) introduced 150–250-W average-power YAG-laser-based systems. These systems had the capacity to produce deep-engraved marks that can still be read after applications of paint and grease cover the marked component.

Mask-marking applications development (see Photo 10) centered on engineering advances to allow marks to be varied for sequential coding. Techniques such as automotive digit changes proved economically viable. Leading the development of mask changes were A-B Lasers and Lasertechnics. A-B Lasers broke new ground with a device that enables routine periodic changes in bar code marks.

Photo 9. Robot-manipulated, fiberoptic-delivered YAG laser beam for deep section marking. (Lumonics)

Photo 10. Lasertechnics was one of the leaders in developing mask changes for mask-marking applications. (Lasertechnics)

Fiberoptic-beam-delivery techniques for scanned beam marking found applications in industries such as automotive, where code information has to be placed in locations that are difficult to access.

Drilling

High-brightness YAG lasers led the applications news in drilling technology. Improved beam divergence characteristics of these lasers leads to better focusability and, consequently, the higher pulse energy density necessary for deep-hole drilling. Continued commercialization of the slab YAG laser in this area is expected to produce industry interest.

Development work with fiberoptic-delivered drilling beams progresses toward the eventual implementation of the technique in multiaxis, three-dimensional turbine blade drilling. Researchers at Coherent General, working with improved robot manipulation of fibers, have greatly improved hole-to-hole accuracy.

Drilling large quantities of small, shallow holes at rates synchronized to workpiece motion has been shown to be feasible. After demonstrating drilling on-the-fly at IMTS '90 (see Photo 11), Raycon Corp. backed away from marketing the system, citing market conditions and the need for increased product and process development.

End users such as those companies drilling thousands of holes in turbine engine components value high-speed drilling at rates of 40 holes/s. As this industry shifts to hotter engine design, cooling hole requirements in components of the engine not being drilled today will increase the need for high-speed, accurate, and repeatable laser drilling.

Surface treatment

When the editors last looked at laser surface treatment, we were disappointed to see that these high-technology applications of laser processing continued to meet user resistance. The metallurgical advantages of laser treatments were usually offset by equipment cost and system complexity in industries where labor skills are generally low.

The 1992 perspective is, however, changing rapidly as several breakthrough applications in 1991 seem poised for growth into the mid-90s. The first of these, and potentially the biggest in terms of market segment interest, is laser cladding of automobile exhaust valves.

First publicized by Avco Everett Metalworking Lasers and Fiat in the mid-1970s, laser cladding of exhaust valves languished as a result of potential customer indifference to the technical and cost benefits associated with the production of

Photo 11. System for drilling on-the-fly at 40 holes/s. (Raycon)

metallurgically bonded, high-density, low-dilution coatings produced from laser melting of powdered alloys.

Researchers in Europe, and now in the US, have finally managed to create user awareness of this process, and several projects are under way to replace conventional plasma arc cladding with the more cost-effective laser process. Systems are expected to be installed and operating in 1992. At EMO '91, Sulzer Brothers, in conjunction with Amysa-Yverdon SA, demonstrated its newly developed laser-induction technology. This process combines the preheating effectiveness of an RF induction coil with the phase control of laser energy to effect alloy melting.

Also exhibited at EMO was the surface melting of cast-iron automobile camshaft lobes using the rapid solidification cooling characteristics of laser-produced heat. Citroen Industrie demonstrated this process with a 5-kW CO_2 laser on a system of the company's own design. Meanwhile, Salzgitter Oberflachen Technik GmbH was preparing to build a camshaft-melting system under contract from Volkswagen for 1992 delivery.

As in the case of valve cladding, the developers at Avco were again way out in front in this technology by producing the first ledaburite layers on camshafts for Volkswagen in the late 1970s.

Like a phoenix rising from the ashes of user indifference, laser shock hardening returned to the scene as a viable technique for shallow case hardening. Wagner Loss Technologies, using a pulsed YAG laser developed at Battelle Institute (Columbus, OH), proposed to take this precision hardening process to market in 1992. This leads the editors to wonder about how many other "old" laser applications once judged negatively will, because of more enlightened acceptance of laser processing in general, prove cost-effective in the 90s. Several examples of such processes are laser surface alloying, laser plate bending, and in-line paper slitting.

Miscellaneous applications

An application that caused considerable comment in 1991 was laser paint stripping of aircraft fuselages. In this process, a high-power pulsed CO_2 laser ablates away layers of paint to expose untouched, bare-metal surfaces for repainting.

The US Navy has contracted with International Technology Associates and UTIL to build two large, robotically manipulated paint strippers. This is seen as the tip of the iceberg, with commercial aircraft and other potential paint-stripping applications (such as offshore drill rigs) looming as big markets for this application.

Future trends

If the editors are correct that the 1990s will be the decade for industrial lasers, then certain trends should be becoming evident at this juncture to point the way toward a bright future. Indeed, a review of contributions from our annual survey presents a few directions this vital technology may take.

- In the area of marketing, it is suggested that industry awareness and acceptance of industrial laser technology will remain a major marketing factor. In the US, an initial step already has occurred: The formation of a supplier industry organization, the Laser Systems Product Group, in conjunction with the Assocation for Manufacturing Technology (NMTBA). A steering committee is directing this new group in activities to expand market interest. In the UK, a joint government/industry activity called "Make It With Lasers" is carrying the message of the cost benefits of laser processing to industry in that country.
- Training and technical support for laser purchasers is expected to become increasingly important for successful implementation of the more sophisticated laser-processing systems, such as five-axis cutters.
- Traditional supplier lines will become more blurred as consolidation, buyouts, and joint ventures produce a growing group of importers into domestic markets. A case in point is the expected expansion of Japanese cutting-system suppliers into Europe via such methods.
- Laser technology concerning better beam quality, reduction in physical size, improvement in conversion efficiency, and selling prices will accelerate in the next two years. Among the more interesting advances are: microwave-excited CO_2 lasers; near-diffraction-limited output in 500-W YAG lasers; slab YAG lasers at 1 kW average power and high efficiency (greater than 5%); higher average power (less than 50 W), diode-pumped solid-state lasers; and commercial 50-kW CO_2 units.
- Further out are commercial versions of the carbon monoxide, chemical oxygen iodine laser, and higher power copper-vapor units.
- In the applications field, users can expect high-speed laser cutting to replace die stamping in auto applications, widespread welding of coated metal for auto bodies, and increasing use of tailored blanks in body fabrication.
- Thick-plate cutting and bevel-edge preparation with multi-kilowatt CO_2 lasers will become common. Cutting thicknesses up to 20 mm will be achievable with usable edges. This will effectively reduce the market for plasma arc cutters.
- Five-axis heat treating systems will extend this laser process to more complex shapes currently not able to be processed by induction techniques.
- Finally, expect continued decreases in selling prices as the supplier industry consolidates and becomes more competitive and responsive to new market sectors.

SECTION 1

Laser material processing data & guidelines

Readers who use this book as a reference source for processing data should be aware that the information presented was, in most cases, developed from laboratory experiments. This means the process performed was done under controlled conditions wherein the laser output, beam delivery, beam/material interaction, and ambient conditions were all ideal for laser processing. These conditions produce optimum processing results that may be superior to those experienced on the shop floor. Therefore, data presented should be viewed as relative and approximate. Many readers find the data serves well as a guideline for potential results.

General processing

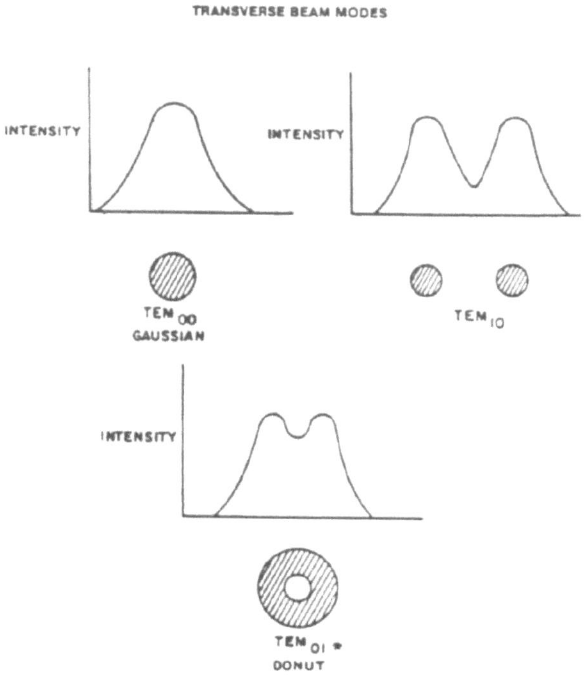

Figure 1. Energy intensity profiles for various CO_2 laser output modes. (Source: *Industrial Laser Annual Handbook*, 1986)

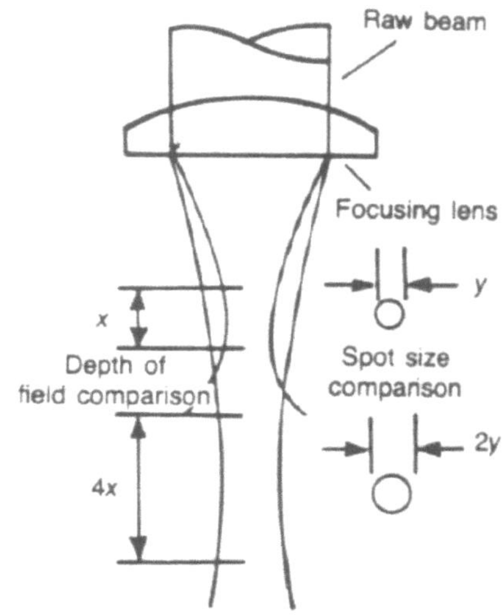

Figure 2. The choice of an appropriate lens entails a compromise of effective, usable power. While shorter focal lengths provide smaller spot diameters with a higher energy concentration, they have a much more restrictive depth of field. (Source: Spectra-Physics Industrial Laser Division)

Table 1. CO_2 laser processing. (Source: *Industrial Laser Annual Handbook*, 1990)

Material \ Process	Weld	Cut	Drill	Heat Treat	Comments
Acrylic	–	E	G-E	–	
Aluminum	F-G	F-G	P-F	P	Reflectivity & thermal conductivity restrain processes, remove oxides
1000 Series	F-G	F-G	F	–	Pure metal welds well. Thick section difficult to cut
2000 Series	P-G	G	–	–	2219 no cracks, 2024 requires 4047 filler to avoid cracking
3000 Series	F	G	–	–	
4000 Series	G	G	E	–	
5000 Series	F-G	G-E	G-E	–	5456 & 5086 good weldability, 5052 fair weld & cut
6000 Series	P-F	G-E	P	–	6061 requires filler metal, some magnesium boil-off
Beryllium Copper	P-F	–	–	–	High % alloys weld best
Brass	P	F-G	–	–	Cu-Zn brass outgasses preventing good welds
Carbon Composite	–	P-F	P	–	
Carbon Steel	E	G-E	F-G	F-G	Heat treat success depends on carbon content
Low	E	E	–	F	Carbon content too low for good heat treat

Table 1. (continued)

Material \ Process	Weld	Cut	Drill	Heat Treat	Comments
Medium	G-E	E	F	E	
High	F	G	–	E	Brittle welds at higher carbon content > 0.04%
Cast Iron (grey)	–	–	–	G	Requires absorbent coating
Cast Iron (nod.)	–	–	–	G	Requires absorbent coating
Ceramic, Al_2O_3	P-F	G-E	G-E	–	
Ceramic, SiC	P	P-F	F	–	
Ceramic, Si_3N_4	P	P	F	–	
Coated Steels	–	G	–	–	Coating degradation at cut edge possible
Copper	F-G	F-G	–	–	Thin-section welding < 0.120″ good with O_2 assist
Delrin	–	E	E	–	
Fabric, nat.	–	E	–	–	
Fabric, syn.	–	E	–	–	
Felt	–	E	–	–	
Fiberglass	–	P-F	P-F	–	Edge quality problems
Formica	–	E	–	–	
Free Machining	P	–	–	–	Presence of sulfur can cause porosity
Galvanized Steel	F-G	G	–	–	Zinc depletion during weld process. Controlled gap successful
Glass	–	G-E	G-E	–	Depends on coefficient of thermal expansion. Pyrex E quartz P
Hastalloy X	P-F	G	–	–	Requires high pulse rate to prevent hot short cracking
HSLA	G	G	–	–	
Inconel	E	G	P-F	–	
Kevlar	–	E	E	–	Slight char
Kovar	F-G	E	–	–	Electroless nickel plating causes cracking
Laminates	–	F-G	F-G	–	Depends on epoxies used
Leather	–	E	E	–	
Lexan	–	E	–	–	
Molybdenum	P-F	F	–	–	Brittle welds, acceptable if high strength not required
Monel	G	–	–	–	Ductile welds
Nickel	G	–	G	–	Ductile welds
Nylon	–	E	E	–	
Paper	–	E	E	–	Control of byproducts necessary in high-speed processing
Plastic	–	G-E	G-E	–	
Plexiglass	–	E	E	–	
Plywood	–	E	G-E	–	Expect some edge char
Polycarbonate	–	F-G	F-G	–	Discolor on cut edge
Polyethylene	F	E	G	–	
Polypropylene	–	E	E	–	

Table 1. (continued)

Material \ Process	Weld	Cut	Drill	Heat Treat	Comments
Rene 41	–	G	–	–	
Rubber	–	F-E	F-E	–	Best results in sulfur or green state
Silicon Carbide	–	F	–	–	Subject to cracks
Silicon Nitride	–	F	–	–	Subject to cracks
Spring Steel	–	E	–	–	
Stainless Steel	G-E	G-E	F-G	N/A	With appropriate consideration to metallurgical restraints
304	G-E	G-E	GE	N/A	Slag present after cutting, 303 cracks after welds
316	G-E	G-E	–	N/A	
410	G	G	–	N/A	High carbon alloys require pre- and post-heat treat
Steel, alloy 4140	–	–	–	E	Requires absorbent coating
Steel, alloy 4340	–	–	–	E	Requires absorbent coating
Steel, tool	–	E	–	–	
Stellite	G	G	–	–	
Tantalum	F-G	F	F	–	Special precautions against oxidation
Titanium	G-E	G	E	E	Ductile welds, special precautions against oxidation
Tungsten	P-F	–	F	–	Brittle welds, requires high energy
Waspalloy	–	E	–	–	
Wood	–	G-E	G-E	–	Cut edges show char
Zircalloy	G	F-G	–	–	
Zirconium	F-G	–	–	–	Ductile welds, special precautions against oxidation

P = Poor, F = Fair, G = Good, E = Excellent

Table 2. Solid-state laser processing. (Source: *Industrial Laser Annual Handbook*, 1990)

Material \ Process	Weld	Cut	Drill	Comments
Aluminum	F-G	G-E	G-E	Depending on thickness holes 0.05″ deep
1000 Series	E	G	G	Depending on thickness
2000 Series	G-E	G-E	G	2024 requires filler (4047)
4000 Series	E	F-G	F-G	Thickness-dependent, welding requires filler
5000 Series	G-E	G-E	G	5052 & 5653 require filler metal
6000 Series	G-E	E	E	6061 requires filler (4047) or (718)
7000 Series	F-G	G	–	7075 requires filler (4047)
Berilyum Copper	G	–	–	High alloy percentages weld better
Brass	P-F	F-G	G	Zinc depletion, holes 0.04″ deep
Carbon Steel	F-G	F-G	G	
Low	G	–	G	
Medium	G	F-G	G	
High	P-F	G-E	G-E	Short hot cracking, brittle welds
Ceramic	–	G-E	E	Better suited for CO_2 lasers
Copper	G-E	G-E	G-E	High reflectivity causes uneven results in welding, holes to 0.04″
Diamond	–	F-G	E	
Diamond (Poly. cyst.)	–	G	–	
Free Machining	P-F	–	–	
Gallium Arsenide	–	E	E	0.025″ thickness
Gold	E	G-E	E	Cut pure & yellow gold
Graphite	–	–	G	Holes 0.112″ deep
Hastalloy X	G-E	G-E	G-E	Requires high pulse rate to prevent cracking
Inconel	F-E	G-E	G-E	Depending on thickness
Iron/nickel alloy	E	–	–	
Kovar	G-E	G-E	–	Phosphorous in elect. nickel plate causes cracks
Molybdenum	–	G	G	0.064″ at 4.8 in/min. Welds may be brittle
Monel	G	–	–	Ductile welds
Nickel	G	–	G	Holes 0.08″ deep. Good ductile welds
Nickel Silver	G	G	G	Cuts 0.06″–0.08″ thick, holes 0.07″ deep
Nimonic	–	–	G	
Plastic	–	P-F	P-F	Drilling poor on thermoplastics, cutting poor on most
Pyrographite	–	G	–	
Sapphire	–	G	G-E	0.02″ thickness
Silicon	–	F-G	G-E	Success depends on thickness

Table 2. (continued)

Material \ Process	Weld	Cut	Drill	Comments
Silicon Nitride	–	G	G	
Silver	E	–	G	Holes 0.06″ deep
Stainless Steel	G-E	G-E	G-E	Cutting excellent to 0.1″, good to 0.2″, fair to 0.8″
304	G-E	G	G	Welding excellent to 0.060″, good ductile welds
316	G	G-E	G	
410	F	–	–	Develops hardened structure
440C	P	–	–	Pre- and post-heat treat recommended
17-4PH	F	–	–	Post-weld heat treat
Tantalum	G	G-E	E	Ductile welds, precaution against oxidation
Titanium	G	G	E	Inert atmosphere required, drill up to 0.125″, cut up to 0.1″
Tungsten	–	E	G	
Waspalloy	F-E	G-E	G-E	Depending on thickness
Zircalloy	–	E	G	
Zirconium	G	P	G	

P = Poor, F = Fair, G = Good, E = Excellent

Laser welding

The energy in a focused spot of laser light can, when absorbed by a metal's surface and converted into heat, cause melting and subsequent rapid cooling of the melt to form a fusion joint. Laser welding, a process in use for more than 20 years, represents about 25% of the applications for solid-state (Nd:YAG) lasers and about 15% of the applications for high-power CO_2 lasers. During the first half of this decade, welding applications are expected to grow at an increasing rate annually. (Source: *Industrial Laser Annual Handbook*, 1990)

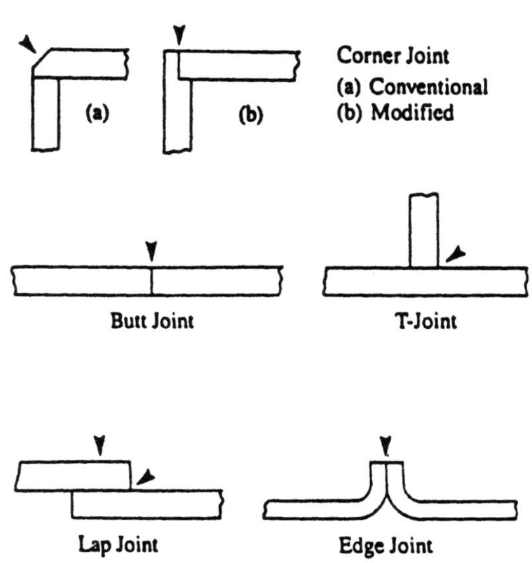

Figure 3. Joint designs for laser welding. Arrows show directions in which the laser beam may be incident. (Source: VanderWert, *Industrial Laser Annual Handbook*, 1986)

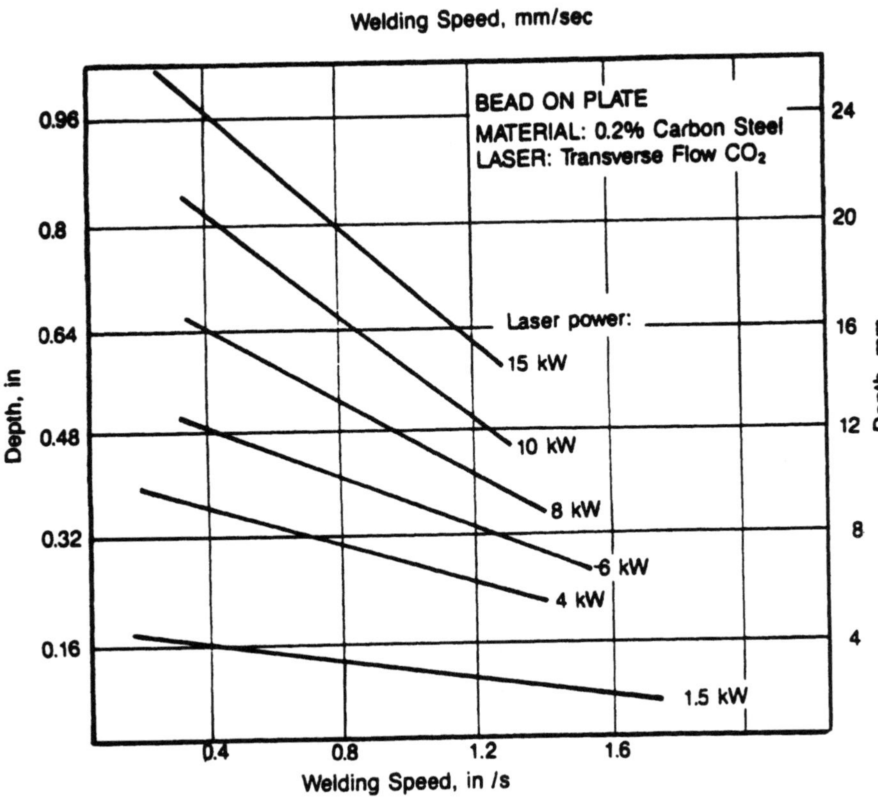

Figure 4. Welding carbon steel with a transverse-flow CO_2 laser. This family of curves of welding depth vs. speed for laser powers in the range of 1.5–15 kW was generated on the basis of information provided by a cross section of industry sources. (Source: *Industrial Laser Annual Handbook*, 1990)

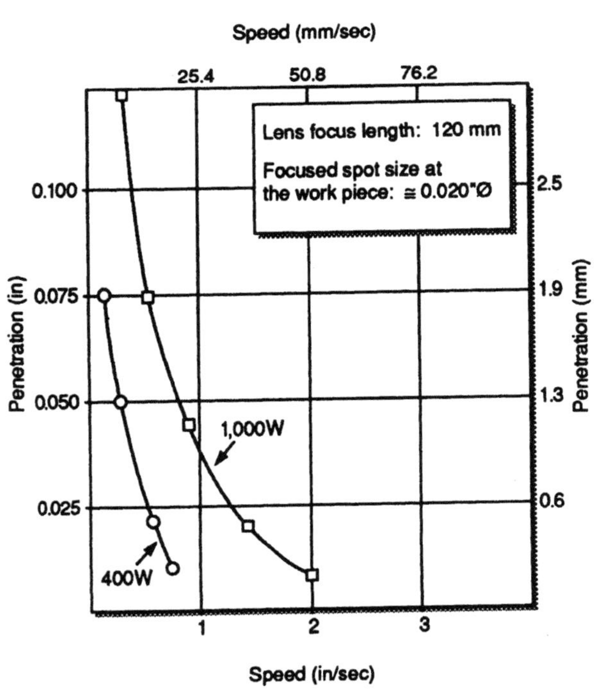

Figure 5. Nd:YAG lasers welding stainless steel (304). (Source: Lumonics Ltd.)

Table 3. Laser welding joint tolerances. (Source: *Industrial Laser Annual Handbook,* 1990)

Joint Type (see Figure 3)	Tolerances Gap, max.	Vertical Mismatch, max.
Butt	0.10 T*	0.25 T
Corner-conventional	0.25 T	
-modified	same as for butt joint	
T-Joint	0.25 T	
Lap Joint	0.25 T	
Edge Joint	same as for butt joint	

* T refers to the thickness of the thinnest section

Table 4. Comparison to competitive processes. (Source: *Industrial Laser Annual Handbook,* 1990)

Characteristics	Laser	Electron beam	Resistance spot	Gas-tungsten arc
Heat generation	Low	Moderate	Moderate	Very High
Weld quality	Excellent	Excellent	Good	Excellent
Weld speed	High	High	Moderate	Moderate
Initial costs	High	High	Low	Low
Operating/ maintenance costs	Moderate	Moderate	Moderate	Low
Tooling costs	Low	High	High	Moderate
Controllability	Very good	Good	Low	Fair
Ease of automation	Excellent	Good	Fair	Fair
Range of dissimilar materials	Wide	Wide	Narrow	Narrow

(Source: adapted from M.M. Schwartz, "Laser Welding," *Metals Joining Manual,* McGraw-Hill, 1979, Chapter 2.)
(Note: This information is very application specific.)

Table 5. Crack sensitivity rating of metals. (Source: *Industrial Laser Annual Handbook,* 1990)

(in increasing order)
Hastelloy B2
Hastelloy S
Hastelloy C4
Hastelloy B to Inconel 600
304 Stainless Steel
Inconel 625
Inconel 718
316 Stainless Steel
310 Stainless Steel
304 Stainless Steel to Inconel 718
Modified Inconel 600
Hastelloy X
330 Stainless Steel

Table 6. Laser spot weld parameters. (Source: *Industrial Laser Annual Handbook,* 1990)

Material	ΔH_{298T_b} J/mm³	ΔH_v J/mm³	I W/cm²	D mm
Al	8.11	29.09	4.5×10^6	$0.41t^{1/2}$
Cu	12.64	42.00	8.2×10^6	$0.25t^{1/2}$
Fe	19.02	48.00	8.6×10^5	$0.09t^{1/2}$

ΔH_{298T_b} is the energy required to heat 1 mm³ of a metal from room temperature (298 K) to its boiling temperature.
ΔH_v is the energy required to vaporize 1 mm³ of metal.
I is calculated laser intensity required to raise the metal surface to the boiling point in 0.1 ms, assuming reflection coefficients of 90% for Al and Cu and 50% for Fe. Al and Cu are difficult to weld because of their high reflection coefficients.
D is the melt pool penetration depth in a conduction weld.
t is the laser pulse duration (in ms). It is assumed that the metal surface has reached its boiling point before time t.

Laser cutting

The focused energy of a laser beam can, when absorbed by a material's surface and converted to heat, cause melting and vaporization of most metals and nonmetals. The process, usually accompanied by some type of gas to provide additional heat (if oxygen) or to eject cut byproducts (if inert) is known as laser cutting.

Laser cutting is the largest material-removal process in number of units installed and one of the most extensively documented cost-effective industrial laser processes.

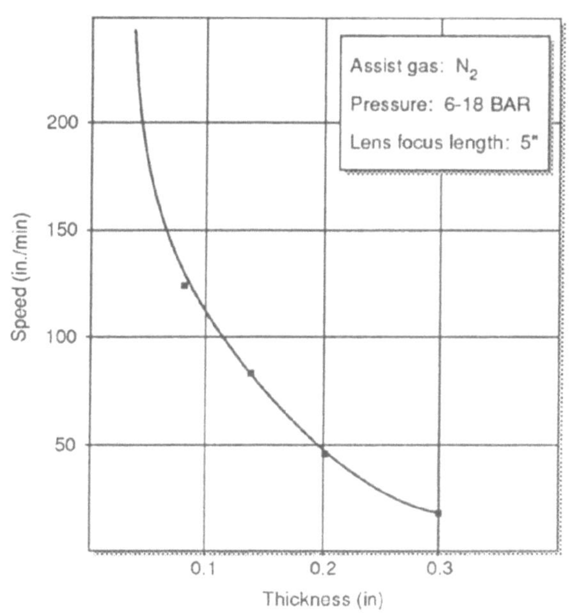

Figure 6. 1500 W CO_2 cutting of stainless steel (304). (Source: Data averaged from industry sources.)

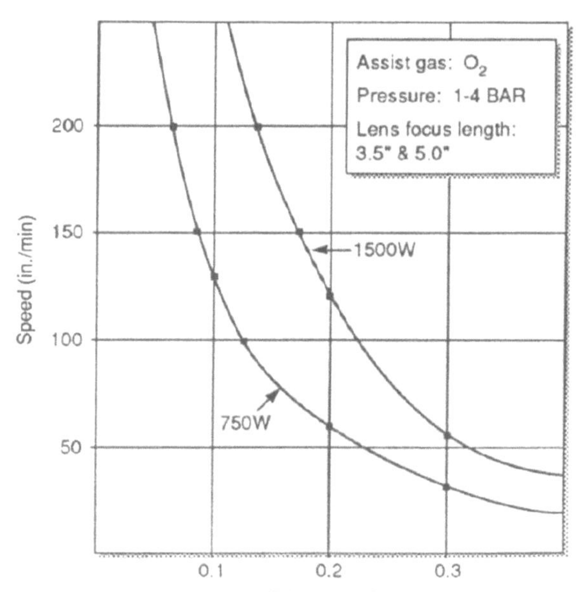

Figure 7. CO_2 cutting of mild steel at 750–1500 W. (Source: Data averaged from industry sources.)

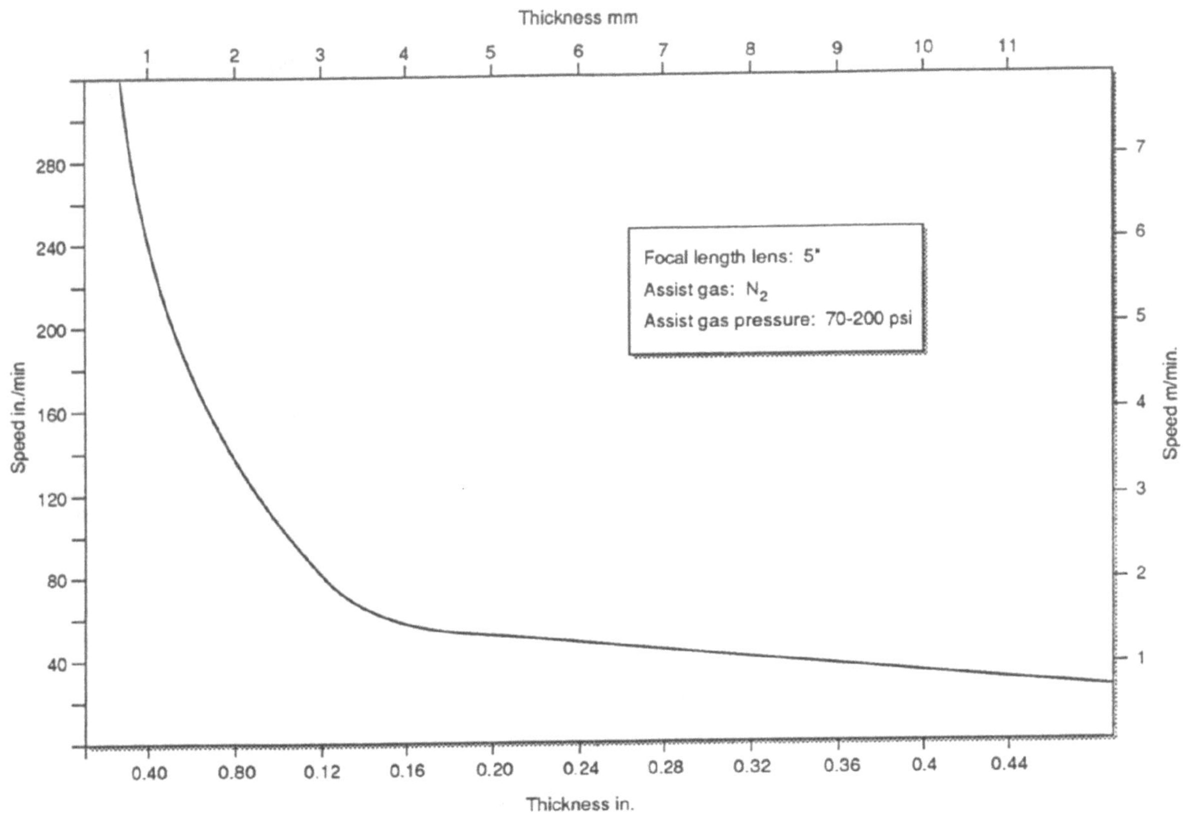

Figure 8. Aluminum cutting with 1500 W. (Source: PRC Laser Corp.)

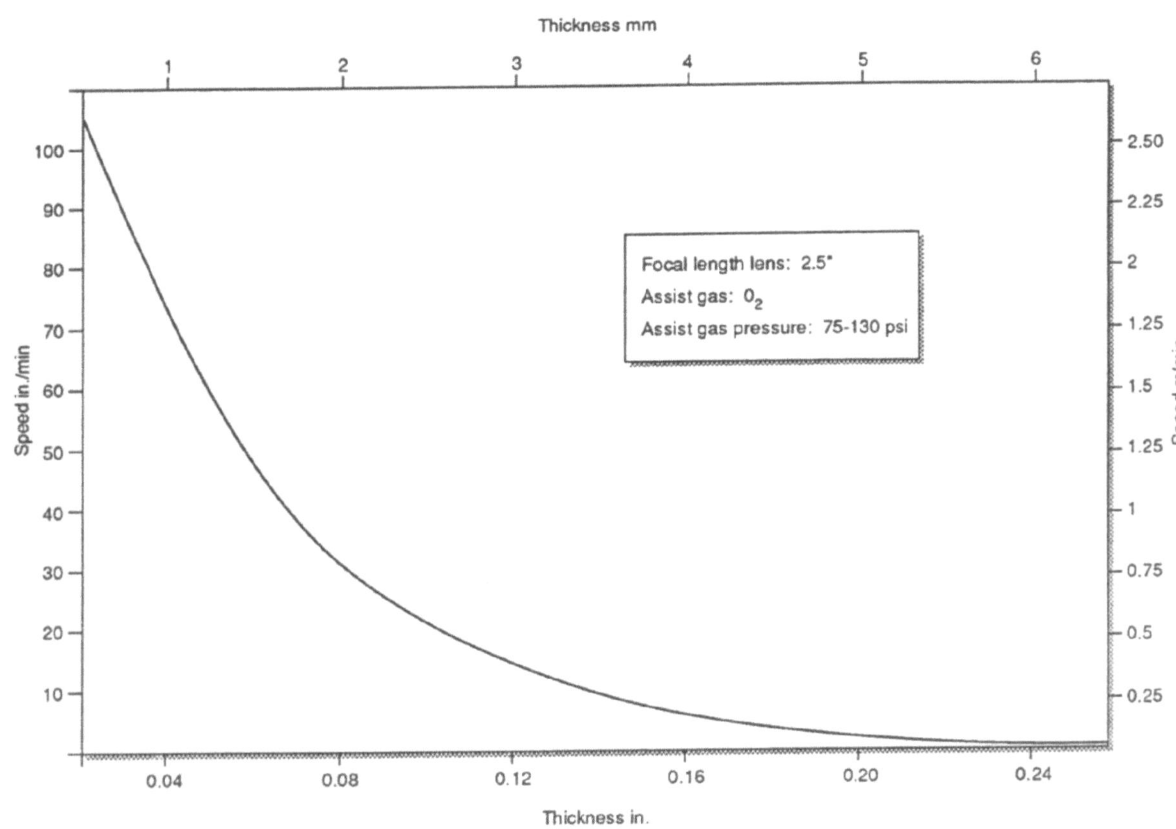

Figure 9. Copper cutting with 1500-W CO_2 laser. (Source: PRC Laser Corp.)

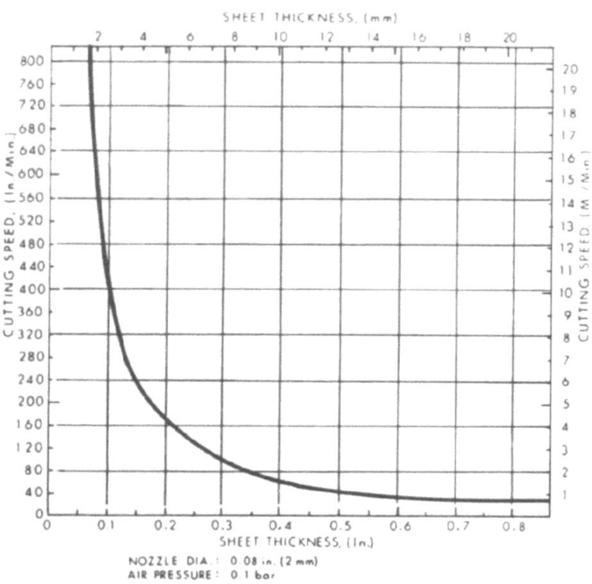

Figure 10. CO_2 cutting speeds for acrylic at 500 W are plotted against sheet thickness. (Source: *Industrial Laser Annual Handbook*, 1990)

Table 7. Pulsed Nd:YAG laser cutting (1 kW average power). (Source: *Industrial Laser Annual Handbook*, 1990)

Material	Thicknesses (*d*) (mm)	Cut Speed (*V*) (mm/s)	*P/d* (W/mm)
Mild Steel (CR4)	0.7	167	1429
Stainless (304)	0.3	233	3333
Stainless (304)	0.9	133	1111
Stainless (304)	5.0	13.3	200
Nimonic (C263)	6.0	9.2	167
Aluminium	6.2	10.8	161
Copper	3.0	10	333
Copper	10.0	0.67	100
Armour plate	12.0	1.33	83
Kevlar	5.5	5	182

Table 8. Typical cutting speeds for titanium alloys using a CO_2 laser. (Source: *Industrial Laser Annual Handbook*, 1990)

	Thickness		Laser Power	Cutting Speed	
	in	mm	W	in/min	m/min
Copper	0.04	1	1500	120	1.5
Copper	0.08	2	1500	40	0.5
Copper	0.04	1	1000	80	1.0
Brass	0.04	1	1500	200	3.0
Brass	0.08	2	1500	80	1.5
Brass	0.04	1	1000	120	2.0

(cutting gas: oxygen)

NB: Cutting speeds (if any) are highly dependent on surface condition and alloy content.

Table 9. Typical cutting speeds for copper alloys using a CO_2 laser. (Source: *Industrial Laser Annual Handbook*, 1990)

Thickness		Laser Power	Cutting Gas	Cutting Speed	
in	mm	W		in/min	m/min
0.06	1.6	500	Argon	60.0	1.5
0.12	3.2	500	Argon	12.0	0.3
0.12	3.2	500	Oxygen	80.0	2.0
0.25	6.4	500	Oxygen	40.0	1.0
0.25	6.4	1000	Oxygen	80.0	2.0
0.10	2.5	1000	Oxygen	240.0	6.0

Table 10. Typical cutting speeds for selected polymers using a 500-W CO_2 laser.
(Source: *Industrial Laser Annual Handbook*, 1990)

Thickness		Polyethylene		Polypropylene		Polystyrene		Nylon		ABS		Polycarbonate		PVC	
in	mm	in/min	m/min	in/min	m/min	in/min	m/min	in/min	m/min	in/min	m/min	in/min	m/min	in/min	m/min
0.04	1	430	11.0	670	17.0	750	19.0	780	20.0	830	21.0	830	21.0	1100	28.0
0.08	2	160	4.0	275	7.0	290	7.4	315	8.0	320	8.2	320	8.2	430	11.0
0.12	3	87	2.2	160	4.0	165	4.2	190	4.8	200	5.0	200	5.0	250	6.4
0.16	4	60	1.5	110	2.8	120	3.0	140	3.5	140	3.6	140	3.6	170	4.3
0.20	5	47	1.2	79	2.0	90	2.3	100	2.6	105	2.7	105	2.7	125	3.2
0.24	6	39	1.0	63	1.6	71	1.8	79	2.0	83	2.1	83	2.1	98	2.5
0.28	7	31	0.8	51	1.3	63	1.6	63	1.6	67	1.7	67	1.7	79	2.0
0.31	8	24	0.6	43	1.1	47	1.2	47	1.2	51	1.3	51	1.3	67	1.7
0.35	9	20	0.5	35	0.9	39	1.0	39	1.0	43	1.1	43	1.1	55	1.4
0.39	10	16	0.4	27	0.7	35	0.9	31	0.8	35	0.9	35	0.9	47	1.2
0.47	12	12	0.3	16	0.4	27	0.7	20	0.5	24	0.6	24	0.6	39	1.0

NB (a) Cutting speeds can be changed dramatically by changes in molecular weight, degree of crystallinity, and porosity.
(b) As a first approximation, cutting speeds and maximum material thickness can be assumed to vary in a linear manner with laser power (between 100 and 1500 W).

Table 11. Rule-of-thumb cutting speeds for thermoplastics with a CO_2 laser at 500 W.
(Source: *Industrial Laser Annual Handbook*, 1990)

Thickness	in	.04	.08	.12	.16	.20	.24	.31	0.4	0.5
	mm	1	2	3	4	5	6	8	10	12
Cutting	in/min	750	300	200	150	100	80	50	30	20
Speed	m/min	20.0	8.0	5.0	3.5	2.5	2.0	1.2	0.8	0.5

Table 12. Cutting speeds for selected thermoset plastics and fiber-reinforced materials with a CO_2 laser.
(Source: *Industrial Laser Annual Handbook*, 1990)

Material	Thickness		Cutting Speed		Laser Power
	in	mm	in/min	m/min	(Watts)
Formica	0.06	1.6	312.0	7.8	400
	0.06	1.6	560.0	14.0	1200
	0.12	3.0	112.0	2.8	400
Phenolic Resin	0.12	3.0	116.0	2.9	400
	0.24	6.0	44.0	1.1	400
Fiberglass	0.06	1.6	208.0	5.2	450
(glass-reinforced	0.06	1.6	600.0	15.0	1200
epoxy resin)	0.13	3.2	96.0	2.4	400
	0.18	4.5	60.0	1.5	400

NB Cutting speeds for materials such as fiberglass depend on the relative proportion of glass, resin, and trapped air in the material.

Table 13. Cutting results for wood and wood-based products using a CO_2 laser. (Source: *Industrial Laser Annual Handbook*, 1990)

Type	Thickness		Cutting Speed		Laser Power
	in	mm	in/min	m/min	W
Poplar	0.4	10	200	5.0	500
Douglas Fir	0.4	10	140	3.5	500
Yellow Pine	0.4	10	128	3.2	500
Walnut	0.4	10	152	3.8	500
Cherry	0.4	10	172	4.3	500
Scotch Pine	0.4	10	132	3.3	500
Beech	0.4	10	160	4.0	500
Teak	0.4	10	140	3.5	500
Mahogany	0.4	10	124	3.1	500
Oak	0.4	10	116	2.9	500
Ash	0.4	10	104	2.6	500
Ebony	0.4	10	48	1.2	500
Chipboard	0.65	17	48	1.2	1000
Chipboard	0.6	15	68	1.7	1000
Medium	0.5	12	120	3.0	1000
Density	0.6	15	88	2.2	1000
Fiberboard	0.7	18	68	1.7	1000
Hardboard	0.16	4	320	8.0	500
Plywood	0.65	17	56	1.4	1000
Plywood	0.5	12	160	4.0	1000
Plywood	0.5	12	56	1.4	500
Plywood	0.35	9	112	2.8	500

NB Generally use high pressure (4–6 bar) air.

Table 14. Cutting data for full-penetration profiling of ceramic materials using a CO_2 laser. (Source: *Industrial Laser Annual Handbook*, 1990)

	Thickness		Cutting Speed		Laser Power
	in	mm	in/min	m/min	W
Glass	0.04	1	60	1.5	500
	0.08	2	40	1.0	500
	0.12	3	20	0.5	500
Alumina	0.04	1.0	56	1.4	500
	0.08	2.0	24	0.6	500
	0.08	2.0	80	2.0	1000
Silica	0.04	1.0	24	0.6	1200
Ceramic Tile	0.25	6.3	24	0.6	1200

Table 15. Cutting speeds for miscellaneous materials using a CO_2 laser. (Source: *Industrial Laser Annual Handbook*, 1990)

Material	Thickness		Cutting Speed		Laser Power
	in	mm	in/min	m/min	W
Asbestos Board	0.25	6.0	60	1.5	400
Asbestos Cement	0.2	5.0	48	1.2	500
Carpet (auto)	0.14	3.5	1800	45.0	1300
Carpet	0.25	6.0	720	18.0	400
Carpet	0.25	6.0	1400	35.0	1200
Cotton	0.6	15.0	40	1.0	500
Felt	0.25	6.0	1600	40.0	1000
Felt	0.25	6.0	760	19.0	400
Lead	0.04	1.0	80	2.0	500
Leather	0.06	1.6	720	18.0	1200
Paper (airmail)	0.002	0.05	40000	1000.0	500
Paper (bond)	0.005	0.12	24000	600.0	500
Paper (cardboard)	0.008	0.2	3600	90.0	500
Paper (cardboard)	0.18	4.6	360	9.0	350
Wool (worsted)	0.03	0.7	2000	50.0	500
Zinc	0.04	1.0	80	2.0	500
Zirconium	0.12	3.0	240	6.0	500

Table 16. Nd:YAG cutting data. (Source: *Industrial Laser Annual Handbook*, 1990)

Material	Thickness		Cutting Speed		Laser Power
	in	mm	in/min	m/min	W*
Stainless Steel 304	0.02	0.5	80.0	2.0	230
	0.04	1.0	36.0	0.9	230
	0.12	3.0	20.0	0.5	230
	0.2	5.0	10.0	0.25	230
	0.4	10.0	4.0	0.1	230
	0.64	16.25	1.0	0.025	230
	1.08	27.4	0.12	0.003	230
Stainless Steel 316	0.12	3.0	5.0	0.12	100
	0.16	4.0	4.0	0.10	100
Mild Steel	0.1	2.5	22.0	0.56	350
	0.2	5.0	5.0	0.13	350
	0.3	10.0	0.4	0.01	350
Aluminum	0.04	1.0	28.0	0.7	230
	0.06	1.5	18.0	0.45	320
	0.1	2.5	11.0	0.27	170
	0.16	4.0	4.0	0.10	135
Tungsten	0.08	2.0	1.2	0.03	180
Titanium	0.04	1.0	40.0	1.0	120
	0.12	3.0	12.0	0.3	120
Silicon	0.016	0.4	16.0	0.4	100
Hastelloy X	0.08	2.0	18.0	0.45	120
	0.16	4.0	4.0	0.10	120
	0.24	6.0	2.0	0.05	120
Gold	0.12	3.0	0.6	0.015	190
Si-Nitride	0.1	2.5	2.8	0.07	300
	0.2	5.0	1.6	0.04	300

*NB: Nd:YAG lasers differ from CO_2 lasers in that the fastest cutting speeds can often be obtained at output powers far less than the maximum-rated average output. The results shown here for stainless steel describe the performance of a laser rated at 400 W max mean power working at 230 W. At powers in excess of this, the divergence of the beam increases and the beam becomes more difficult to focus to an intense spot needed for cutting. All cuts here were carried out using the lasers in their pulsed mode. Pulse frequencies for laser cutting generally lie in the range of 10 to 100 Hz and pulse lengths are of the order of 0.2–2.0 ms. Oxygen is generally used as the cutting gas, except where it may damage the workpiece (for example, Ti alloys). In these cases an inert gas can be employed.

(Information gathered from several sources including: Lumonics, Rugby, U.K.; Raytheon, Burlington, Mass.; LASAG, Thun, Switzerland; and Van Dijk, *Industrial Laser Annual Handbook* 1987 and 1988).

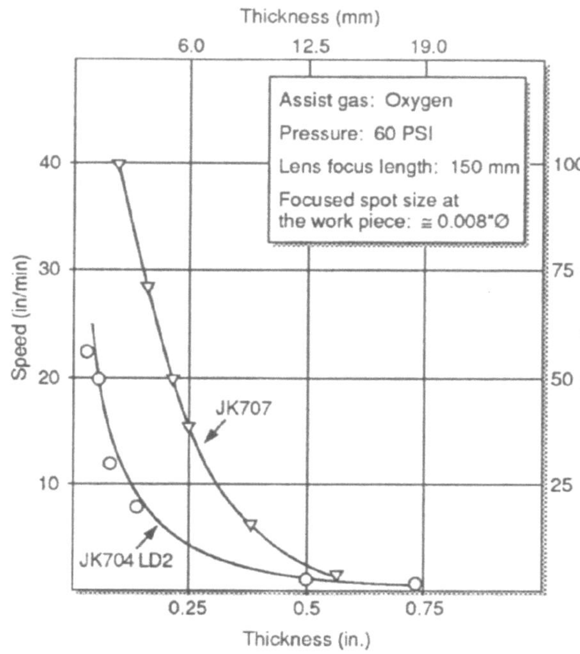

Figure 11. Nd:YAG lasers cutting stainless steel (304). (Source: Lumonics Ltd.)

Laser drilling

Laser hole drilling is the process resulting from the impingement of a focused laser beam onto the surface of a material. Single or multiple pulses of laser energy are used to vaporize the material to produce a hole with slight taper in most materials. Trepanning, a technique to produce high-quality holes, employs rapidly sequenced laser pulses around the hole edges.

Table 17. Laser drill process limits.
(Source: *Industrial Laser Annual Handbook*, 1990)

Laser	Diameter	Depth	Metal	Nonmetal
	lower upper (inches)	(inches)		
Ruby	<0.001 to 0.060	1.0	G	M
Nd:glass	<0.001 to 0.060	1.0	G	M
Nd:YAG	<0.001 to 0.060*	1.0	E	M
CO_2	<0.005 to Open*	0.500	E	E

M = Marginal G = Good E = Excellent
* The maximum diameter for trepan drilling, however, is limited only by the operational range of the work positioning system and can be much larger.

Laser cladding

The process of melting alloying elements onto the surface of a metal to improve its wear and corrosion resistance is known as laser cladding. In this process, a broad-area beam melts gravity-fed powder or preplaced alloys causing them to flow over a substrate surface and solidify to form a dense, homogeneous, low-dilution protective layer.

Table 18. Comparison of cladding processes from information supplied by C-E Industrial Lasers, now Trumpf Industrial lasers. (Source: *Industrial Laser Annual Handbook*, 1990)

	ADHESION		QUALITY				OTHER	
PROCESS	Interface Bond (CLAD)	Interface Bond (SUBSTR)	Surface Finish	% Dilution	Coating Homo-geneity	Porosity	Deposition Rate (lb/h)	Dimensional Control
Flame Spray	Liquid	Solid	Good	1-10	Fair	5%	1-6	Fair
Weld Overlay	Liquid	Solid	Poor	15-30	Poor	High	Low	Poor
Fuse Weld	Sol/Liq	Solid	Good	1-30	Fair	Low	Low	Fair
MIG	Liquid	Liquid	Fair/Good	1-50	Poor	Med	1-6	Fair
TIG	Liquid	Liquid	Fair/Good	1-50	Poor	Med	1-8	Fair
Plasma TR Arc	Liquid	Liquid	Fair	5-25	Fair	Low	1-15	Fair
Plasma Spray	Liquid	Solid	Good	5-30	Good	5-10%	1-15	Fair/Good
Detonation	Solid	Solid	Very Good	Very Low	Good	Low	High	Good
CO_2 Laser	Sol/Liq	Liquid	Very Good	0-90*	Very Good	0-5%	1-10	Very Good

* *In laser applications, the degree of dilution is controllable making surface alloying possible.*

Laser heat treating

Laser heat treatment in the form of transformation hardening produces hard, fine-grained structures with controlled case depth under accurate process control. The process utilizes a broad-area, low-power-density beam directed onto the suitably prepared surface of a metal.

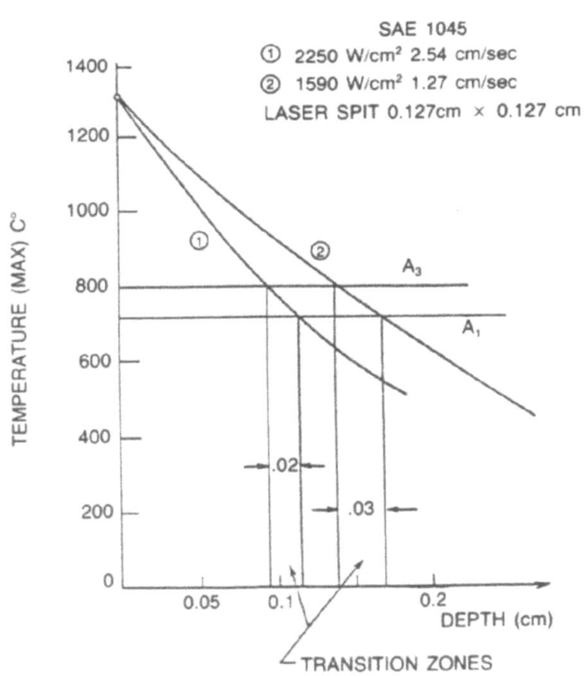

Figure 12. Influence of processing parameters on heat penetration in laser surface transformation hardening. (Source: Belforte, EOSD 14, 12, pp. 48-50, Dec. 1982)

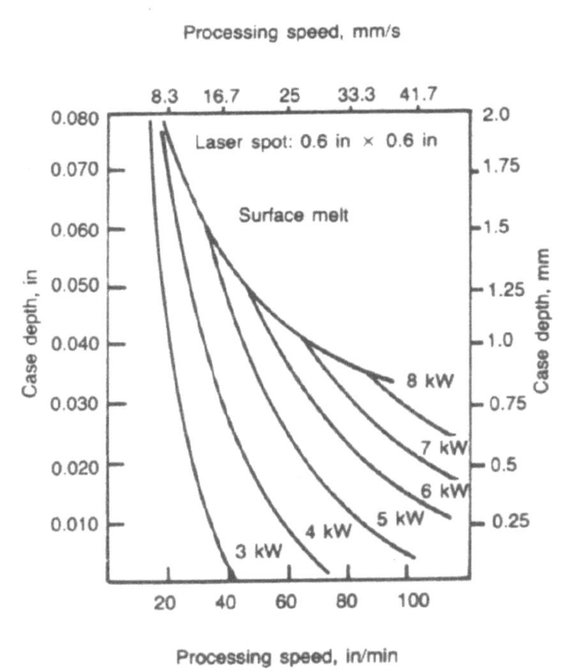

Figure 13. Laser surface hardening of SAE 4140 steel. (Source: *Industrial Laser Annual Handbook*, 1990)

Laser safety

Laser safety is an important aspect of industrial laser material processing. Proper precautions to ensure safe operation of the laser system should be standard practice. Excellent information on laser safety is available from the Laser Institute of America (Orlando, FL). The following tables present useful information; however, readers are advised to seek advice from their plant safety officer on other safety aspects.

Table 19. Basic laser bioeffects.
(Source: *Industrial Laser Annual Handbook*, 1990)

Laser type	Wavelength (μm)	Bioeffect Process	Tissue Effected: Skin	Cornea	Lens	Retina
CO_2	10.6	Thermal	×	×		
Hydrogen Fluoride	2.7	Thermal	×	×		
Erbium-YAG	1.54	Thermal	×	×		
Nd:YAG[a]	1.33	Thermal	×	×	×	×
Nd:YAG	1.06	Thermal	×			×
GaAs (diode)	0.780–0.840	Thermal	b			×
He-Ne	0.633	Thermal	b			×
Argon	0.488–0.514	Thermal/ Photochemical	×			× c
XeFl	0.351	Photochemical	×	×	×	
XeCl	0.308	Photochemical	×	×		

a Wavelength @ 1.33 μm more common in some research Nd:YAG lasers has demonstrated simultaneous cornea/lens/retina effects in biological research studies.
b Power levels not normally sufficient to be considered a significant skin hazard.
c Photochemical effects dominate for long term exposures to retina (exposure times > 10 s).

Table 20. Maximum permissible exposure limits as specified by ANSI Z-136.1 (1986).
(Source: *Industrial Laser Annual Handbook*, 1990)

Laser Type	Wavelength (μm)	MPE Level (W/cm^3) Exposure Time: (s) 0.25	10	600	3×10^4
CO_2	10.6	—	100×10^{-3}	—	100×10^{-3}
Nd:YAG[a] (CW)	1.33	—	5.1×10^{-3}	—	1.6×10^{-3}
Nd:YAG (CW)	1.064	—	5.1×10^{-3}	—	1.6×10^{-3}
Nd:YAG Q-switched[b]	1.064	—	17×10^{-6}	—	2.3×10^{-6}
GaAs (diode)	0.840	—	1.9×10^{-3}	—	610×10^{-6}
He-Ne	0.633	2.5×10^{-3}	—	293×10^{-6}	17.6×10^{-6}
Krypton	0.647	2.5×10^{-3}	—	364×10^{-6}	28.5×10^{-6}
	0.568	2.5×10^{-3}	—	31×10^{-6}	18.6×10^{-6}
	0.530	2.5×10^{-3}	—	16.7×10^{-6}	1.0×10^{-6}
Argon	0.514	2.5×10^{-3}	—	16.7×10^{-6}	1.0×10^{-6}
XeFl	0.351	—	—	—	33.3×10^{-6}
XeCl	0.308	—	—	—	1.3×10^{-6}

a Nd:YAG operating at less common 1.33 μm wavelength.
b Repetitively pulsed @ 11 Hz, 12 nsec pulses with pulse energy of 20 mJ/pulse.

Table 21. Nominal hazard zone distance values for various lasers.
(Source: *Industrial Laser Annual Handbook*, 1990)

Laser Type	Exposure Criteria	Hazard Range (m) Diffuse	Lens-on-laser	Direct
Nd:YAG	8 hr	1.4	11.3	1410
	10 s	0.8	6.3	792
CO_2	8 hr	0.4	5.3	390
	10 s	0.4	5.3	390
Argon	8 hr	12.6	1.7×10^3	25.2×10^3
	0.25 s	0.25	33.3	240

Laser Criteria Used for Nominal Hazard Zone Distance Calculations

Laser Parameter	Nd:YAG	CO_2	Argon
Wavelength (μm)	1.064	10.6	0.488
Beam power (W)	100.0	500.0	5.0
Beam divergence (mrad):	2.0	2.0	1.0
Beam size at aperture (mm):	2.0	20.0	2.0
Beam size at lens: (mm):	6.3	30.0	3.0
Lens focal length: (mm):	25.4	200.0	200.0
MPE: 8 hr: (μW/cm^2):	1.6×10^3	1.0×10^5	1.0
MPE: 10 s: (μW/cm^2):	5.1×10^3	1.0×10^5	—
MPE: 0.25 s: (μW/cm^2):	—	—	2.5×10^3

Table 22. Main decomposition products from laser-cut nonmetallic materials.
(Source: *Industrial Laser Annual Handbook*, 1990)

Decomposition Products	Materials Polyesters	Leather	PVC'S Wt (%)	Kevlar	Kevalar Epoxy Spectra Composite	
(a) FTIR analysis						
Acetylene	0.3-0.9	4.0	0.1-0.2	0.5	1.0	0.05
Carbon Monoxide	1.4-4.8	6.7	0.5-0.6	3.7	5.0	0.3
Ethylene	1.5-5.4	4.3	0.7	—	0.3	2.9
Methane	0.2-0.4	0.8	0.0-0.1	—	0.3	0.2
Hydrogen Chloride	—	—	9.7-10.9	—	—	—
Hydrogen Cyanide	—	—	—	1.0	1.3	—
C_6-C_8 alkenes	—	—	0.7-0.8	—	—	—
(b) GC analysis						
Benzene	3.0-7.2	2.2	1.0-1.5	4.8	1.8	—
Benzonitrile	—	—	—	0.5	0.7	—
Biphenyl	—	—	0.01-0.09	0.0	--	—
Chlorobenzene	—	—	0.01-0.04	—	—	—
Ethyl-benzene	0.1-0.4	0.1	0.01-0.6	0.1	—	—
Naphthalene	—	—	0.00-0.01	—	—	—
Nitric Oxide	—	—	—	—	0.2	—
Nitric Dioxide	—	—	—	0.6	0.5	—
Phenyl-acetylene	0.2-0.4	trace	—	0.1	—	—
Styrene	0.1-1.1	0.3	0.01-0.05	0.3	—	—
Toluene	0.3-0.9	0.1	0.06-0.1	0.2	0.2	—
Vinyl chloride	—	—	0.02	—	—	—
l-Decene	—	—	—	—	—	0.3
l-Heptene	—	—	—	—	—	0.2
l-Hexene	—	—	—	—	—	0.8
l-Octene	—	—	—	—	—	0.1

Laser vs. nonlaser process comparisons

Table 23. Laser vs. nonlaser process comparisons: Welding. (Source: *Industrial Laser Annual Handbook*, 1990)

Laser	**Electron Beam**	**Laser**	**TIG**
Advantages:	*Advantages:*	*Advantages*	*Advantages*
Welding in air	Deep penetration	Rate	Thickness
Equipment reliability (better than 80%)	High weld quality	Low total heat input	*Disadvantages:*
High throughput	Weld high-conductivity metals	Narrow heat affected zone	Rate
Multiple workstations	Fast welding rates	Single-pass butt welding	Part clamping
Weld magnetic metals	Dissimilar metals	Simple fixturing	Interpass cleanup
Simplified fixturing	Low distortion	Narrow gap filler	Bead width
Multiaxis manipulation	Welds high-carbon steel	Bead width	Thermal distortion
Flexibility for production	Weld steel with Ph and S	*Disadvantages:*	Penetration
Disadvantages	*Disadvantages*	Thickness	
Penetration limited to 1.0 in.	Up-time	Equipment cost	
Limited success in highly reflective metals	Process in protective environment		
High equipment cost	Lower throughput		
	High equipment cost		
	Complex fixturing required		
	Complicated multiaxis manipulation		
	Part cleaning		

Laser	**Resistance**
Advantages:	*Advantages:*
Noncontact	Equipment cost
Fast rate	Part clamping
Dissimilar metals	*Disadvantages:*
Low total heat	Electrode wear
Minimum weld seam	Joint access
Strength	Rate
Complex shapes	Part distortion
Disadvantages:	Dissimilar metals
Part fit-up	Heat-sensitive components
Equipment cost	Joint efficiency
	Rectilinear shapes
	Contact process

Table 24. Laser vs. nonlaser process comparisons: Metal cutting. (Source: *Industrial Laser Annual Handbook*, 1990)

Laser *Advantages:* No dies needed Rapid design change Cut complex shapes Cut three-dimensional shapes Minimum set-up time Reduced scrap Narrow kerf width Cut nonmetal patterns Cut tempered steel Good edges Short delivery time Cut large sheet without handling Noise *Disadvantages:* Single thickness Slow rate on large volumes Equipment cost	**Punch Press** *Advantages:* Rate on high volume Low cost on volume Regular, repetitive shapes *Disadvantages:* Multiple dies for complex shapes Tooling cost Tooling design and fabrication time Set-up time Steel must be annealed Noise Thin stainless plate	**Laser** *Advantages:* Accuracy Edge quality Narrow kerf Minimum HAZ No part distortion Low cutting noise Reduced metal waste Cut nonmetal templates *Disadvantages:* Slower rate in thicker (0.250 in.) metals High equipment cost Thick metals >0.5 in.	**Plasma Arc** *Advantages:* Faster on thicker (0.280 in.) metals Cut multiple layers Lower equipment cost Cut thicker>0.5 in. *Disadvantages:* Accuracy Edge quality Wide kerf width Larger HAZ Noise Operating cost (considerable) Can only cut metals
Laser *Advantages:* Rate Edge quality Reduced scrap No part distortion Complex shapes Narrow kerf Maximum part nesting *Disadvantages:* Single thickness Equipment cost	**Nibbling** *Advantages:* Low equipment cost *Disadvantages:* Scalloped edges Slow on complex shapes Tool wear Surface finish Scrap Part nesting	**Laser** *Advantages:* Narrow kerf (0.020 in.) Low maintenance More compact nesting Faster cutting rates Three-dimensional cutting *Disadvantages:* Equipment cost Small HAZ Limited with low thermal conductivity metals	**Abrasive Fluid Jet** *Advantages:* Material thickness Fluids present on metals Cut all metals No HAZ No part distortion *Disadvantages:* Pump maintenance (1000 h) Fluid must be filtered to remove chemical Noise (80 dB or more) Must protect adjacent surface from rebounding abrasive User education to high-pressure plumbing Equipment cost Slow cutting rates Multiaxis cutting
Laser *Advantages:* Rate Cut paper patterns *Disadvantages:* Edge quality Metal thickness	**Wire EDM** *Advantages:* Edge quality Metal thickness Accuracy *Disadvantages:* Fixturing Electrode wear Wire costs	**Laser** *Advantages:* Rate Reduced set-up time Flexibility Reduced scrap *Disadvantages:* Equipment cost Edge quality	**NC Milling** *Advantages:* Accuracy Edge quality No HAZ Lower equipment cost *Disadvantages:* Rate Set-up time Tool wear Machined chips Cutting fluids

Table 25. Laser vs. nonlaser process comparisons: Nonmetal cutting. (Source: *Industrial Laser Annual Handbook*, 1990)

Fabrics **Laser** *Advantages:* Speed Edge quality *Disadvantages:* Fumes	**Water Jet** *Advantages:* No fumes No edge discoloration *Disadvantages:* Edge quality Can leave threads	**Fabrics** **Laser** *Advantages:* Speed Flexibility Edge quality Accuracy *Disadvantages:* Fumes	**Knives** *Advantages:* Throughput Multilayer *Disadvantages:* Knives dull Accuracy Dust Lack of flexibility
Wood **Laser** *Advantages:* Speed *Disadvantages:* Edge char	**Water Jet** *Advantages:* Clean edges Thickness *Disadvantages:* Speed Wet surface	**Rubber** **Laser** *Advantages:* Tight tolerance Low cost in low volume *Disadvantages:* Equipment cost	**Die Punch** *Advantages:* Low cost in high volume *Disadvantages:* Poor tolerances Edge quality
Plastic **Laser** *Advantages:* No swarf Polished edge High accuracy Narrow kerf Speed Part intricacy *Disadvantages:* Equipment cost Fumes Single layer cutting	**Router/Sawing** *Advantages:* Low cost Cut multilayer *Disadvantages:* Large kerf Excess swarf Poor edge quality Labor intensive	**Plastic** **Laser** *Advantages:* Speed Edge quality *Disadvantages:* Fumes Edge quality certain materials	**Water Jet** *Advantages:* Stacked parts Edge quality *Disadvantages:* Operating cost

Table 26. Laser vs. nonlaser process comparisons: Metal drilling. (Source: *Industrial Laser Annual Handbook*, 1990)

Laser *Advantages:* Speed No tool wear/breakage Flexibility Lower operating costs Holes of any shape Drill oblique holes Drill nonmetals Fixturing minimal Less down-time Drill on curved surfaces Noncontact drilling *Disadvantages:* High equipment cost Surface quality Taper in direct drill holes Heat affected zone Thickness >0.750 in. Recast layer Hole size control	**EDM** *Advantages:* Surface quality Low equipment cost Hole depth No taper Small HAZ Constant hole size *Disadvantages:* Consumable cost high Electrodes need frequent replacement Slow drilling rate Multiple set-up Hole shape limited	**Laser** *Advantages:* Hole quality Small hole size No drill breakage No burrs Process rate No scrap Drill on irregular surfaces Drill hard metals *Disadvantages:* Hole taper Depth limit Equipment cost	**Mechanical Drill** *Advantages:* Deep holes Straight holes Low cost *Disadvantages:* Hole quality Minimum hole size Drill wear Drill breakage Burrs Scrap management Set-up time Speed

Table 27. Laser vs. nonlaser process comparisons: Transformation hardening. (Source: *Industrial Laser Annual Handbook*, 1990)

Laser	**Induction**
Advantages:	*Advantages:*
Minimal part distortion	Fast process rates
Selective hardening	Deep case obtainable
No quenchent needed	Lower equipment cost
Thin case capability	Coverage area
Harden inaccessible locations	High efficiency
Case depth controllable	*Disadvantages:*
Eliminates post processing	Downtime for coil change
Improves fatigue life	Fabrication of complex coils
Complex structures	Coil replacement
Disadvantages:	Quenchent
Equipment cost	Part distortion
Process rate	Limited part geometry
Coverage area	Thermal load
Absorbent coatings necessary	

Table 28. Laser vs. nonlaser process comparisons: Cladding. (Source: *Industrial Laser Annual Handbook*, 1990)

Laser	**Plasma Arc**
Advantages:	*Advantages:*
Low heat input	Produces thick layers
Thin layers	High deposition rates
Low dilution	Low equipment cost
Low porosity	Covers large areas
Higher hardness	*Disadvantages:*
Reduced alloy cost	High heat input
Small heat affected zone	Part distortion
Low cost of powder alloys	High dilution
Automatic alloy feed	Low hardness
Less post process cleanup	Excess alloy to be removed
High cooling rate	Cost of feed materials
Complex clad alloys	Tight process control
Tight process control	
Disadvantages:	
Initial equipment investment	
Process rates	

Table 29. Laser vs. nonlaser process comparisons: Marking. (Source: *Industrial Laser Annual Handbook*, 1990)

Laser	**Ink Jet**	**Laser**	**Stamping**
Advantages:	*Advantages:*	*Advantages:*	*Advantages:*
Permanent marks	Relatively small investment	Noncontact	Low initial investment
Noncontact (pressureless)	Mark some materials that lasers can't	Nondestructive	Low operating cost
Mark directly on product	High contrast mark	Controlled depth	*Disadvantages:*
Mark curved or irregular surfaces	Compact marking head	Easily automated	Limited applications
Wide range of applications	Low capital cost	Flexible marking	Difficult to serialize
No consumable inks	High speed	High mark quality	Surface must be relatively flat
No curing time	Programmable	Mark irregular shapes	Can be destructive
No clean-up	Computer controlled	Rapid rates	Quality of mark
Rapid set-up	*Disadvantages:*	Multiple fonts	Readability
Mark rate	Susceptible to damage by heat, light, chemicals, and friction	Automatic serialization	Heavy material stress
Can serialize easily	Consumables must be inventoried	Easy layout change	Relatively slow
Can make small marks	Often requires curing process/time	No moving parts	Difficult to change data
Low running cost	Clean-up require	Low maintenance	Die wear
Versatility	Surface must be clean	*Disadvantages:*	
Precise	Surface must be flat	Large initial investment	
Easy system integration	Quality of mark is rate dependent		
Mark quality	Difficult to serialize		
Low maintenance	Consumable cost		
Disadvantages:	Marks can smear		
Large initial investment			
Will not mark all materials			
Low contrast on some materials			
Equipment comparatively large			

Table 30. Laser vs. nonlaser process comparisons: Wire stripping. (Source: *Industrial Laser Annual Handbook*, 1990)

Laser	**Mechanical**
Advantages:	*Advantages:*
Noncontact	Simpler with large standard-type wires
Good with microwires	Low cost
Precise insulation removal	*Disadvantages:*
Good on twisted pairs	Not possible with microwave
Disadvantages:	Damage conductors
Slow on thick wires	
Slow on thick insulation	
Equipment cost	

Table 31. Laser vs. nonlaser process comparisons: Other applications. (Source: *Industrial Laser Annual Handbook*, 1990)

Photo CVD	**Thermal CVD**
Advantages:	*Advantages:*
Spatial control	Cost
Precise dopant control	*Disadvantages:*
Speed	Dopant control
Localized structures	Speed
Disadvantages:	Problem to generate localized structures
Cost	
Photolithography	**UV lamp**
Advantages:	*Advantages:*
Higher photon fluxes	Simplicity
Resolution	Resist technology
Speed	*Disadvantages:*
Microstructures below 0.5 μm	Low speed
Flash on fly	Resolution
Disadvantages:	Smallest microstructure 0.8 μm
Equipment cost	
Resist technology	

SECTION 2

Review of laser processing

Modeling of laser material processing

Dr. John C. Ion
Laser Processing Laboratory
Lappeenranta University of Technology
Finland

Mathematical models that describe the principles of laser materials processing are reviewed in this article. The fields covered include: process variables, temperature fields, phase transformations, physical and chemical properties, and structure-property relationships. Charts are presented that illustrate how the choice of process variables defines the dominant mechanism of processing, which characterizes the various methods of laser treatment.

For individual treatments, the use of models for optimization of process variables and material properties is discussed. The methods described show considerable potential for automatic parameter selection and process control systems, and in assessing the suitability of new alloys for novel forms of laser processing.

Introduction

Models of laser materials processing relate the process variables (the beam and material properties) to the product characteristics (the geometry and properties of the processed region) by describing mathematically the changes that occur during the treatment.

Models have many uses:

- Treatments can be simulated, reducing the need for expensive testing during optimization of a process.
- Model results provide valuable input for on-line process control systems.
- A flexible model based on sound physical principles is a useful tool for designing new materials and novel processes.

Few complete process models exist, due to the large number of variables involved and the complexity of many treatments. However, the principles of most industrial laser processes are understood well enough for descriptions to be formulated, either theoretically or empirically.

The simplest models are mathematical relationships derived from empirical data—curve fits and "rules of thumb." By incorporating a sound physical basis into a model, its range of application can be extended, and practical analytical solutions can often be derived through the use of sensible approximations. The most detailed models require few assumptions and are more flexible, but may require a powerful mainframe computer for a numerical solution. The detail in the model and the method by which the formulation is solved depend on the complexity of the problem, the reliability of input data, the accuracy required in the results, and the computing power and time available.

Models that have a practical application to the most common methods of laser materials processing are reviewed here. It is not the intention to analyze models in detail and compare results—which can be difficult due to the accuracy with which some process variables are known—but rather to outline the principles on which they are based and direct the reader to the appropriate source for more information.

The process variables are considered first, since these determine the temperature fields induced in the material and, consequently, the processing mechanism(s) involved: heating, melting or vaporization. A process may involve more than one mechanism—for example, a deeply penetrating weldment is formed by heating, melting, and vaporization. A mechanism may be fundamental to a number of different processes—for example, melting is the dominant factor in glazing, cladding, and alloying. Models of temperature fields and microstructural transformations are therefore classified in terms of process mechanisms to illustrate their relevance to a wide range of treatments. Finally, these component models are drawn together into specific practical applications for some of the most common industrial processes. The review concentrates on metallic alloy processing using continuous wave CO_2 lasers, although many of the principles described are relevant to other treatments based on the heating effect of a power beam.

Process variables

A laser beam impinging on a material is characterized by its power, wavelength, modes (spatial and temporal energy distributions), polarization, divergence, cross-sectional shape and size, and traverse rate. The response of the material depends on its absorptivity to the beam, thermal conductivity, density, specific heat capacity, phase transformation temperatures, specific latent heats of transformation, and its dimensions. Data for most engineering materials are available in

Table 1. Symbols and units

a	thermal diffusivity ($m^2 s^{-1}$)
A	fraction of incident beam power absorbed (-)
c	specific heat capacity ($J kg^{-1} K^{-1}$)
d	depth of treatment (m)
D	aperture diameter (m)
e	base of natural logarithms (2.718)
E	energy (J)
f	focal length (m)
j	constant
k	constant
L_m	volumetric latent heat of melting ($J m^{-3}$)
L_v	volumetric latent heat of vaporisation ($J m^{-3}$)
q	beam power ($J s^{-1}$)
r_B	beam radius or half-width (m)
t	time (s)
T	temperature (K)
T_m	melting temperature (K)
T_o	initial temperature (K)
T_v	boiling temperature (K)
v	beam traverse rate ($m s^{-1}$)
λ	thermal conductivity ($J s^{-1} m^{-1} K^{-1}$)
ρ	density ($kg m^{-3}$)

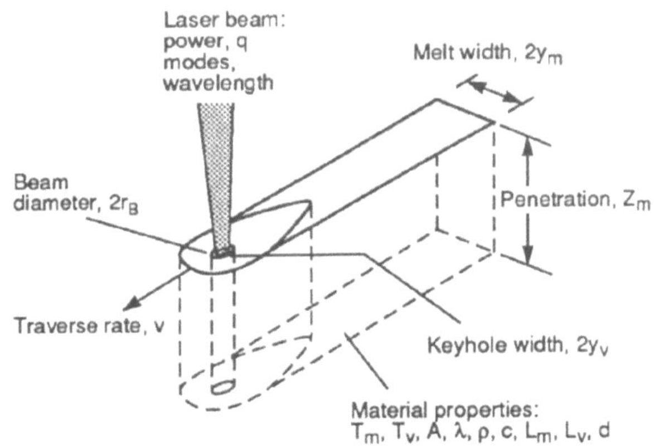

Figure 1. Schematic illustration of laser keyhole welding showing the principal process variables.

handbooks and databases.[2] Many material properties are dependent on other process variables; for example, thermal properties vary with temperature, which introduces non-linearities into equations, complicating modeling. The main process variables are listed in Table 1 and illustrated in Figure 1.

The wavelength of laser light, λ, is related to the energy of the lasing transition, E, by:

$$\lambda = \frac{hc}{E} \tag{1}$$

where h is Planck's constant and c is the velocity of light. The beam mode describes the energy distribution across a section of a beam produced in a stable optical resonator; for example, TEM_{00} denotes a Gaussian energy distribution about a central peak. The diameter of such a beam is often defined by the points at which the power density has fallen to a fraction (normally $1/e$ or $1/e^2$) of the peak. Unstable resonators used in higher power lasers produce beams with annular (often referred to as TEM_{01*}) and multimode energy distributions.

To compare the properties of different beams, the quality can be defined in terms of the K-factor, which expresses the focusability (spot size and focal length) of a beam in terms of a Gaussian beam. The minimum theoretical diameter, $2r_B$, to which a laser beam can be focused is[3]:

$$2r_B = \frac{4\lambda f}{K\pi D} \tag{2}$$

where f is the focal length and D is the aperture of the focusing optics. K is 1 for a Gaussian beam and less than 1 for other beam modes. An analogous system uses the M^2 notation, where $M^2 = 1/K$. Aberrations introduced by the focusing optics increase the practical minimum spot size. Relationships can be derived for other properties of a focused beam, such as depth of focus.[4] The beam can also be formed into a variety of distributed heating patterns for surface treatments with suitable optics.

By combining the main process variables into independent groups, an analysis can be simplified and the nature of processing characterized more easily. In Figure 2, beam parameters are combined to form power density and beam interaction time, such that practical operating regimes for various types of laser processing can be defined.[5] Beam parameters can be combined with material properties to form dimensionless groups and linked with analytical heat flow models to construct an operating chart (see Fig. 3).[6] A model-based approach using dimensionless variables allows the thermal cycles to be characterized in terms of all the relevant process parameters, and can thus be applied to a range of materials and processes. Once the processing method has been established, it is then of practical interest to be able to predict the geometry of the processed zone, for which a knowledge of the temperature fields induced is required.

Temperature fields

The laser beam is normally regarded as a well-defined heat source that is restricted to the material surface in some treatments or focused to penetrate the material in others. The governing differential equation of heat conduction is:

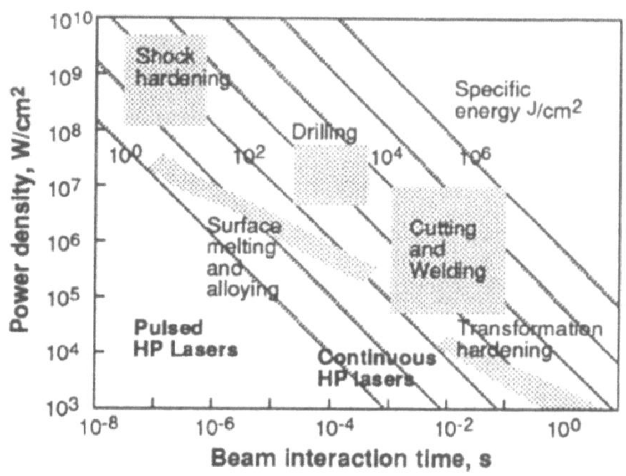

Figure 2. Practical operating regimes for various methods of laser materials processing.[5]

$$\frac{\partial T}{\partial t} = \frac{\partial}{\partial x}\left(\frac{\lambda}{\rho c}\cdot\frac{\partial T}{\partial x}\right) + \frac{\partial}{\partial y}\left(\frac{\lambda}{\rho c}\cdot\frac{\partial T}{\partial y}\right) + \frac{\partial}{\partial z}\left(\frac{\lambda}{\rho c}\cdot\frac{\partial T}{\partial z}\right) + q_v \quad (3)$$

where T is temperature; t is time; x, y, and z are coordinates; λ is thermal conductivity; c is specific heat capacity; and q_v is the ratio of the rate of heat supply per unit time and volume.

Surface heating

Equation (3) can be solved numerically by integration[7–14] and finite difference[15–17] methods to give temperature fields around a variety of moving and stationary surface heat sources. Such solutions are flexible and can take into account variations of material properties with temperature, but they may require considerable computing power and time for execution.

Equation (3) may also be solved analytically, by making approximations such as treating the substrate as a semi-infinite body, with material properties which do not vary with direction or temperature, and assuming heat flow to be steady state. Solutions for temperature fields around moving point[18,19] or disc[20–23] heat sources then yield equations for the thermal cycles in the material. Expressions for peak temperature, heating and cooling rates, and thermal dwell times can then be obtained, which indicate that

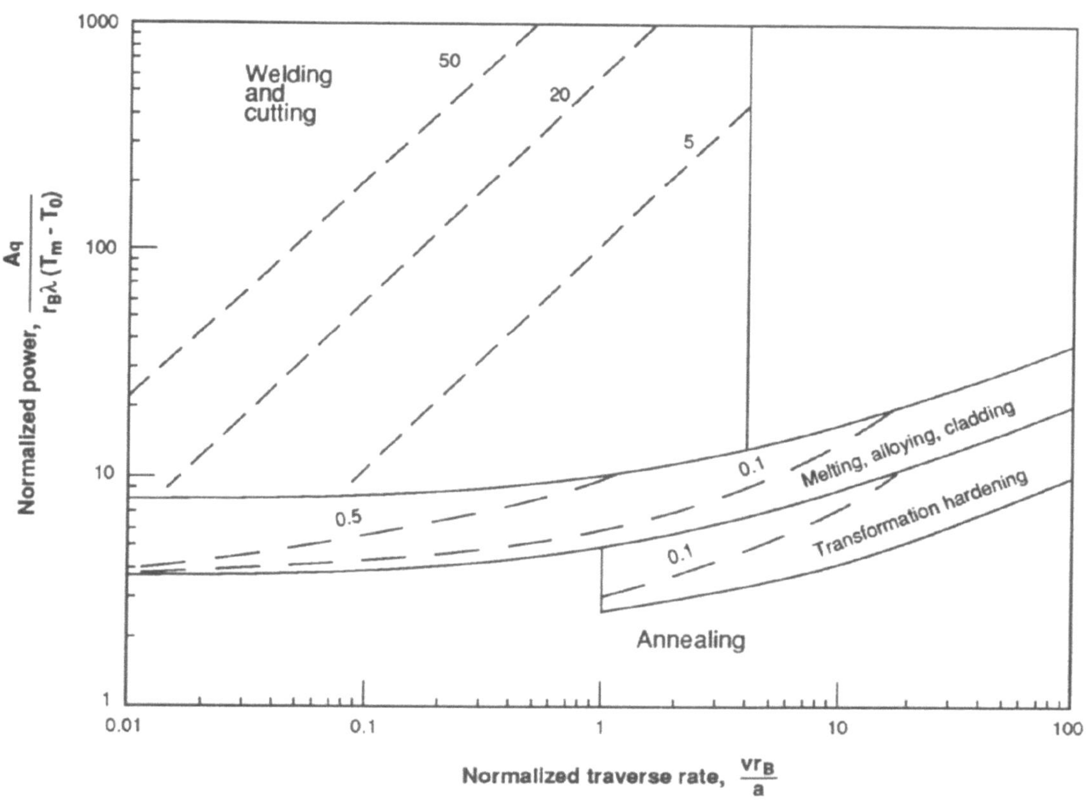

Figure 3. Theoretical operating regimes for laser materials processing (solid lines). The broken contours represent normalized processing depth (d/r_B) and are constructed using analytical heat flow models.[6]

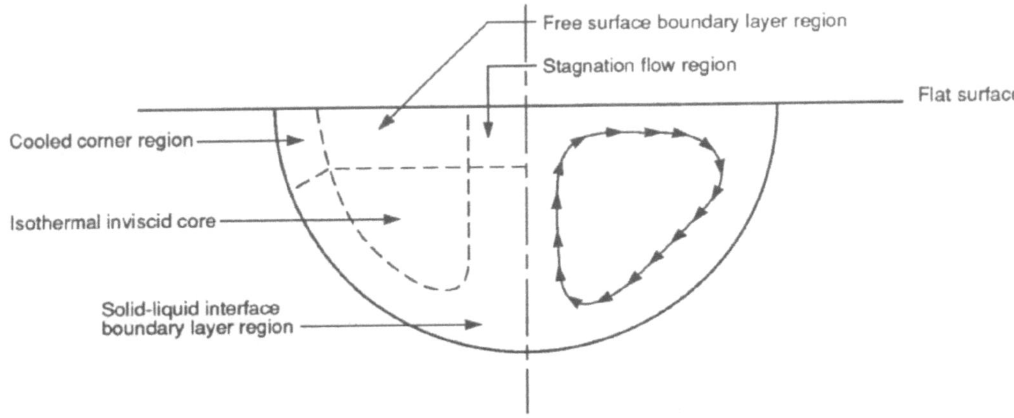

Figure 4. Schematic illustration of flow regimes in a laser-melted pool, which can be modeled numerically.[31]

heating rates of the order $10^4 K/s$ and thermal cycles less than 1 s are common for laser treatments. Such analytical solutions yield explicit relationships between process variables and temperature fields that can be manipulated readily and that provide a reliable basis for models of solid-state structural transformation processes such as transformation hardening, carburizing, and annealing.

Surface melting

Surface melting is the basis of laser glazing, remelting, alloying, particle injection, and cladding. Improved corrosion and/or wear properties can be achieved in ferrous and nonferrous alloys through microstructural refinement, reduced segregation, and the production of nonequilibrium phases. The melt volume can be estimated roughly by substituting the melting temperature isotherm into the temperature field models of the previous section, with an appropriate correction for the latent heat of fusion. However, numerical solutions to models for convectional heat and mass transfer[24–31] are required if the shape of the molten pool and patterns of fluid flow, such as those in Figure 4, are to be predicted.

Dimensionless groups such as the Peclet number (the ratio of the rate of convection to the rate of conduction) and the Prandtl number (the ratio of kinematic viscosity and thermal diffusivity) give indications of the relative importance of various processing mechanisms. Such studies have explained the role of trace elements in influencing convectional heat transfer driven by surface tension, and compositional homogenization.

Surface vaporization

A pulsed laser beam can be used to selectively remove the surface of a material, which may also give rise to shock hardening in some materials. Models that consider vaporization[32,33] and a combination of vaporization and liquid expulsion[34–36] have been used to quantify the rate of materials removal in terms of the process variables.

Keyhole processing

Suitable optics allow a laser beam to be focused to a spot, with a power density in excess of 100 kW/mm^2, sufficient to vaporize engineering materials. Under such conditions a stable, deeply penetrating "keyhole" of vaporized material can be formed and traversed through the material to produce deep, narrow cuts and welds, shown schematically in Figure 1.

By approximating the keyhole to a moving, through-thickness line heat source, analytical[18,19,22,37–48] and numerical[49,50] solutions to the heat conduction problem allow temperature fields to be predicted. The line can be expanded to a cylinder[7,51–53] or a triangular prism[54] to represent the geometry of the keyhole and the weld bead, respectively. Such solutions provide a means of estimating idealized cutting and welding performance, but a complete treatment requires the consideration of all process variables.

Microstructure and properties

Solid-state transformations

Solid-state transformations control the microstructure and properties of the heat-affected zone (HAZ) produced by transformation hardening, welding, and homogenization treatments. Continuous cooling transformation (CCT) diagrams provide some information for microstructural prediction in ferrous alloys. However, rapid laser thermal cycles may result in incomplete transformation on heating and nonequilibrium distributions of alloying elements, which should be taken into account.

Analytical[21,55] and numerical[56] models of ferrite/pearlite transformation on heating and austenite transformation on cooling have been developed. These models help explain observations that coarse phases in steels (particularly carbides) and phases that require long-range diffusion for transformation may be retained during rapid laser processing.

Such models allow the microstructure to be predicted as a function of the process variables and provide a means of predicting physical and chemical properties. Similarly, diffusion models allow the extent of carburization[57] and homogenization[58] to be characterized.

Solidification microstructures

Fine solidification microstructures can be produced in laser welds and surface remelting processes, with superior physical and chemical properties to the base material. Equilibrium and metastable phase diagrams[59] provide a guide to possible solidification sequences, although phase transformations occur under conditions far from equilibrium.

Conventional methods for predicting solidification microstructures have been adapted to take into account the rapid cooling rates of low heat input processes. For example, the duplex ferrite-austenite region of the Schaeffler diagram for stainless steels has been modified to describe laser welding.[60,61] In some cases, new solidification mechanisms have been identified.[62] Laser weldments in an aluminium alloy[63] have shown that primary and secondary dendrite arm spacings can be predicted using theoretical extrapolations from the lower cooling rates found in large castings.

Very thin surface layers produced by laser melting of metals may solidify with a variety of nonequilibrium microstructures, depending on the composition, cooling rate, solidification velocity, and temperature gradient. Diagrams with these properties as axes have been compiled for a variety of metals, including spheroidal graphite cast iron[64] and Fe-B alloys[65] (in which amorphous microstructures are possible). Extended solid solutions produced by cladding with a Ni-Hf alloy powder have also been modeled by considering temperature and concentration fields.[66] Such methods allow surface microstructures to be engineered through appropriate selection of the process variables.

Properties

Hardness changes in the HAZ determine many of the important properties of laser-processed material. CCT diagrams for conventional heat treatment of ferrous alloys give estimates of the hardness to be expected for rapid cooling cycles. Empirical phase-property relationships have been used to express hardness as a function of microstructure and chemical composition for laser transformation hardening of ferrous alloys.[21] Empirical equations for hardness prediction following rapid thermal cycling have also been developed.[67]

Similar techniques developed for arc welding[22] have been applied successfully to HAZ hardness prediction after laser welding.[68] By considering coarsening and dissolution of strengthening precipitates in age-hardenable aluminium alloys,[69–71] the degree of softening following thermal cycling may be calculated. Residual stresses induced during laser transformation hardening[72] and surface melting[73] of ferrous alloys have also been investigated, with the aim of predicting conditions under which a desirable compressive stress state can be produced at the surface for enhanced fatigue properties. Models for corrosion and wear resistance involve many variables, and are difficult to formulate in terms of the process variables.

Process models

Transformation hardening

Some of the methods described in the previous sections have been combined into charts for laser transformation hardening,[16,21,55,74–76] an example of which is shown in Figure 5.

Figure 5 can be used to select process variables for transformation hardening of a particular steel to a given depth, without surface melting. The volume fraction of martensite (f_m) and the hardness to be expected (H_v) can also be estimated. Simplifications and other semi-empirical methods have also been used to predict case depths under certain conditions.[77] Such methods allow parameters to be optimized rapidly with respect to the process variables or the material properties. Models have also been incorporated into adaptive control systems that ensure a uniform hardened case, even for hardening of complex geometries, based on monitoring the surface temperature.[78] Simple economic models of transformation hardening have shown that components with a low treatment area:weight ratio can be hardened more economically using laser processing than using a conventional treatment such as gas carburizing.

Surface glazing, cladding, and alloying

Process models for surface treatments involving melting comprise components for heat flow by conduction and convection, convectional mass transport, solidification sequences, phase transformations, and structure-property relationships. Models have been constructed to show the effect of process variables on melt depth,[79] cooling rate,[80] liquid state homogenization,[25] and surface rippling[24] during laser melting. These models are valuable for a variety of processes, such as selecting parameters for desensitizing stainless steel by laser melting.

Models of laser cladding have been presented in the form of charts from which process variables for particular coverage rates can be estimated.[81,82] Process models, expert analysis, and monitoring systems form the basis for the development of parameter optimization[83] and adaptive control[84] systems for laser cladding. Investigations into laser surface alloying have identified dimensionless process parameters,[85,86] which control the geometry and composition of alloyed surfaces. Processes involving an alloy addition contain another set of parameters, which must be included in a complete process model.

Welding

The physics of keyhole welding are complex, involving energy absorption by the vapor and the cavity walls, plasma formation, and equilibrium between vapor pressure, surface tension, and hydrostatic pressure. The effects of process variables such as beam polarization, plasma formation and suppression techniques, assist, and shielding gases are diffi-

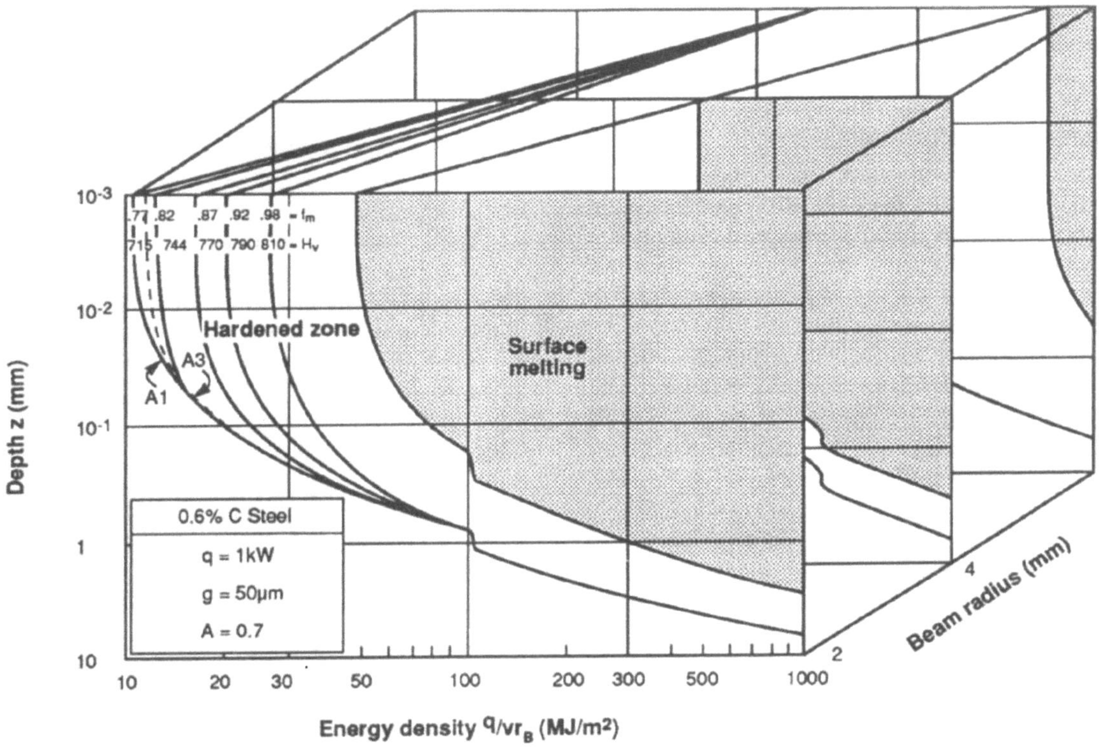

Figure 5. Laser transformation hardening diagram constructed for an 0.6 wt% C steel using analytical models of heat flow and phase transformations and empirical structure-property relationships.[21]

cult to describe quantitatively. Exact process models are therefore complex and specific in nature.

Many of the principles mentioned above have been modeled using numerical methods,[15,87–96] from which relationships between process variables can be obtained. The characteristic "nail head" bead profile often found in laser welds can be reproduced using a balance of line and point sources to represent the heating effects of, respectively, the keyhole and the surface plasma.[97–100] Thermal analysis of a pulsed heat source[101] also provides valuable information about processing rates and penetration achievable for pulsed processing.

Experimental laser welding data have been summarised in previous volumes of the *Industrial Laser Annual Handbook* for a large number of materials. These suggest that the maximum processing rate, v, depends on the power, q, and the material thickness, d, according to:

$$v = k\, q\, d^{-j} \tag{4}$$

where k and j are constants, k is determined by factors that influence the process efficiency, and j generally lies between 0.8 and 2. Models based on simple energy balances normally yield integer values for j, depending on the assumptions made about the weld bead geometry. Within certain operating ranges such models are useful for parameter selection, but factors such as keyhole instability and plasma formation may introduce new processing mechanisms, which require additional modeling.

In contrast to arc welding, standards that define process variables in terms of weld properties for particular applications are not yet available for laser welding. In some cases, the codes designed for arc welding have been adapted, with allowance for the high hardness values found in laser welds in steels due to the high cooling rate. However, the *Industrial Laser Annual Handbook* and others[48,102] present various empirical methods to assess process variables for successful laser welding of particular joints.

Cutting

Empirical data for laser cutting of ferrous and non-ferrous alloys, polymers, wood-based products, and ceramics have shown that the minimum power, q, required for cutting various materials of thickness, d, with a kerf width, s, at a speed, v, is described by[103,104]:

$$q = k\, d^{0.21}\, s^{0.01}\, v^{0.16} \tag{5}$$

where k is a constant for a given material. Due to the complexities of penetration processing outlined in the previous

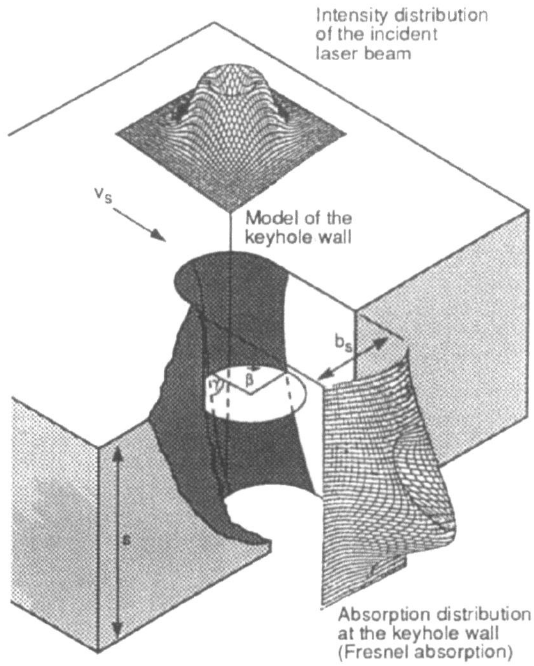

Figure 6. Schematic illustration of the laser-beam/workpiece interaction during laser cutting.[106]

section, coupled with the variety of cutting mechanisms that may be operating (such as melt shearing, vaporization, or chemical degradation), consistent process models are difficult to formulate. Work is being done to optimize process variables[105] and improve cut edge quality, and a self-consistent three-dimensional model of laser cutting based on the laser-material interaction (see Fig. 6) that allows process parameters and cut quality to be optimized for metals processing has been developed.[106]

Conclusions

Mathematical models of heat flow and structural change provide a sound basis for understanding the physical processes that occur during laser materials processing. Complete process models have been developed for simpler treatments such as transformation hardening, and many of the components of more complex processes such as cutting and welding can be modeled or determined empirically. Much can be achieved by adapting predictive methods for conventional processes to the shorter time scales of laser treatments. Models are now used to optimize a process with regard to productivity, economics, material properties, etc., and are rapidly becoming an integral part of on-line process control and expert systems.

With continued growth in the capacity and data processing rates of computers and the sophistication of process sensors, mathematical models will play an increasing role in intelligent processing of materials using lasers. A materials engineer may soon be able to exploit such technology to design a component, select a material and microstructure, and devise a laser-based adaptively controlled fabrication scheme to fully use the advantages offered by laser materials processing.

Acknowledgments

I would like to thank my colleagues in the Laser Processing Laboratory at Lappeenranta University of Technology and Dr H.R. Shercliff and Prof. M.F. Ashby of Cambridge University Engineering Department for many productive discussions.

References

1. C.J. Smithells, *Metals Reference Book*, 5th ed., Butterworths, London, 1976.
2. A.L. Edwards, "For Computer Heat Conduction Equations A Compilation of Thermal Properties Data", *UCRL-50589*, Feb. 24 1969.
3. O. Gregersen and F.O. Olsen, *Proc. ICALEO '90*, Laser Institute of America, Toledo, 1990.
4. J.T. Luxon, in *Industrial Laser Annual Handbook*, pp.38-48, PennWell Publishing Co., Tulsa, 1986.
5. E.M. Breinan, B.H. Kear, and C.M. Banas, *Physics Today, 29, 44* (1976).
6. J.C. Ion, H.R. Shercliff, and M.F. Ashby, to be published.
7. H.S. Carslaw and J.C. Jaeger, *Conduction of Heat in Solids*, 2nd ed., Clarendon Press, Oxford, UK (1959).
8. Y. Arata and I. Miyamoto, *Trans. Jap. Weld. Soc., 3, 163* (1972).
9. H.E. Cline and T.R. Anthony, *J. Appl. Phys., 48, 3895* (1977).
10. M. Lax, *J. Appl. Phys., 48, 3919* (1977).
11. T.W. Eager and N.S. Tsai, *Weld. J., 62, 346s* (1983).
12. O. Sandven, *Proc. Materials Processing Symposium (ICALEO'83)*, Laser Institute of America, Toledo, p.208 (1984).
13. M. Davis, P. Kapadia, J. Dowden, W.M. Steen, and C.H.G. Courtney, *J. Phys. D: Appl. Phys., 19, 1981* (1986).
14. M.K. El-Adawi and E.F. Elshehawey, *J. Appl. Phys, 60, 2250* (1986).
15. J. Mazumder and W.M. Steen, *J. Appl. Phys., 51, 941* (1980).
16. S. Kou, D.K. Sun, and Y.P. Le, *Metall. Trans., 14A, 643* (1983).
17. K. Inoue, E. Ohmura, K. Haruta, and S. Ikuta, *Trans. JWRI, 16, 277* (1987).
18. D. Rosenthal, *Weld. J., 20, 220s* (1941).
19. D. Rosenthal, *Trans. ASME, 68, 849* (1946).
20. N. Rykalin, A. Uglov, and A. Kokora, in *Laser Machining and Welding*, Chapter 3, Pergamon Press, Oxford, UK, 1978.
21. M.F. Ashby and K.E. Easterling, *Acta Metall., 32, 1935* (1984).
22. J.C. Ion, K.E. Easterling, and M.F. Ashby, *Acta Metall., 32, 1949* (1984).
23. M. Bass, in *Physical Processes in Laser-Material Interactions*, pp. 77–115, Plenum Press, New York, 1983.
24. T.R. Anthony and H.E. Cline, *J. Appl. Phys. 48*, 3888 (1977).
25. T.R. Anthony and H.E. Cline, *J. Appl. Phys. 49*, 1248 (1978).
26. J.M. Dowden, M. Davis, and P. Kapadia, *J. Appl. Phys. 57*, 4474 (1985).
27. S. Kou and Y.H. Wang, *Metall. Trans. 17A*, 2265 (1986).

28. C.L. Chan, J. Mazumder, and M.M. Chen, *Mat. Sci. Tech. 3*, 305 (1987).
29. A. Paul and T. DebRoy, *Proc. Conf. Modeling and Control of Casting and Welding Processes*, TMS/AIME, Warrendale, (1988).
30. T. Zacharia, S.A. David, J.M. Vitek, and T. DebRoy, *Weld. J. 68*, 499s–509s (1989).
31. J. Mazumder, M.M. Chen, C.L. Chan, R. Zehr, and D. Voelkel, *Proc. 3rd European Conf. Laser Treatment of Materials* (ECLAT '90), AWT, Germany, p.37 (1990).
32. J.F. Ready, *J. Appl. Phys. 36*, 462 (1965).
33. J.G. Andrews and D.R. Atthey, *J. Inst. Math. Appl. 15*, 59 (1975).
34. M.V. Allmen, *J. Appl. Phys. 47*, 5460 (1976).
35. C.L. Chan and J. Mazumder, *J. Appl. Phys. 62*, 4579 (1987).
36. C.L. Chan and J. Mazumder, *J. Appl. Phys. 63*, 5890 (1988).
37. M.F. Ashby and K.E. Easterling, *Acta Metall. 30*, 1969 (1982).
38. R.J. Grosh, E.A. Trabant, and G.A. Hawkins, *Q. Appl. Math. 13*, 161 (1955).
39. C.M. Adams, *Weld. J. 37*, 210s (1958).
40. C.M. Adams, W.G. Moffatt, and P. Jhaveri, *Weld. J. 41*, 12s (1962).
41. W.F. Hess, L.L. Merrill, E.F. Nippes, and A.P. Bunk, *Weld. J. 22*, 377s (1943).
42. Z. Paley, J.N. Lynch, and C.M. Adams, *Weld. J. 43*, 71s (1964).
43. E.G. Signes, *Weld. J. 51*, 473s (1972).
44. D.E. Schillinger, I.G. Betz, and H. Markus, *Weld. J. 49*, 410s (1970).
45. A.A. Wells, *Weld. J. 31*, 263s (1952).
46. P. Jhaveri, W.G. Moffatt, and C.M. Adams, *Weld. J. 41*, 12s (1962).
47. M.J. Bibby, J.A. Goldak, and G.V. Shing, *Canad. Metall. Q. 24(1)*, 101 (1985).
48. D.T. Swift-Hook and A.E.F. Gick, *Weld. J. 52*, 492s (1973).
49. S. Kou, *Proc. Conf. Modeling of Casting and Welding Processes*, pp.129–138, Metall. Soc. AIME, New York (1981).
50. S. Kou, *Weld. J. 61*, 175s (1982).
51. J.N. Gonsalves and W.W. Duley, *Can. J. Phys. 49*, 1708 (1971).
52. U.C. Paek and F.P. Gagliano, *IEEE J. Quant. Electron. QE-8*, 112 (1972).
53. R. Trivedi and S.R. Srinivasan, *J. Heat Transfer (Trans. ASME) 96*, 427 (1974).
54. T. Hashimoto and F. Matsuda, *Trans. Nat. Res. Inst. Metals (Jpn) 7*, 96 (1965).
55. W.B. Li, K.E. Easterling, and M.F. Ashby, *Acta Metall. 34*, 1533 (1986).
56. D. Farias, S. Denis, and A. Simon, *Proc. Euro. Scientific Laser Workshop on Mathematical Simulation*, Sprechsaal Publ. Group, Coburg, p.31 (1989).
57. P. Canova and E. Ramous, *J. Mat. Sci. 21*, 2143 (1986).
58. Y. Nakao and K. Nishimoto, *Trans. Japan Weld. Soc. 17*, 84 (1986).
59. M. Rappaz, *Int. Mat. Rev. 34*, 93 (1989).
60. S.A. David and J.M. Vitek, *Int. Mat. Rev. 34*, 213 (1989).
61. Y. Nakao, K. Nishimoto, and W.P. Zhang, *Trans. Jpn. Weld. Soc. 19*, 100 (1988).
62. H.K.D.H. Bhadeshia, S.A. David, and J.M. Vitek, *Mat. Sci. Tech. 7*, 50 (1991).
63. G.K. Solbakk, J.C. Ion, and J.E. Tibballs, *Proc. 1st Nordic Laser Materials Processing Conference*, Senter for Industriforskning, Oslo, 1987.
64. H.W. Bergmann, *Surf. Eng. 1*, 137 (1985).
65. A. Bloyce, I. Hancock, and H.W. Bergmann, in *Laser/Optoelektronik in der Technik*, p. 417, Springer Verlag, Berlin (1986).
66. A. Kar and J. Mazumder, *Acta Metall. 36*, 701 (1988).
67. Y. Arata, K. Nishigichi, T. Ohji, and N. Kohsai, *Trans. JWRI 8*, 1 (1975).
68. E.A. Metzbower, *Weld. J. 69*, 272s (1990).
69. H.R. Shercliff and M.F. Ashby, *Acta Metall. 38*, 1789 (1990).
70. H.R. Shercliff and M.F. Ashby, *Acta Metall. 38*, 1803 (1990).
71. O.R. Myhr, PhD Thesis, Technical University of Norway (1991).
72. W.B. Li and K.E. Easterling, *Surf. Eng. 2*, 43 (1986).
73. M.R. James, D.S. Gnanamuthu, and R.J. Moores, *Scripta Metall. 18*, 357 (1984).
74. H.R. Shercliff and M.F. Ashby, Cambridge University Engineering Department Report CUED/C-Mat/TR134 (1986).
75. W.M. Steen and C. Courtney, *Met. Technol. 6*, 456 (1979).
76. Y. Arata, K. Inoue, and S. Matsumura, *Proc. Materials Processing Symposium (ICALEO '83)*, Laser Institute of America, Toledo, p.100 (1983).
77. O. De Pascale, C. Esposito, P. Boffi, and M. Lepore, *Materials Chemistry and Physics 19*, 205 (1988).
78. A. Drenker, E. Beyer, L. Böggering, R. Kramer, and K. Wissenbach, *Proc. 3rd European Conf. Laser Treatment of Materials (ECLAT '90)*, AWT, Germany, p. 283 (1990).
79. S.C. Hsu, S. Chakravorty, and R. Mehrabian, *Metall. Trans. 9B*, 221 (1978).
80. L.E. Greenwald, E.M. Breinan, and B.H. Kear, *Proc Conf. Laser-Solid Interactions and Laser Processing 1978*, AIP, New York (1979).
81. W.M. Steen, *Proc. NATO ASI Laser Surface Treatment of Metals*, p. 369, Martinus Nijhoff, Dordrecht (1986).
82. L.J. Li and J. Mazumder, *Proc. Symp. Laser Processing of Materials*, TMS-AIME, PA (1984).
83. L. Li, W.M. Steen, and R.D. Hibberd, *Proc. 3rd European Conf. Laser Treatment of Materials (ECLAT '90)*, AWT, Germany, p. 355 (1990).
84. L. Li, W.M. Steen, R.D. Hibberd, and D.J. Brookfield, *Proc. Int. Cong. Lasers and Laser Processing*, vol. 1279, SPIE, The Hague (1990).
85. T. Chande and J. Mazumder, *Metall Trans. 14B*, 181 (1983).
86. T. Chande and J. Mazumder, *J. Appl. Phys. 57*, 2226 (1985).
87. P.G. Klemens, *J. Appl. Phys. 47*, 2165 (1976).
88. T. Miayazaki and W.H. Giedt, *Int. J. Heat and Mass Transfer 25*, 807 (1982).
89. J.K. Kristensen, L.H. Hansson, and F.L. Smidth, *Proc. Conf. Electron and Laser Beam Welding*, Pergamon Press, Oxford (1986).
90. N. Hamada, O. Ichiko, and H. Soga, *Trans. ISIJ 26*, B-120 (1986).
91. J. Dowden, P. Kapadia, and N. Postacioglu, *J. Phys. D: Appl. Phys. 22*, 741 (1989).
92. R. Peretz, *Optics and Lasers in Engineering 7*, 69 (1987).
93. M.C. Tsai and S. Kou, *Proc. Conf. Power Beam Processing*, ASM International, Ohio (1988).
94. W.H. Giedt and L.N. Tallerico, *Weld. J. 67*, 299s (1988).
95. R. Peretz, *Optics and Lasers in Engineering 11*, 27 (1989).
96. D. Schuöcker, *Proc. 3rd European Conf. Laser Treatment of Materials*, AWT-Arbeitsgemeinschaft, Germany, p. 55 (1990).
97. W.M. Steen, J.M. Dowden, M.P. Davis, and P. Kapadia, *J. Phys. D: Appl. Phys. 21*, 1255 (1988).
98. R. Akhter, M. Davis, J. Dowden, P. Kapadia, M. Ley, and W.M. Steen, *J. Phys. D: Appl. Phys. 21*, 23 (1989).
99. A. Tiziani, A. Zambon, F. Bonollo, and M. Cantello, *Proc Conf. Laser-5*, IITT International, Gournay-sur-Marne, France, p.75 (1989).

100. G. Molino, M. Penasa, G. Perlini, A. Tiziani, and A. Zambon, *Proc. 3rd European Conf. Laser Treatment of Materials (ECLAT '90)*, AWT, Germany, p. 195 (1990).
101. P. Ravi Vishnu, W.B. Li, and K.E. Easterling, *Report TULEA 1990:17*, Luleå University of Technology, Sweden (1990).
102. C.J. Dawes and M.N. Watson, *Proc. Materials Processing Symposium (ICALEO '83)*, Laser Institute of America, Toledo, p. 73 (1983).
103. I. Belic and J. Stanic, *Optics and Laser Technol. 19*, 309 (1987).
104. I. Belic, *Optics and Laser Technol. 21*, 277 (1989).
105. Y. Aberkane and M. Dumas, *Proc. 3rd Int. Conf. Lasers in Manufacturing (LIM-3)*, IFS Publications, Bedford, UK, p. 77 (1986).
106. D. Petring, P. Abels, and E. Beyer, *Proc. Int. Cong. Optical Science and Engineering*, SPIE vol. 1020, Washington (1988).

Laser-beam quality and brightness in industrial applications

K.D. Hachfeld
Coherent General Inc.
Sturbridge, MA, USA

Industrial lasers have been successfully cutting, welding, drilling, heat-treating, etc., for many years. Volumes have been written about the associated methods, process rates, and quality. This data has been collected at corporate research laboratories, technology centers, laser job shops, and, of course, laser manufacturers. Most of this has been done with relatively loose knowledge about details of the laser beam emanating from high-power industrial lasers.

Typically, laser manufacturers declare the rated laser output power, the beam size as it leaves the laser, and the beam divergence, along with some "fluffy" comment about how good the beam quality is. But this is rapidly changing.

Intuition and simple computer modeling of Gaussian laser beams convince us that if we can only focus the beam to a smaller spotsize on the workpiece, we will have more power density on the material. That should give us smaller kerf widths, faster cutting rates, and, possibly, other benefits. But is it as simple as that?

So far, no mention has been made of workpiece material properties, the dynamics of the cutting process, or geometry, all of which strongly impact process rates and quality. To simplify analysis of the laser-machining process, we need to separate the condition of the laser beam as it travels from the laser to the workpiece from what happens once the beam hits the material. Material interactions are difficult to model accurately, though such work continues to be done with improving success.[1–4] On the other hand, the laser beam has been understood for decades, yet not adequately modeled because real beams are more complex than we generally want to accept.

For example, acrylic laser burns have served the high-power CO_2 laser community well in that they have shown qualitatively all the defects of real beams and have given approximate beam sizes.[5] This data has been good enough to indicate that beams are rarely close to the diffraction-limited TEM_{00} model we have in mind, though they may look essentially Gaussian. That means the size of the focused beam on the workpiece is not usually what the user expects, unless real measurements have been made right at the focus. However, this region also has the greatest potential for destroying the measuring instrument.

In the last few years, instruments have been developed that not only probe the raw laser beam but also the laser focus and give the intensity distribution, beam size,[6] and beam quality[7–9] of high-power CO_2 lasers. These instruments have given the laser designer new opportunities to evaluate each laser's suitability for cutting, welding, and other industrial applications. These instruments also indicate problems faced by the laser-machine integrator. This paper attempts to answer questions relating to laser-beam quality and brightness and what impact these features have in the workplace.

Laser-beam quality

Real industrial laser beams are not usually the textbook TEM_{00}, TEM_{01^*}, or higher order modes derived from empty stable-resonator theory. They can come close, but they are more likely to be the superposition of several of these modes, with each mode contributing some of the total power emitted by the laser. In addition, there are likely to be circular diffraction ripples impressed on the Gaussian intensity distribution, which originate at any location where the beam passes through and touches a hard-edged aperture. CO_2 lasers often show the ripples close to the laser, but less so further away. There may also be a low-level "pedestal" of laser power stretching significantly beyond the edges of the beam that is likely to add to the heat-affected zone on the workpiece.

The laser designer strives for beam perfection. He aims to get as much power in the TEM_{00} mode—which really does have a Gaussian intensity distribution and is really at the physical limit imposed by nature—but he rarely gets it.

To begin to understand beam quality, imagine that the beam diameter can be measured anywhere along a beam and compared with the mathematical model:

$$W(z) = W_{02} \cdot \left\{ 1 + [\lambda M^2 \cdot (z-a)/(\pi \cdot W_{02}^2)]^2 \right\}^{1/2} \qquad (1)$$

where $W(z)$ is the beam radius at distance z from the laser; W_{02} is the smallest beam radius, at the beam waist; a is the location of the beam waist from the laser; λ is the laser wavelength; and M^2 is the beam quality figure of merit.

The envelope of such a beam is illustrated in Fig. 1, where a distinction is made between a real laser beam having a mode radius W and the theoretical best TEM_{00} mode with radius w. The (bold) real-beam envelope is said to have lower beam

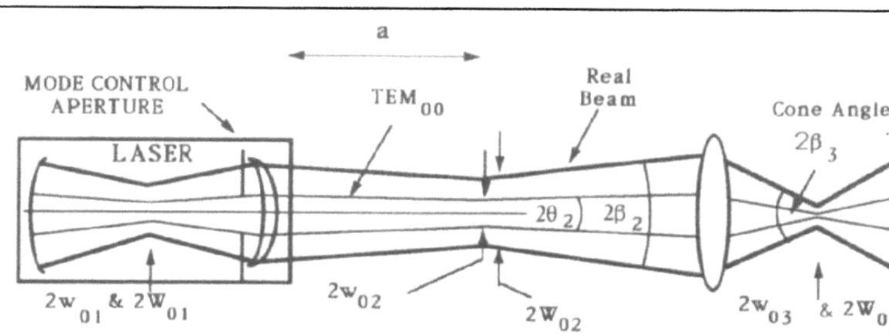

Figure 1. Spatial relationship between modes of different order from the same source.

quality than the TEM_{00} beam because it is consistently larger all the way to the workpiece. The space between the laser and the lens represents a simple beam-delivery system on a machine. The space between the lens and its focused beam diameter $2.W_{03}$ is the "hot" end of the machine, where the workpiece would normally be placed.

The mode control aperture indicated in the laser in Fig. 1 represents any mode-size-limiting mechanism the laser designer has built into the product. As such, it influences the actual size and mode purity of the emitted laser beam. In principle, the aperture size can be varied to give either the ideal TEM_{00} mode or a larger beam that could be a combination of modes. The essence is that the TEM_{00} mode will always be "nested" within the real beam.

We can define a dimensionless beam-quality figure of merit M^2 as:

$$M^2 = (\pi/4\lambda) \cdot (2W_0 \cdot 2\beta) \tag{2}$$

where $\pi = 3.14$ and 2β is the real mode's total beam divergence, in contrast with 2θ, the TEM_{00} mode's total beam divergence. Each is measured between the $1/e^2$ intensity points.

Here, equation 2 and subsequent equations are written in terms of $2W_0$, the focal beam diameter, and 2β, the full beam divergence. This was done because that is what is usually measured, and it eliminates confusion. Written this way, there is an obvious link with much of the laser literature, which has standardized on using the radius W and beam divergence half-angles θ and β.

In solid-state laser circles, $(2W_0 \cdot 2\beta)$ is known as the millimeter-milliradian product. In practice:

$$M^2 = 0.74 \cdot (2W_0 \cdot 2\beta) \quad \text{for } \lambda = 1.06\ \mu\text{m (for Nd:YAG)},$$

$$M^2 = 0.074 \cdot (2W_0 \cdot 2\beta) \quad \text{for } \lambda = 10.6\ \mu\text{m (for } CO_2\text{)}.$$

In both cases, $2W_0$ is in millimeters and 2β is in milliradians.

In principle, the lowest and best value M^2 can have is $M^2 = 1$, corresponding to the lowest order TEM_{00} mode. However, typical high-power lasers have values of M^2 between 1.2 and 200, depending on the design technology.

While M^2 is rapidly becoming accepted in the USA as *the* Quality Figure of Merit, an alternative, K, is being offered in Europe.[7,9] K is the inverse of M^2. Hence, K is 1 for TEM_{00} and becomes less than 1 for lower beam quality.

Another useful way of looking at M^2 is that higher order modes from the same laser source scale linearly with M, relative to the TEM_{00} mode. That means:

$$2W = M \cdot (2w) \tag{3}$$

everywhere along the beam, including the focal point or waist, where W becomes W_o. Consequently, the total beam divergence 2β for the real beam is given by:

$$2\beta = M \cdot (2\theta) \tag{4}$$

and:

$$2\theta = 2\lambda/\pi w_o \tag{5}$$

Thus, if a design change made to a given laser reduces its M^2 value, the new beam can have smaller dimensions and a smaller beam divergence.

An example would be if we put an aperture into a YAG laser. A high-mode-order beam results with the aperture wide open. M^2 will be high ($M^2 = 100$ at 300 W is not unusual), as will the beam diameter and divergence. Conversely, with the aperture closed to the point where only the TEM_{00} mode operates, then $M^2 = 1$ and beam diameter and divergence will be the smallest possible–but the laser power will also be typically low.

It is instructive, when comparing the beam quality emitted by different lasers, especially lasers with different wavelength, to look at the component values of M^2. Generally, when looking only at beam quality, the lower the value of M^2, the better. The significance of low M^2 to the laser machine designer and user is that:

1. A given diameter of beam remains more collimated over a greater distance. Hence, larger machines with more travel can be built while retaining the same diameter of optics. This can be seen from the "range of collimation b," also known as the "confocal parameter," defined by:

$$b = 2\pi \cdot (2W_{02})^2/4 \cdot \lambda M^2 \tag{6}$$

The range of collimation is indicated in Figure 2 by the distance between planes A and B, at which the beam

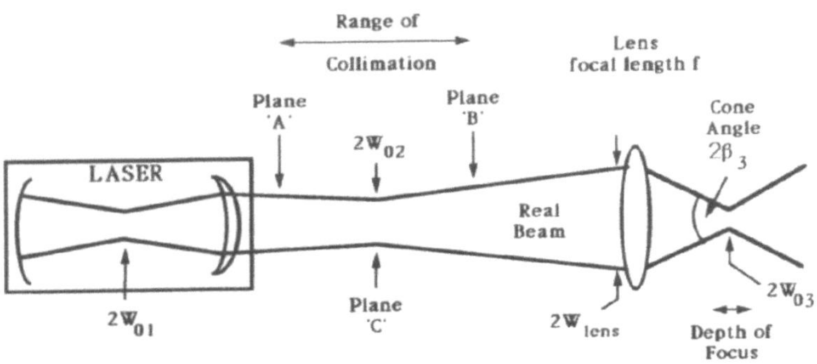

Figure 2. A simple laser-beam delivery system.

diameter is 41% larger than at plane C, the waist. This becomes:

$b = 1.48 \, . \, (2W_{02})^2/M^2$ (for Nd:YAG)
$b = 0.148 \, . \, (2W_{02})^2/M^2$ (for CO_2)

In each case, b is in meters and $2W_{02}$ is in millimeters.

2. Smaller focal spot diameters can be achieved on the workpiece, resulting in greater material penetration for the same laser power. A good approximation for the spot size on the workpiece is given by:

$$2W_{03} = 4 \, . \, \lambda M^2 \, . \, f \, / \, \pi \, . \, 2W_{lens} \quad (7)$$

which assumes that the beam incident on the lens is approximately collimated—a reasonable assumption.

In practice:

$2W_{03} = 1.35\times10^{-3} \, . \, M^2 \, . \, f \, / \, (2W_{lens})$ (for Nd:YAG)
$2W_{03} = 13.5\times10^{-3} \, . \, M^2 \, . \, f \, / \, (2W_{lens})$ (for CO_2)

In each case, f, $2W_{lens}$ and $2W_{03}$ are in millimeters.

3. The depth of focus (DOF) can be increased, dependent on machine design details. Expressed in terms of the focal spot diameter, which the operator usually specifies, we have:

$$DOF = 2\pi \, . \, (2W_{03})^2/4 \, . \, \lambda M^2 \quad (8)$$

For practical purposes:

$DOF = 1{,}480 \, . \, (2W_{03})^2/M^2$ (for Nd:YAG)
$DOF = 148 \, . \, (2W_{03})^2/M^2$ (for CO_2)

In each case, DOF and $2W_{03}$ are in millimeters.

At the limits of this range, the power density on the workpiece is half that at the focus.

4. The "cone angle $2\beta_3$" is small, making it easier to feed the beam down long, narrow gas nozzles or confined spaces:

$$2\beta_3 = 4 \, . \, \lambda M^2 \, / \, \pi . 2W_{03} \quad (9)$$

Practically:
$2\beta_3 = 1.35 \, . \, M^2 \, / \, 2W_{03}$ milliradians
$2\beta_3 = 0.077 \, . \, M^2 \, / \, 2W_{03}$ degrees (for Nd:YAG)

For CO_2:
$2\beta_3 = 13.5 \, . \, M^2 \, / \, 2W_{03}$ milliradians
$2\beta_3 = 0.77 \, . \, M^2 \, / \, 2W_{03}$ degrees

In each case, $2W_{03}$ is in millimeters.

Laser-beam brightness

Laser-beam brightness is the parameter that places laser light apart from light created by other means. It is a measurable combination of laser-beam power and quality with basic units of watts per square centimeter per steradian.

As such, it can be thought of in terms of a beam that is focused on a workpiece where the watts per square centimeter represent the familiar power density incident on the material. The steradian represents the three-dimensional solid angle Φ of the cone of light striking the material. Φ is related to the two-dimensional cone angle β (in radians) of Fig. 1 by:

$$\Phi = \pi \, . \, \beta^2 \quad (10)$$

Brightness B is defined in terms of the concepts given above:

$$B = (4/\pi)^2 \, . \, P \, / \, (2W_0 \, . \, 2\beta)^2 \quad (11)$$

where P is the laser beam power.

Equation 6 can be rewritten in terms of M^2 as:

$$B = P \, / \, (\lambda M^2)^2 \quad (12)$$

In practical terms:

$B = 89.0 \, (P \, / \, (M^2)^2)$ for $\lambda = 1.06$ µm (Nd:YAG)
$B = 0.89 \, (P \, / \, (M^2)^2)$ for $\lambda = 10.6$ µm (CO_2)

In each case, B is brightness in MW/cm^2-steradian, which is 10^6 W/cm^2-steradian, and P is laser power in watts.

This shows that brightness will be high for a given beam M^2 value if the beam power is high. Alternatively, for a fixed laser-beam power, the brightness will be high if M^2 is low. Highest brightnesses are achieved when the laser power is high at the same time M^2 is low. This results in high power densities for small beam cone angles incident on the workpiece.

Applying these concepts to Fig. 1, we conclude that the TEM_{00} beam has higher brightness than the real beam if they

Figure 3. Beam quality vs. output power for various Nd:YAG lasers.

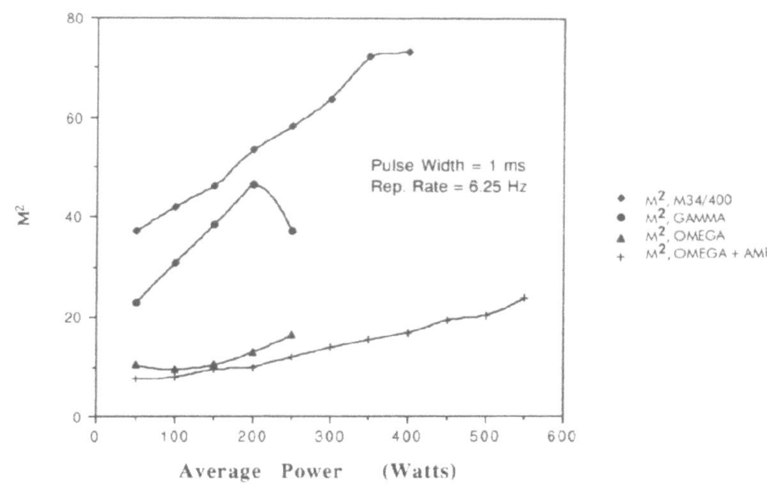

have equal power. Therein lies the catch! Generally, the laser designer makes every effort to achieve low M^2 values while maintaining high power because that leads to greatest application versatility. However, sometimes the design is pitched toward a specific group of applications, based on input researched by laser applications engineers, knowing that the laser application sets the requirements for the laser-beam power and quality.

Note that beam quality is constant everywhere along a beam, providing the beam quality has not been degraded by poorly mounted mirrors or distorted lenses. So, with reference to Fig. 1, M^2 values measured at the workpiece using values of W_{03} and β_3 should equal those measured at the laser using values of W_{02} and β_2.

Equation 6 indicates that brightness is also constant everywhere along a beam, if in addition there is no loss of laser power as the beam passes from the laser to the workpiece. Thus, the definitive comparison between laser designs is given by the brightness value under similar test conditions.

Further understanding is gained by also looking at a plot of M^2 vs. output power. Figure 3 illustrates that laser beam quality is not fixed for all values of laser output power. It is dynamic. High brightness, illustrated in Fig. 4, is obtained by M^2 remaining low as the power rises.

The value of the brightness concept is that it can give a rapid (back of the envelope) answer to how much power density will fall on the workpiece for a specified cone angle β_3. Only the brightness number for the laser has to be known because the concept is independent of any beam-delivery details. Equation 11 can be rewritten to give the power density I_3 on the workpiece, independent of wavelength, as:

$$I_3 = (\pi/4) . (2\beta_3)^2 \ . B \qquad (13)$$

For easy implementation, $I_3 = 0.78 \times 10^{-6} . (2\beta_3)^2 . B$, with $2\beta_3$ in milliradians, and $2.39 \times 10^{-4} . (2\beta_3)^2 . B$, with $2\beta_3$ in degrees. In each case, I_3 is in MW/cm^2 and B is in MW/cm^2-steradian.

This makes the assumption that the beam-delivery optics have large enough diameters to transmit the beam and that there is negligible distortion.

For comparison, Table 1 gives measured M^2 and brightness values for some high brightness lasers. It shows that the beam quality of high-power CO_2 lasers come much closer to the theoretical diffraction limit ($M^2 = 1$) imposed by their wavelength than do high-power Nd:YAG lasers. This should not be viewed negatively, as the difference is almost entirely due to the difference in laser wavelength. Instead, there is more opportunity for improvement in spot size and range of collimation with Nd:YAG than with CO_2 lasers.

The table also shows that the high mean brightness values of CO_2 lasers are due to their high average powers, which cannot be matched by Nd:YAG lasers. On the other hand, Nd:YAG lasers outshine CO_2 lasers when pulsed, as indicated by the peak brightness and peak power values. Given these differences, it is easy to see why Nd:YAG lasers can drill and cut 1-in.-thick metals at low speed, while CO_2 lasers cut thinner material at much higher speed.

A word of caution: Laser designs are becoming more esoteric in an attempt to improve efficiency and brightness—in some cases at the expense of beam symmetry. Some laser applications do benefit from beams that are not round, but they tend to be very specific. In general, a beam with circular symmetry and the same values of M^2 in two perpendicular planes—say, the horizontal and vertical planes—will be easiest to use.

Mode matching with a telescope

We have now looked at the laser end and the workpiece end in terms of beam quality and brightness, which has allowed us to compare the ability of two different lasers to irradiate the target. What should be noted is that two different lasers having the same M^2 and B values fitted to the same beam-delivery system do not automatically produce the same beam

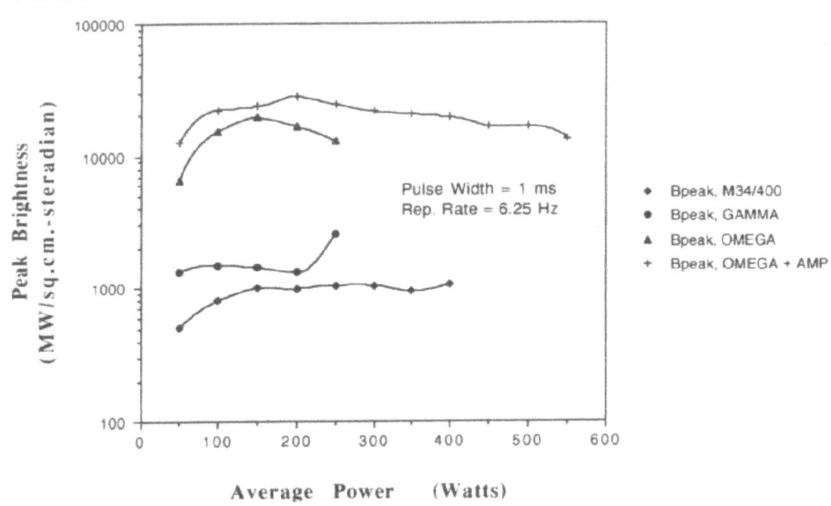

Figure 4. Peak brightness vs. output power for various Nd:YAG lasers.

on the target. Laser X may have a large beam diameter at the laser with low beam divergence, and vice versa with Laser Y. Thus the beam on the focusing lens and the workpiece will probably be different in size. This does not invalidate anything said so far.

To assure laser interchangeability, the machine designer should run a computer analysis of the beam propagation for each laser and fit a suitable telescope between the laser and focusing lens. Still all the equations given above remain valid.

Laser-beam dynamics

It is worth thinking about the mechanisms that change a laser beam and its quality as it propagates to the workpiece. Different mechanisms affect the user in different ways, and some of these can be managed by the operator.

As the laser beam is bounced off beam-bending mirrors, passes through various gases, or is transmitted through laser rods, an output coupler, or a cover slide, it becomes defocused and distorted. It is important to know that the defocusing component only changes the size and location of the focal point on the workpiece. While inconvenient, it is still possible for focus to be recovered with no change in beam quality or brightness, if the appropriate adjustments have been built into the beam-delivery system. However, added distortion reduces the brightness, tends not to be recoverable, and negatively impacts the application.

Typically, as the output is varied, the beam from a solid-state laser (Nd:YAG) will change more than one from a gas laser (CO_2).

Table 1. Brightness and M^2 for several lasers

Laser	Laser wavelength (microns)	Mean O/P power (watts)	Peak O/P power (watts)	M^2	Brightness Mean (MW/cm^2-sterad)	Brightness Peak (MW/cm^2-sterad)
Nd:YAG						
Gamma (40J, 1ms)	1.06	250	40,000	38	15	2465
Omega (40J, 1ms)	1.06	250	40,000	17	77	12,318
CO_2						
Arrow (CW)	10.6	1100	—	1.8	302	—
Arrow (1.2ms, 125 Hz)	10.6	500	5500	1.8	137	1511
Vulcan (CW)	10.6	3000	—	2.2	552	—

Brightness = 89.0 (Power)/$(M^2)^2$ for Nd:YAG lasers
Brightness = 0.89 (Power)/$(M^2)^2$ for CO_2 lasers
Brightness in MW/cm^2-steradian
Power in watts

Intrinsic distortions

Lasers are not very efficient. Therefore there is a large difference between how much power is put into a laser and how much power comes out in the form of a laser beam. The difference nearly always gets wasted as heat and has to be removed. Heat-removal mechanisms vary according to the laser design, but often result in temperature gradients being established across the laser rod or laser gas, through which the beam passes. This is the principal mechanism affecting the size of the beam as power into the laser is varied. Careful design limits this effect, but the user has little influence on it.

Extrinsic distortions

To some degree, extrinsic distortions are under the control of the laser user and relate mostly to optics cleanliness. In order of significance, they are:

1. Contamination, such as dirt, on the output coupler, focusing lens, windows, cover slides, and other "transmitting" optics. Metal spattered from the workpiece back onto the focusing lens absorbs some laser power and heats the optic. Its dimensions and refractive index change, resulting in a change of focal spot size and location. Hence, the kerf width and penetration of the workpiece change. This is as appropriate to high-power CO_2 lasers as it is to Nd:YAG lasers, where the use of lowest absorption materials is recommended.
2. Beam-delivery mirrors also increasingly absorb laser power as they become dirty or damaged. This has less influence on the application because the mirror can only change its curvature. However, if the laser is part of a complex machine with many mirrors in the beam-delivery path, the cumulative effect may become dominant. Face-mounted mirrors must have near perfect pads to mount against or the mirror surface and beam will be distorted.
3. The gas in the beam-delivery tubes, all the way from the laser to the workpiece, should be filtered, free of water, oil, and solvent vapors. In addition, care must be taken when dusting off CO_2 laser optics with Freon and other fluorocarbon gases because they also absorb laser radiation. Lack of attention here results in beam "blooming," an arbitrary increase in beam size and stability, giving poor results on the workpiece.

Good maintenance pays dividends, not only because the beam brightness is maintained at the workpiece, but because focal spot size changes and focus position shifts will be limited.

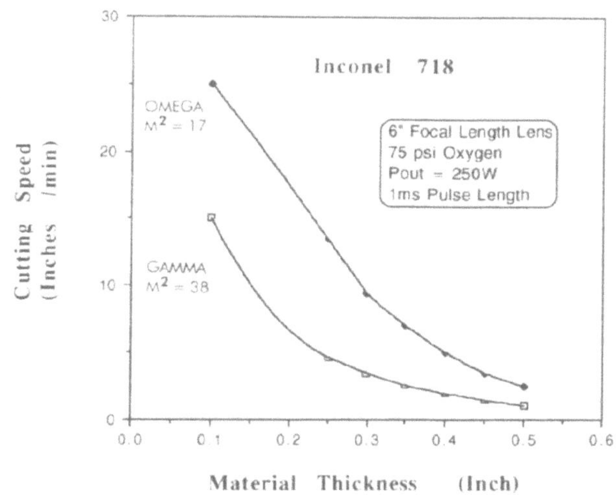

Figure 5. Cutting speed vs. material thickness for two Nd:YAG lasers.

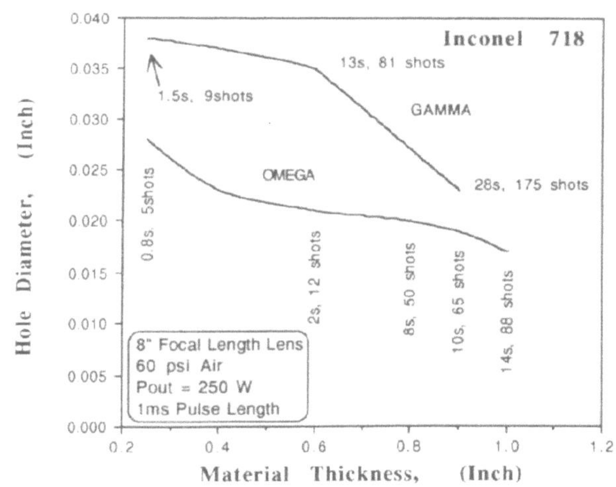

Figure 6. Percussion drilling rates and hole sizes for two Nd:YAG lasers.

Performance and beam brightness

The relationships of cutting speed and drilling efficiency with brightness are not simple to obtain experimentally because different lasers having the same beam-delivery system will normally produce different spot sizes unless a telescope is fitted. Nonetheless, Figures 5 and 6 show typical machining improvements. For example, these figures indicate that Omega drills Inconel 718 twice as efficiently as Gamma, though it is five times as bright. Hence the relationship between brightness and drilling efficiency is not linear, which is not very surprising considering the complexity of the activity within the material.

Some general observations about high-brightness lasers can be made:

- Higher brightness lasers make machines cut faster and drill deeper until the speed is limited by removal of melted material.
- Improved cutting rates are achieved because smaller spot sizes are available, and this can be exploited in thin stock.
- Thicker material can be machined, principally because the cone angle is reduced, when maintaining the same spot size on the workpiece.
- Making the spot very small on thick stock has no merit because the kerf will be so narrow that the material has no opportunity to exit.
- Further benefits in machining rates can be obtained through helping liquid and gaseous material exit the kerf by changing assist gas pressure, nozzle shape, and kerf width.

Conclusion

The concept of laser-beam brightness, taken together with laser-beam quality, can be used in a simple manner to compare attainable power densities on the workpiece. This does not require detailed knowledge of the beam-delivery system. As such, it allows comparison between lasers of different manufacture.

Interchangeability between lasers having the same M^2 and B values is only guaranteed if a mode-matching telescope can also be fitted.

Brightness and quality concepts only deal with the beam to the point of impact with the target. Workpiece mechanical and thermal properties, delivery geometry, and customer-imposed constraints will eventually limit process rates.

Acknowledgments

The author is pleased to acknowledge experimental help and many discussions with colleagues in the laser industry that have contributed to this article, as well as Mrs. P. Gross' editorial assistance.

References

1. S. Ramanathan and M.F. Modest, "Effect of Variable Properties on Evaporative Cutting with a Moving CW Laser," *ASME Proceedings HTD-135*, 101–108 (1990).
2. S.Y. Bang and M.F. Modest, "Multiple Reflection Effects on Evaporative Cutting with a Moving CW Laser," *ASME Proceedings HTD-137*, 65–72 (1990).
3. F.O. Olsen, "Theoretical Investigations in the Fundamental Mechanisms of High Intensity Laser Light Reflectivity," *Proc. SPIE 1020, High Power CO_2 Laser Systems and Applications*, Hamburg (Sept 1988).
4. F.O. Olsen, "Cutting Front Formation in Laser Cutting," *Annals of the CIRP 38* (Jan. 1989).
5. D. Whitehouse and C.J. Nilsen, "Plastic Burn Analysis for CO_2 Laser Beam Diagnostics," Laser Materials Processing Symposium, ICALEO 90, Boston, MA (Nov. 1990).
6. G.C. Lim, "Laser Beam Quality," *The Industrial Laser Annual Handbook, 1989 Edition*, pp. 96–116.
7. P. Loosen, et. al, "Diagnostics of High-Power Laser Beams," *Proc. SPIE 1024, Conference on Beam Diagnostics and Beam Handling Systems*, Hamburg (1988).
8. W. Woodward, "A New Standard for Beam Quality Analysis," *Photonics Spectra* (May 1990). Also, information from Coherent Inc., Components Group, 2301 Lindbergh St., Auburn, CA 95603.
9. M.W. Sasnett and T.F. Johnston Jr., "Beam Characterization and Measurement of Propagation Attributes," *SPIE 1414*, 21 (1991).
10. Gregersen and F. Olsen, "Beam Analysing System for CO_2 lasers," Laser Materials Processing Symposium, ICALEO 90, Boston, MA (Nov. 1990).

High-radiance near-infrared lasers for material processing

Ken Manes
Advanced Applications Laser Program
Lawrence Livermore National Laboratory
Livermore, CA

Background: Technology transfer initiative

The US Congress has, through its enactment of new laws during the 1980s, made it clear that it wants taxpayer-funded research and development to have a greater impact on the country's ability to compete economically and technologically in world markets. The Department of Energy's contractor-operated laboratories, such as the University of California's Lawrence Livermore National Laboratory (LLNL), have been singled out as important centers of technology that can be better leveraged to the benefit of US industry. At LLNL, a technology-transfer program has been implemented to facilitate the process of interface between US industry and the Laboratory.

In recent years, US industries have been moving to automate many of their production-line processes. In the automobile industry, for example, the use of robots and lasers have made it possible to increase productivity, improve quality, and make the change to new models at lower cost. In Japan, advances in the fields of lasers and robotics have afforded Japanese automakers a potential advantage over their US counterparts. To compete more effectively, the US industry must have laser-based welding, cutting, and drilling capabilities that are more accurate, reliable, and flexible than existing systems.

Of all of LLNL's many areas of competence, its ability to fully comprehend how lasers can be applied to solve real problems and then develop advanced laser systems is probably one of its most valuable and important capabilities. Under the laboratory's technology-transfer program, industry is becoming convinced that LLNL's staff and facilities can be accessed in a timely manner, and that the participating company's proprietary data and LLNL's intellectual property, resulting from the project, can be protected and effectively applied to the task.

Introduction

Substantial experience with solid-state laser machining of metals, plastics, and ceramics has been accumulated in recent years. Excellent accounts of this emerging technology are generating industrial interest as it becomes clear how economically superior near-infrared laser sources wedded to fiber optical distribution systems can be compared to e-beam or even CO_2 laser welders.[1]

Solid-state lasers have been used successfully in the automobile industry for almost two decades.[2] For example, Nd:YAG lasers having output power levels of only 350 W drill holes angled at 35° in hardenable steel (Rc 60 hot forged powder metal alloy) transmission components such as gears. These holes are 7.5-mm deep × 1.5-mm in diameter and are cut in 3.5 sec using the commonly available medium power lasers, without the assistance of an oxidizing gas. A more recent account of the potential of laser materials processing in the automobile industry—containing many more examples and 146 references—has been prepared by General Motors Research Laboratories.[3]

With the advent of face-pumped zig-zag slab lasers, greatly improved beam quality has dramatically extended the drilling performance of Nd:YAG lasers. General Electric Aircraft Engines uses Nd:YAG lasers in turbine engine manufacture, and thus actively supports internal research and development of these systems. The CF series (wide-body) engines use 80,000 laser-drilled transpiration cooling holes; the smaller CFM series (commuter) engines have about 40,000 cooling holes. GE engineers are increasing the numbers and depths of these holes in superalloy steels while calling for similar holes in ceramics and composites in their new engine designs. Early this year, GE announced in several popular magazines 1000 W performance from its face-pumped lasers.[4] Dramatic examples of face-pumped laser performance attributable to their high power, superior beam quality, and fiberoptic compatibility have been reported[5–7]:

- Holes drilled at oblique angles with single laser pulses "on the fly" in 1.5-mm-thick steel sheet at the rate of five/s.
- 5-cm-deep holes less than 1 mm in diameter drilled in stainless steel.
- A single 0.5 mm fiber, 50 m in length, is 90% efficient in transmitting 40 J/2 ms laser pulses at 900 W average power.

GE's research group is not alone in having recognized the potential of face-pumped lasers. The promise of virtually unlimited clean energy from controlled fusion has motivated research into inertial confinement fusion (laser fusion) at LLNL for almost 20 years. During that time, seven generations of Nd:Glass lasers for fusion experiments have been built at LLNL. The latest of these, NOVA, is capable of delivering up to 100 kJ in 1 ns (or 100 TW) to laser fusion target assemblies.

As now envisioned, an ICF electric power plant will need a driver delivering 3–5 mJ per 5–7-ns pulse at 10 Hz. LLNL has maintained an ICF laser-driver research effort throughout its ICF program history, and during the past several years face-pumped solid-state lasers have become serious contenders for practical ICF drivers.

Zig-zag slab laser development has been supported by DARPA and SDIO at LLNL for over four years. During the past two years, diode pumping technology for these devices has made rapid progress under both DoD and LLNL internal sponsorship. Thermo-optical design codes, optical propagation codes, laser materials development, laser pump source development, and integrated tests of laser hardware at LLNL generally meet or exceed the standards of performance of corresponding technologies worldwide.

Material processing laser requirements

In their 1988 review of laser surface treatment, A. Bloyce and T. Bell give examples of many of the emerging laser surface engineering processes beginning to have a strong impact on manufacturing.[8] Fig. 1, adapted from that review, indicates the laser irradiance and pulse duration regimes required for some of these processes. We have added drilling, cutting, and welding of mild steel to the figure as well as a lower irradiance application, laser-assisted hydraulic mining. Such laser material processing applications represent one of the largest segments of the laser systems market. In 1989 alone, laser materials processing system sales reached approximately $1.2 billion.

Unfortunately for US competitiveness in the manufacturing arena, the USA has steadily lost market share since the invention of the laser. In 1989, about 43% of this market was Japanese, another 26% was European, and the remaining 31% was supplied by the USA.

Beyond the sort of generic requirements suggested by Fig. 1, manufacturers around the globe are beginning to identify more specific requirements that, to one extent or another, are already being met by lasers but could be accomplished more cost-effectively by a diode array or a diode array pumped high-radiance device. Several such requirements sets are outlined below. In addition to being optical-fiber-compatible and requiring kilowatt power levels, each application has its unique specifications:

Punching cooling holes in aircraft engine turbine parts:

- ≥30 J per pulse
- 1 to a few ms pulse duration; may be bursts
- High radiance desired
- 5–30 Hz, cut "on the fly"

Welding mild sheet steel:

- 1- to a few-mm-thick stainless steel
- ≥10 cm/s welding and ≥20 cm/s cutting
- Stand-off distances of 10s of cm
- Beam quality ≈ 10 × diffraction limited
- Estimated power requirements: 2–4 kW

Machining composites and ceramics for aircraft:

- Al:Carbon composites probably require ablative pulses
- Ceramics may require 2ω or 3ω short pulses
- Pulse shape and format flexibility needed

Laser-assisted hydraulic drilling:

- ≥1 kW/cm^2 must be delivered to the mining face
- Laser beam must traverse approximately 1 cm of water
- Off-road vehicle mounted for field mobility

One day soon we'd like to add to this list an ICF driver laser for commercial power production capable of delivering 3–5 mJ/pulse at 10 Hz. To be sure, the laser diode's demonstrated electrical efficiency must be maintained as production is fully automated and scaled up, and laser diode pump cost must be reduced to pennies/peak watt. Note that a solid-state ICF driver laser can become cost-competitive with alternative power generation technologies *only* if volume production of laser diodes of order 10^8 peak watts per year greatly reduces the current price of laser diode pump arrays. Laser materials processing could make up a significant part of this volume.

Calculated laser requirements for sheet-metal cutting and welding

To weld sheet metal, a hot plasma line source (keyhole) should be established in the workpiece by drilling almost through the material at high temperature (≈ 20,000° K) with a focused laser beam. Drilling proceeds after an initial few hundred nanosecond transient period, during which the plasma is formed and the reflectivity of even polished metal surfaces becomes very low.[9] As the laser beam moves relative to the workpiece, molten metal flows around the keyhole and resolidifies to form the weld. The plasma formation and drilling process is most efficient over a narrow range of irradiance, 5–50 MW/cm^2.[10] Care must be taken to avoid the formation of unwanted oxides, and the strength of the uncontrolled melt region may be lower than that of the rest of the material. Cutting is a similar process except that molten metal is typically blown out of the kerf by an assist gas (Fig. 2).

We have scaled the recently reported welding performance of three different Nd:YAG industrial laser systems to arrive at an estimate for the laser power and beam quality required of a fiber-compatible solid-state laser to match the demonstrated performance of the RS850 CO_2 laser system. At IMTS'90, the claim was made that this laser had been able to

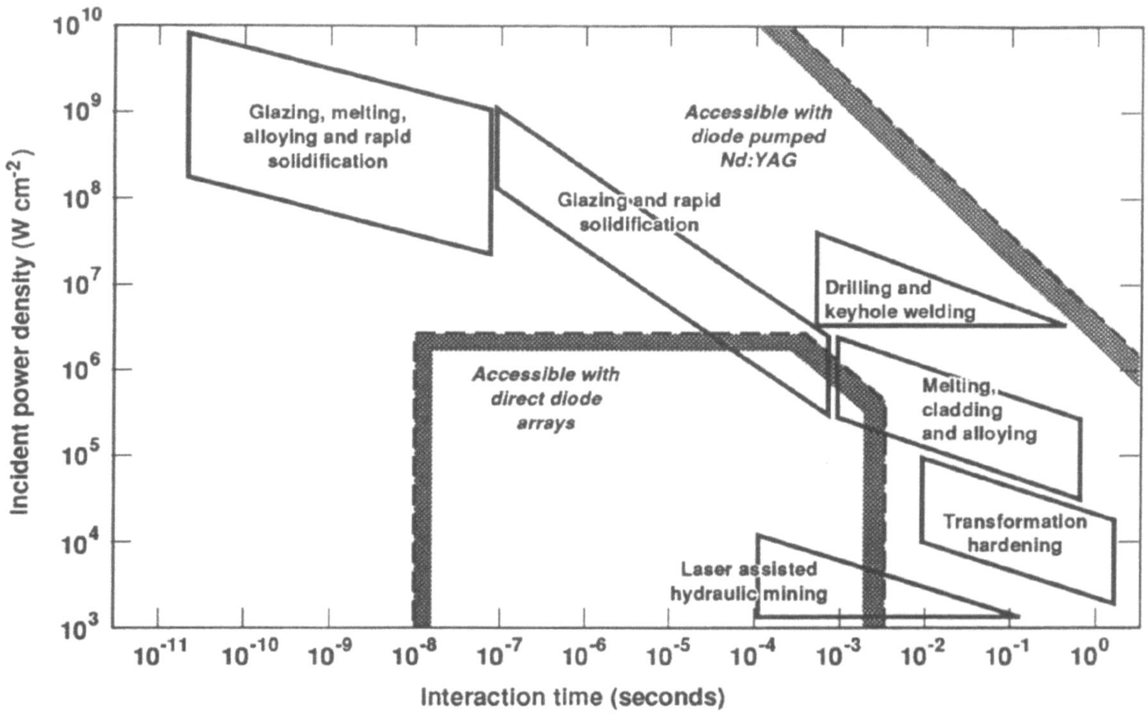

Figure 1. Optical-fiber-delivered multi-kW solid-state lasers can access the treatment regimes of interest for laser materials processing. Adapted in part from reference 8.

lap weld two 1-mm-thick plates of steel together with a bead width of ≈1 mm at almost 200 in./min. Based on our calculations and laboratory experience, what follows are notional solid-state lasers that should be able to meet this goal.

It is desirable to stay about 20 cm away from the work surface, presumably using a 20-cm focal-length optic. Since practical focusing systems that can be conveniently carried by robotic arms can accommodate 2-cm-diameter focusing optics with ease, we have selected 1 cm as our baseline beam diameter at the focus lens. The relation between the size of a partially coherent laser's focal spot and the size of a coherent patch or speckle on the focusing lens is a simple one. If we arrange that the beam entering the final focusing lens be collimated—i.e., have an average radius of curvature of ∞—then the intensity spot $1/e$ radius on the workpiece will be:

$$W_0 = \frac{2f}{k\,L_A} = .01 \text{ cm}$$

where $f = 20$ cm, $k = 2\pi/\lambda$ and the lateral coherence length or speckle radius in the lens plane, L_A, is .07 cm. Many welding laser manufacturers report their laser's radiance in terms of the number of mm-mrad achieved at a given output power. The full angle subtended by a partially coherent beam that meets our specifications would be:

$$\theta(L_A) = \frac{2}{\pi}\frac{\lambda}{L_A} = 1 \text{ mrad}$$

If we require our beam to be about 1 cm across at the lens, then we need a beam quality of about 10 mm-mrad, and more than a factor of 2 increase in this number would probably not be tolerable.

Data from Martek Lasers Inc. on its MM1800 three-rod laser[11] show this laser delivers low powers in a beam 5 mm in diameter that diverges into an angle of 15 mrad (a 75 mm-mrad beam) and 1800 W into a beam about 2.5 mm in diameter that diverges into an angle of about 25 mrad (63 mm-mrad). Lumonics Inc. reports its model JK706 rated at 1000 W produces a 10-mm diameter beam that diverges into 10 mrad for a 100-mm mrad beam.[11] Both these manufacturers use a rod technology that does not scale favorably to higher radiance. It is unlikely, therefore, that a rod system can satisfy our requirements.

Zig-zag slab lasers achieve significantly better beam quality. The first commercially available laser of this type is the Munich Laser Systems model P500. This device produces a 6-mm-×-6-mm square beam that diverges into 5 mrad for a 30 mm-mrad laser. GE researchers have reported ≤10 mm-mrad beams at 1020 W and LLNL experience has been

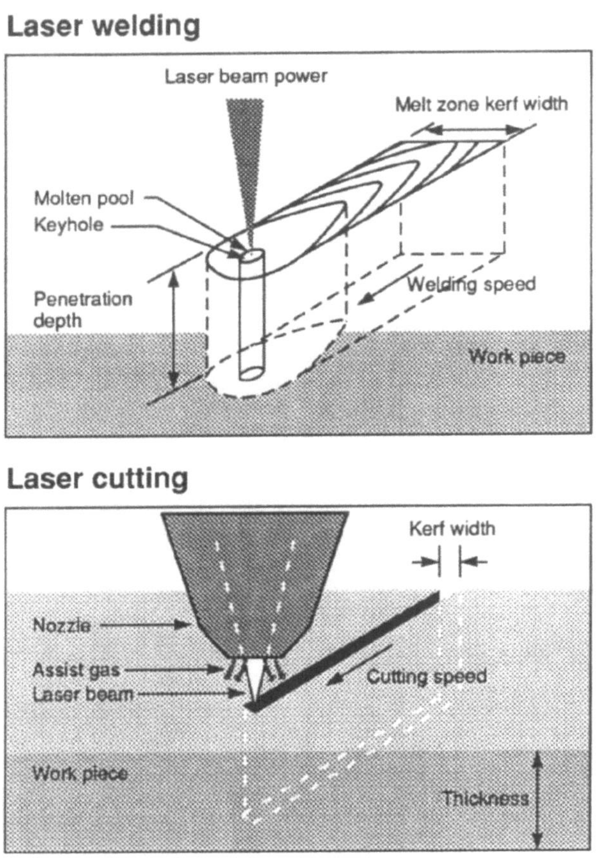

Figure 2. A comparison of welding and cutting shows their similarities. In laser welding, energy is coupled into a plasma keyhole at high temperature (≈20,000°K); in laser cutting, energy coupling is similar to welding, an assist gas is used to blow molten metal out of the kerf, and a reactive gas (O_2) can improve cutting speed.

encouraging.[5–7] We conclude that our beam quality requirements can best be met by a face-pumped slab Nd:YAG laser.

In the absence of a truly comprehensive theoretical treatment of keyhole welding, the most reliable way to estimate the power required is to scale from recent performance data accumulated with high power Nd:YAG industrial lasers. The Martek MM1800 rod laser has been used to weld monel at 20 in/min with a weld depth of 5.3 mm and a kerf width of about 3 mm.[1] From their quoted divergence and because their tests were conducted with *f*/5 optics, we can estimate their spot $1/e^2$ intensity radius as 150 to 200 μm. To meet our assumed requirements, we want about half this spot radius using an *f*/20 focusing system and intend to attain it by exploiting the much higher radiance of a slab laser. To scale the Martek data 5.3 mm to 2.0 mm depth, we can safely assume that the speed can increase by this ratio. We expect that the MM1800 could weld 2-mm-thick sheet steel at 53 in./min or 2.24 cm/s. It would not be able to do so using *f*/20 optics, however.

The laser that meets our requirements will interact with less material since its focal spot is half as large. We get another speed multiplier because of the smaller kerf, and we might choose to use the elliptical focus normally produced by slab lasers to avoid overlap problems at high speeds. A three-fold increase in speed owing to the smaller kerf is expected; so with 1800 W we should be able to weld two 1-mm sheets of a typical steel such as AKDQ/CRS 1008/1010 together at a speed of 159 in./min or 6.73 cm/s. The CO_2 laser manages 200 in./min; to match it, we arrive at a required solid-state laser power of:

$$P_L = 1800 \frac{200}{\frac{5.3}{2}\frac{3}{1}20} = 2300 \text{ W}$$

A second estimate comes from scaling cutting data made with 2-mm-thick galvanized iron sheet using the MLS P500 slab laser operated at 300 W. This laser was able to cut this particular material at 800 mm/min or 1.33 cm/s. Unfortunately, no information is given on kerf widths, however, scaling to the required rate of 8.5 cm/s would require:

$$P_L = 300 \frac{8.5}{1.33} = 1900 \text{ W}$$

A final estimate may be made using data taken with the Lumonics JK706 oscillator-amplifier laser. This device has a beam quality lower than the rest, 100 mm-mrad. When used to weld Type 304 stainless steel to depths between 1.5 and 2.0 mm, this laser achieved a speed of 60–80 cm/min at a kerf width of 2.5 mm. Scaling the width down to 1 mm sets a range of laser powers required:

$$2500 \le P_L \le 3400 \text{ W}$$

This last estimate is the highest because it is based on data taken with the laser having the poorest beam quality. Lumonics notes that its laser welds are typically conduction-loss dominated. The JK706 is the only model the company produces that delivers enough power density to the workpiece to do keyhole welding. The JK707 model is identical to the JK706 except for a patented low-divergence resonator that improves beam quality to about 20 mm-mrad but lowers output power to 600 W. This system has cut 2-mm-thick Type 304 stainless at 2.5 cm/s with kerf widths between 0.5 and 1.5 mm. Scaling this data to the required speed yields:

$$P_L = 600 \frac{8.5}{2.5} = 2040 \text{ W}$$

Note that the extrapolation of the JK707 20 mm-mrad laser data is close to the P500 30 mm-mrad laser-based estimate. The laser power required to meet our requirements is therefore taken to be in the range of 1900–2300 W.

As a check on this number, we can compare it to the recent parameterization of CO_2 laser welding of steels by L. Mannik and S.K. Brown.[13] They derive a formula that should be valid at the speed of interest, v = 8.5 cm/s:

$$P_L = (3.83d + 0.69dv) \times 1000 = 1940 \text{ W}$$

Table 1. HAP solid-state lasers ranked by radiance[11]

Laser (source)	Average power (watts)	Beam quality (mm-mrad)	Radiance $MW/cm^2/sr$
NEC (5 rods)	2000	224	6.5
M34, CGI	250	74	7.4
NEC (3 rods)	800	120	9.0
JK-706, Lumonics	1000	100	16.2
Omega Adv. Rod, CGI	250	35	33.1
P-500, MLS	500	30–50	50.0
MM 1800, Martek	1800	63	73.5
JK-707, Lumonics	600	20	243.2
GE FPL, GE CRD*	500	5	3242.0[5–7]
Vortek-YAG, LLNL*	2000	10	3242.0 (design)
Diode-YAG, LLNL*	275	4.5	4000.0
MPSSL, LLNL*	300	3	5404.0 (short runs)

*Laboratory models—not currently available.

where $d = .2$ cm penetration depth. Their focal spot radius of .0093 cm was close to our requirement, but their observed kerf width of 0.6 to 0.75 mm at our required speed is smaller in proportion to the lower absorption of their 10.6 μm wavelength compared to the 1.06 μm wavelength of Nd:YAG.

Mannik and Brown also give convincing evidence that higher-order spatial modes would lead to significantly higher values of P_L/d to maintain the same welding speed because the larger focused spot sizes reduce the power density on the workpiece. Beam-quality effects are clearly measurable even with nearly diffraction-limited CO_2 laser beams. The solid angle subtended by the our near-infrared beam will be:

$$\Omega = \frac{1}{\pi}\left(\frac{\lambda}{L_A}\right)^2 = 7.3 \times 10^{-7}\, sr$$

which would make the average radiance of a 2000 W laser with a 0.5-cm radius spot at the focusing lens:

$$R = \frac{2000}{\pi.5^2\Omega} = 3.2 \times 10^9 \frac{\text{W}}{\text{cm}^2 sr}$$

We can assess the difficulty of our task by comparing this value of radiance to the theoretical maximum:

$$R_{\max} = 4\frac{2000}{\lambda^2} = 7.1 \times 10^{11} \frac{\text{W}}{\text{cm}^2 sr}$$

and to the values achieved by the industrial lasers already mentioned in this article.

At 5000 W, the RS850 CO_2 laser can achieve the requirement as long as its beam quality is held within 2.4 × the diffraction limit. Stacked up against this impressive performance, the Martek MM1800 is a 70-×-diffraction-limited laser that reaches an average radiance of $R = 5.2 \times 10^7$—about 61 times smaller than required. The MLS P500 delivers a radiance of $R = 7 \times 10^7$, about 45 times too low. The JK706 reaches a radiance of only $R = 1.6 \times 10^7$, about 200 times too low, while the JK707 comes closest to the mark with $R \approx 2.8 \times 10^8$, still about 12 times too low. In all cases, the stated assumptions are used based on published specifications; improved performance may be achievable under special conditions. What we seem to require is a face-pumped slab laser with a beam quality of about 10 times the diffraction limit or better that is capable of delivering 1900–2300 W.

Fig. 3 shows an LLNL simulation that contends that these parameters can be met by a single LLNL crystal laser using Nd:YAG. Table 1 lists commercially available lasers, ranked according to radiance, along with three LLNL designs that, as of this writing, are just beginning to be tested. A very similar GE face-pumped laser system—reported extensively at ICALEO'90—comes very close to satisfying our requirements and is listed in the "laboratory device" group because

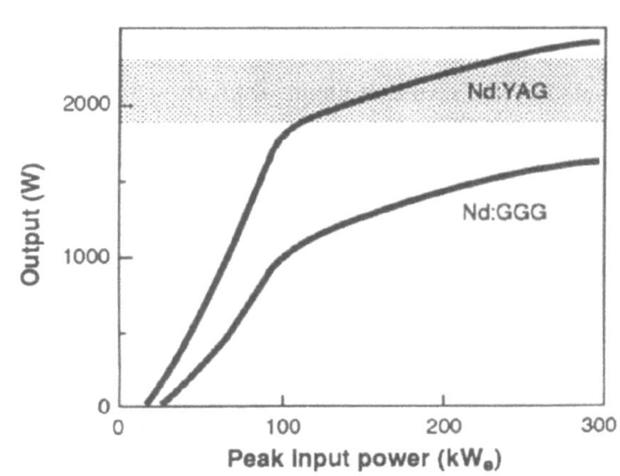

Figure 3. Calculated performance of LLNL Nd:GGG and Nd:YAG slab lasers.

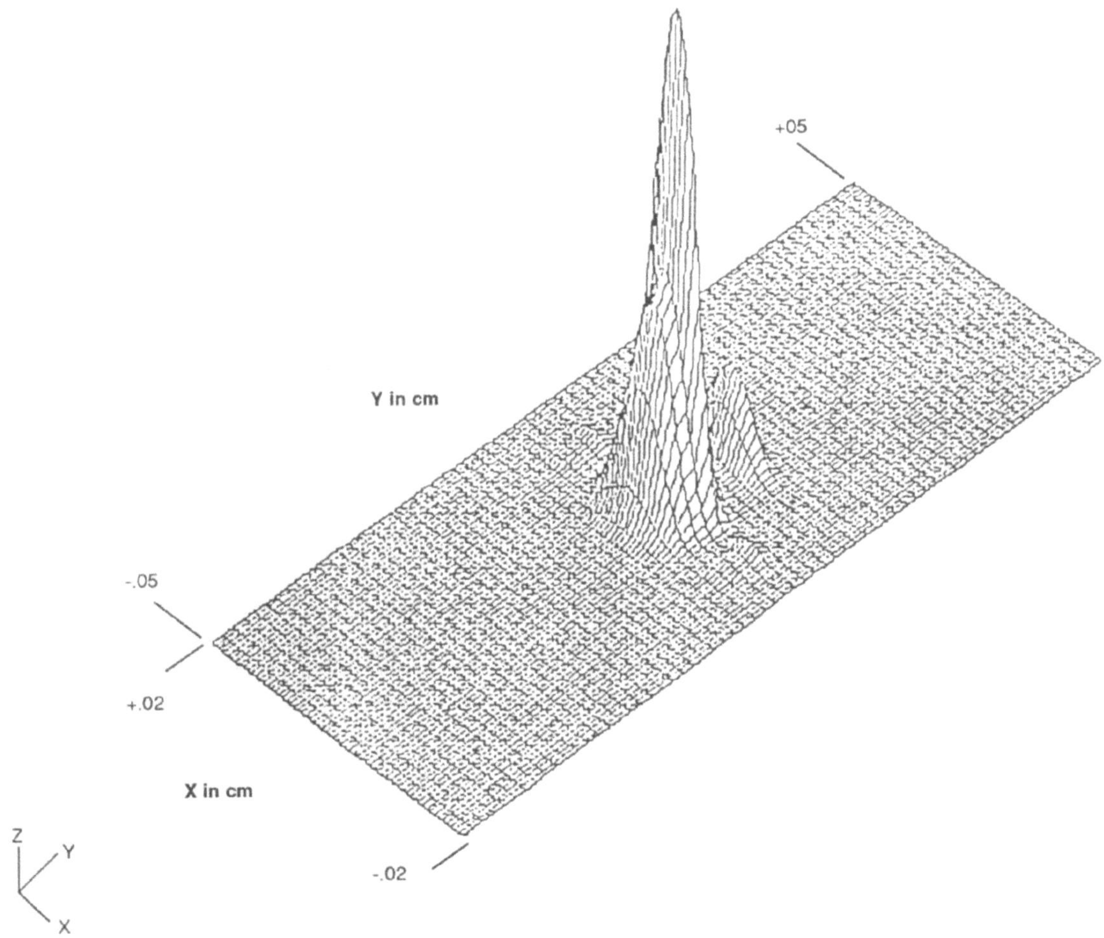

Figure 4. Far-field beam profile calculated for LLNL Nd:GGG slab laser. FWHM is 200 μm, peak irradiance is 51 MW/cm^2.

it has not been brought to market; however, it has been evaluated by GE's Aircraft Engine Division with very favorable findings.[5–7,14]

LLNL has performed interferometric measurements of a Nd:GGG slab. This slab—pumped by Vortek arc lamps—achieved 490 W beams with smooth near-field profiles in early tests. We have completed initial resonator calculations that include these measured aberrations; an example can be found in Fig. 4. The calculated far-field beam profile shown in Fig. 4 exceeds our welding radiance requirement by a factor of more than 4.

Since radiance is a conserved quantity in an ideal passive optics system, even a fiber-delivered slab laser beam may retain its high brightness with good design, but a poor-quality input beam can never be made more coherent by fiber transport alone. Fig. 5 is a plot of kerf width vs. speed for 2-mm-thick steel that shows the performance of each laser mentioned. The MM1800 comes closest to meeting our speed requirement but is unable to meet our 20-cm stand-off requirement because of its lower beam quality without resorting to much larger, heavier, lower *f* /# optics that severely limit the depth of focus.

Once inside the keyhole, the GE drilling demonstrations suggest that the laser pulse forms an index guide.[6] Visible plasmas emerging from the hole have ion temperatures in the neighborhood of 2 ev. Wall temperatures, and thus the radiation temperature in the hole, are closer to the vaporization temperature of the metal—about 2000°K, or less than 10% of the ion temperature. Charge neutrality is assured so the electron density mimics the ion density, but temperatures need not be the same in this non-local thermodynamic equilibrium plasma. Electron temperatures are probably several ev, and the dominant absorption mechanism is likely to be classical collisional absorption or inverse Bremsstrahlung. The spatial

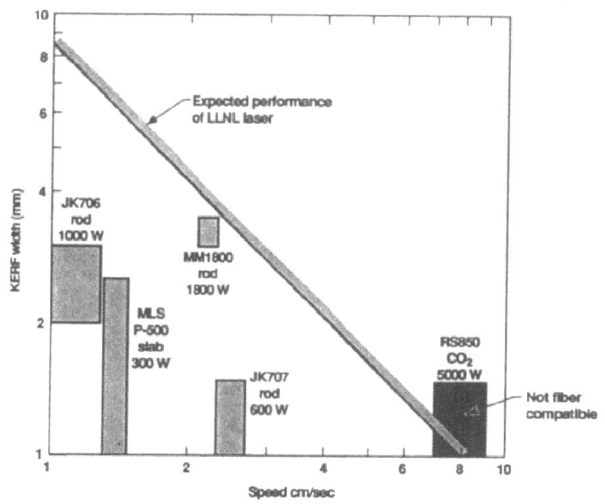

Figure 5. This comparison of commercial lasers with the RS850 speed does not restrict them to meet the stand-off requirement. The RS850 laser, unfortunately, is not fiber compatible because of its 10 μm wavelength.

damping rate or absorption coefficient for inverse Bremsstrahlung is highly density- and temperature-dependent:

$$\kappa_{ib} = \frac{Zn_e^2}{T_e^{3/2}(1 - n_e/n_c)^{1/2}}$$

where n_e is the electron density, T_e is the electron temperature, Z is the average atomic number of the wall material and $n_c \approx 10^{21}$ electrons/cc is the critical density for 1μm light. Inverse Bremsstrahlung is thus strongest for low temperatures, high densities, and high Z wall materials.

The picture that emerges is of a channel in pressure equilibrium with cool high-density electrons leading to low index of refraction near the cool walls and hotter lower density electrons along the axis of the hole. Such an electron density distribution forms an index guide that would tend to keep a high-quality laser beam from interacting with the walls.

Inverse Bremsstrahlung in plasma gradients at other than normal incidence is known as resonance absorption. Microwave propagation in the ionosphere suffers from this effect and has received much theoretical attention over the years. In a plasma having a steep quasilinear gradient, the resonance absorption on reflection is given quantitatively by:

$$\varphi(\tau) = 2.31\tau \exp(-\frac{2\tau^3}{3})$$

where $\tau = (k L_e)^{1/3} \sin\theta_i$.[15] The predicted absorption fraction when the laser beam bounces from the wall will be given by $0.5\varphi^2$. A preliminary index-guiding calculation for 300 μm keyholes suggests that the our notional beam focuses in the hole in about 1 mm, after which it diverges to bounce from the keyhole plasma near the wall at a depth of almost 2 mm at an interaction angle of about .04 radians. Such a shallow incidence angle, θ_i, keeps the absorption down to a few percent per bounce, even if the electron density scale length, L_e, is 10s of μm.

If deeper keyholes are desired, this guiding phenomena continues to deposit energy in the walls at a rate sufficient to maintain the wall temperature at the vaporization point in the face of thermal conduction losses radially. It takes about 1 kW/cm of hole depth to maintain a steel wall in this equilibrium state. As a result, the maximum hole depth, limited by wall losses, for a train of 10 J 1 msec laser pulses of the quality we require would be about 10 cm.

Efficiency, reliability, and ultimate cost are strongly dependent on technology. The first HAP solid-state industrial lasers built for sheet steel welding may well be arc-lamp-pumped. By far the most attractive option from the user's point of view would be one that gives him a compact, efficient, and maintenance-free material processing laser system. Laser-diode-pumped Nd:YAG potentially offers these features to a greater extent than any previous laser.

LLNL solid-state lasers

Continuous wave Nd:YAG lasers are commonplace, but HAP crystalline lasers that can compete effectively with CO_2 laser systems, such as those needed for sheet-metal welding, require improved technology.

Depending on the crystal size used, its temperature, and the available pump power, the repetition rate of the laser amplifier can range up to more than 6000 Hz. Thermal-optical stress calculations show that a Nd:YAG slab $10 \times 40 \times 200$ mm^3 can achieve a gain of 10 at 100 Hz. An oscillator designed for high repetition rates or cw operation might employ $5 \times 40 \times 200$ mm^3 slabs and should have usable gains >2 at 3000 Hz using vortex stabilized water wall arc lamps. Improved performance is possible using large-area laser-diode pumping, as discussed below.

At 500 W of output, the slab laser should already be quite a challenge to electric discharge machining manufacturing. We expect improved performance in the near future as new slab laser models—at around 1 kW of output from a single Nd:YAG head—are introduced into the industrial laser market. A barrier exists at this level chiefly because conventional flashlamp technology is limited to wall loadings of about 300 W/cm^2. Lamp life falls dramatically above a wall loading of about 400 W/cm^2.

Though there are gains to be made by increasing the specific power loading in Nd:YAG, conventional lamps are unequal to the task. This is what caused LLNL to search for higher-power pump sources. Vortek lamps currently survive for several hundred hours at wall loadings of 1.5 kW/cm^2. Laser diode arrays not only are more efficient than lamps but also deliver their power at the correct wavelength to pump Nd-doped garnets, thereby minimizing thermal effects. Scaling up boule growth has allowed the fabrication of larger garnet slabs, but without better pumps their potential cannot be realized.

At LLNL, two Vortek arc lamps have been modified to pulse close to the optimum power density and at arbitrary pulse repetition frequency and duty factor. Photo 1 is our Vortek lamp test bed showing a Nd:GGG slab mounted in a cooling fixture. In a few years, however, we contend that diode-pumped solid-state lasers will carry the field and that wavelength conversion of these devices will offer the manufacturing engineer great latitude in working modern industrial materials.

The key issues associated with developing high-average-power diode pump sources include cost, reliability, and thermal management. LLNL has developed a reliable, low-cost packaging technology based on silicon microstructures that is ideally suited for high-average-power applications.[16] The basic package design (see Photo 2) features LLNL-designed microchannel coolers bonded in a unique LLNL process to Al:GaAs laser diode arrays.[17] Current fabrication rates stand at about eight of these packages per day and can be expanded several fold depending on demand. These packages routinely achieve peak powers of 90 W and average powers of 20 W. When stacked as shown in Photo 3, they make compact, high-intensity and high-average-power optical pump sources for solid-state lasers.

To meet our welding requirements, we propose to optimize diode array performance for high-duty-factor optical output, ≈0.2, with about 200 ms duration pulses coming at 1000 Hz of 0.8083 μm light and 360 W/cm^2 peak irradiance. Conceding that laser-diode-pumped solid-state devices would be highly desirable, critics often cite exorbitant cost as the reason for not even considering this pumping option. LLNL experience points emphatically in the opposite direction. Current LLNL fabrication costs stand at less than $10/peak watt and should drop rapidly as volume increases.

A high-power diode-pumped solid-state laser test bed (see Photo 4) complements our Vortek arc-lamp-pumped test bed. The compact device shown used 10 diode pumping packages

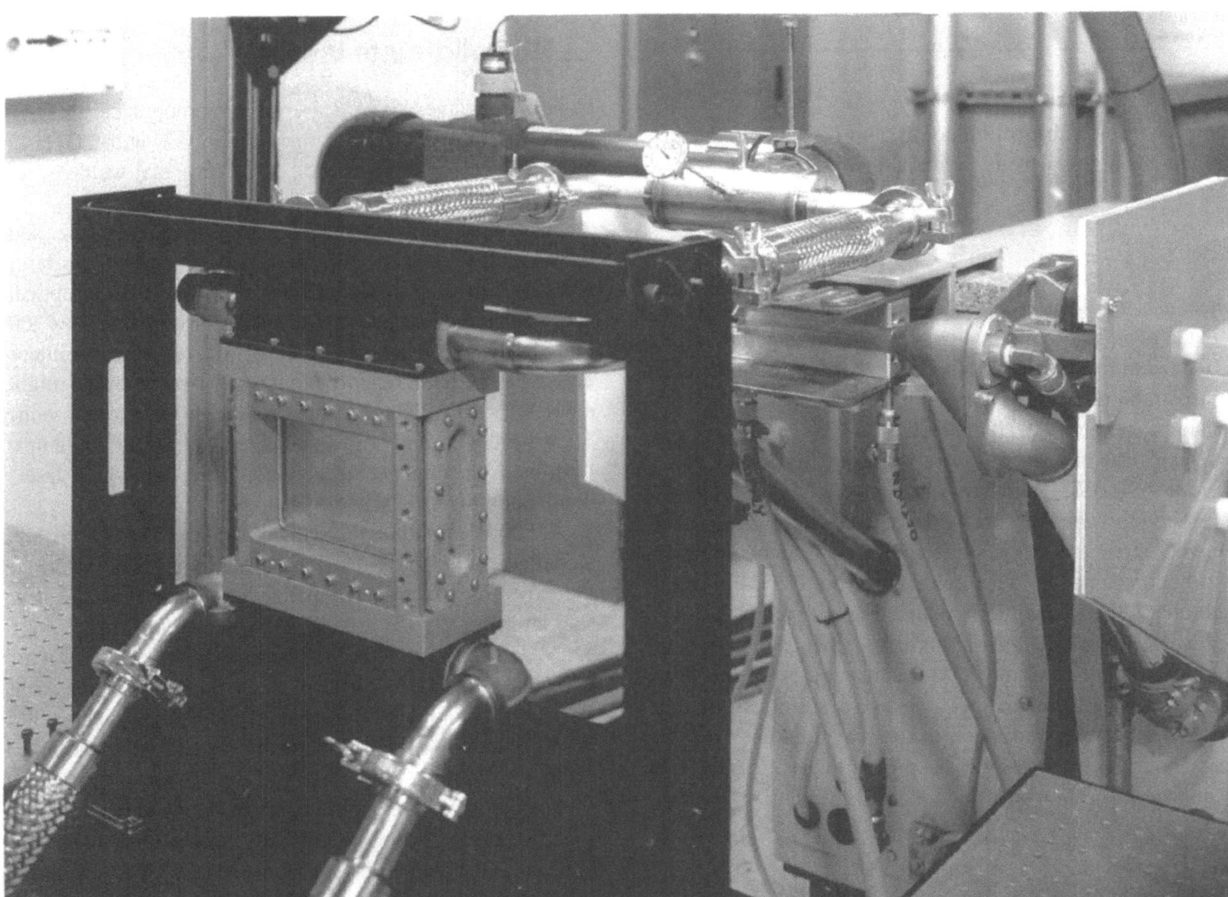

Photo 1. Vortek arc-lamp-pumped crystal slab laser test facility.

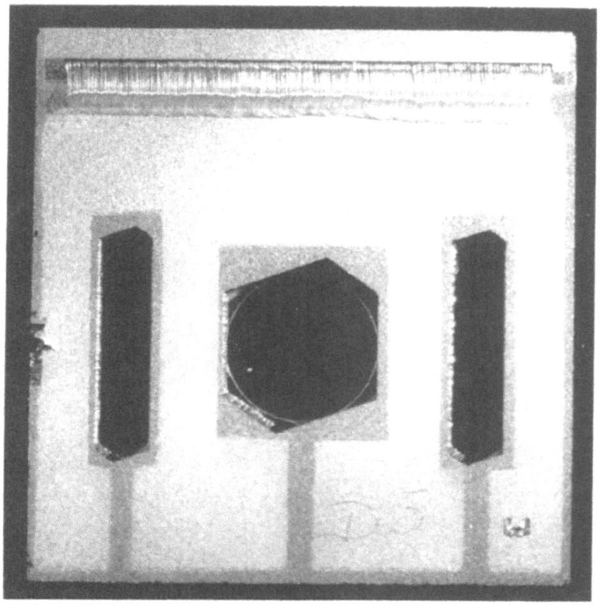

Photo 2. High-average power diode pump array element employs silicon microchannel cooler.

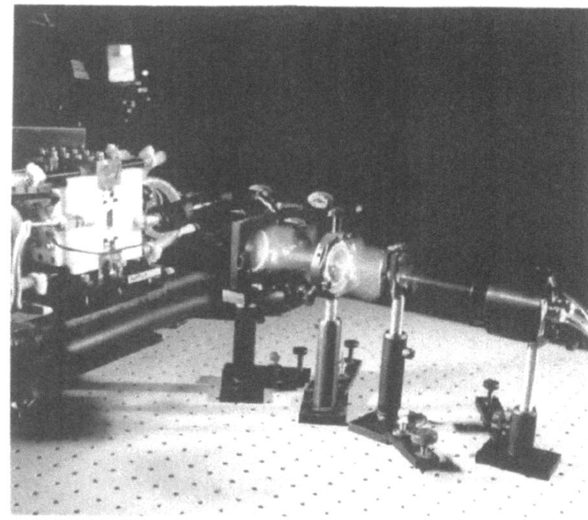

Photo 4. This LLNL diode-pumped Nd:YAG slab laser delivers 40 W at 1.064 μm for 200 W of diode pump power at 0.808 μm. The same laser reached an average power of 275 W when fitted with more powerful pumping arrays. The infrared laser beam has been converted to green by a nonlinear crystal.

to produce a 40 W laser beam at 1.064 μm. With 32 cm^2 of pumping diode laser arrays, this same laser reached an average power of 275 W. As the figure shows, the infrared beam can also be frequency doubled in a nonlinear crystal to produce green light. Life testing so far has shown no degradation after several billion shots; however, we feel the most important near-term task to be addressed is to operate this device at over 200 W average power and conduct life tests. LLNL's diode array package should operate for several thousand hours without any maintenance. It is tolerant of rapid cycling on and off; a diode-pumped solid-state laser's turn on transient period of 10s of seconds will be dominated by thermal equilibration of the crystalline slab itself.

LLNL remains a major R&D center for HAP crystalline lasers in the USA, and quite likely is the only laboratory capable of building a diode-pumped HAP crystalline laser suited to industrial needs in the next year. There is sharp competition in this rapidly developing field, and a failure to apply this understanding may have significant negative consequences for US industry. Allied Signal Corp. has grown 90

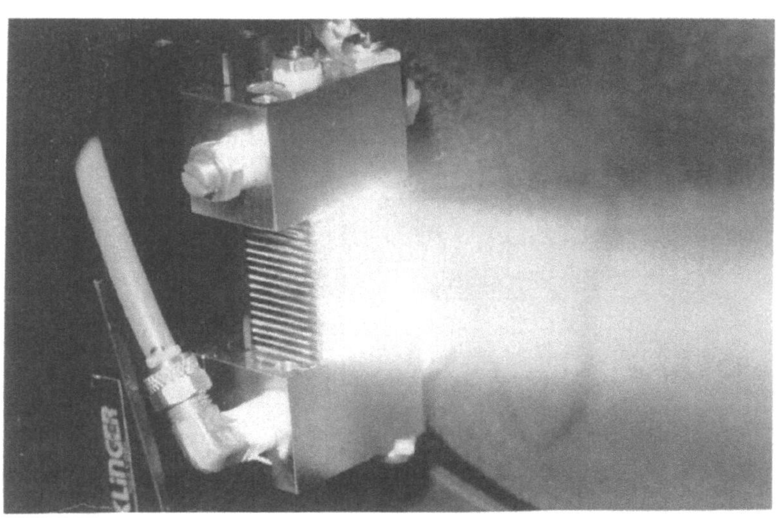

Photo 3. When stacked, the LLNL cooled diode packages make compact efficient solid-state laser pumps. This array delivers a peak power of 1.45 kW with a duty factor of 0.2 in 100-μsec pulses.

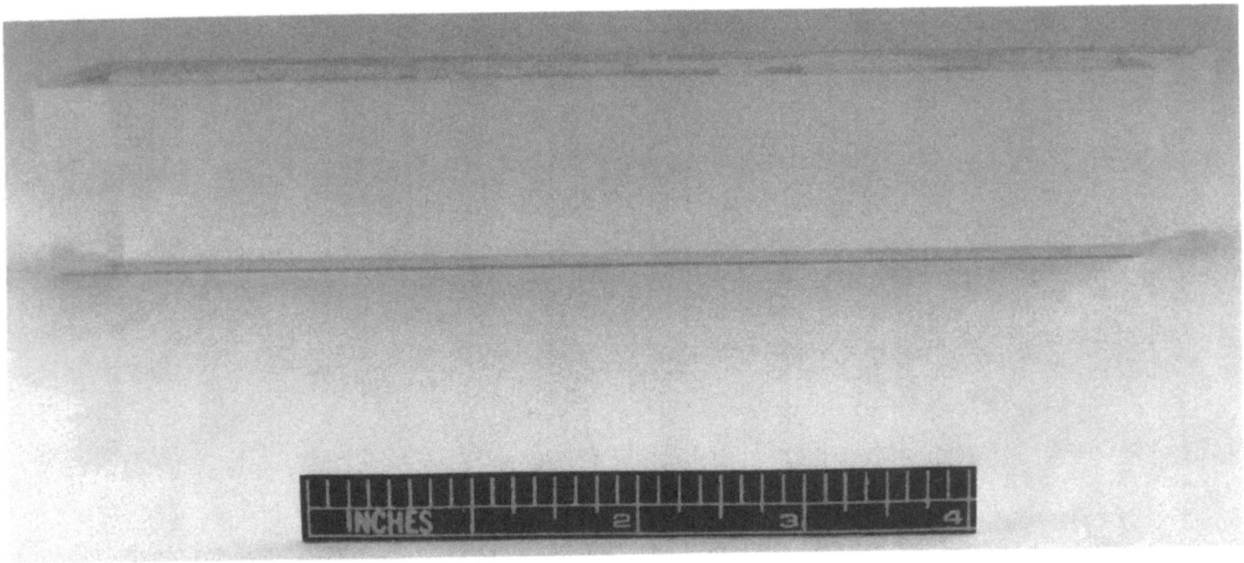

Figure 6a. This Nd:YAG zig-zag slab is $.7 \times 4 \times 20\ cm^3$.

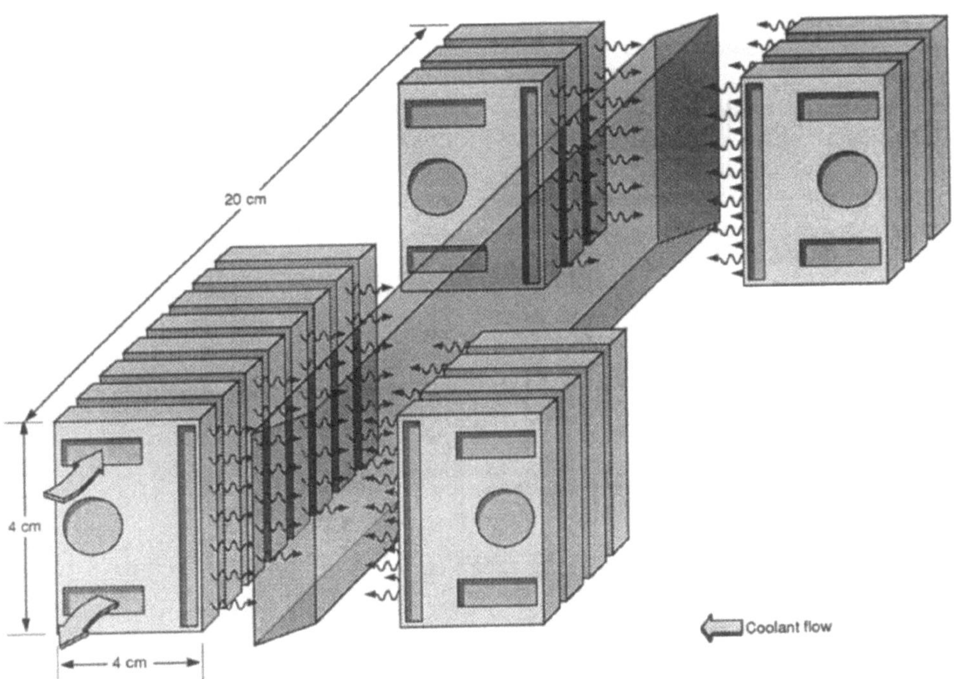

Figure 6b. Conceptual design for a 2-kW average power, 8-kW peak power diode-pumped solid-state laser.

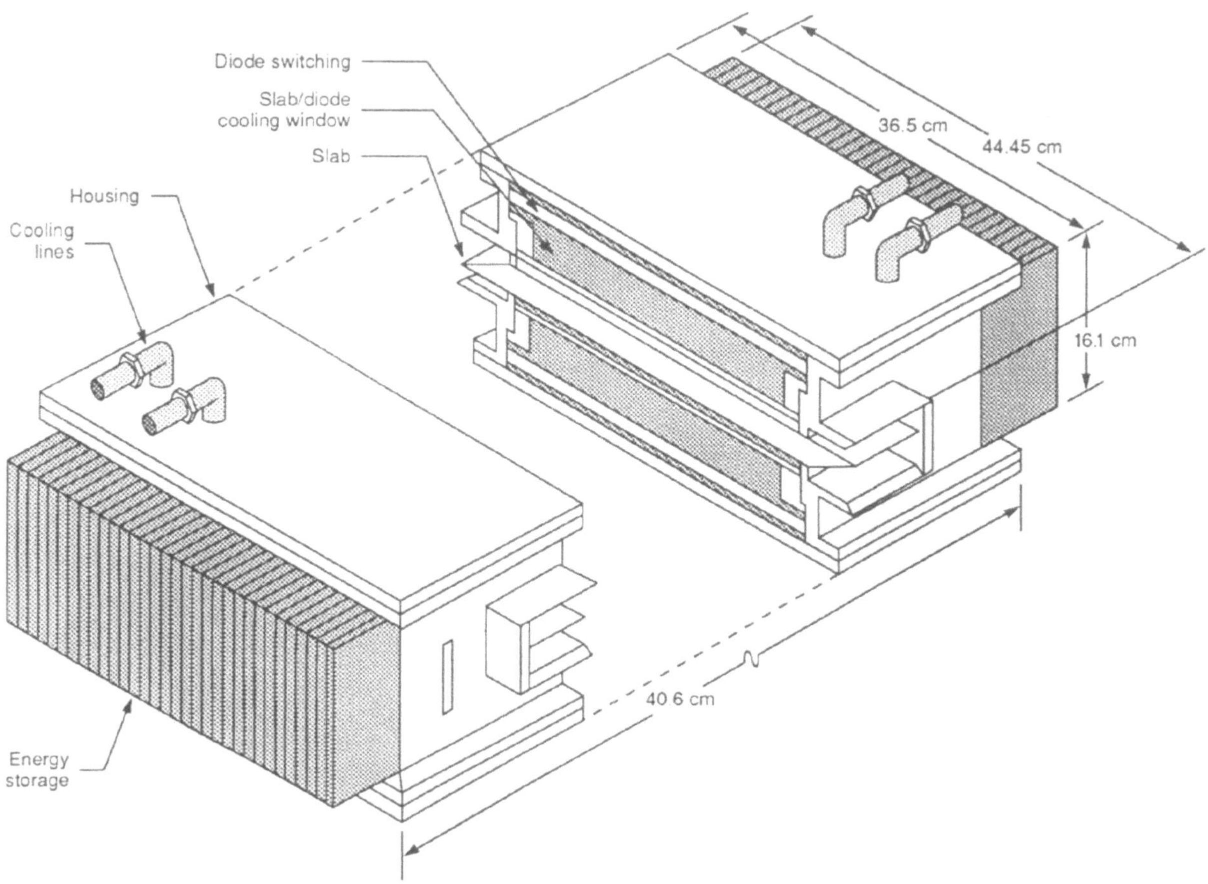

Figure 7. Diode-pumped GGG amplifier.

× 200 mm boules of Nd-doped GGG and similar sized boules of co-doped Nd:Cr:GSGG. They are capable of scaling flat interface YAG crystal growth to comparable sizes. Fig. 6a shows a Nd:YAG crystal slab grown recently by Allied, which already meets the dimensions required for a practical welding laser, $10 \times 40 \times 200$ mm^3.[18] Fig. 6b contains a conceptual design for a 2-kW average power, 8-kW peak power diode-pumped solid-state laser using these new slabs, which might package into a unit like that sketched in Fig. 7 with a footprint of only 40 × 50 cm.

In summary, an industrial laser whose optics and power supply would fit into a volume roughly equal to that used by current, much-lower-power solid-state lasers is on the horizon. The possessor of this technology will be able to cut and weld sheet-steel structures such as automobile bodies with significantly lower per-unit cost and improved reliability than will a competitor constrained to today's methods, including CO_2 lasers.

Work for this article was performed by the Lawrence Livermore National Laboratory under the auspices of the US Department of Energy, Contract W-7405-Eng-48.

References

1. G. McFadden and D.M. Filgas, "Designing Nd:YAG Lasers for Higher Powers," *Lasers & Optronics 9*, 32 (August 1990).
2. D.M. Roessler, "US Automotive Applications of Laser Processing," *The Industrial Laser Annual Handbook*, ed. by D. Belforte and M. Levitt, PennWell Publishing, Tulsa, OK (1988).
3. D.M. Roessler, "Update on Laser Processing in the Automotive Industry," *GMR-6893* issued 12/12/89.
4. J. Keller interview with M. McLaughlin of GE, "GE Squeezes 1,000-W Beam From Face-Pumped Laser," *Military & Aerospace Electronics 1* (April 1990).
5. J.P. Chernock, GE Research Ctr. Schenectady, NY, 1990.
6. T.J. Rockstroh, "Application of slab-based Nd:YAG lasers at GE Aircraft Engines," ICALEO'90.
7. A.L. Ortiz Jr., "On-the-fly drilling with a fiber delivered face pumped laser beam," ICALEO'90.
8. A. Bloyce and T. Bell, "Laser Surface Engineering," *The Industrial Laser Annual Handbook*, ed. by David Belforte and Morris Levitt, PennWell Publishing, Tulsa, OK (1988).
9. M. von Allmen et al., "Absorption Phenomena in Metal Drilling with Nd-Lasers," *JQE QE-14*, (Feb. 1978).
10. M. von Allmen, "Laser Drilling Velocity In Metals," *J. Appl. Phys. 47* (Dec. 1976).

11. *Lasers & Optronics 1990 Buying Guide 8*, 292; *Laser Focus World Buyers Guide 1991 26*, 155.
12. "Commercial slab laser cuts better than conventional YAGs," *Laser Focus World 26* (August 1990).
13. L. Mannik and S.K. Brown, "A Relationship Between Laser Power, Penetration Depth and Welding Speed in the Laser Welding of Steels," *Journal of Laser Applications 2*, 3 & 4 (1990).
14. J.P. Chernock, "Characteristics of a 1 kW YAG Face-Pumped Laser," ICALEO'90.
15. V.L. Ginsberg, *Propagation of Electromagnetic Waves in Plasmas*, Gordon and Breach, N.Y. (1960) or Pergamon, N.Y. (1964), Chs. 4 and 6.
16. D.B. Tuckerman and R.F. Pease, *IEEE Electron Device Lett EDL-2*, 126 (1981).
17. R. Beach et al., "High-reliability silicon microchannel submount for high average power laser diode arrays," *Appl. Phys. Lett. 56 (21)*, 2065–2067 (May 1990).
18. Grown by E. O'Dell and R. Morris, Allied-Signal Inc.

Physical mechanism and modeling of deep-penetration laser welding

Dieter Schuöcker
Technical University
Vienna, Austria

Introduction

Deep-penetration laser welding is one of the most important applications of high-power lasers. It has experienced rising application all over the world because it allows joining workpieces with high speed, good seam quality, and minimum distortion of the workpiece due to the relatively low heat input.

Most recently, the application of laser welding in industry is strongly stimulated by the fact that lasers with a beam power of 10–20 kW and very good beam quality are now on the market. Using these lasers, an excellent seam profile can be obtained that resembles e-beam welds, with their low width-to-thickness ratio.

The theoretical description of deep-penetration welding is an important task, providing deeper insight into the physical mechanism of this process. Only a correct description of the relevant phenomena leads to a mathematical treatment with an appropriate agreement between theoretical and experimental results. In addition to the scientific aspect, theoretical work is also of technical importance because it allows the prediction of processing results without experimental investigations that take much time and can thus be quite expensive.

There are two basic approaches to the theoretical treatment: simulations and models. Simulations try to describe the processes under consideration as precisely as possible and use a complicated set of differential equations that must be solved using point-by-point procedures. Simulations yield results that in general agree very well with experimental findings and are, therefore, most appropriate for gaining a better understanding of the process. Nevertheless, simulations are not very well suited to a fast prediction of processing results because the solution of the differential equations take much time.

The second approach takes the limitations of simulations into account and uses simplified models. In this approach, only those phenomena that are of first-order importance are considered, thus simplifying the mathematical treatment. A further simplification is obtained with an approximative mathematical treatment of the few main phenomena included in the analysis by abandoning differential equations and replacing them with simpler approximative formulae.

It should be mentioned that these two methods of mathematical treatment very often approach each other. Simulations must be simplified step-by-step by introducing certain approximations to obtain resolvable equations. Models must take into account more and more phenomena (for example, of second- or third-order importance) to obtain a reasonable agreement with experimental values.

Physical mechanism

If a laser beam with a sufficiently high intensity moves over the desired contour where two pieces have to be joined, evaporation takes place under the beam's focus. If the absorbed laser power is high enough, a channel filled with vapor is built up and extends over the full depth of the workpiece (see Fig. 1). Due to the movement of the heat source, the channel takes on an elongated shape, so that the length can be up to 10 times larger than the width. The channel is also wider at the front (i.e., direction of beam motion) and narrower at the back and thus is called a "keyhole." The fraction of the laser beam that enters the keyhole is absorbed to a great degree due to a combination of Fresnel absorption at the walls of the keyhole and absorption by the particles in the vapor.[1–3]

The latter energy is also transported by collisions to the walls. Therefore, the walls of the keyhole are heated and reach a temperature near or above the boiling point, thus providing the necessary vaporization. Due to the latter, the energy input to the wall is reduced to some extent by evaporation cooling.[4] The keyhole is surrounded by a molten region, where the width of this region determines the width of the welding seam. Due to the movement of the beam and keyhole in the welding direction, solid material is molten at the front side of the melt pool, flows on both sides of the keyhole to the back side, accumulates there and finally resolidifies, thus joining the two pieces.

Due to the high energy absorbed by the vapor particles from the laser beam, ionization takes place and a plasma is formed. This strongly luminous plasma (see Photo 1) is typical for deep-penetration welding.[1] The keyhole is held open continuously due to an equilibrium between the vapor pressure and the dominating hydrodynamic pressure at the

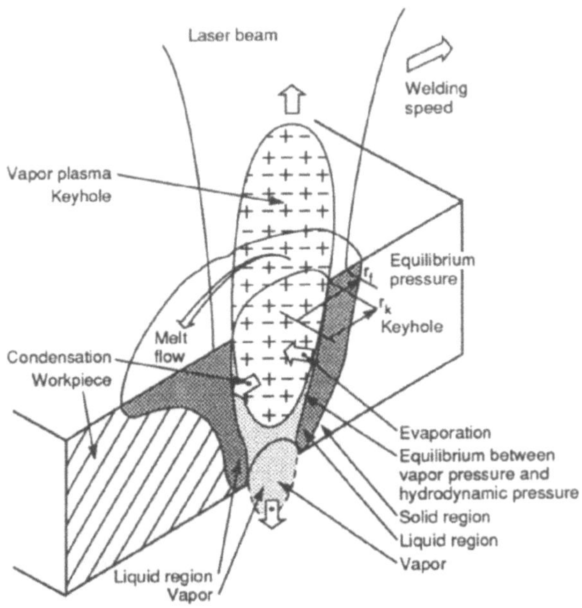

Figure 1. Mechanism of deep-penetration laser welding.

Photo 1. Luminosity during deep-penetration laser welding (steel, 4-kW CO_2 laser, IST-TU Vienna).

walls of the keyhole. The latter results from a fluid circulation caused by spatial variations of surface tension due to the strong temperature decrease from the wall of the keyhole to the melting isothermal region (thermocapillary flow).[5] This fluid circulation takes place in vertical-radial planes.

Previous theoretical work

As early as 1973, eight years after the invention of the CO_2 laser, Swift-Hook and Gick from the United Kingdom published the first model on laser welding.[6] They started from the assumption that the laser beam that scans over the surface of the workpiece can be treated as a moving line source. With the further assumption that the width of the weld seam must be equal to the maximum lateral extension of the melting isothermal, they obtained a relationship between absorbed laser power, seam width, and welding speed. Since they neglected important phenomena such as vaporization and keyhole formation, their model yields only partial agreement with experimental results and applies mainly to conduction welding.

Only three years later, Klemens, also from the UK, presented a much more sophisticated model that became a basis for a lot of subsequent work.[7] The model is based on the assumption of a circular keyhole with vertical walls that is held open by a balance between vapor pressure within the keyhole, surface tension, and hydrodynamic pressure in the melt surrounding the keyhole. Absorption of radiation is assumed to take place only by the vapor particles and is analyzed in detail under consideration of ionization. The temperature distribution is determined from the solution of the heat-conduction equation. The theory yields predictions on absorption, melt-pool geometry, weld seam, and welding speed.

Up to that point all models of laser welding neglected the vertical variations in the workpiece entirely. However, in 1977, Cline and Anthony, again from the UK, published a model that concentrated on these vertical variations.[8] This model assumes 100% absorption and deduces an exponential decrease of the temperature in the vertical direction from the solution of the heat conduction equation. With the further assumption that the bottom of the keyhole must be at the boiling point, the authors determined the depth of the keyhole.

These early models of laser welding were able to provide analytic solutions for the temperature distribution and all other quantities of interest by neglecting the temperature dependence of all material properties involved in the welding process. Since these dependencies are quite strong, Mazumder and Steen considered these dependencies in a work published in 1980,[9] but had to pay for this improvement with the complication of numerical step-by-step solutions of the heat conduction equation. The calculation is carried out for a Gaussian beam intensity distribution and starts with the assumption that the wall of the keyhole must be at the boiling point. The authors consider attenuation of the laser beam in the vertical direction, heat conduction, convection, and radiation. This highly sophisticated theory comes very close to the physical reality and thus gives good insight into the physical mechanism of the process (see Fig. 2).

These models of deep-penetration welding treated heating of the workpiece because this is the most obvious phenomenon that contributes to welding. None of the models, however, treated the viscous flow of liquid material in detail, although this phenomenon is as necessary for welding as heating.

The first treatment of the latter phenomenon appeared in 1986, in an article written by Davis, Kapadia, and Dowden

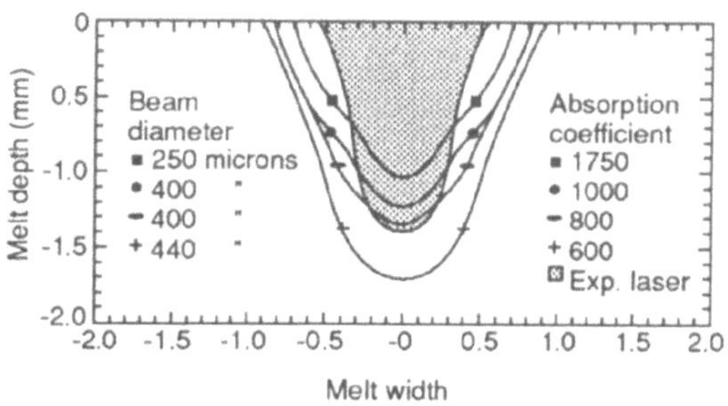

Figure 2. Variation in laser melt widths with beam diameter and absorption coefficient at a laser power of 1570 W, reflectivity of 0.8, and welding speed 33.5 mm/s (Mazumder and Steen 1980[9]).

from Essex, UK.[10] They started from the assumption of a circular keyhole and a molten region that is also circular but not concentric to the keyhole due to the movement of the beam. Vertical variations were neglected. It is assumed that the wall of the keyhole is at the boiling point and the molten region is limited by the melting isothermal. With these assumptions and the further assumption that only horizontal melt flow takes place, Navier-Stokes equations are solved, thus yielding a description of the melt flow during welding. With these results presented by a stream function, the model gives good insight into the mechanism of material transport in the liquid region (see Fig. 3).

Also in 1986, the same authors together with Postacioglu published their most sophisticated theory of deep-penetration welding.[11] This theory also accounted for the vertical variations, especially in and around the keyhole. The model starts with the assumption that laser power is absorbed by vapor filling the keyhole. The latter is held open by an equilibrium between vapor pressure and surface tension. A nonviscous vertical flux of vapor in the keyhole and a viscous vertical flow of molten metal around the keyhole are assumed. The mathematical treatment includes the solution of the Navier-Stokes equations, as well as mass, energy, and particle-conservation equations. Absorption of radiation by vapor in the keyhole is treated with a two-flux model.

This most sophisticated model, which considers the largest number of phenomena contributing to laser welding relative to all previous models, yields a wealth of information on the welding process. For instance, it is possible to determine the vertical cross section of the keyhole. The same authors also provided (somewhat later in 1986) an analysis of the vertical cross section of the weld seam under consideration of upwelling phenomena such as vapor and melt flow perpendicular to the surface of the workpiece (see Fig. 3).[12]

It is well known that the vertical cross section of the weld seam shows a wide, more or less circular section near the upper surface of the workpiece and a narrow section with parallel walls at the bottom of the workpiece. It was a fine idea of Steen, Dowden, Davis, and Kapadia to combine a moving point source and a moving line source to describe the

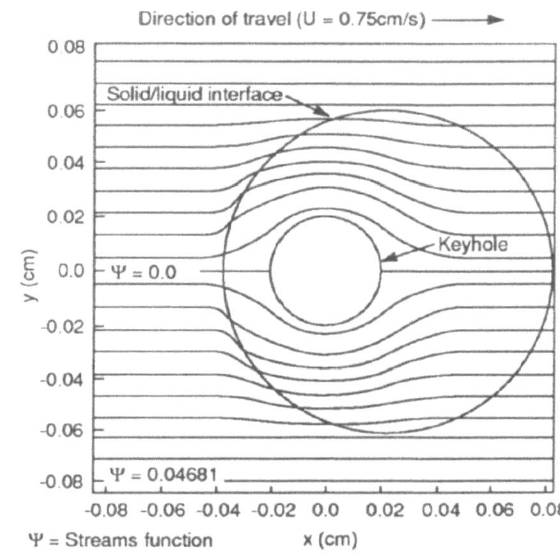

Figure 3. Contour map of the streams function. The contours are multiples of 0.0629 cm/s (after Davis, Kapadia, and Dowden[10]).

laser beam absorbed in the keyhole.[13] The temperature distribution resulting from this combined source model yields a vertical shape of the weld seam that agrees very well with that found from experimental investigations.

So far, most of the models described here treated the absorption mechanism in a simplified manner or neglected it entirely by assuming 100% absorption. But at ILT in Aachen, FRG, Herziger, Beyer, Loosen, Poprawe, and coworkers pro-

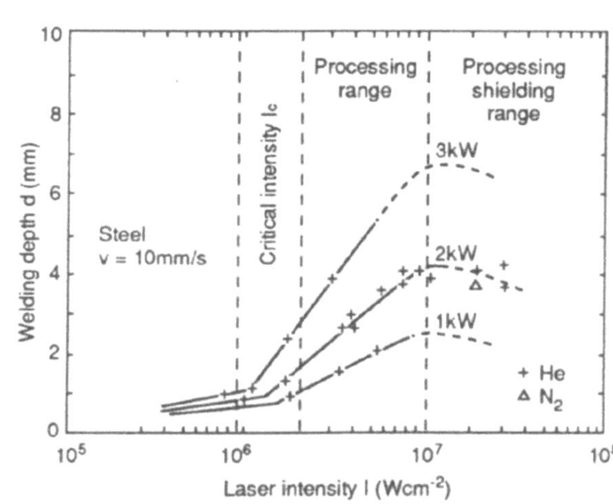

Figure 4. Welding depth dependent on laser intensity, divided into useful and ineffective ranges. The experimental results confirm theoretical work by Herziger, Beyer, and coworkers.[1] (Herziger et al., ILT, Aachen).

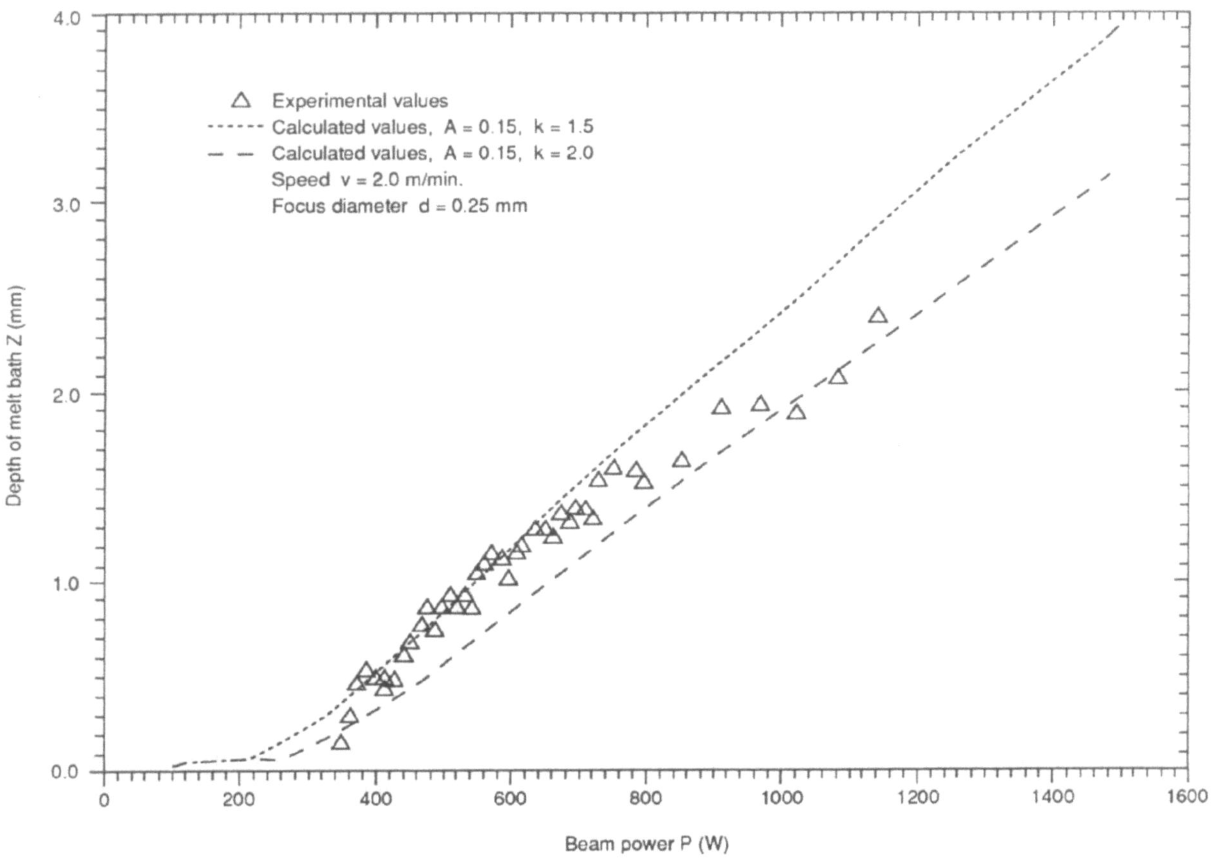

Figure 5. Welding depth dependent on laser power calculated for given absorption (*A*) and two different heat conductivities (*K*) compared with experimental values. Effects of the plasma have been neglected. (After Beck, Dausinger, and Huegel.[2])

vided very detailed theoretical studies of the absorption process in the keyhole plasma (e.g., by inverse Bremsstrahlung) and related phenomena.[1] This theoretical work was extremely successful in predicting optimum processing conditions (see Fig. 4).

In 1989, Beck, Dausinger, and Huegel published an improved theory of absorption and keyhole formation.[2] This theory considered not only absorption by plasma particles but also by the Fresnel mechanism due to multiple reflections. This work gives excellent insight into the mechanism of keyhole formation and elucidates the role of multiple reflections (see Fig. 5).

In 1991, the Essex group published a comprehensive treatment of the thermocapillary flow in the molten pool. They found agreement with experimental evidence if they used a nonlinear boundary-layer model that takes into account thermal conduction and fluid circulation. Somewhat later, the same group studied waves appearing at the surface of the melt pool due to mechanical flow instabilities.[5,14]

This overview shows that the existing work in deep-penetration welding can be divided into two model groups. The first covers self-consistent models that are highly sophisticated and treat laser welding with a high degree of correspondence to physical reality, but they also require very complicated analytic solutions and sometimes allow only step-by-step solutions. The other group treats only specific aspects, such as plasma formation or keyhole formation, and so cannot be regarded as self-consistent laser-welding models. Thus there appears to be a need for a simple but self-consistent model of deep-penetration laser welding that can be described by some relatively simple formula.

Simplified model

The model developed in Vienna starts with the assumption that the keyhole is fully established and is maintained in a stable way during the welding process. For the sake of simplicity the actual shape of the keyhole—with its wider front portion and long, narrow back portion—is replaced by a simpler model keyhole with oval shape. The radius (r_K) of this model keyhole and its length (l_K) are determined by the temperature (T_w) at the wall of the keyhole, the net energy input to the keyhole (P_{net}), and welding speed (v). These quantities are initially regarded as unknown and must be determined by the model calculation. Therefore the characteristic dimensions of the keyhole, such as radius and length, are also initially unknown and must result from the model calculation. The relationship between these dimensions and the temperature at the wall of the keyhole, net energy input to the keyhole, and welding speed can, for instance, be determined with the help of the moving line source model.[4]

The absorbed laser power depends on a first-order approximation mainly on the overlapping of the cross section of the laser beam and the keyhole. This is because approximately 100% of that part of the beam that enters the keyhole is absorbed, and the remainder is nearly totally reflected by the surface of the workpiece. If a simple model beam with constant power (P_L) and radius (r_F) is assumed, the absorbed laser power can be calculated with the overlapping fraction (η) of the beam cross section. This fraction depends on the beam radius (r_F), keyhole radius (r_K), and, to a lesser degree, keyhole length (l_K)—neglected in the following):

$$P_{abs} = \eta(r_F, r_K, l_K) \cdot P_L \tag{1}$$

A detailed calculation of η based on the moving line-source approximation is described in reference 4.

The energy gain of the workpiece derived from the absorbed laser power is reduced somewhat by vapor leaving the keyhole at its upper and lower opening. Under the assumption of thermal equilibrium between the evaporating walls of the keyhole and the vapor in it, the mass loss by evaporation dependent on the wall temperature (T_w) can be calculated from the Clausius-Clapeyron equation. The latter dependence is described by the function $f(T_w)$. With the energy carried away per evaporated particle, E_{evap} and with the cross section of the keyhole (F_K) dependent on keyhole radius and length, an equation for the evaporation power loss is obtained:

$$P_{evap} = E_{evap} \cdot F_K(r_K, l_K) \cdot f(T_w) \tag{2}$$

The net energy gain of the workpiece is thus given by the absorbed laser power reduced by evaporation heat loss:

$$P_{net} = P_{abs} - P_{evap} \tag{3}$$

This net energy gain takes place at the walls of the keyhole and is carried away from the keyhole by heat conduction. Therefore, heat-conduction loss into the bulk of the workpiece must be equal to the net energy gain. This depends on the temperature at the wall of the keyhole (T_w), the geometry of the keyhole, the thermal properties of the workpiece, and, last but not least, the welding speed, an important property of all moving heat sources.

An analytic description of these heat-conduction losses can be obtained by using the moving line source model already mentioned (for details, see reference 4). In principle, the conduction heat loss can be written with the function $g(r_K, l_K, v)$, which describes the dependence on the keyhole geometry and the welding speed (K = thermal conductivity, d = thickness of the workpiece):

$$P_{cond} = K \cdot d \cdot T_w \cdot g(r_K, l_K, v) \tag{4}$$

Figure 6 (solid line) shows an example of the net energy gain calculated as the difference between absorbed laser power and evaporation heat loss, depending on the radius of the keyhole for an arbitrary value of the temperature at the wall. It points out that the net energy gain is relatively low for a small keyhole because, in this case, only a minor portion of the beam enters the keyhole.

With rising radius of the keyhole, the net energy gain increases because a rising fraction of the beam enters the keyhole. If the keyhole radius approaches the beam radius, all the laser power enters the keyhole and a maximum of the net energy gain is obtained. Finally, with further increase in the keyhole radius, a slight decrease of the net energy gain appears because the evaporation losses then prevail over the absorbed laser power due to the larger evaporating surface. The net energy gain described here depends also on the welding speed, but only slightly. A hidden parameter is the length of the keyhole, which is determined by the welding speed and keyhole radius.

Figure 6 also shows the net energy gain given by the heat conduction losses (broken line), dependent on the keyhole radius. Due to the strong influence of welding speed on heat conduction losses, different curves are obtained for the various welding speeds.

In general, the curve for the net energy gain as a difference between absorbed laser power and evaporation loss and the curve for net energy gain given by the heat-conduction loss intersect in two points. At these points, the energy balance is in equilibrium, which means the net energy gain obtained from the absorbed laser power minus evaporation loss is equal to heat-conduction loss into the bulk of the workpiece. Unfortunately, two intersections mean there is an unstable situation; but if the welding speed is continuously increased, the two curves will finally touch each other, thus yielding one single intersection point, corresponding to a stable solution of the energy balance.

From this point, the net energy gain of the workpiece, the welding speed, the keyhole radius, and the length of the oval keyhole can be determined easily for an arbitarily chosen value of the temperature at the wall of the keyhole. So, a further relationship is needed to determine the temperature at the wall of the keyhole. This can be found from a description of a balance of the forces that appear at the wall of the keyhole: vapor pressure, surface tension, and hydrostatic pressure in the melt around the keyhole.

It is generally assumed that surface tension is the most important phenomenon to balance vapor pressure in the keyhole. Nevertheless, A. Kaplan[15] showed that the hydrodynamic pressure caused by the thermocapillary flow (see reference 7) due to the strong decrease of the temperature near the keyhole and related variation of surface tension is strongly dominating. The flow mentioned here takes place in a radial direction, streaming away from the keyhole toward the liquid-

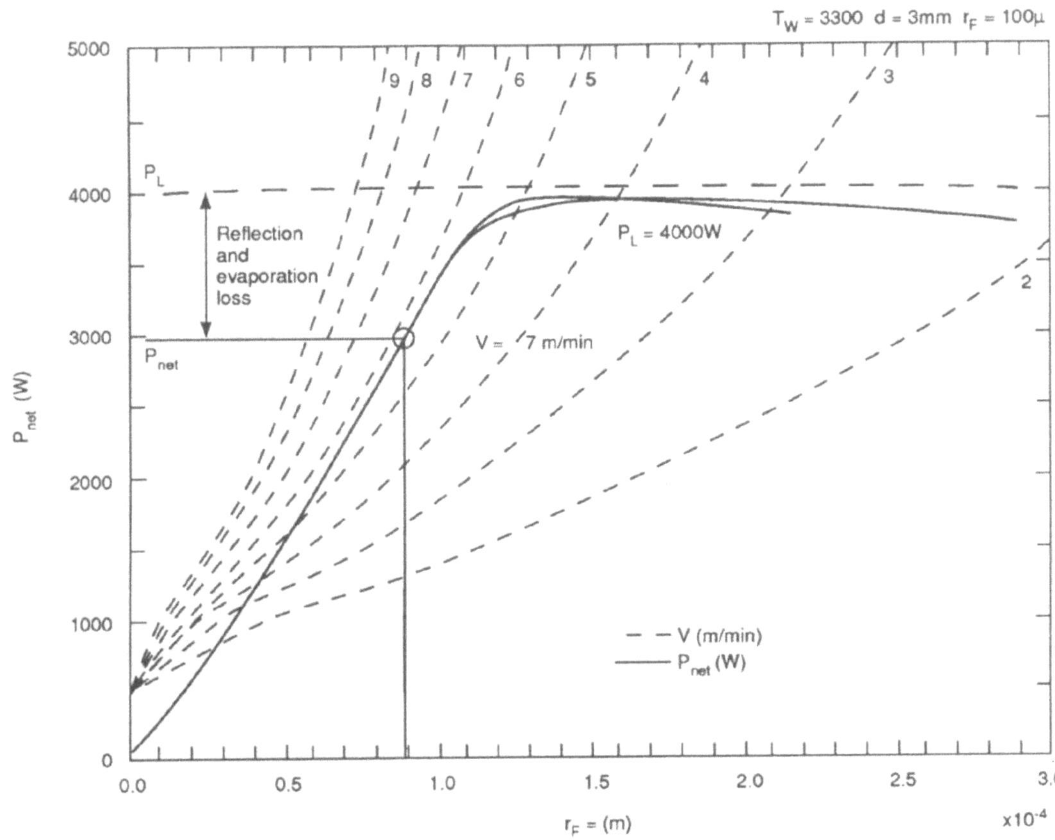

Figure 6. Net energy gain (P_{net}), calculated from difference between absorbed laser power and evaporation heat loss (solid line) and heat conduction loss (broken line) dependent on the keyhole radius (r) and welding speed (v) for a laser power of 4 kW and a focus radius of r = 100 μm. Steel, d = 2 mm.

solid-boundary, where it is then bent to the vertical direction. In the middle of the workpiece, the flow returns in the horizontal direction back to the keyhole wall, where the flow then becomes vertical again, thus forming a toroidal flow field.

The solution of the Navier-Stokes equation under strongly simplified boundary conditions yields the maximum flow speed, appearing at the surface of the melt in the radial plane through the x axis. Assuming that the melt flow width is determined in principle by the average distance between keyhole wall and melting isothermal, this speed can be expressed as:

$$V_{max} = \frac{1}{\mu} \cdot \frac{d\sigma}{dT} (T_b - T_m) \qquad (5)$$

where μ is dynamic viscosity, $\frac{1}{\mu} \cdot \frac{d\sigma}{dT}$ is temperature coefficient of the surface tension, T_b is the boiling point, and T_m is the melting temperature.

For typical values, an average flow speed of $v_{max}/2$ = 35 m/s is obtained. This flow speed determines the hydrodynamic pressure P_{dyn}:

$$P_{dyn} = \rho \cdot \frac{v^2}{2} \qquad (6)$$

where ρ = density.

After some averaging operations for the vertical and horizontal directions, a typical value of $P_{dyn} = 3.10^5\ Pa$ is obtained. Under quasi-steady state conditions, all forces acting on the wall of the keyhole must be in equilibrium. Therefore, the vapor pressure (P_{dyn}) must equal the hydrodynamic pressure (P_{evap}), thus yielding an equation for the temperature (T_w) at the wall of the keyhole:

$$P_{evap}(T_w) = P_{dyn} \qquad (7)$$

For typical values mentioned above, T_w = 3500 K. If the diagrams shown as an example for T_w = 3300 in Fig. 7 are calculated for the above value of T_w, final results for the welding speed and all the relevant quantities can be obtained. These data can also be used to determine the width of the weld seam. As an example, for laser welding of 8 mm steel with a beam characterizied by $P_L = 10$ kW, $r_F = 100$ μm, seam width $b = 0.5$ mm has been obtained, a value that agrees quite well with experimental evidence obtained with the Rofin-Sinar RS 10.000 CO_2 laser.

This model uses strongly simplified approaches and describes welding speed and seam width for given laser power, focus radius, thermal properties, and workpiece thickness. Very few analytic equations are used. Although these equations are relatively simple formulas and not point-by-point solutions of differential equations such as the heat-conduction equation and Navier-Stokes equation, the order of magnitude of the numerical results seems to be quite reasonable and agrees with observed results.

The model contains two approximations. First, it is assumed that beam power is absorbed 100% in the keyhole. Secondly, it is also assumed that no variations take place in the vertical direction in the workpiece. From a physical point of view, the first shortcoming is more interesting because the appropriate analysis of plasma formation could refine the model considerably and include the description of the influence of the shielding gas. The extension to three-dimensionality seems not to be very urgent because in practice welding is usually carried out with a laser power greatly above the minimum laser power needed for full penetration of the workpiece. Only in this case can high welding speed and excellent weld quality be achieved.

References

1. G. Herziger, et al., numerous papers during the last 20 years, including an overview lecture by Herziger, "Laser material processing," *Gas Flow and Chemical Lasers*, p. 55, Plenum Press, N.Y. (1984).
2. M. Beck, F. Dausinger, and H. Huegel, "Studie zur Energieeinkopplung beim Tiefschweissen mit Laserstrahlung," *Laser u. Optoelektronik 21* (3), pp. 80–84 (1989).
3. U. Del Bello and M. Cantello, "Heavy section laser welding," Europ. Industrial Laser Forum, March 1991, The Hague.
4. D. Schuoecker, "Modelling of the laser welding process," 3rd European Conference on Laser Treatment of Materials, ECLAT 90, Erlangen (1990).
5. N. Postacioglu, P. Kapadia, and J. Dowden "A theoretical model of thermocapillary flows in laser welding," *J.Phys.D: Appl.Phys.24*, pp. 15–20 (1991).
6. D.E. Swift-Hood and A.E.F. Gick, "Penetration welding with Lasers: Analytical study indicates that present laser beam welding capabilities may be extended tenfold," *Welding Research Supplement*, p. 492 (1973).
7. P.G. Klemens, "Heat balance and flow conditions for electron beam and laser welding," *J. of Appl.Phys. 47*, 5, p. 2165 (1976).
8. H.E. Cline and T.R. Anthony, "Heat treating and melting material with a scanning laser or electron beam," *J. of Appl.Phys. 48*, 9, p. 3895 (1977).
9. J. Mazumder and W.M. Steen, "Heat transfer model for cw laser material processing," *J.Appl.Phys. 51*, 2, p. 941 (1980).
10. M. Davis, P. Kapadia, and J.Dowden, "Modelling the Fluid in Laser Beam Welding," *Welding Research Supplement*, p. 167-2 (1986).
11. J. Dowden, N. Postacioglu, M. Davis, and P. Kapadia, "A keyhole model in penetration welding with a laser," *J.Phys.D: Appl.Phys. 20*, p. 36 (1987).
12. N. Postacioglu, P. Kapadia, M. Davis, and J. Dowden, "Upwelling in the liquid region surrounding the keyhole in penetration welding with a laser," *J.Phys.D: Appl.Phys. 20*, p. 340 (1987).
13. W.M. Steen, J. Dowden, M. Davis, and P. Kapadia, " A point and line source model of laser keyhole welding," *J.Phys.D: Appl.Phys. 21*, p. 1255 (1988).
14. N. Postacioglu, P. Kapadia, and J. Dowden, "Theory of the oscillation of an ellipsoidal weld pool in laser welding," *J. Phys.D: Appl.Phys.*, accepted for publication.
15. A. Kaplan, private communication.

Potential and challenges in laser welding structural steels

Jens Klæstrup Kristensen
Advanced Welding Centre, Force Institutes
Copenhagen, Denmark

Introduction

The large industrial area of welding has so far seen only limited use of lasers. A substantial effort is now being put into this area, and it is predicted that laser welding will make a breakthrough in industrial use during the next few years.

The need for a new welding process has been obvious for some time. The traditional welding processes—with regard to quality, efficiency, health, safety, and, perhaps most importantly, the ability to automate—are pressed against limits imposed by the processes themselves. Lasers offer obvious welding advantages, such as high speed, low distortion, single-pass welds in great thickness, and ease of automation. However, laser sources with the necessary power capability and beam quality have only recently been developed.

Limits to arc welding

Today, arc-welding processes—such as stick electrodes (SMAW), MIG/MAG, and submerged arc welding (SAW)—are by far the most important. The dominating trend within arc welding has for many years been the substitution of the stick electrodes by the more productive MIG/MAG welding processes, the latter offering a continuous wire feed. These processes in their simpler forms are further substituted for by more sophisticated variants, such as flux-cored powder-filled wire (FCAW) or welding using pulsed power sources. The key phrases in this development have been:

- Increased productivity.
- Increased quality.
- Increased ability to weld in position.
- Increased control of the thermal cycle.
- Increased health and safety properties.
- Automation and robotization.

Although this trend has been clear for about two decades, the speed of acceptance has been very slow, and the development is still far from finished. The automation and robotization of the arc-welding processes have also been quite slow and the acceptance level achieved very low, compared to other areas of materials processing such as machining. This appears to be due to arc-welding limitations, many of which may be related to the fundamental mechanisms of the process itself.

The large molten pool in arc welding has the following consequences:

- Limitations in the maximum obtainable welding speed.
- Limitations in the ability to weld in position.
- Limitations in productivity due to the multipass nature.
- Limitations in automation, especially for multipass welds.

The control of the individual weld-pass geometry is a key problem in the automation and robotization of arc welding. The number of independent variables is extremely large and the physical phenomena involved very complex, making mathematical modeling or regression analysis extremely difficult to perform.

Benefits of laser welding

Laser welding of structural steels will have a tremendous impact on the entire concept of fabrication. The high accuracy with which the different elements can be welded will greatly influence assembly methods and make it possible to introduce new design concepts, which up to now have been impossible to manufacture due to the limitations connected with traditional welding techniques.

Shipbuilding is an excellent example of an industrial area in which laser welding shows great potential. It is therefore not surprising that the few published investigations concerning laser welding of steel in greater thickness are concentrated in this area. D.R. Martyr investigated the potential of the laser-welding and -cutting process with regard to shipbuilding, giving special reference to wide gap joints (up to approximately 2.5 mm).[1] The possibility of filling such wide joints was demonstrated, some mechanical testing results presented, and the design of a production cell sketched. S.J. Brooke demonstrates the possibility of performing one-sided T-butt-joint welds with filler wire under normal production tolerances.[2] Promising static and dynamic mechanical tests were reported. Findings and discussions similar to those already mentioned are presented by P. Seyffarth in reference 3.

The introduction of laser welding to shipbuilding can be compared to the change from riveting to welding almost 50 years ago. In shipbuilding, a majority of the total price comes from the materials and the engine. The remaining cost, on which the real competition between shipyards takes place, is dominated by the cost of welding. A 50,000-ton tanker, for example, has more than 500,000 m of weld line. Of this, 80% is in subsection and section assembly, a potential area for laser welding. Although this corresponds to only 50% of the weld hours, a dramatic increase in productivity is within reach by application of laser welding. A corresponding reduction in filler metal costs and labor costs as well as the ability to hold production tolerances are also attractive features of laser welding.

The demands for laser joint preparation are much greater than with traditional welding processes, which has been considered a major drawback. As will be discussed later, recent investigations appear to show that laser joint preparation demands are actually obtainable using the preparation techniques used today (such as flame cutting). This is because the precision of this process for reasons other than welding has increased dramatically during the last decade.

Thermal cycles in laser welding

It is well known that laser welding in thick sections is performed as "deep-penetration welding" or "keyhole welding." The actual mechanism of the keyhole formation and movement are very complex. The influence of the plasma is especially difficult to describe precisely. From a rather simple argument, however, one may obtain a crude estimate of the equilibrium geometry of the keyhole, as well as the thermal cycles involved.

Figure 1 shows the calculated equilibrium keyhole geometry for a beam weld. The model used is based on the equilibrium between the vapor pressure in the cavity and the forces due to the surface free energy (surface tension). It does not, however, consider in detail the influence of the plasma, which is taken into account only through an efficiency factor.

From the calculated keyhole wall temperature, we may numerically calculate the thermal cycles sensed by the material in different distances from the beam axis. Typical results are shown in Figure 2. Analytical solutions to the high-velocity, two-dimensional heat source may also be used to estimate the thermal cycles.

For welds in steel, the cooling time from 800° to 500°C is often considered to determine the degree of hardening; it is assumed that the hardness of a given weld may be directly correlated with this value (e.g., see reference 5 for further discussion). From the high-velocity-limit analytical solution to the moving line heat source, the result shown in Figure 3 for the cooling time may be deduced. Compared to traditional arc welding, the cooling times are significantly shorter, especially for high welding speeds. The effect of this will be discussed later.

It is also interesting to compare the size of the molten pool in laser welding to that in arc welding. This is done in Figure 4, in which a TIG weld in steel with a speed of 5 mm/s is compared to a laser weld of the same speed. The molten pools are seen from above. It should be noted that the laser weld penetrates approximately 20 mm into the steel plate.

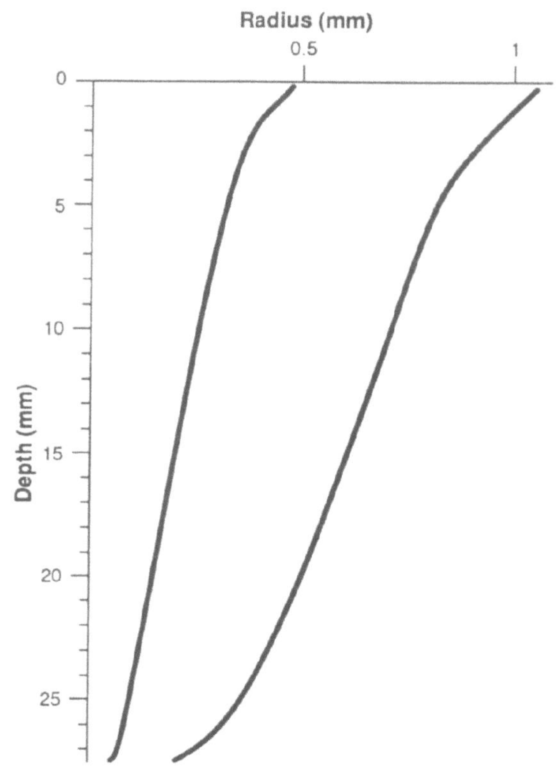

Figure 1. The calculated equilibrium cavity geometry for a weld in austenitic stainless steel using a Gaussian distributed beam. The inner contour represents the liquid-vapour interphase, the outer the liquid-solid interphase. Although the simulation disregards plasma effects, the geometry approximately represents a 15-kW, 10 mm/s laser weld.

Challenges to laser welding

Compared to arc welding, laser welding offers a number of possibilities:

- Full-penetration, one-pass welds.
- High welding speed.
- Low heat input.
- Low distortion.
- Relatively simple automation.
- High quality.
- Reduced amount of filler material.
- Highly improved health and safety properties.

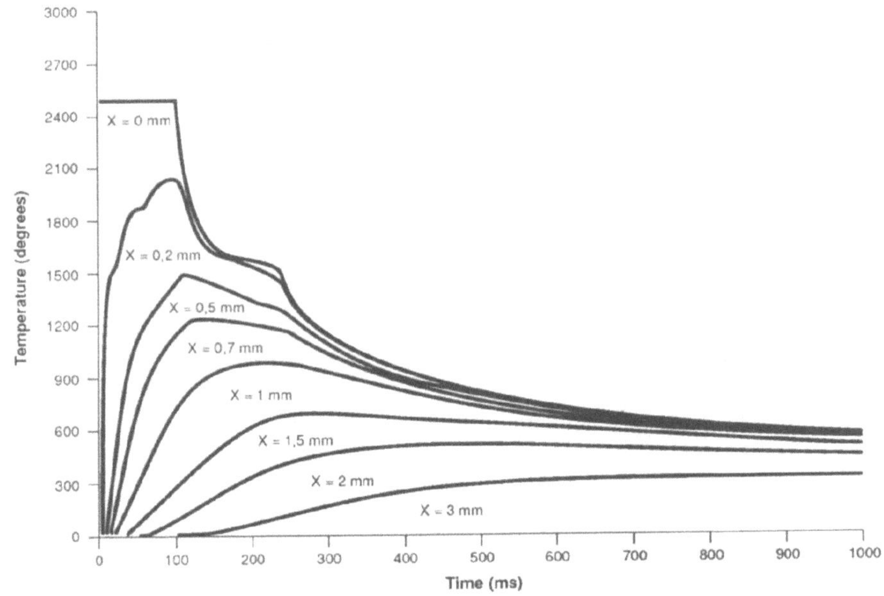

Figure 2. Typical thermal cycles corresponding to Figure 1; x is the distance to the cavity.

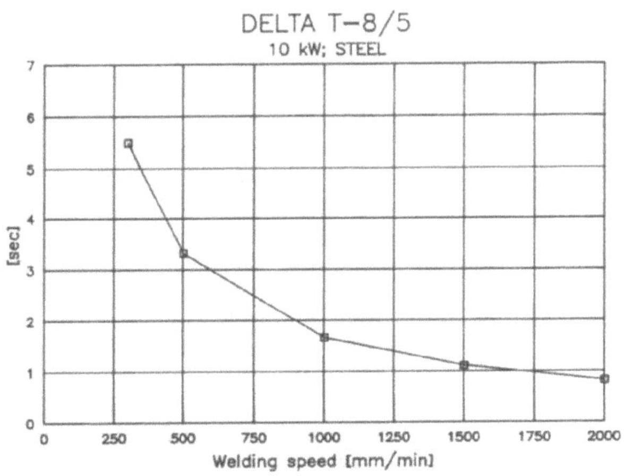

Figure 3. The cooling time from 800° to 500°C as a function of welding speed for a 10-kW autogenous laser weld in steel.

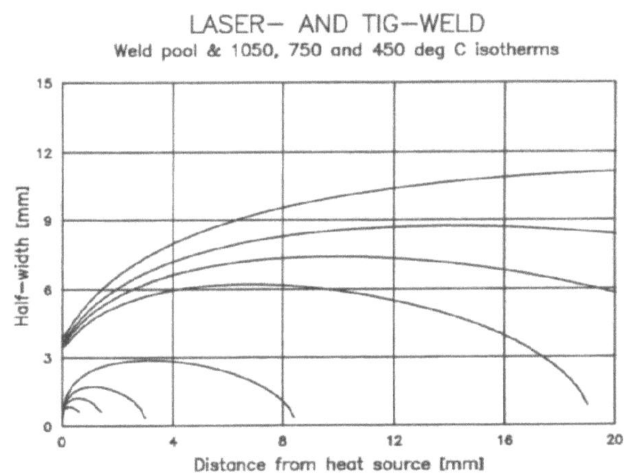

Figure 4. Comparison of the molten pool size and isotherm distribution of a TIG weld in steel with a speed of 5 mm/s to a laser weld of the same speed. The molten pools are seen from above.

Still, several drawbacks have so far restricted the use of laser welding. These disadvantages, which are also the challenges to further development, may be grouped into two factions: equipment and processes, and materials.

Equipment and process-related challenges

The drawbacks to laser-welding equipment and processes include:

- High capital cost for high-power lasers.
- Restrictions in the obtainable power and/or beam quality.
- Low overall electrical conversion efficiency.
- CO_2 laser wavelength not suited to fiberoptic transmission.
- Enhanced requirements in joint preparation.
- Enhanced requirements for fixturing.
- Enhanced requirements for seam tracking.

As already mentioned, an impressive amount of international development effort has been directed toward improving the size and quality of the laser sources and beam-delivery

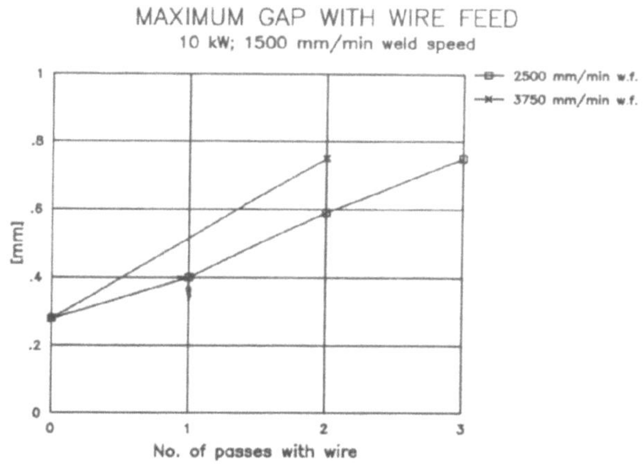

Figure 5. The maximum gap width obtainable with wire feed as a function of the number of passes. Zero passes correspond to no filler wire used.

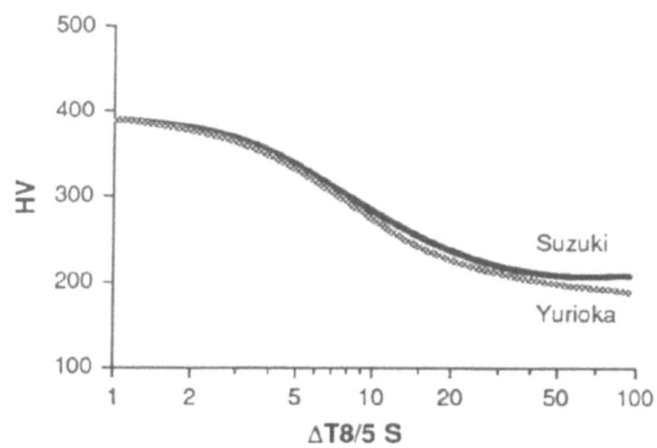

Figure 6. Calculated relationship between the cooling time from 800° to 500°C and the hardness of the heat-affected zone for the steel of the composition in Table 1. The two curves represent different formulas.

system. There is no doubt that high-power laser sources of sufficient beam quality for deep-penetration welding will be available at an acceptable price in the near future. Off-line programming, seam tracking, and other automation and robotization needs will certainly be available as well. Since these tools are not specific to laser welding, the results of broad development work can be utilized.

Process technologies such as plasma suppression and filler wire addition are not mentioned here because these techniques are considered available already. Filler wire addition may be used to improve the possibilities for gap filling as well as to control the weld metal metallurgy (to be discussed later).

Figure 5 shows the possibilities for welding steel having a joint gap. As can be seen, a maximum gap upon approximately 0.3 mm may be accepted without filler wire addition; a wider gap requires some additional material. The gap-filling ability for a wire is actually very close to what would be expected from simple volume arguments, although the actual mechanism undoubtedly is somewhat more complicated.

Materials-related challenges

Hardness levels

The thermal cycles associated with laser welding are generally much faster than those in traditional welding processes—a natural consequence of the low total heat input and high welding speed. However, it also means steels will obtain a local maximum hardness that is somewhat higher than normally observed in welding.

It was stated earlier that it is common in welding to consider the cooling time from approximately 800° to 500°C in determining the hardness of a steel. In traditional welding, this refers to the heat-affected zone (HAZ), which is the zone of interest. In laser welding, however, the interesting zone is the weld metal itself because the weld metal composition can only be controlled to a limited extent, whereas the HAZ is extremely narrow. Thus, because the weld metal's microstructure is quite unique, the empirically deduced formulas correlating cooling time and hardness must be used with great care.

Figure 6 shows two examples of the empirically deduced relationship between cooling time and hardness for a typical normalized steel. The composition of this steel is shown in Table 1. Measured and calculated hardness values for the same steel are further compared in Figure 7. It is obvious that the HAZ hardness corresponds well with the calculated value, while the weld metal hardness shows a lower value for low welding speed due to the difference in its microstructure.

In laser welding, the deformations (and, consequently, the stresses) are smaller and more locally distributed than in arc welding. In addition, the hydrogen level can easily be controlled to almost zero. Thus an important requirement for hydrogen-induced cracking can be avoided, which effectively decouples hardness level from cracking tendency.

In general, one has to consider the properties of a more or less martensitic structure. Thus the importance of hardness level—or, alternatively, the amount of martensite and its

Table 1. Parent metal composition [%]

C:	0.127	Cr:	0.091	Cu:	0.258
Mn:	1.330	Nb:	0.019	B:	0.0006
Si:	0.383	Mo:	0.004	W:	0.013
S:	0.005	V:	0.004	Ti:	0.002
P:	0.011	Ni:	0.082	Sn:	0.013
Al:	0.039	Co:	0.015	O:	0.0017

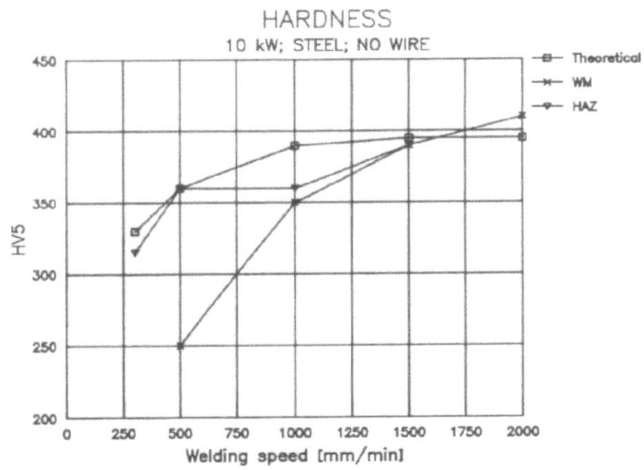

Figure 7. Measured and calculated laser weld hardness in the steel mentioned in Figure 6. The theoretical curve is calculated from the thermal cycle.

microstructure and composition—with respect to cracking tendency must be reevaluated.

Fracture toughness and tensile test ductility

Traditionally, Charpy impact tests and (occasionally) CTOD tests are used to measure fracture toughness. Scattered results have been observed in the work thus far performed on evaluating local fracture toughness of the structure in a laser weld. This is because, during its dynamic development, the crack does not remain in the structure in which the notch/fatigue crack is placed.

This phenomena has been reported several times[1,2,9] and is thoroughly discussed by E.R. Wallach et al.[6-8] It has been found that fracture deviation may occur in Charpy V-notch testing of welds narrower than approximately 4 mm, and that the use of fatigue-cracked Charpy specimens or specimens with side grooves are of limited value when the weld becomes even narrower. In some cases, CTOD testing can be used with better results than Charpy testing, presumably because of the sharper crack initiation and slower deformation rate. Further, two routes are proposed for achieving a good toughness in beam welding: a reduction in carbon equivalent, which is beneficial; and the introduction of oxygen, which promotes the formation of acicular ferrite due to the nucleation ability of the oxides.

For high temperatures, the crack often deviates into the parent metal, while for lower temperatures it stays in the weld metal. The driving force for the deviation seems to be the very large variation in hardness or strength, where the crack tip minimizes its potential energy through the deviation. The amount of plastic strain the weld metal undergoes also seems important. The transformation temperature observed in Charpy testing of laser welds will in many cases not be the transformation temperature of the weld metal but rather the temperature at which the cracks stops deviating into the parent metal.

Although laser welds often fulfill the requirements for Charpy testing, the results published so far seem to indicate that for some combinations of steel and welding data, the local fracture toughness in the weld metal or HAZ of a laser weld may be limited to values not accepted for bulk materials. In the same way, the ductility—as measured by longitudinal tensile test specimens exclusively made up of weld metal—may show a low value. Thus the crucial point to investigate is which demands one must apply for local brittle zones in the narrow laser weld. This case is very similar to local brittle zones in arc-welded multipass steel weldments.

The influence on the properties of a weldment including local brittle zones continue to be discussed intensively in arc welding.[10-15] Much evidence seems to indicate that a weld metal with mechanical strength higher than that of the parent metal is able to protect local brittle zones against failure. This result is very interesting to laser weldments, where the weld metal is highly overmaching.

New steels

As in the 1940s and 1950s when the change from riveting to welding took place, many of the problems referred to in this article can be solved through the development of new steel types. The idea is to use modern steel production technology to produce very lean steels at the level of strength relevant for shipbuilding. This is not done today because normalized steels satisfy present demands.

Fatigue and corrosion properties

From work with traditional welding techniques it can be expected that the fatigue properties of a welded joint are, to a very high degree, determined only by the sharpness of induced geometrical faults. Thus laser-welded joints will be expected to show fatigue properties that are not very different from traditional welded joints.

The only corrosion-induced problems that could be unique to laser-welded structures seem to be stress corrosion cracking and corrosion fatigue cracking. From our present knowledge, these phenomena in typical media will only be relevant for high-strength steels. Because laser-welded joints may possess high hardness locally in the weld metal, in some cases it may be relevant to test laser-welded structures for this effect.

Controlling weld metal properities

In certain cases, weld metal hardness must be controlled at a level lower than normal in laser welding. Such is the case when, for example, the environment in which the weld is going to be used contains appreciable amounts of hydrogen. In the same way, there are cases in which the fracture toughness of the weld metal or its ductility will have to be controlled to a higher level than normal in autogenous laser

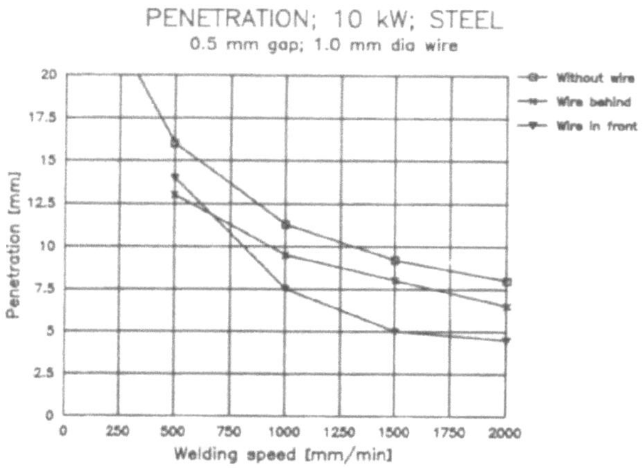

Figure 8. Penetration depth with and without wire addition as a function of welding speed, using 10-kW laser weld in steel.

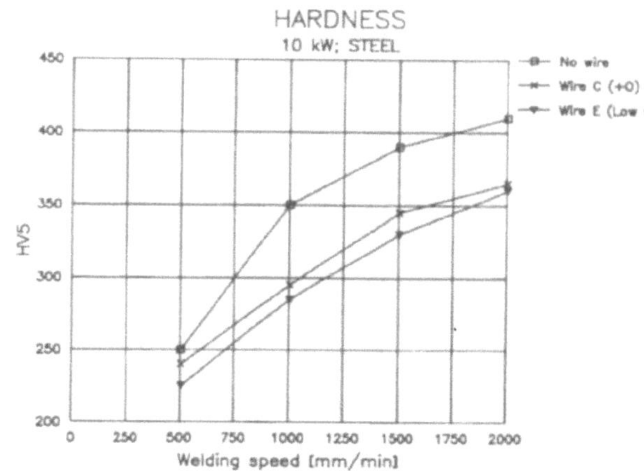

Figure 9. Measured hardness values of a 10-kW laser weld in the steel with the composition shown in Table 2 using different filler wire compositions. The wire is fed in behind the beam axis.

welding. As already discussed, different routes seems possible for doing so:

- At all welding speeds, a reduction in carbon content is beneficial, as the properties of the martensite formed to a first approximation is determined by the carbon level. Other elements, such as manganese, will also influence the properties, but the dependence is not totally clear.
- The carbon equivalent is relevant for medium and low welding speed, where the microstructure need not be fully martensitic. The carbon equivalent basically describes the hardenability of the steel.
- The introduction of oxygen is supposed to promote the formation of acicular ferrite due to the nucleation ability of the oxides. Titanium oxides are known to be especially stable.

The references at the end of this article contain a few examples of the investigations into the properties of laser-welded steels.[9,12,16] Some of our findings will be shown below.

Table 2. Filler wire composition [%]; Dia. 1.0 mm powder-filled wire

Wire A:	0.07 C	0.50 Si	1.10 Mn	—	0.03 O
Wire B:	0.07 C	0.50 Si	1.10 Mn	0.05 Ti	0.03 O
Wire C:	0.07 C	0.50 Si	1.10 Mn	—	0.08 O
Wire D:	0.07 C	0.50 Si	1.10 Mn	0.05 Ti	0.08 O
Wire E:	0.003 C	—	—	—	—

A normalized steel with the composition shown in Table 1 was welded with and without filler wire. The composition of the wires are shown in Table 2. The experimental details were:

- CO_2 laser with 10 kW continuous power and polarization perpendicular to the weld line.
- I-joint with 0.5 mm gap; machined edges.
- Wire fed into the weld pool at a 45°C angle, either in front of or behind the beam axis.
- Plasma-suppressing helium jet and argon shielding gas are used.
- Focus point position optimized for maximum penetration in bead-on-plate welds.

Figure 8 shows the penetration depth obtained with and without filler wire. Without filler wire the welds just passed the demands of Charpy testing at 20°C, while the pure weld metal elongation was measured to approximately 10%. Wires C, D, and E showed significant reductions in hardness level; the results are shown in Figure 9. However, hardness is only one characteristic of the weld. Demands of Charpy tests, weld metal tensile tests, or longitudinal bend tests will still limit the maximum welding speed. For example, in the present case, the allowable hardness will be limited to approximately 325 μm, if the minimum weld metal ductility is demanded to be 20%.

Discussion

In contrast to the arc-welding processes that dominate today, laser welding structural steels shows great potential, thanks to such attractive properties as high welding speed, low distortion, and ease of automation. Still, a number of challenges remain to be addressed, in process and equipment and in metallurgy. While solutions to the process-dependent challenges seem reasonably straightforward, the material-dependent challenges call for a reevaluation of test methods and acceptance levels.

References

1. D.R. Martyr, *Weld. Rev.*, 106 (May 1987).
2. S.J. Brooke, *Proc. of 5th Int. Conf. on Lasers in Manufacturing*, 165 (1988).
3. P. Seyffarth, *Shiffbauforchung 28*, 22 (1989).
4. J.K. Kristensen and T. Kristensen, *Proc. of The Int. Coll. on Welding; Beam Technology, Arc Welding Technology*, Aachen, Germany (1989).
5. B. de Meester, *Weld. in the World 28*, 48 (1990).
6. E.R. Wallach, M. Ohara, P.L. Harrison, and M.N. Watson, *Proc. of The Int. Conf. on Power Beam Technology*, ed. by J.D. Russel, The Welding Institute, p. 43-1 (1986).
7. M. Ohara and E.R. Wallach, *Proc. of The Int. Conf. in Tokyo Electron and Laser Beam Welding*, The International Institute of Welding (IIW), 237 (1986).
8. B.E. Hall, M.T. Harvey, and M.T. Wallach, *Proc. of The 2nd Int. Conf. on "Power Beam Technology*, ed. by J.D. Russel, The Welding Institute, 139 (1990).
9. M.N. Watson and I.M. Norris, *Proc. of The Int. Conf. on "Power Beam Technology*, ed. by J.D. Russel, The Welding Institute, P351 (1988).
10. B. Lian, R.M. Denys, and Van De Walle, "An experimental Assess ment on the Effect of Weld Metal Yield Strength Overmaching in Pipeline Girth Welds," 3rd Int. Conf. on Welding and Performance of Pipelines, London, UK (November 1986).
11. R.M. Denys, "Fracture Testing of Weldments in Relation to Local Brittle Zones," *Proc. of Int. Sem. on Modern High Strength Steels in Marine Structures*, Delft, The Netherlands (Oct. 1987).
12. C.P. Debel, "Local Brittle Zones", Ris National Laboratory, Metallurgy Dept. Rep. 1/21 (June 1988).
13. R.M. Denys, "The Future of Wide Plate Testing in Welding Research," *Proc. of Welding 90*, ed. by Kocak, IITT International (1990).
14. B. Petrovski and S. Sedmark, "Evaluation of Crack Driving Force for HAZ of Mismatched Weldments using Direct J-Integral Measurements in Tensile Panels," *Proc. of Welding'90*, ed. by Kocak, IITT International (1990).
15. P. Dong and J.R. Gordon, "The Effect of Under- and Overmatching on Fracture Prediction Models," *Proc. of Welding '90*, ed. by Kocak, IITT International (1990).
16. M. Hill, J.H.P.C. Megaw, A. Aitchison, and J.H. Aubrey, *Proc. of The 2nd Int. Conf. on Power Beam Technology*, ed. by J.D. Russel, The Welding Institute, 108 (1990).
17. E.A. Metzbower, P.E. Denney, F.W. Fraser, and D.W. Moon, *Weld. J.*, 39 (July 1984).

Laser welding in the pipeline industry

V.E. Merchant
The Laser Institute
Edmonton, Alberta, Canada

Introduction

The construction of pipelines, in the plant or in the field, is an expensive process that requires reproducible high-quality welds because of the possible environmental and economic effects of weld defects, particularly those leading to breaks. This requirement has been accentuated by the shift to high-strength low-alloy (HSLA) steels for pipelines, with increasingly stringent demands on welding procedures. Rothwell et al.[1] report that "As new materials and design approaches for the pipeline construction industry are developed and economic pressures increase, methods which revolve around craft skills and arbitrary workmanship standards become less appropriate . . . given the intrinsically repetitive nature of cross-country pipeline welding, new, mechanized and automatic approaches will be sought, in which the quality of welds, and the way in which it is assessed, are related to prior engineering decision, rather than to individual craft performance in the field."

Many major pipeline companies have been investigating alternatives to the common shielded metal arc (SMAW) and automated gas metal arc welding (GMAW) procedures, in order to decrease costs. In the early 1980s, the forecast for very large pipeline construction jobs, in particular that from Prudhoe Bay in Alaska to the continental US, led to an examination of the need for high-speed, highly automated welding processes. Laser welding was considered worth investigating as a candidate process. Considerable effort has been spent investigating laser welding for high-strength steels used for pipelines and developing equipment to apply the technology to the field environment.

Laser welding is a keyhole, or deep-penetration, welding process, in which the focused power from the high-power laser forms a vaporized channel partially or completely through the thickness of the material. As the laser is scanned along a butt joint, material at the front of the beam is melted, flows around the side of the hole, and resolidifies at the rear of the hole. This forms an autogenous joint without the use of filler material.[2] Photo 1 compares the cross section of a laser weld (1a) to a double-pass submerged arc weld (1b) in 0.375-in. wall-thickness HSLA steel for pipeline applications. The submerged arc weld (SAW) is the spiral weld used to manufacture the pipe from a roll of steel.

Photo 1 clearly shows the narrow weld and heat-effected zone produced in the low-heat-input deep-penetration process. By comparison, in the SAW, a double-V groove-weld preparation is filled with large amounts of filler metal. Similar preparations are required for SMAW or GMAW. The principle advantage of laser welding is speed. It has been projected that laser welding has the capability of producing high-quality welds at a speed that will increase the overall production rate.

This article is a summary of past and present laser pipe-welding research. The development of the machines and procedures required for laser pipeline welding is described in the following section. The laser development program at Majestic Laser Systems Ltd. (Edmonton, Canada), with which the author is most familiar, is presented as a means of familiarizing the reader with details about a laser-welding system.

Field-welding pipelines

Majestic Laser Systems Ltd., founded in the early 1980s, intended to bring to the pipeline construction industry the PIE laser technology developed at the University of Alberta and described briefly in a following section. Engineering development of the pipeline laser-welding system was divided into three phases:

- Laser to generate a powerful beam of infrared radiation.
- Beam-delivery system, to deliver the beam to the pipe.
- A vehicle to transport the laser and the beam-delivery system to the pipeline right-of-way.

In addition, investigations were undertaken to determine the conditions for producing a reliable and reproducible weld in various thicknesses of modern HSLA steels.

Majestic Laser Systems ceased activity in July of 1985. The prime reason for the company's failure was the decrease in pipeline construction projects. When the company and its activities were initiated in 1981, the long and very expensive pipeline from Prudhoe Bay to the continental US was envisioned. But by 1985, all environmental concerns about the possibility of an Exxon Valdez-type disaster had been stifled and the much shorter pipeline from Prudhoe Bay to Valdez was constructed using conventional technology. In addition,

Photo 1. Comparison of (a) single-pass laser weld and (b) double-pass submerged arc weld in 0.375-in. wall-thickness pipe.

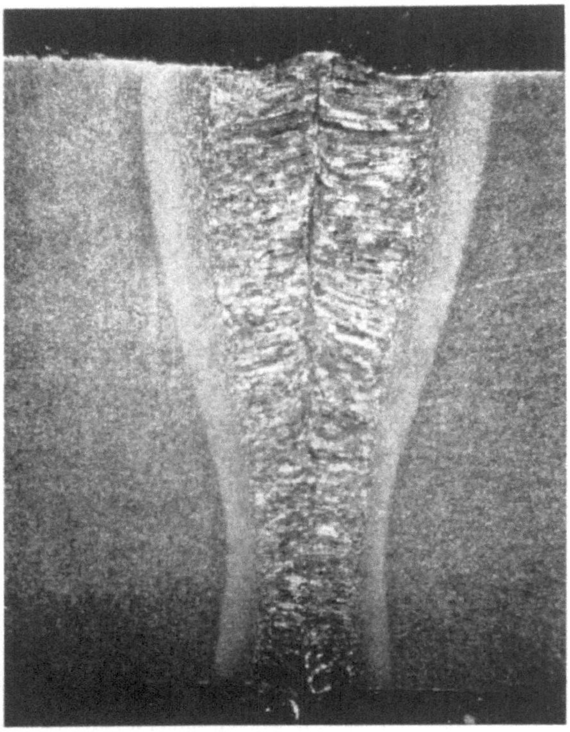

(a)

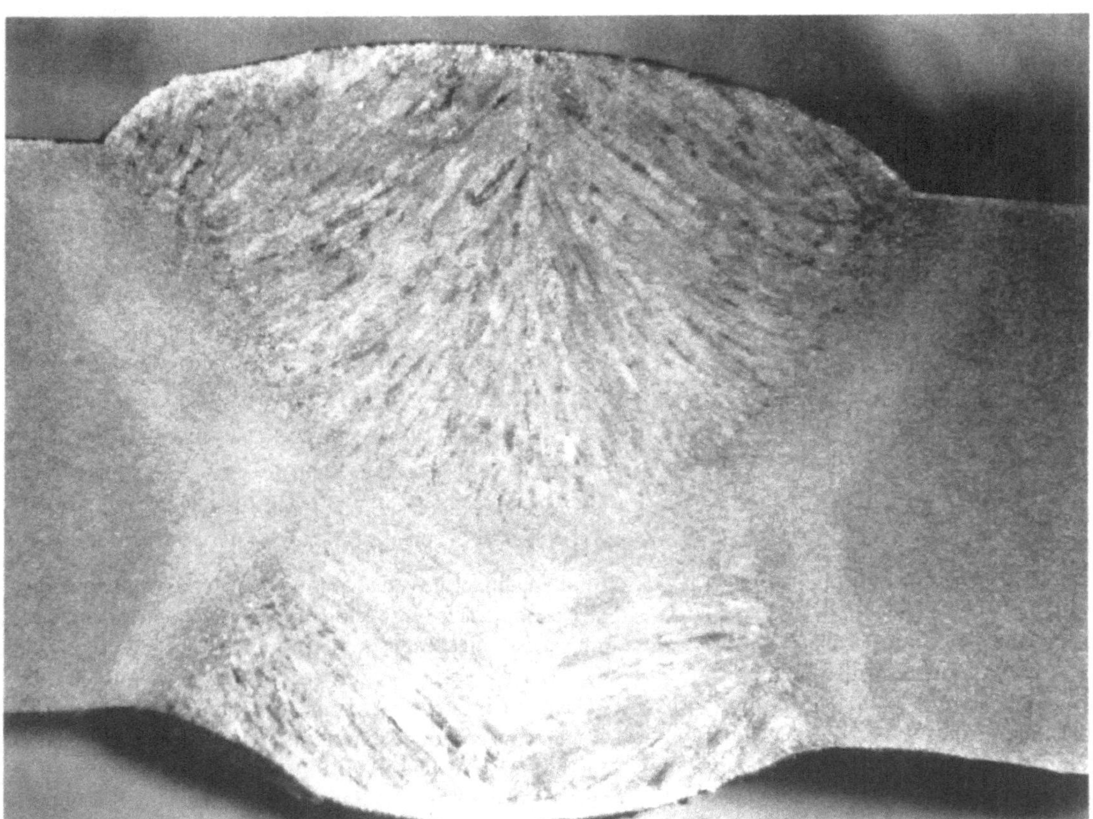

(b)

Photo 2. The 20-kW laser developed by Majestic Laser Systems Ltd. for incorporation into an off-road vehicle for pipeline welding in the field.

the price of oil had become very low, halting many oil-exploration projects and all pipeline-construction projects. When no pipelines were being built, no new pipeline welding processes were needed.

The following sections provide more detail of each of the development phases of the laser system developed by Majestic. Note that the beam-delivery system was only partially constructed at the time company activity was terminated, and construction of the vehicle had not commenced.

Laser development

The objective of the laser development phase of this project was the construction of a highly controllable, rugged, compact, 20-kW laser. Controllability was required to relieve the operator on the pipeline right-of-way of concerns about details of laser operations, such as mirror alignment and cooling-water temperature. The operator—a welder, not an engineer or laser specialist—would simply push a button to produce a laser beam to complete his job. All control and monitoring operations would have to be done remotely and automatically, without operator interaction.

The second requirement was for a rugged laser, of sufficient durability to withstand the jolting and jarring as it was moved across the crudely graded roads in the pipeline right-of-way. This motion would take place, in general, in a "ready-to-operate" mode as the laser was moved from one welding site to the next. It was anticipated that the laser would be used in an approximately 5-min-on, 5-min-off cycle. The laser would have to have the ability to produce a good-quality mode while operating with a considerable side-to-side or lengthways tilt due to uneven terrain.

A third requirement of the laser was compactness, since the laser system, including power supplies and cooler, had to be transported to the site in a vehicle. The overall vehicle length was limited to one-half the length of pipes being welded. Vehicle width was limited by regulations governing travel on public highways.

The power requirement for this project was for a 20-kW laser, due to the perceived need to weld 1-in. steel. Laser systems of this power rating were not commercially available at that time, though a laser design featuring lower power levels—known as photo-initiated impulsively enhanced electrically excited lasers (PIEs)—had been developed at the University of Alberta.[3,4] (The PIE acronym was in response to TEA and COFFEE—Transversely Excited Atmospheric lasers and Continuously Operating Fast Flowing Electrically Excited lasers, respectively.) The PIE lasers used a series of high-repetition rate pulses to preionize or condition the laser discharge region to absorb a large amount of power from a continuous power source. Specialized pulsed high-voltage circuits were developed to generate the pulses.[5] This technology was projected to be capable of meeting Majestic Laser Systems' requirements. Lasers with similar excitation schemes had been developed elsewhere.[6–8]

A PIE laser was designed by Majestic Laser Systems to fulfill the criteria of power, ruggedness, and compactness. Details about the laser (see Photo 2) as constructed at Majestic Laser Systems are presented in references 9 and 10.

Beam-delivery system

It was envisioned that the laser would be installed in a vehicle that would be driven up to the location of a joint and placed approximately parallel to the pipe. A motorized system of mirrors mounted on rails attached to a clamp would be extended from the vehicle and clamped onto the pipe. When the laser was turned on, the mirror system would move in such a way as to keep the circumference of the pipe (see Fig. 1). At the completion of the weld, the clamp would be released from the pipe and the system retracted into the vehicle.

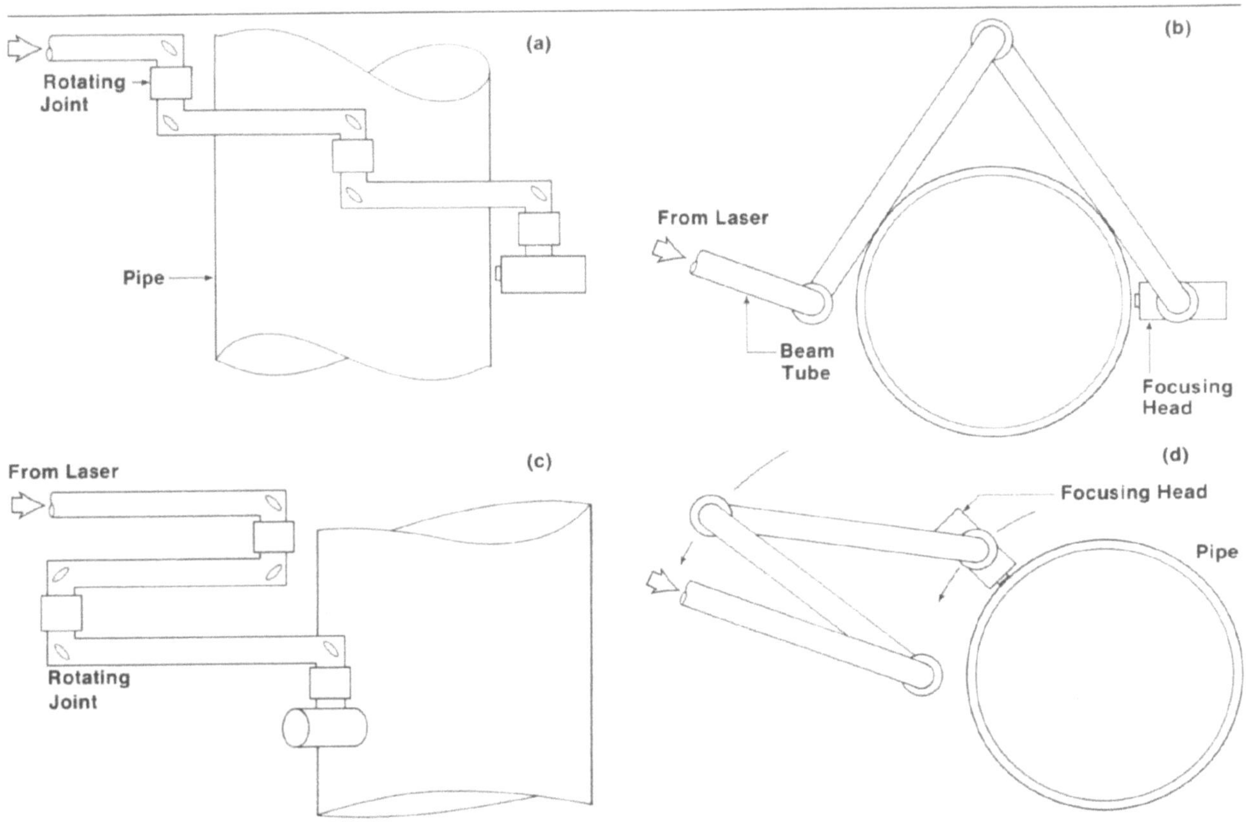

Figure 1. A proposed beam-delivery system for laser welding of a stationary pipeline from a stationary laser, using a moving beam system. (a) and (c) give top views; (b) and (d) end views. With the laser in the nine o'clock position, (a) and (b) show welding in the three o'clock position and (c) and (d) welding in the 11 o'clock position. Each joint consists of two mirrors; the upper joint is slaved to the motion of the welding head.

To accommodate the high laser power, all mirrors in the delivery system would be water-cooled. The delivery system would include the focusing head that would concentrate the laser power on the pipe seam; the gas shields required for plasma suppression and protection of the molten metal from atmospheric contamination during the welding process; and a commercial seam tracker to ensure that the beam hit the seam. Majestic Laser Systems designed and fabricated the focusing head and a compensating elbow mount. The focusing head contained a rotating joint and transverse slides that enabled the head to move longitudinally to follow the seam and radially to account for irregularities in the pipe. The compensating elbow, at the intersection of two beam tubes, adjusted the angle of the mirror to automatically bisect the angle between the beam tubes.

The beam-delivery system was envisioned to be contained inside a flexible, accordion-like tent that would be automatically extended and retracted simultaneously with the system itself. This cover was intended to ensure that the welding operation would be protected from the ambient environment, in particular from variations in temperature that may effect the accuracy of the system and from winds that would effect the gas shielding. Similar shields had been found to be necessary for successful implementation of automated GMAW.

The vehicle

A vehicle would be needed to move the laser along the pipeline right-of-way and to provide a controlled environment for the laser. The beam-delivery system would be contained inside the vehicle and would be deployed when needed. The vehicle would also carry the diesel electric power supply to generate the electricity required to operate the laser and accessories, and a closed-cycle refrigerator to dissipate the excess energy generated by the laser.

A commercially available off-road vehicle could be modified to carry the laser and associated equipment. Two personnel were expected to be involved, a driver and a welder-operator. The welder-operator would have to ensure alignment of the clamp and position of the welding head before the welding operation commenced. It was envisioned that conventional peripheral equipment would be used to the greatest extent possible, including preheat torches and the internal pipeline clamp. The welding procedure development would include an assessment of the ability of "end-prep" machines to prepare the ends of the pipes to sufficient accuracy for laser welding.

Welding procedure and development

For welding in the field, it had been envisioned that the laser would initially perform a tack weld with shallow penetration and subsequently a deep-penetration weld through the material thickness. Once the tack weld had been performed, the internal pipe clamp could be disengaged and used to prepare the next joint while the deep-penetration pass was being completed. This technique would lead to the fastest possible construction rate with a single-laser system.

After Majestic's laser was operational but before its operation and controls were optimized, a welding development program was initiated. Much of this work was bead-on-plate on a pipe rotating under a stationary beam; however, some work was on pipes with a gap or a mismatch at the joint. Focusing and gas-shielding conditions were found that allowed single-pass welding on 9.5-mm wall-thickness steel at 90 in./min. Some results of this welding development work have been published.[11] The development work was limited to welds on a rotating pipe and was not extended to circumferential welds on a stationary pipe in the duration of Majestic Laser Systems. The concept of tack welding was not verified.

Other laser pipe-welding developments

Offshore applications

Concurrently with the development of the laser pipeline welding machine for dry-land construction, several offshore construction companies had been investigating laser use for offshore construction using lay barges. Among these companies was Brown and Root (Houston, TX), which had been active in international offshore construction and was particularly interested in laser welding for the J-bend configuration in which only one welding station was practical. An engineering design study showed that the welding station should incorporate three lasers, one on either side of the pipe and one spare.

Contracts for welding development work were awarded to United Technology Research Center, which had 15-kW laser capability, and Culham Laboratories of the UK Atomic Energy Agency, where there was 10-kW capability. Much of the work was performed in the flat position, with the pipe rotating on a horizontal axis. However, toward the end of the project, some work was performed with a rotating mirror system. Welding procedures were developed to confine the liquid metal when laser welding in the overhead position, as required when performing a 360° weld on a horizontal pipe. Some of this work has now been published.[12,13]

In a benchmark publication,[12] it was concluded that in the flat and overhead welding positions, the stable welding limits are established by the maximum weight of molten material that can be supported by surface tension. The data presented in this paper lead to the conclusion that the maximum wall thickness of stationary pipe that could be welded by a continuously operated laser beam deflected by a mirror system around the pipe was about 13 mm. It might be possible to weld thicker pipes using a repetitively pulsed laser, in which case the metal would solidify before it would have time to move.[2] In other words, the liquid metal would be stabilized by inertial forces rather than primarily surface tension forces. This hypothesis has not been tested in practice, however.

Fairey Engineering, a UK firm, was contracted to design the beam-delivery system for the offshore environment. This company's design was similar in principle to that envisioned by Majestic Laser Systems.

Brown and Root apparently intended to use a laser on a lay barge in 1985. Their activity was approximately simultaneous with that at Majestic Laser Systems, but neither was aware of the other. Both activities were terminated in 1985, with the fall in the price of oil and in oil exploration activity.

Soviet pipe-welding activity

Since 1977, laser-welding activity has been ongoing at the All Union Scientific Research Institute of Pipeline Construction in Moscow.[14–17] As early as 1979, an investigation of laser welding pipeline steels was reported,[15] using powers up to 21 kW in a variety of welding positions. With a laser power of 21 kW, 15-mm wall-thickness steel was welded at speeds of 1.5 m/min. Few details of the welding results were given, but the publications indicated that the welds had a small reinforcement and contained undercutting to a depth of 0.5 mm, except those performed in the overhead position. Some contained defects, particularly micropores and pores with diameters 0.1 to 0.2 mm. Nevertheless, all bend-test specimens were reported to be deformed to 180° without fracture, and the majority of tensile test specimens failed in the parent material rather than the weld material. There are no reports, however, of this technology moving from the laboratory to field conditions.

Italian pipe-welding activity

In the 1970s, Italsider, an Italian manufacturer of pipeline steels, participated with United Technologies Research Corp. in a study of laser welding of pipeline steels.[18] Single-pass and dual-pass welds were performed in material of 12-mm and 26-mm thickness with powers of 10–15 kW. Initial results reported in the literature showed welds with poor visual, metallographic, and radiologic characteristics, specifically the presence of porosity, shrinkage cracks, and high hardness values.

Nevertheless, surprisingly good mechanical properties were observed with tensile strengths of many welds exceeding that of the base metal, and several welds with Charpy impact shelf energies above that of the base metal. The behavoir was attributed to fusion zone purification, a phenomenon previously noted in laser welding in which the beam preferentially vaporizes nonmetallic impurities that are generally present in the form of oxides and sulfides. The published study recommended improvements in cleaning and shielding practices, and use of alloys favorable to autogenous welding.

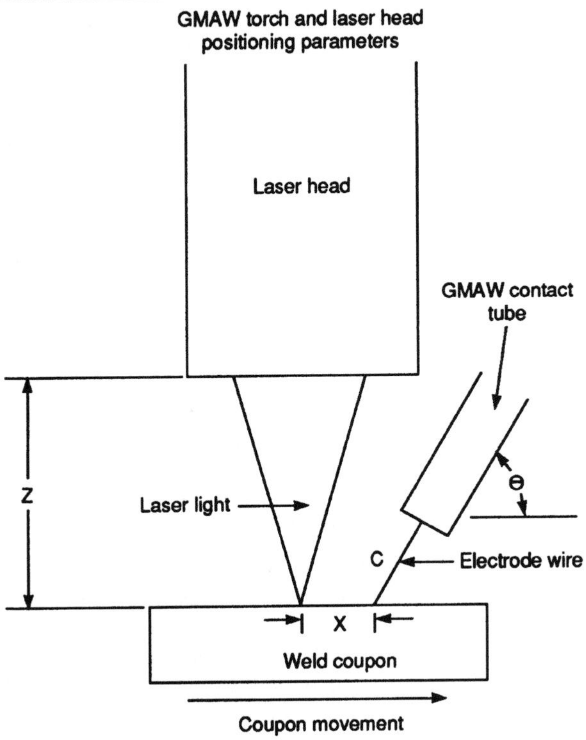

Figure 2. Schematic view of welding head used for laser-assisted arc welding, incorporating laser-focusing mechanism and GMAW torch.

New concepts in laser pipe welding

Laser-assisted arc welding

As already mentioned, laser welding is a keyhole, or deep-penetrating, welding process that produces joints at high speed and without the use of filler material. With lasers being installed in factories for pipe welding, the possibility of laser pipeline welding in the field or offshore is a topic that can be expected to be raised again. However, when this occurs, it may be appropriate to evaluate innovative processes such as laser-assisted gas metal arc welding (LAGMAW)[19] instead of simple laser welding. In laser-assisted arc welding, the relatively inexpensive power from an electric arc is combined with the expensive and highly controllable power from the laser, as shown schematically in Fig. 2.

The combined laser and arc process has several advantages over each process individually; some of these advantages would have potential benefit in pipe welding. The power from the laser serves to anchor the arc and prevent wandering. In thin materials, this would potentially allow welding at higher speeds than that achieved with the arc process alone. If the arc power is deposited in the keyhole produced by the laser, the combined welding process simulates the effect of a higher power laser and could weld thicker steel than that achieved by laser welding alone. The arc power can serve to preheat the steel and enhance welding in materials such as aluminum that are normally too reflective to be readily welded with a laser alone. Since a filler metal is used in the LAGMAW process, the tight fit-up requirements normally needed for laser welding are relaxed. The combined process can weld a given thickness of steel with a lower heat input than can each process individually. This may have metallurgical and mechanical advantages such as greater impact strength.

Moreover, it is expected that the laser power will anchor the arc to the bottom of the groove, which may allow welding with a groove far narrower than that possible with gas metal arc welding alone. Because the groove is narrower, less filler metal and less welding time is required and dramatic increases in productivity over that obtained with gas metal arc welding may result. However, further research is needed to confirm these expectations and to ensure that use of the narrow groove does not result in defects such as lack of fusion.

For a given thickness of steel, the combined laser-arc process may lower the laser-power requirements and hence capital cost of a welding project, as compared to the cost to be incurred using laser welding alone. If it is confirmed that a narrower groove and less filler metal can be used to produce high-quality welds, the combined laser-arc process may increase the productivity of a welding project as compared to that when using arc welding alone. However, the process is not sufficiently developed for field application at the present time.

Laser-assisted electric resistance welding

Another new concept in pipe welding, applicable to plant applications, is laser-assisted electric resistance welding (ERW).[20] In ERW, a high current flows between two electrodes, one mounted on either side of a butt joint. The higher resistance at the butt causes melting and fusion of the two pieces of material. However, the current flows preferentially at the metal surface, rather than being equally distributed throughout the material thickness. Thus the butt joint has a wider fusion zone at both the upper and lower parts of the seam, as shown in Fig. 3.

In experimental trials, this situation was rectified by directing a laser between the two pieces of metal as they were shaped into a butt (Fig. 4) and positioning the electrodes close to the point where the laser energy was absorbed. This re-

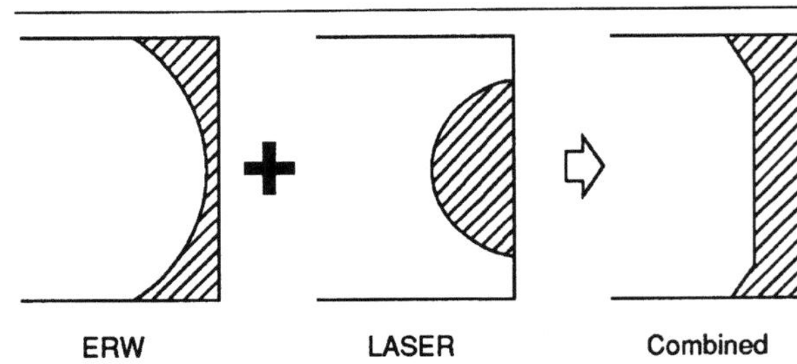

Figure 3. The concept of uniform heating produced by laser-assisted electric resistance welding.

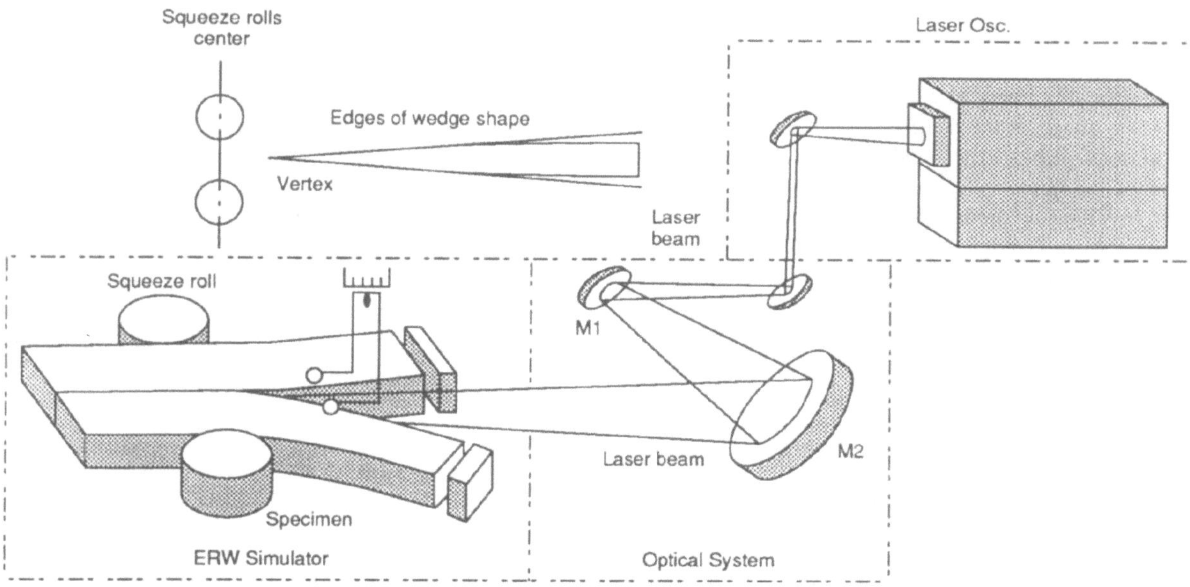

Figure 4. Schematic view of welding configuration used in laser-assisted electric resistance welding.

sulted in a melt zone of relatively uniform thickness over the thickness of the pipe, as shown in Fig. 3. Morever, the occurrence of lack-of-fusion defects was reduced. Notice this effect was caused by relatively low laser powers, for example 3.5 kW, as compared to the power in the electric current, which was 207 kVA.

Figure 5. Set up for high-speed laser welding of pipe, with focusing of the beam into the wedge formed by shaping of pipe from sheet metal. Welding plated pipes is possible.

Welding with polarization-enhanced absorption

Another departure from the keyhole welding process using lasers has been investigated by Beyer et al.[21] at the Fraunhofer Institute. In this case, the laser was also focused into the convex-shaped "funnel" or wedge formed when a piece of steel is being formed into pipes (Fig. 5). The laser was used without an auxiliary power source.

It is well-known that absorption of optical radiation striking a metal surface at a glancing angle is a function of laser polarization. In this case, the polarization of the laser was chosen so that the laser energy was absorbed in the seam at the apex of the wedge. The power of the laser was high enough (20 kW) that both metal surfaces melted. The forming operation causes a joint to form by squeezing the two molten pieces of metal together. Solidification takes place immediately, since the beam can no longer continue to heat the metal at that location. This process is claimed to be considerably faster than the keyhole process, relatively insensitive to misalignment of the beam and potentially useful for producing pipe from preplated steel. This laser-welding process is to be used in the Hoesch Rohr AG pipe mill[22] beginning in 1992.

Conclusion

The task of developing a laser pipeline-welding system remains to be completed. While lasers have demonstrated hundreds of thousands of hours of production operation for a variety of difficult applications in factories, laser operation in the rough-and-ready oil-patch environment is another matter. Have current lasers demonstrated reliability to withstand the rigors of cross-country or maritime environments, where downtime might cost a million dollars a day? Can laser mirrors that direct the laser beam to the workpiece remain functional in the muddy and windy environment of a pipeline right-of-way, the greasy and salt-laden atmosphere of an offshore lay-barge, or the thermal gradients close to a large mass of steel preheated to welding temperature? Can the mirrors of the complicated beam-delivery system remain aligned in a nonstop production environment? These questions must be answered.

This article has described in some detail one attempt to produce a laser pipe-welding system, along with several other projects in lesser detail. Variations on the laser-welding process potentially useful for pipe manufacture and joining have been described. While the impetus for pipe-welding machines largely disappeared with the collapse of the pipe market accompanying the fall of oil prices in 1985, the impetus for developing laser pipe-welding processes and the potential financial savings that can be made on repetitive fabrication processes still exists. The next few years will see a number of laser installations in the more friendly factory environment. These installations will provide valuable experience.

Acknowledgments

A partner in Majestic Laser Systems Ltd. was Majestic Contractors, North America's largest dry-land pipeline construction contractor. Financial support for design and construction came from the Industrial Research Assistance Program of the National Research Council of Canada. In the welding development work at Majestic Laser Systems, valuable assistance was provided by Fred Seaman, then with The Illinois Institute of Technology Research Institute.

References

1. A.B. Rothwell, D.V. Dorling, and A.G. Glover, "Welding Metallurgy And Process Development Research for The Gas Pipeline Industry," *Advanced Joining Technologies, proceedings of the International Institute of Welding Congress-Joining Research July 1990*, T.H. North, ed., Chapman and Hall, London, pp. 175–192 (1990).
2. C.M. Banas, "High Power Laser Welding," *The Industrial Laser Annual Handbook*, M. Levitt and D. Belforte, eds., PennWell Press, Tulsa OK (1985).
3. K.H. Nam, H.J.J. Seguin, and J. Tulip, "Operational Characteristics of a PIE CO_2 Laser," *IEEE Journal of Quantum Electronics QE-15*, 1, pp. 44–50 (Jan. 1979).
4. S.K. Nikumb, H.J.J. Seguin, V.A. Seguin, and H. Reshef, "Gain and Saturation Parameters of a Multikillowatt PIE CO_2 Laser," *J. Phys.E: Sci. Instrum 20*, pp. 911–916 (1987).
5. V.E. Merchant, H.J.J. Seguin, and J. Dow, "A High-power, High-Repetition Rate Pulser for Photo-Impulse Ionized Lasers," *Review of Scientific Instruments 49*, p. 1631 (1978).
6. N.A. Generalov, V.P. Zimalov, V.D. Kosynkin, Yu. P. Raiser, and D.I. Roitenburg, "Steady Externally Sustained Discharge with Electrodeless Pulsed Ionization in a Closed Loop Laser," *Fizika Plazmy 3*, pp. 626–643 (1977).
7. A.E. Hill, "Continuous Uniform Excitation of Medium-Pressure CO_2 Laser Plasmas by Means of Controlled Avalanche Ionization," *App. Phys. Lett. 22*, pp. 670–673 (1973).
8. J.P. Reilly, "Pulser/Sustainer Electric-Discharge Laser," *J. Appl. Phys. 43*, p. 3411 (1972).
9. V.E. Merchant, "Development of a New 20 kW CO_2 Laser," *Laser Focus* (May 1985).
10. V.E. Merchant, M.R. Cervenan, and H.J.J. Seguin, "An Industrial Quality 20 kW Infrared Laser," Lasers '85 Conference, Las Vegas (Dec. 1985).
11. V.E. Merchant, M.R. Cervenan, and H.J.J. Seguin, "New Developments in High-Power Laser Welding," *Welding for Challenging Environments*, Proceedings of an October 1985 conference published by The Welding Institute of Canada.
12. J.H.P.C. Megaw, M. Hill, and S.J. Osborne, "Girth Welding of X-60 Pipeline with a 10 kW Laser," Culham Laboratories Preprint #CLM-P773, Culham Laboratory, Abington, Oxfordshire (1986).
13. J.S. Foley and C.M. Banas, "Laser Welding Stability Limits," *Proc. ICALEO '87, Focus on Laser Materials Processing*, S.L. Ream, ed., IGS Publication.
14. A.M. Belen'kii et al, "Laser Beam Welding with Dagger-shaped Penetration," *Svar. Proizs. 1977*, 11, pp. 23–24; translated in *Welding Production 1977 24*, 11, pp. 9–10.
15. I.A. Shmeleva et al, "The Properties of Welded Joints Produced with a High-power Laser Beam," *Svar. Proiz. 1977*, 11, pp. 13–15; translated in *Welding Production 1979 26*, 11, pp. 1720.
16. E.S. Lur'e, I.A. Shmeleva, and V.S. Smirnov, "Thermophysical Processes in Laser Welding Pipeline Steels," *Svar. Proiz 1986*, 7, pp. 31-35; translated in *Welding Production 1986*, 7.
17. I.A. Smeleva and E.S. Lur'e, "Effect of the Parameters on the Penetration Depth in Laser Welding Pipe Steels," *Avtomaticheskaya Svarka 1987 40*, 8, pp. 61–62; translated in *Welding International 1988*, 7, pp. 594–595.
18. C. Parrini, C. Banas, and A. DeVito, "Laser Welding of Pipeline Steels," *Welding of HSLA (microalloyed) Structural Steels* (1976).
19. K.H. Magee, V.E. Merchant, and C.V. Hyatt, "Laser-Assisted Gas Metal Arc Weld Characteristics," presented at the 1990 International Symposium on Applications of Lasers and Electro-Optics (ICALEO '90) and to be published in the conference proceedings.
20. K. Minamida, H. Takafuji, N. Hamada, H. Haga, and N. Mizuhashi, "Wedge shape welding with multiple reflecting effect of high-power CO_2 laser beam," *The Changing Frontiers of Laser Materials Processing, Proc. of ICALEO '86*, C.M. Banas and G.L. Whitney, eds., Springer Verlag IFS Publications Ltd., UK, in association with the Laser Institute of America, pp. 97–104.
21. K. Behler, E. Beyer, G. Herziger, and O. Welsing, "Using the Beam Polarization to Enhance the Energy Coupling in Laser Beam Welding," *Laser Materials Processing, Proc. of ICALEO '88*, G. Bruck, ed., Springer Verlag IFS Publications Ltd., UK, in association with the Laser Institute of America, pp. 98–105.
22. "Is Laser Pipe Welding Coming Soon?" *Industrial Laser Review 3*, 9, p. 21 (Feb. 1989).

Tailored welded blanks: A new alternative in automobile body design

J-C. Mombo-Caristan,* V. Lobring,* W. Prange, and A. Frings****
*Thyssen Steel Technical Center, Detroit, MI
**Thyssen Stahl AG/Forschung, Duisburg, FRG

Introduction

The automobile body is a complex manufacturing engineering design. It must combine structural integrity with corrosion resistance, weight reduction, dimensionality controls, nesting accuracy, scrap management, and manufacturability at reasonable cost. It must meet safety and health-standard regulations with respect to recycling and crash tests.

The challenge for each generation of vehicle is to optimize all of these factors. Automobile body frames can be broken down into more than 200 different formed parts produced from different steel grades, thicknesses, coatings, and geometries, each of them fulfilling a defined role when assembled. From soft tooling and die tryout through production, dies are designed and presses aligned to form, deep-draw, pierce, trim, and flange each part. Also, controls are made for quality and dimensionality nesting before parts are assembled.

The large number of parts gives us an idea about the multiplicity of processes and assembly lines required to manufacture cars and trucks, causing not only increased capital investment but also increased personnel and variable costs.

To cut these costs, the application of tailored welded blanks (TWBs) becomes a universally attractive alternative.[1–3] A tailored welded blank can be composed of several flat steel sheets of different strengths, thicknesses, and coatings, welded together before being formed into a single panel. Of all available welding techniques, laser welding promises great potential.

Tailored welding techniques

Tailored welding techniques have to be chosen carefully to guarantee a successful forming operation, acceptable corrosion resistance level, and strength and fatigue resistance requirements. Of all welding techniques (including resistance welding types, MIG, TIG, plasma arc, etc.), laser welding yields the following superior results:

- A narrow weld seam that enables corrosion resistance throughout the heat-affected zone. This is made possible by the long-range zinc sacrificial action zone on galvanized material.[4]
- A material's formability reduced only within a narrow heat-affected zone that barely alters the macroscopic formability of the panel as a whole and thus enables successful deep-drawing operations.[5,6]
- A laser weld seam, stronger than the base material, that establishes a smooth joint between the two blanks and enables forming with minimum die wear.

Although laser welding incorporates many technical advantages, mash seam welding is a well-accepted alternative (see Table 1). Volkswagen (VW) and Volvo are reported to have implemented, in production, multigauge side rails and pillars made from mash-seam-welded TWBs.

Spatially distributed mechanical and metallurgical properties must meet predetermined requirements when formed as an automobile body panel. Figure 1 illustrates the variety of potential material combinations that can be tailored for specific needs. A very well-known application is the side-rail application (see Photo 1). Finite element analysis (FEA) comparing current designs and new TWB concepts can yield thinner gauge material on the TWB for equal strength and fatigue resistance characteristics, thus reducing the weight of the part. Typical parts of interest in the automobile industry are grouped into six general TWBs (see Figure 2).

Laser-welding lines at Thyssen

To support its steel sales activities, Thyssen Steel has been developing techniques that bring laser-welding technology from job-shop laboratory stages to industrial production.[1] Thyssen TWBs are produced from manufacturing lines with proven licensed systems supported by five US patents. The first generation of a Thyssen-designed and -built laser-welding line was based on a system with moving optics and fixed blanks. Producing 500,000 TWBs per year for a floor-panel application used by a European car manufacturer, this line has been running constantly in three-shift production in Duisburg, FRG, since 1985 (see Photo 2).

The second-generation line is called "Konti-machine" because the blanks are moving continuously[7,8] during the welding process while the optics remain fixed (see Photo 3). Konti-machines allow more flexibility with regard to a

Table 1. Comparison of laser and mash seam welding.

Laser welding (butt weld)	Mash seam welding (lap weld)
1. There is no physical contact with the material.	1. It is based on a technology known for decades: resistance welding.
2. It performs equally well with all types of coatings, with dissimilar steel grades, thicknesses, and geometries.	2. It does not require an accurate edge preparation.
3. The weld seam is narrower than the zinc's long-range sacrificial cathodic action zone, thus permitting corrosion resistance to be effective throughout the heat-affected zone.	3. It does not require shielding gas.
4. It performs well in the steel industry environment with reduced maintenance scheduling.	4. On the other hand, it produces a 10% thickness increase due to the material overlap, required frequent electrode maintenance; and the weld seam is wider than the zinc's long-range sacrificial action zone.

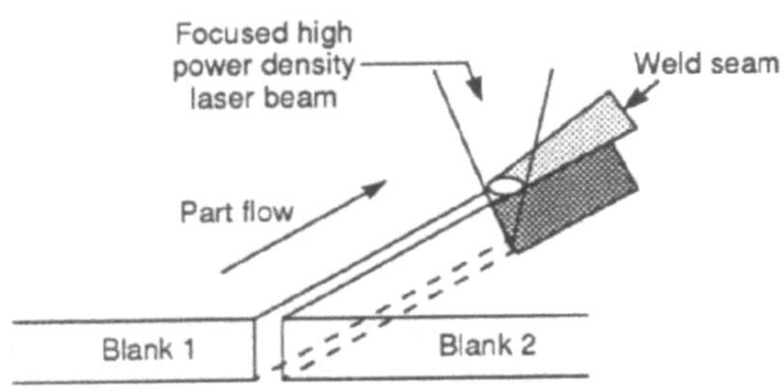

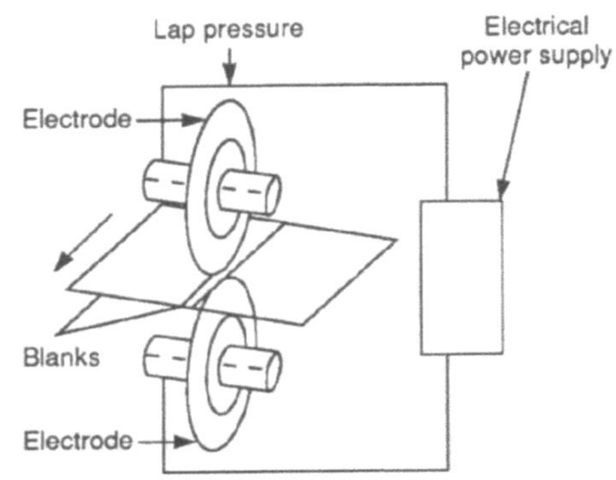

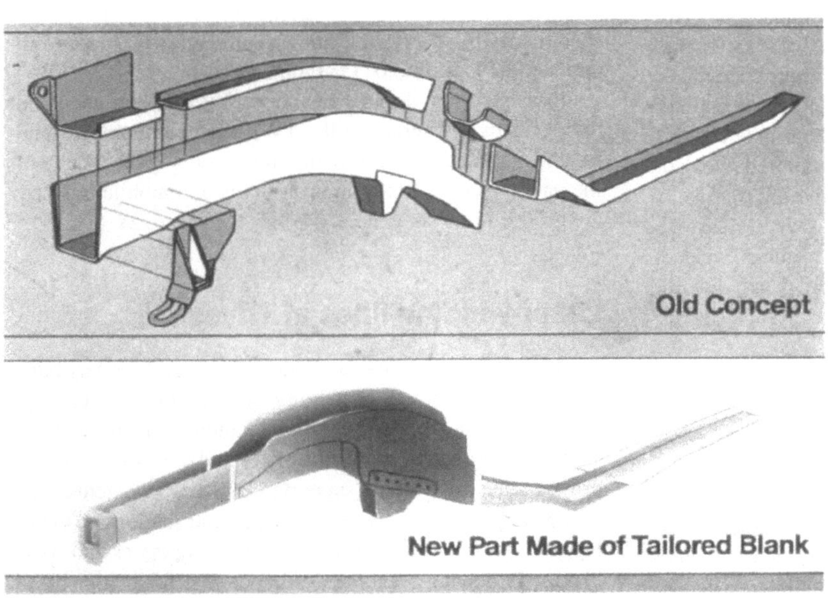

Photo 1. Comparison of traditional and TWB designs of a side rail. In the old concept, six differently formed parts are handled and assembled together. In the new concept, a single TWB made out of three different material strengths and gauges is stamped. Finite element analysis comparing both concepts indicate that the TWB can yield thinner gauge material with equal stiffness and fatigue resistance characteristics.

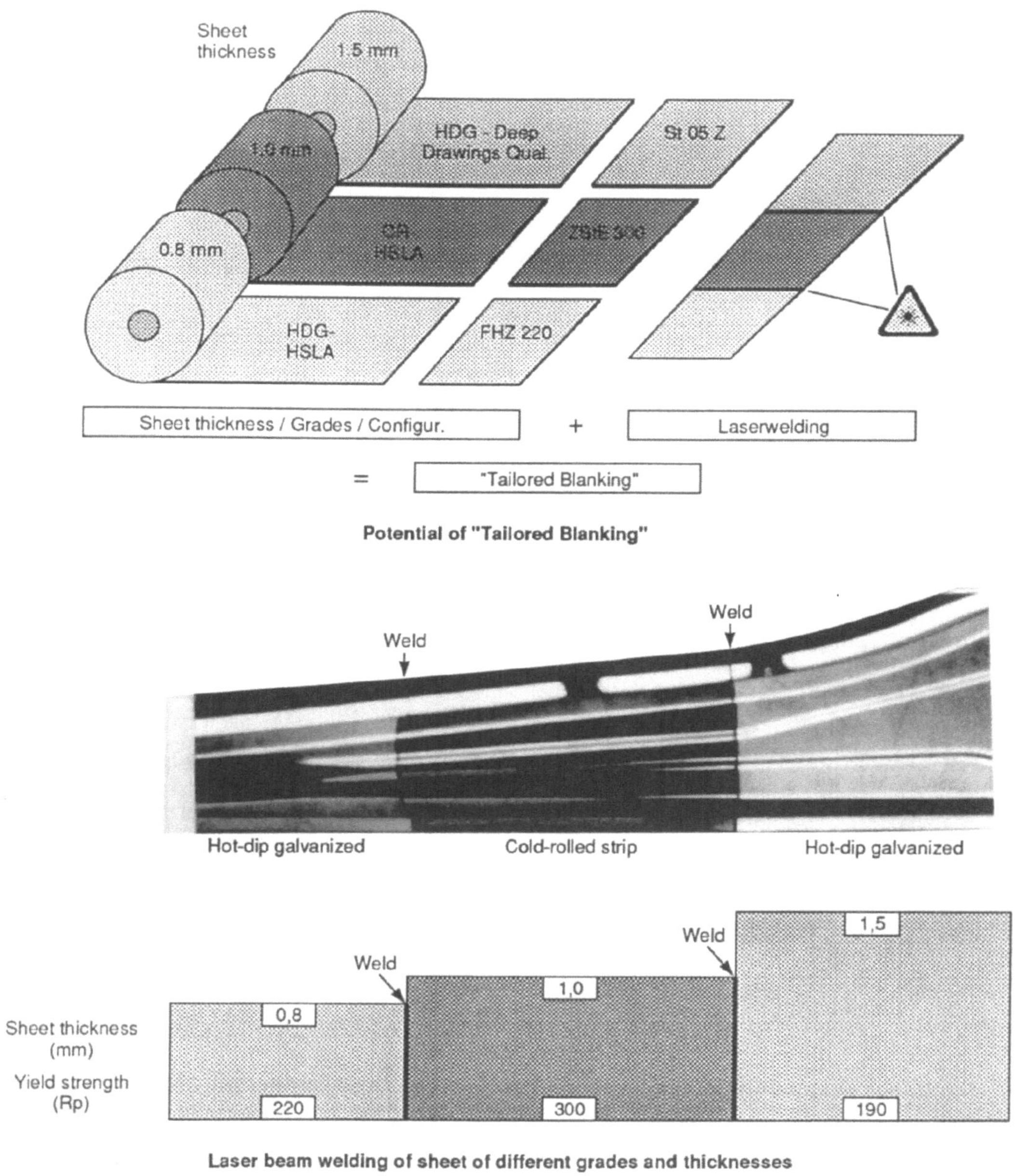

Figure 1. The variety of potential material combinations that can be tailored for specific needs.

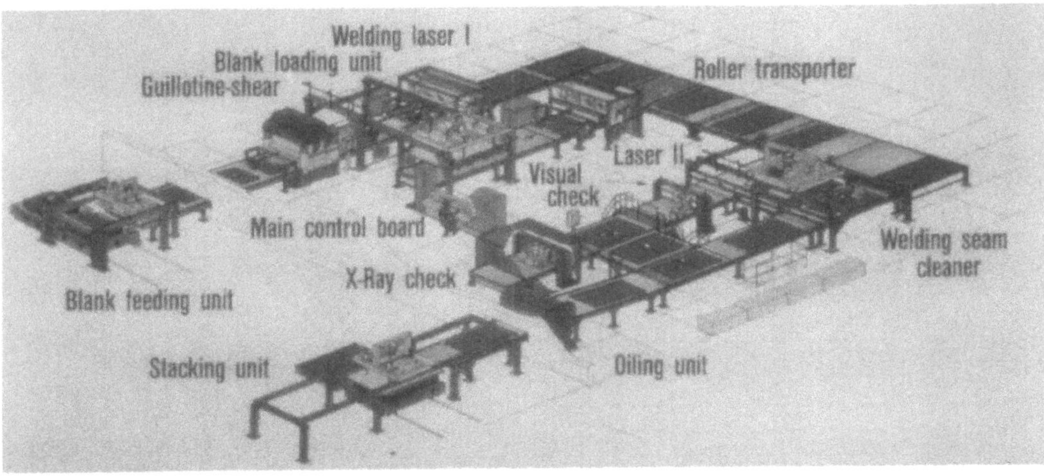

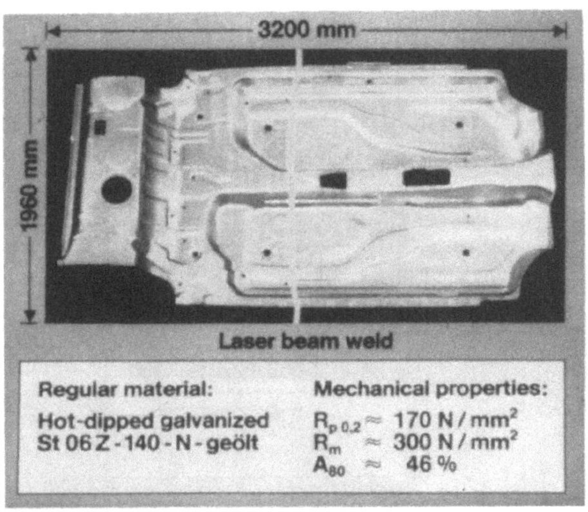

Photo 2. The first commercial scale laser-welding installation at Thyssen in Duisburg, FRG. With this first-generation type of welder (moving optics and clamped blanks), about 500,000 oversized floor panels have been produced per year in three-shift operation since 1985. No finishing is required on the weld seam. The zinc long-range sacrificial action ensure corrosion protection throughout the heat-affected zone. A 10-year rust warranty is in effect for this commercial vehicle.

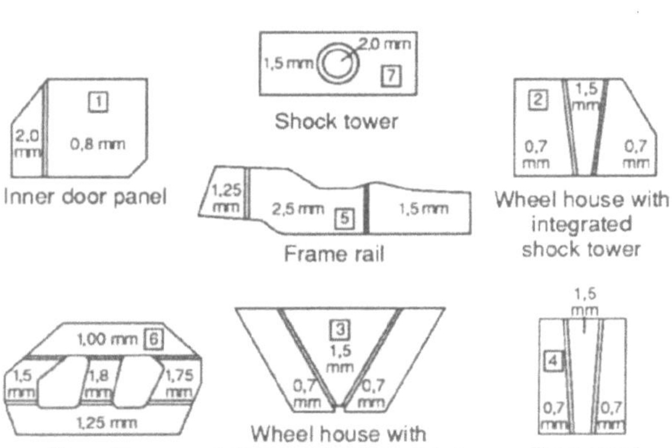

Figure 2. Typical parts made of tailored blanks.

blank's thickness and geometry. Moreover, due to a reduction in material handling, Konti-machines also allow more throughput. The following table summarizes the percentages of cycle time spent on material handling and actual welding:

	Konti-machine	Traditional clamping systems
Welding time	90%	50%
Handling time	10%	50%

This second line, used in production in Duisburg since 1989, has been producing 10,000 TWBs per month for a truck front panel application (see Photo 4). Because many major European automobile manufacturers are already implementing TWB concepts in their car designs, a pilot laser-welding line of this second generation was installed at the Thyssen Steel Technical Center in Detroit, MI, in December 1990 (see Photos 5–8). Several parts have been deep-drawn successfully (see Photos 9–12).

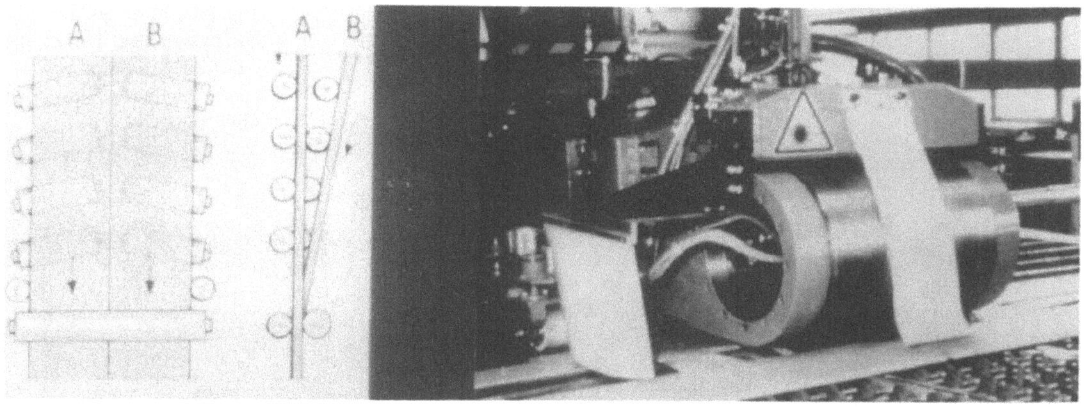

Photo 3. Second generation of laser welder named the Konti-machine because the blanks move continuously in a perfect butt-joint relationship during the welding process while the optics remain fixed. Blank handling is reduced, and 90% of the cycle time during production is spent on actual welding.

Photo 4. Oversized truck front panel made out of two laser-butt-welded steel blanks. The window aperture is reused for another car part. No finishing is required.

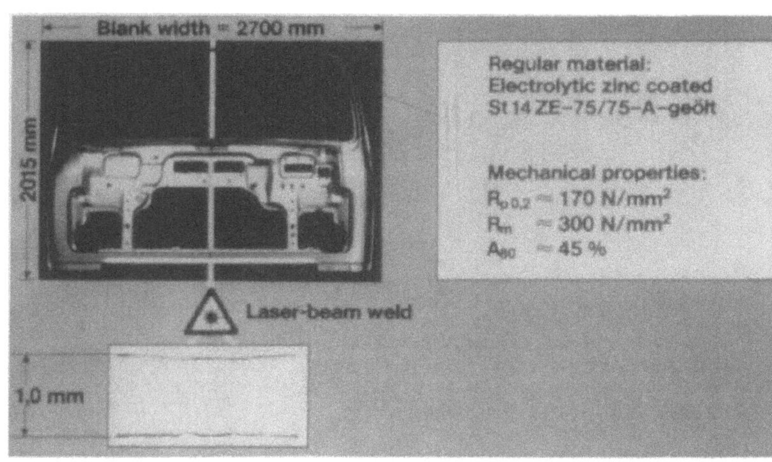

Photo 5. Pilot laser welding line for large-volume production installed at the Thyssen Steel Technical Center in Detroit, MI. Front view of the Konti-machine and downloading roller conveyors.

Photo 6. Side view of the pilot laser-welding line. A laser power of 3 kW with a 200-mm-long focusing mirror yield power density above 10^6 W/cm^2.

Photo 7. Pilot laser-welding line for large-volume production. View of the special shear press looking upstream the Konti-machine. Edge preparation is an essential step for the production of good quality laser-butt-weld seams.

Photo 8. Pilot laser-welding line for large-volume production. One operator supervised the whole system from a central control desk. Another operator trained in reading weld seams examines a suspect weld seam on a specially designed visual inspection table.

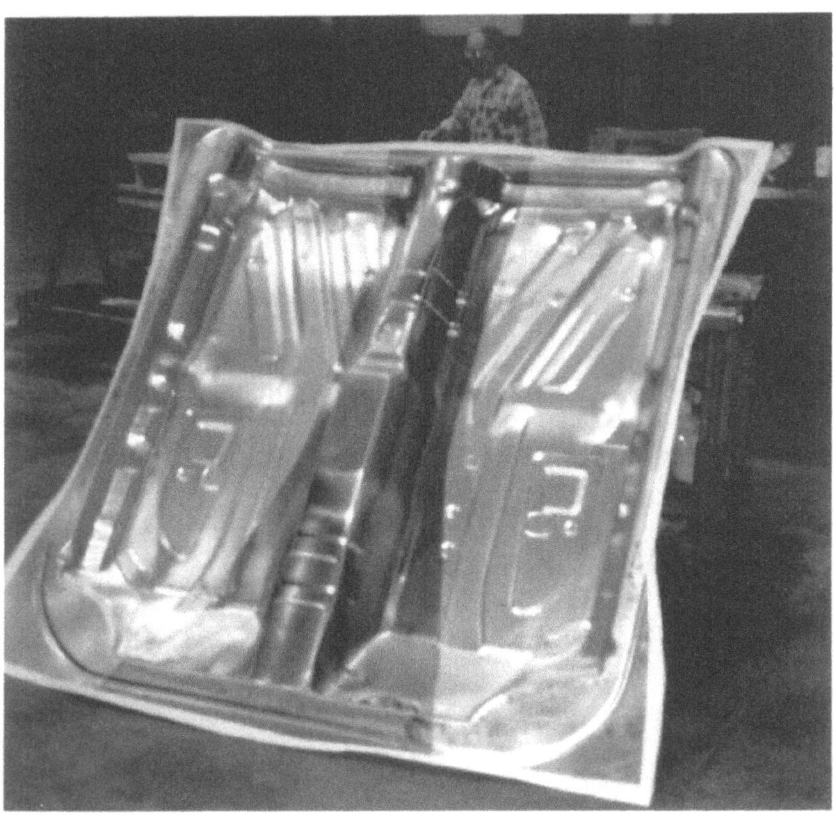

Photo 9. Three-piece floor panel during production and after deep drawing. Two different gauges are used: 0.75 mm for the wings and 1.5 mm for the tunnel area. Several reinforcement plates, such as seat attachment and stick-shift base, are eliminated. Neither sealing nor finishing is required for corrosion protection or sound dampening.

Photo 10. Body-side aperture. Five pieces of three different gauges and strength are laser-butt-welded together before being stamped. Several reinforcement rail assembly are eliminated. Engineering scrap and overall weight are reduced and cost effectiveness is improved.

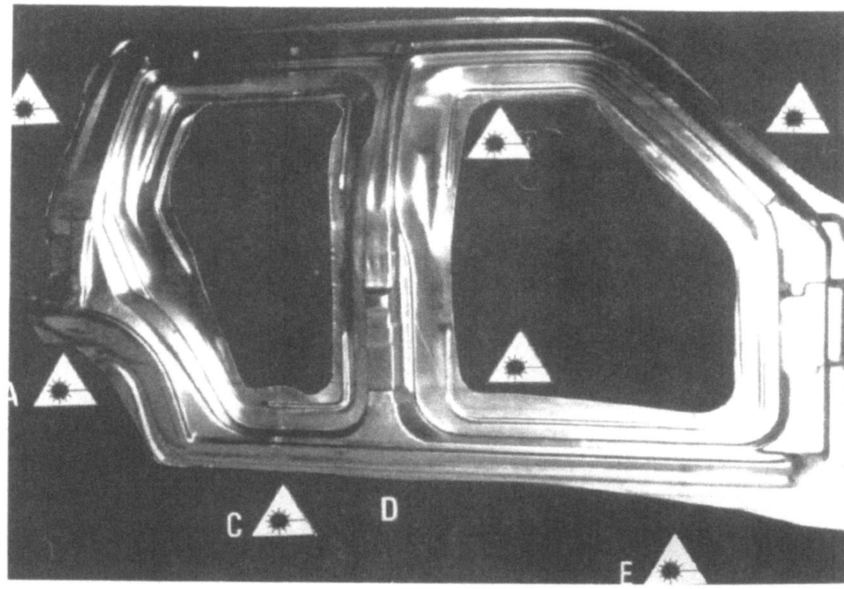

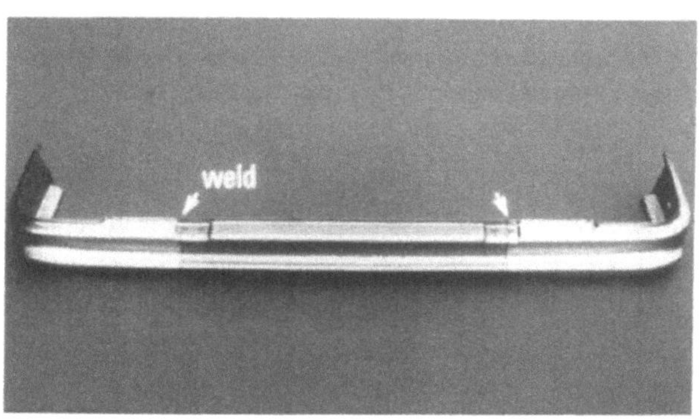

Photo 11. Three-piece bumper made with a stronger middle section.

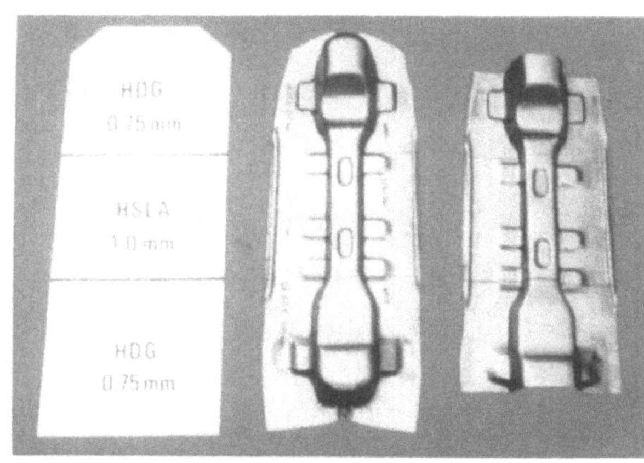

Photo 12. Three-piece floor panel made with a stronger middle transverse section.

Manufacturing quality control

Each of these laser-welding lines has in common the fact that it operates in a genuine industrial environment. These lines are engineered and designed to meet all safety standard regulations (required by the Federal Occupational Safety Health Act—OSHA) and consistent high-quality large-volume requirements. The quality control with these systems include complete statistical process control[9,10] and failure mode effect analysis procedures. Weld-seam geometry monitoring is the focus of special attention because it is a key factor for a successful drawing operation in production.[11] Figure 3 shows a photomicrograph of a laser-weld-seam cross section. Due to a 150% peak hardness increase in the weld zone, a depression at the weld's root side is carefully monitored to minimize die wear during large-volume stamping. During each weld, an in-process monitoring system records the acoustic signal emitted at the weld location and analyzes it in real time.[12] Weld failures are characterized by acoustic distortions and revealed on a screen terminal, along with their exact position on the weld seam (see Fig. 4). A specially designed tiltable table is used to visually inspect each suspect laser weld seam (see Photo 8).

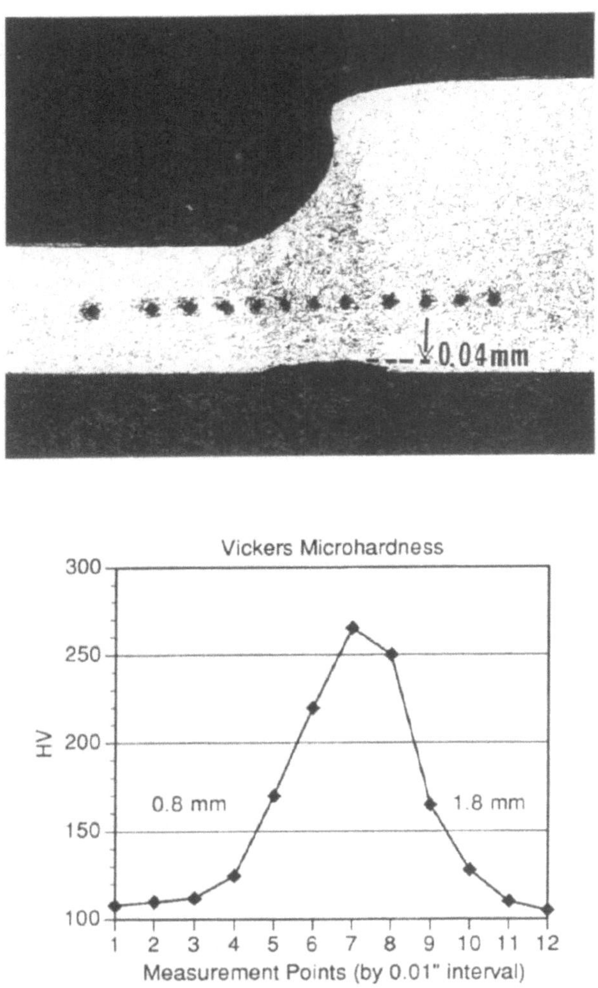

Figure 3. Cross-sectional view of a laser-welded seam (dual gauge 0.8 mm/1.8 mm) and corresponding Vickers microhardness profile. A 0.04-mm weld seam's root depression is monitored to minimize die wear due to hardness increase. Ductility is affected only in the 1-mm-wide seam, but formability of the whole panel is barely affected.

Future of TWBs

In the 1980s, Volvo, Audi, Mercedes, and VW in Europe were the first listed users of TWBs for side rails, truck front panels, floor panels, and reinforced pillars. More developments are on the way with other European automobile manufacturers. In Japan, Toyota started producing laser-welded TWBs[2] for its luxury line of cars and is reported to be implementing this technology in its high-volume production car lines. In the US, General Motors[13] and Ford Motors[14] are getting involved in potential TWB programs for different platforms of cars and truck. Steel companies[15] and satellite stamping plants are ready to propose these new designs at an early stage of involvement. The disadvantage of needing larger dies is offset by a trade-off of technological advantages and important cost savings.

Technological advantages

- The replacement of a discontinuous joint after forming, such as found in spot welding and riveting, by a

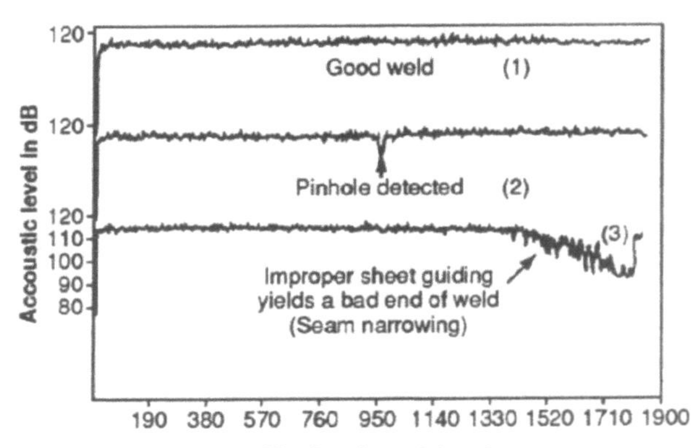

Figure 4. Examples of the acoustic signal recorded during laser welding. Trace (1) shows a good weld seam with a relatively low noise level. Trace (2) shows a weld seam that presents a distortion diagnosed as a pinhole at the visual inspection. Trace (3) shows an imperfect butt joint at the end of the weld.

Table 2. Typical comparative cost study.

Current design	Tailored welded blank design
1. Soft tooling program pricing for each part.	1. Soft-tooling program pricing for a single TWB.
2. Costs of material and transportation.	2. Cost of a TWB.
3. Material blanking: investment + operational + personnel costs.	3. TWB transportation and on-time delivery.
4. Material storage.	4. TWB handling in the stamping plant.
5. Material handling in stamping plant.	5. Die set: investment required for the TWB's size.
6. Die set for each part: investment required for each part size.	6. Stamping presses: investment + operational + personnel costs.
7. Stamping presses: investment + operational + personnel costs.	7. Part handling to the next level assembly line.
8. Material handling to the spotwelding assembly line.	
9. Spotwelding assembly line: investment + operational + personnel costs.	
10. Part handling to the next level assembly line.	

continuous weld seam before forming yields a superior product.

- Weight reduction can be achieved by eliminating reinforcement panels. In addition, a complete FEA shows that TWB designs lead to stronger parts than those produced with the current traditional design.
- The number of dimensionality problems is also reduced when using TWB designs. The homogeneity of a continuous weld joint yields fewer dimensional shifts than would occur if additional spot-welding assembly were required. The reduction of the number of thin-gauge (low-stiffness) panels lowers the scrap rate caused by the handling of low-stiffness panels.
- The manufacturing of panels that are longer than available coil width can be achieved by the laser-welding process. In addition, a continous weld makes obsolete the use of sealant for corrosion protection and sound dampening.
- Extended use of TWBs yields an overall stiffer body-in-white.

Cost savings

- Substantial reduction of engineering scrap material can be achieved by combining intelligent blanking with tailored welding. Optimized tailoring of steel qualities, coatings, and thicknesses can yield substantial material savings.
- The forming of panels usually occurs in four steps: drawing, trimming, piercing, and flanging. Each of these steps necessitates at least one die. Forming a single TWB instead of several panels helps reduce capital investment in stamping plants as well as labor costs.
- TWB technology also contributes to the reduction of assembly-line operations (material storage, handling and transportation, spot welding, and sealing), thus reducing capital investment, personnel costs, and variable costs. For each application, a comparative cost study must be carefully conducted (see Table 2).

Conclusion

TWB technology is a major advantage for the steel industry, as it adds value. It is believed that the laser-welding process is today underutilized as only a welding technique replacement. Further developments will involve extensive computerized finite element analysis for the design of new body-in-white components made out of TWBs (see Fig. 5). The potential for weight reduction could then enhance the success of this technology.

Furthermore, the formability, recyclability, and weldability in large-volume production having been demonstrated, the exposed surface quality seems to be the next challenge for laser-welding technology. Trial studies show that the appearance problem can be overcome by a careful monitoring of the weld-seam geometry.[11] Figure 6 shows a cross section of a laser weld seam for which the root presents a 0.1-mm convexity. This convexity is monitored to give some material to be ground for surface-finishing purposes. Photo 13 shows the encouraging results of these trials. For both cases, the sequence of operation has been:

- Case a: Deep-drawing the TWB, grinding the weld seam, painting.
- Case b: Grinding the weld seam, deep-drawing the TWB, painting.

In both cases, surface quality has passed standard automobile industry requirements. Finishing the part before deep-drawing presents the advantage of an easier automation. It is

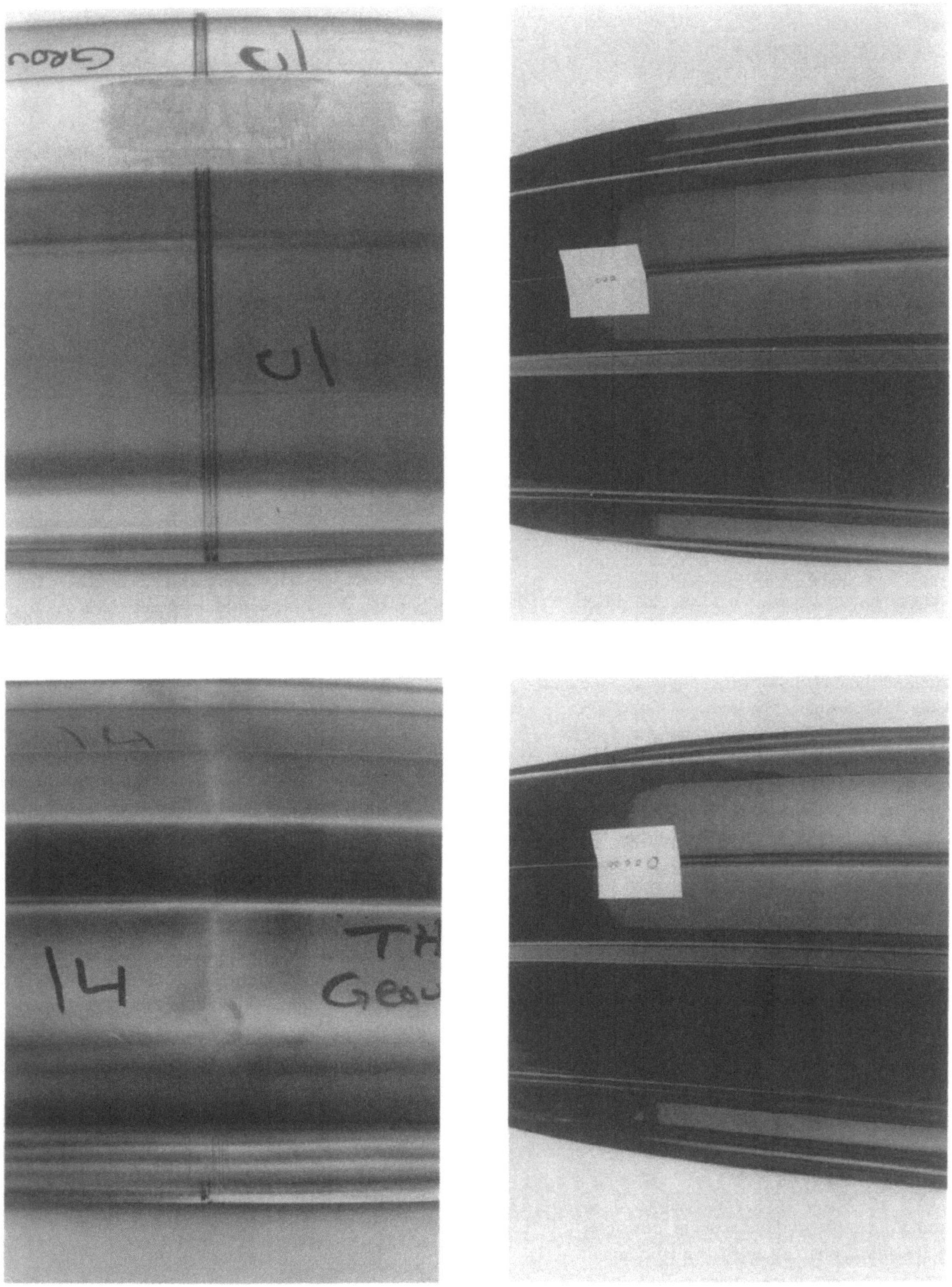

(a)

(b)

Photo 13. Portions of deep-drawn laser-welded panels. In case a, the exposed surface of the panel was ground after being deep drawn. In case b, the panel was ground before being deep drawn. Standard automobile industry painting procedure was then applied.

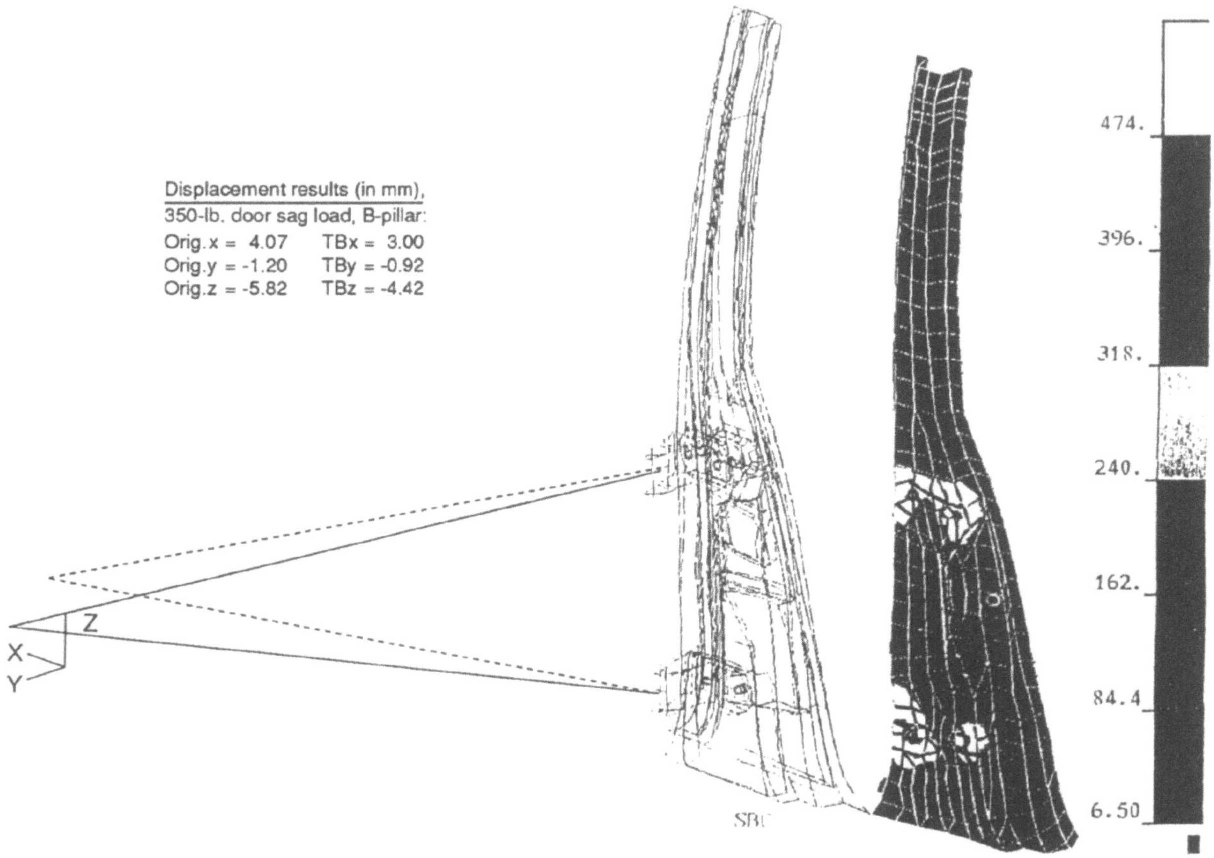

Figure 5. Example of finite element analysis performed on a B-pillar (courtesy of D. Djavairian, The Budd Co., Stamping and Frame Div., Rochester, MI). This test is a 350-lb door sag load applied at the hinge position. The continuous laser weld presents better tensile and bending strength and thus better displacement results than the former concept using spot welds. As a result, using the TWB concept, the blank's thickness can be reduced for an overall lighter body-in-white at equivalent stiffness.

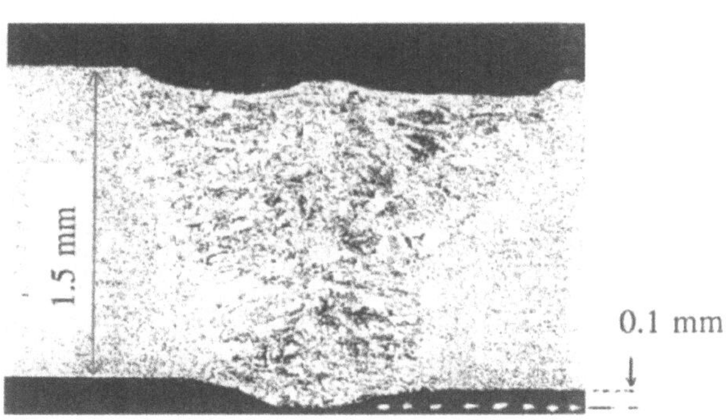

Figure 6. Cross-section profile of a laser weld seam presenting a raised root of 0.1 mm height. A careful assist gas flow and welding speed control are necessary to monitor this raised area; no filler material is used.

important to note that no filler material is used with this technology. Therefore, it is encouraging that the Konti-machine system allows an absolute monitoring of the weld-seam geometry.

References

1. A.J. Selige and W. Prange, "Production and Usage of Laser Beam-Welded Sheet Metal," SAE 870413 (1987).
2. K. Azuma, H. Sugura, F. Matsumi, H. Kato, N. Tanaka, and T. Takasago, "Laser Welding Technology of Joining Different Sheet Metal for One Piece Stamping," ISATA Wien (1990).
3. G. Neiheisel, "High Speed Laser Welding of Deep Drawing Low Carbon Steel," *The Industrial Laser Annual Handbook 1990*, ed. by D. Belforte and M. Levitt (1990).
4. W. Prange, C. Schneider, and A.J. Selige, "Application of Laser Beam Welded Sheet Metal," SAE 890853 (1989).
5. W. Prange, H. Schmitz, C. Schneider, "Tailored Welded Blanks: A Material for New Ways of Design," *ISATA Wien 1*, pp. 342–349 (1990).

6. K. Blumel and C. Schneider, *Improved Formability by Means of Tailored Blanking*, IDDRG Goteborg Vortrag (1990).
7. J. Sturm, "Process for Guiding Sheets to be Butt Welded Together and Apparatus with Means for Conveying and Guiding Sheets or Strips to be Butt Welded Together," US Patent #4,733,815 (1988).
8. J. Sturm and W. Prange, "Apparatus for the Continuous Welding of Strips and/or Sheets," US Patent #4,872,940 (1989).
9. Statistical Methods Office, Ford Motor Co., *Continuing Process Control and Process Capability Improvement* (1990).
10. W. Woodhall and B.M. Adams, "Statistical Process Control," *Handbook of Statistical Methods for Engineers and Scientists*, Ed. H.M. Wadsworth, McGraw-Hill (1990).
11. J-C. Mombo-Caristan and M. Koch, "Seam Geometry Monitoring for Tailored Welded Blank Applications," to be published in 1992.
12. M. Koch, "Qualitatsicherung beim Karosserieblechschweissen," Symposium des Institut fur Laser Techniques, Aach, Feb. 27–28, 1991.
13. A. Wrigley, "GM order blank-fabricating line, *Americal Metal Market*, June 24, 1991.
14. "The Laser Beam: A Future in Autos?" *Proc. WELDEX 91, Welding Design & Fabrication,* Aug. 1991, pp. 40–42.
15. B. Allen, "Laser welded sheet being tested, *American Metal Market-Automotive Steel*, Dec. 17, 1990.

Pulsed welding with radio frequency-excited CO_2 lasers

Ole A. Sandven

Consultant, Trumpf Industrial Lasers Inc.
Sturbridge, MA

Until about 1985, CO_2 lasers used DC excitation of the lasing medium. In principle, this technique is simple and straightforward, but it presents a number of practical and fundamental problems.

First of all, this method requires a very high and therefore potentially dangerous voltage, usually 10–20 kV and sometimes even higher. Secondly, the transfer of energy to the laser medium, which in the CO_2 laser consists principally of a mixture of nitrogen, helium, and CO_2, requires electrodes inside the lasing cavity. These may become a source of contamination by electrode erosion. The discharge power density must be kept fairly low to prevent arcing. Because the discharge is not uniform throughout the cavity, localized hot spots can occur, which in turn can result in electrode erosion and molecular breakdown of the lasing gas. Because the power density per unit volume of the cavity is limited and the power density of the discharge in the cavity uneven, the quality of the output beam may be affected. Finally, when DC excitation is used, the electrodes must be cleaned periodically, which requires the cavity to be opened up, resulting in extra downtime and the possibility of cavity contamination.

The use of AC excitation, particularly in the radio frequency range[1], can circumvent or eliminate many of the problems encountered with DC excitation. Using the typical excitation frequency of 13.56 MHz makes it possible to use electrodes that are external to the laser cavity and separated from the lasing medium by the cavity walls. The cavity walls are made from a dielectric material such as glass, and the high excitation frequency allows the energy transfer to the lasing cavity by capacitive coupling between the external electrodes and the lasing gas. Such an arrangement makes it possible to use much lower operating voltage than that required for DC excitation. Furthermore, contamination of the cavity by the electrodes is eliminated, and periodic electrode cleaning is not necessary. Figure 1 shows the general difference between RF and DC excitation as applied to axial flow CO_2 lasers.

A special advantage of RF excitation is the high-power density and uniformity of discharge that can be obtained in the lasing cavity. This makes it feasible to produce very high-power outputs from relatively small laser cavities and thus reduce space requirements for laser installations. The output beam from a RF-excited laser will also, as a consequence of the uniform discharge, have a low-order mode (TEM_{00} or TEM_{01*}) and low divergence, even at multi-kW output levels. Stability and repeatability of power output and beam mode, low gas consumption, and reliability are other desirable consequences of RF excitation. Equipment cost, however, may be somewhat higher than for DC-excited lasers. Proper shielding of the RF generator and the laser cavity must be provided to prevent EMI radiation.

Because the RF generators is tuned to the load represented by the laser cavity, the variation of output power of the laser in the CW mode is limited to 75%–100% of maximum power. If a lower output is required, this can easily be obtained by high-frequency pulsing at a power on/off ratio that would give the required average power.

A Trumpf TLF 5000 with 5 kW output power was used for the experiments discussed in the following sections.

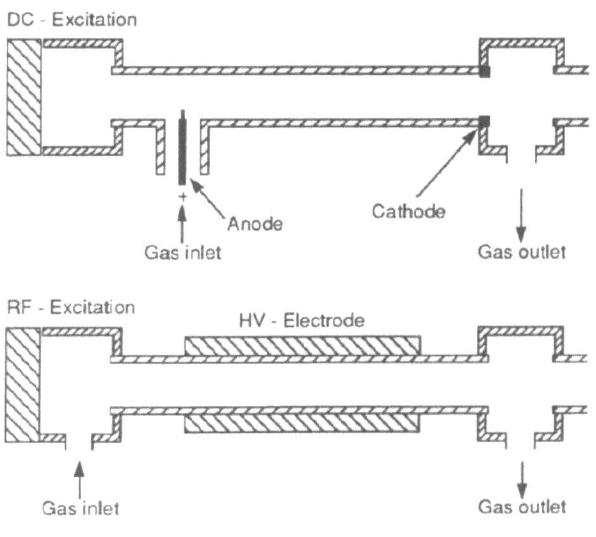

Figure 1. DC and RF electrode configurations in CO_2 gas lasers.

Pulsing of the RF-excited laser

The RF-excited laser can be used in either the CW mode or the pulsed mode (see Fig. 2). Because pulsing is accomplished by switching the RF power on and off electronically, the pulses are essentially gated pulses without pulse amplitude enhancement. And because the response of the lasing medium is extremely rapid at RF-excitation, gated pulsing can be performed over a wide range of frequencies, typically 100 Hz to 100 kHz. (The pulse frequencies should not be confused with the excitation frequency, which is of an entirely different order of magnitude.)

The pulsing is performed with power amplitude of 100% output. The actual average power delivered to the workpiece will thus depend on the duty cycle of the pulsing but will of course always be less than the maximum 100% capacity of the laser in the CW mode. As an example, a 50% duty cycle in the pulsed mode will give 2500 W power from a 5-kW laser. Thus, duty cycle determines the average power level, while pulse frequency can be set independently from 100 Hz to 100 kHz.

At very high pulse frequency, the power will not decay back to the zero line between each pulse, even at very low duty cycles—that is, low ratio between time of power on to time of power off. In the TLF 5000, the maximum frequency at which the power will decay to zero between each pulse peak is about 40 kHz. At this pulse rate, the pulse length is only about 25 µm. The way that solid materials will react to the laser pulses will then be generally indistinguishable from that of exposure to CW laser radiation, although more subtle effects of the pulsing may still be present in the materials response.

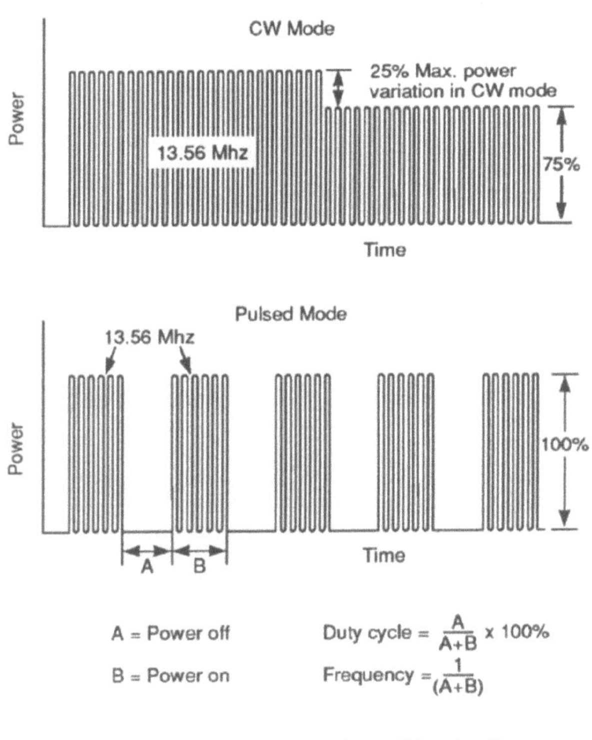

Figure 2. CW and pulsed modes in an RF-excited laser.

At the lower end of the pulse frequency range, a material's response to the pulsed laser radiation becomes affected by the pulsing. At 100 Hz, the pulse lengths is 10 ms, and many of a material's response phenomena will then be rapid enough to keep up with the periodic power change.

In the following discussion, we shall look at some of the effects of pulsing as they pertain to welding at the lower end of the pulse-frequency spectrum. This is only a preliminary discussion of the response to pulsing, as there are undoubtedly other effects of gated pulsing pertaining to welding (as well as other laser processes) that are still unrecognized.

Low-frequency pulsed laser welding

The effect of pulsing in the frequency range of 0.5 kHz–10 kHz on the laser welding of 304 austenitic stainless steel has been investigated.[2,3] The welding was performed using a TLF 5000, equipped with an off-axis mirror focusing head with 150-mm focal length. Bead-on-plate welds were prepared on 0.25-in.-thick plates, using helium shielding gas. The power was kept constant at 2250 W, welding speeds of 50, 100, 150, and 200 in./min were used, and the pulse frequencies at each speed level were in the range of 0.5–10.0 kHz.

Before each run, the duty cycle was adjusted to give the constant average power output of 2250 W. This made the duty cycle change somewhat over the range of frequencies used because the tuning of the RF power generator to the load changes a little with changes in frequency. The variation in the duty cycle was less than 5%, from about 45% to 49%. As already discussed, the duty cycles could not be varied at will without also changing the average power, as the duty cycle is essentially used to set the power level in the pulsed mode of these lasers.

The resulting welds were examined metallographically, and the penetration and bead geometry (expressed as the aspect ratio of the weld bead) determined for each weld. The weld aspect ratio used in welding research is usually defined as penetration divided by weld bead width at midpoint of penetration. But in laser welding, where there often is a tendency to develop weld beads with pronounced "nail heads," this definition may sometimes result in misleading data. Therefore, in this work the aspect ratio was determined as penetration divided by the square root of the weld bead cross-sectional area. (The square root value, rather than the cross-sectional area, was used to obtain a dimensionless number for the aspect ratio.)

The variation in frequency had a marked effect on the appearance of the welds. At low frequency, the top of the weld beads tended to be uneven and, at times, severely undercut. There was also much more weld spatter present at low frequencies than at higher ones. At frequencies below 1.0 kHz, the welds were straight, without "nail heads" and frequently

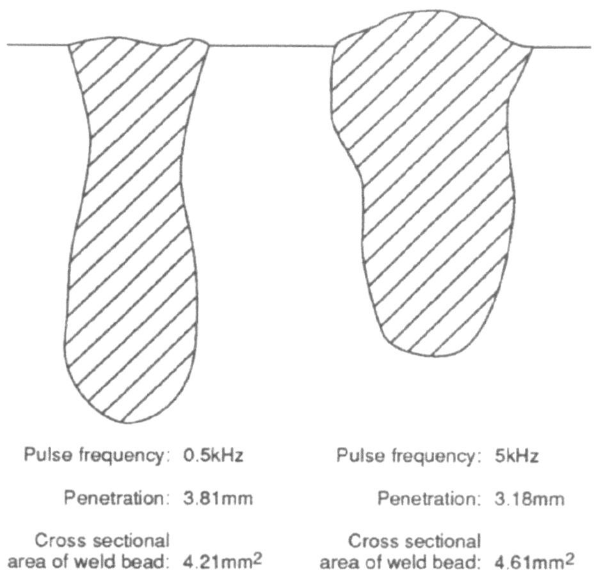

Figure 3. Cross-section of laser weld beads in 304 stainless steel (input of 2250 W pulsed power, mirror optics with 150-mm focal length, and magnification 26X).

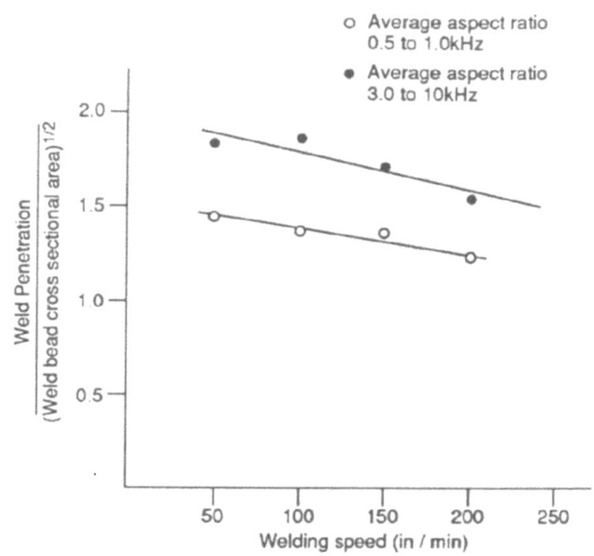

Figure 4. Variation of weld bead aspect ratio with welding speed and pulse frequency.

without any weld reinforcements. At higher frequency, the weld bead cross sections took on a normal shape (see Fig. 3).

The variation in the bead aspect ratio with the pulse frequency is shown in Fig. 4. From a high level at low frequencies, the aspect ratio drops down to a noticeably lower level at higher frequencies. In an intermediate range, from about 1 kHz to about 2.5 kHz, the results were very erratic and nonrepeatable, indicating that some form of transition was taking place in interaction between the laser beam and the workpiece material.

Looking at the variation of the weld penetration with the frequency (see Fig. 5), we can see a behavior similar to that of the aspect ratio variation, namely a high level of penetration at the low end of the frequency range, separated from a lower level at higher frequencies by a transition zone of erratic and nonrepeatable results. The depth of penetration on the high-frequency side was of a magnitude that would be reasonable for a CW laser weld made under similar conditions of power, focal spot radius, and processing speed. The penetration at the low end of the frequency range was, however, some 20%–25% higher than expected.

The reason for the narrowness of the weld bead and the increase in penetration at lower pulse frequencies has not been established, but may be linked to plasma fluctuations. In laser welding, the keyhole will form in the workpiece surface if the power density is sufficiently high (10^6 W/cm^2 or more). A plasma will subsequently be formed above the keyhole by interaction between the laser beam and metal vapor from the keyhole. This will severely reduce the amount of energy delivered to the keyhole, and the resulting reduction of metal vapor pressure above the keyhole will tend to extin-

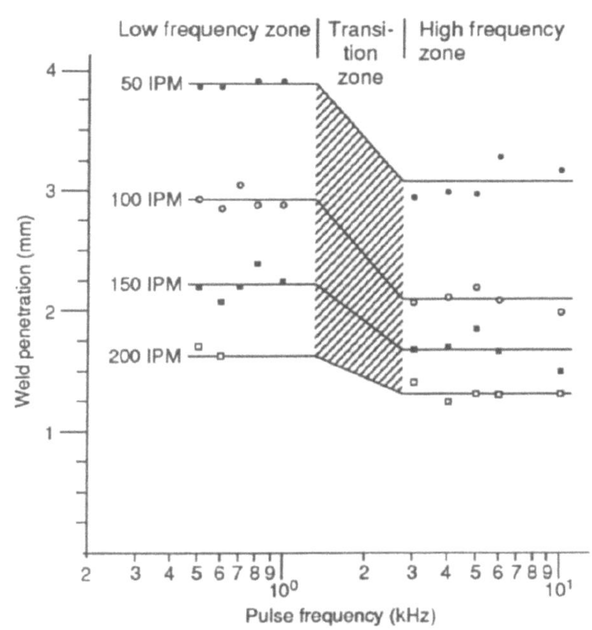

Figure 5. Change in weld penetration with pulse frequency and welding speed in 304 stainless steel (2250 W).

guish the plasma. The process then repeats itself, and the period of the plasma fluctuation is typically of the order of milliseconds. This is the same order magnitude as the pulse lengths used at the low-frequency range in this work, thus presenting the possibility that the pulse frequencies are of the correct magnitude to interfere with the plasma formation.

If this is the case, more energy should be directed to the keyhole at low frequencies, resulting in deep, narrow welds, precisely as observed. The widening of the weld bead and reduction of penetration at higher frequencies could then be assumed to be the result of plasma formation, resulting in the transfer of relatively more energy to the workpiece surface but less to the keyhole. Using low-frequency welding to obtain deeper penetration will, however, require a reduction in the weld spatter and the tendency for formation of undercut weld beads observed in this frequency range.

Pulsed laser welding of galvanized steel

The laser welding of galvanized steel sheets is a potentially important industrial process and has been studied by several authors.[4,5] In general, it is found that in lap welding zinc-coated sheets, the low boiling point of zinc will result in relatively large amounts of zinc vapor during the welding process, and such vapor will tend to form porosity and splatter if the escape of the zinc vapor occurs through the weld zone itself. Furthermore, the interaction between the laser beam and zinc vapor will tend to result in the formation of a plasma above the workpiece surface, which in turn will prevent the laser energy from being deposited in the weld keyhole. This will cause a momentary collapse of the keyhole, followed by the disappearance of the zinc vapor and plasma and the re-establishment of the keyhole. Welding under these conditions gives a generally poor weld with porosity, weld bead sink-in, and weld spatter.

A well-known technique for overcoming such problems in laser welding zinc-coated steel is to provide a gap between the sheets in a lap joint. Such a gap, typically about 0.1 mm wide, provides an alternate escape route for the zinc vapor and reduces the problems associated with it. However, the need for such a gap greatly complicates the process from a practical, industrial standpoint.

Hayden et al[4] found that if the laser welding of galvanized material is performed in a pulsed mode, it is possible to eliminate the need for a gap between the sheets, provided the pulse frequency is correctly chosen. Using a 5-kW TLF 5000 FAF laser at 2.5 kW output, they found that the pulse frequency for optimal welding results depended on the thickness of the zinc coating and the processing speed in such a way that the amount of zinc evaporated by each pulse was constant and approximately 0.4×10^{-3} mm^3. In other words, the number of pulses per unit length of weld for optimal results was proportional to the coating thickness (see Fig. 6). Welds made under these conditions were sound, had properly shaped weld bead, and were mostly free of porosity.

It was assumed that these results were due to an optimal pulse length; individual pulses were too short to evaporate enough zinc to create a plasma that absorbed the beam and resulted in the collapse of the keyhole, but long enough to form the keyhole. At higher coating thickness, more pulses per unit length of weld were needed to evaporate all zinc from the weld zone. Thus the pulse frequency had to be increased to keep the amount of evaporated zinc per pulse constant at constant weld speed. The precise magnitude of the optimal amount of zinc evaporated by each pulse will depend somewhat on the laser system used; that is, average power, focal spot diameter, and so forth.

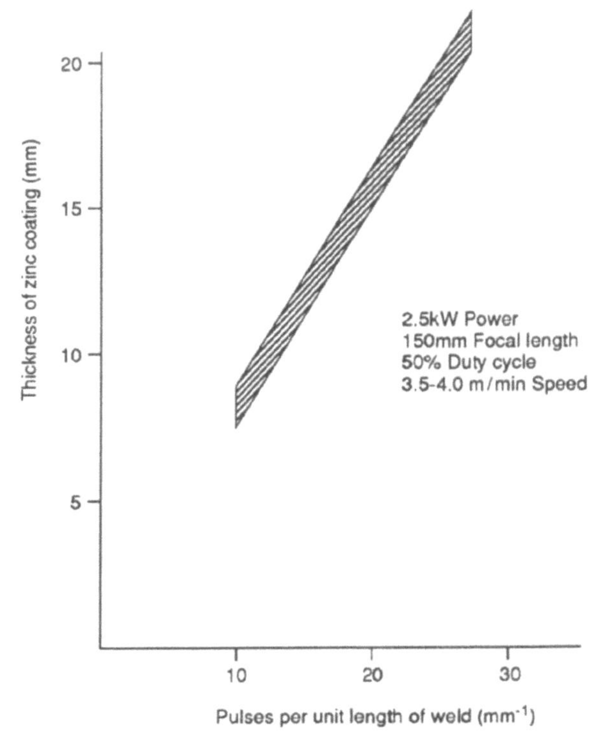

Figure 6. Pulsed laser welding of galvanized steel sheets; envelope of best results (from data by J. Hayden et al.[5]).

Conclusion

The introduction of CO_2 lasers with broad-range pulsing capacity has given the applications engineer an extra processing parameter to work with. The practical utilization of this extra parameter in laser applications is only in its infancy at the present. But as we learn more about the effect of pulsing on various industrial laser processes, new and useful applications of this capability will undoubtedly be found, for both the relatively low-frequency range discussed here and the much higher ranges available in RF-excited lasers.

References

1. R. Wollerman-Wingasse, F. Ackerman, J. Weick, and D. Dechamps, "Multikilowatt RF-excited CO_2 lasers and their application in laser materials processing," *Proc. ICALEO 1986.*
2. O.A. Sandven, "Welding with RF-excited lasers," *ASM Materials Week* (1990).
3. O.A. Sandven, unpublished work.
4. R. Akhter, W.M. Steen, and K.G. Watkins, "Welding Zinc-Coated Steel with a Laser and the Properties of the Weldment," *Journal of Laser Applications 3*, 2.
5. J. Hayden, K. Nilsson, and C. Magnusson, "Laser Welding of Zinc-Coated Steel," *Proc. Int. Conf. Lasers in Manufacturing* (May 1989).

Laser material processing: Effects of polarization and cutting velocity

H.J. van Halewijn
Drukker International B.V.
Cuijk, The Netherlands

Introduction

Laser material processing with linear polarized light gives rise to curved cutting grooves[1] if the cutting direction makes an angle between 0° and 90° with the plane of polarization. In practice, this gives an unacceptable cut quality. To avoid this, the laser beam must be unpolarized, but Nd:YAG laser rods tend to polarize the beam because of stress birefringent effects.[2] If the power of the laser is changed, the direction of the polarization is also changed, in a rather unpredictable way. These effects are eliminated if a Brewster window is introduced in the resonator and a quarter wave plate is mounted outside the laser to get circular polarized light. Unfortunately, if a Brewster window is used, power is reduced 10%–50%, depending on the mode structure and operation cinditions of the laser. Because deviation effects in the groove can disappear for high cutting velocities, it is possible to cut with linear polarized light in every direction.

Diamond and sapphire experiments

The experiments described here were done on two dielectric materials, diamond and sapphire. Both are transparent in the visual region, which makes it possible to see exactly what happens in the laser cut. Table 1 shows some material constants.

The interaction of laser light with diamond results in a rapid graphitization process, so every laser pulse interacts with a thin layer of about 10 μm. This is why graphite is included in Table 1. In Fig. 1, the reflection as a function of incidence with the surface normal of the cutting front is shown for sapphire and graphite. If the laser fluence exceeds 0.140 J/cm^2, the reflectance of graphite alters to lower values for only a few nanoseconds.[3] This fluence level can easily be reached because of a change in the electronic structure, but it will be overlooked here. The $R\pi$ polarization state has a minimum for both materials (the Brewster angle) and is parallel to the plane of incidence. The $R\sigma$ polarization state has no minimum and is perpendicular to the plane of incidence.

Table 1. Some material constants

	Graphite[7]	Diamond	Sapphire
Density [g/cm^3]	2.25	3.51	3.97
Refr. index (real part)	2.15	2.41	1.77
Refr. index (imaginary part)	1.15	—	—
Melt temp. [K]	4700	3500	2053
Diff. coeff. [mm^2/s]*	—	68	2
Heat cond. [W/mK]*	—	150	8
Heat function [W/mm]*	—	1200	10.5

*See references 5 and 6 for definition of these parameters. Heat conduction is the average from room to melting temperature.

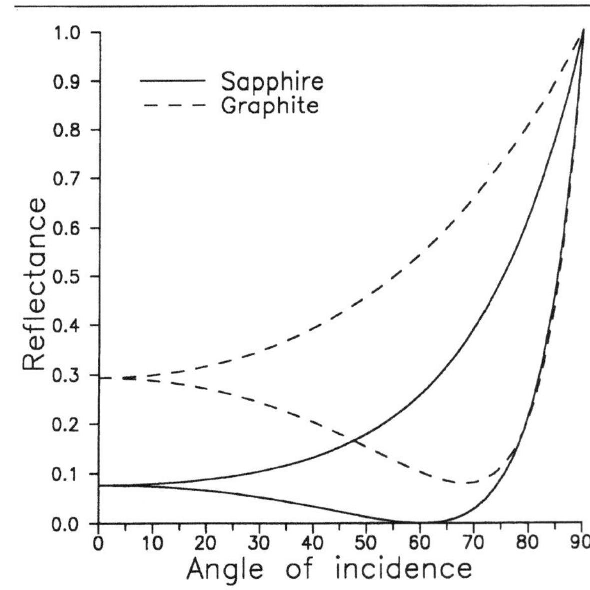

Figure 1. Reflection as a function of the angle of incidence to the surface for sapphire and graphite. $R\pi$ is zero at 60.5° for sapphire and 68° for diamond/graphite. For small and large angles, the difference between $R\pi$ and $R\sigma$ becomes zero, and the aspect ratio becomes 1. The aspect ratio is at maximum around the Brewster angle.

Experimental setup

The laser used for these cutting experiments was a Q-switched Nd:YAG laser with a hemispherical resonator. The lamp current during these tests was constant at 15.5 A, so the beam quality was constant. The laser was operated in the TEM_{00} mode, and the beam was directed through a beam expander and a focusing lens (f = 46.4 mm) to a CNC-controlled xy table with a resolution of 1 µm. The beam was linear-polarized with a Brewster window mounted inside the resonator. The spot size, measured with the knife-edge method, was 7.1 µm (±3%, $1/e^2$) and could be located on the surface of the workpiece with an accuracy of 15 µm. Care was taken in aligning the resonator and choosing the mode selector diameter. After passing through the lens, the power was measured with a Coherent 205 power meter. Throughout the experiments, two powers were used: 1.7 W, T_p = 150 ns and f = 1 kHz; and 4.5 W, T_p = 170 ns and f = 3 kHz.

The experimental procedure can be divided into two parts:

- Cutting grooves were made in both diamond and sapphire with the polarization parallel and perpendicular to the velocity vector. The width, depth, and angle of the cutting front were measured as a function of the velocity.
- When cutting circles in the workpiece, maximum and minimum deviations were measured as a function of the velocity.

Straight cuts

Figure 2 shows the depth as a function of the velocity for two polarization states. The penetration depth for sapphire is higher than for diamond because diamond has a higher heat loss than sapphire in terms of the normalized parameters (see Fig. 3). The depth has a higher value for the polarization parallel to the velocity vector for both materials, although relatively less for diamond.

These values are used to plot normalized cutting velocity and beam power in Fig. 3.[5,6] The melting ratio (MR) is defined as the ratio of the energy delivered by the laser to the energy needed to melt a volume of material equal to the volume of the cut. Theoretically, it can be shown that it is 48% at maximum. The energy transfer efficiency (ETE) is defined as the ratio of the absorbed energy to the laser energy.

It is interesting to note that for higher frequencies or powers, the points for 3 kHz are shifted to lower values of the normalized power and normalized velocity. Because of the shorter cooling time of the material between the pulses and decreasing peak power for 3 kHz, energy transfer becomes more efficient and the process less violent.

Fig. 4 shows the measurements for the angle of incidence at the bottom of th cutting front. For diamond, the angle of incidence is about 60°– 80° for all polarization states and over the whole velocity range (see Fig. 4b). In the case of sapphire, the angle is about 80° for 16 µm/s and decreases smoothly to about 30° for 200 µm/s for both polarization states. A significant difference in the angle of incidence is seen between P = 1.8 W (1 kHz) and P = 4.7 W (3 kHz).

Circular cuts

When circles are cut, deviations arise at the bottom of the cut. Photo 1 shows that for low velocities, the cut has an ellipse-like shape. On the long axis it is curved to the outside, while

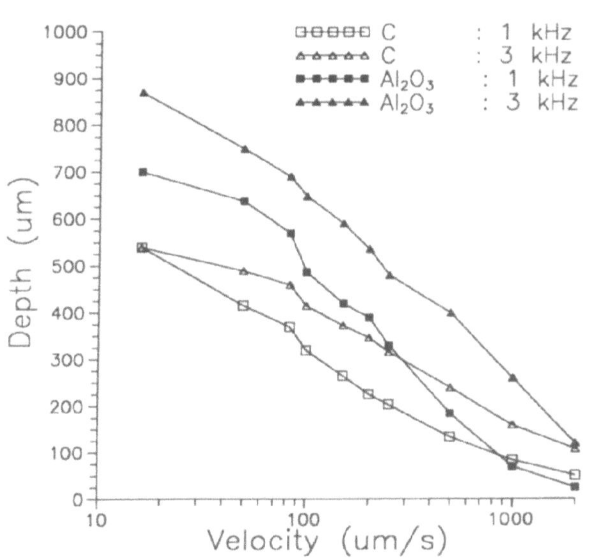

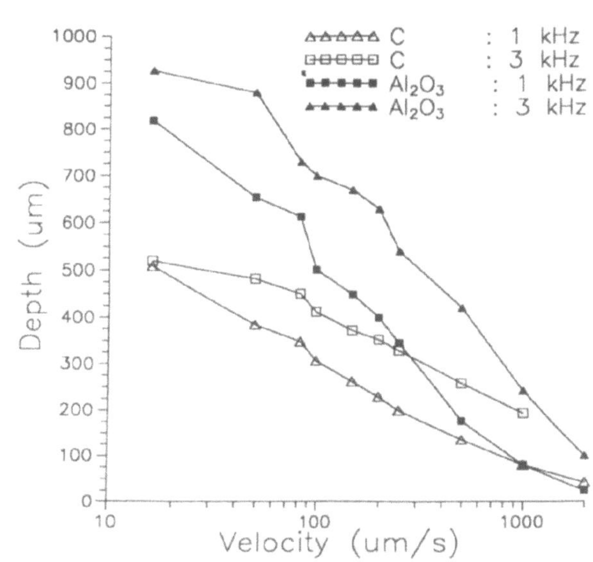

Figure 2. Cutting depth as a function of the velocity for straight cuts, different directions of polarization, and powers.

Figure 3. Normalized cutting velocity as a function of normalized power. If Vnorm = 0.01, the heat loss is significant and reduces for Vnorm > 10. The points in the figure are the average taken between the two polarization states of Fig. 2.

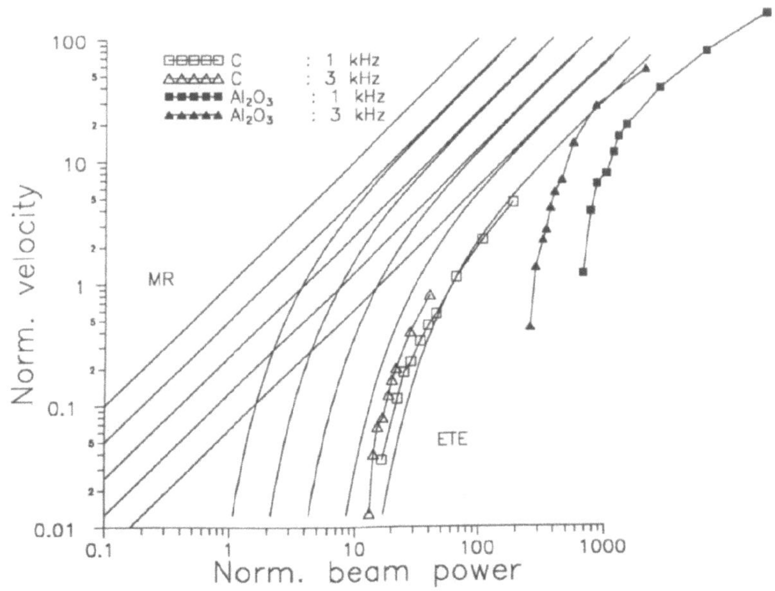

on the short axis it is curved to the inside. The cutting direction is clockwise, and the long and short axes are interchanged if the direction is reversed. If the velocity is raised, the long axis gets shorter and the short axis longer, so the ellipse become a circle. The aspect ratio, defined as the ratio of the long axis to the short axis, is plotted in Fig. 5.

The aspect ratio in Fig. 5 is 3 and decreases smoothly to 1 for diamond in the whole velocity range. The difference between the $R\pi$ and $R\sigma$ components is about the same for angles of 60°– 80°. This means the aspect ratio is a smooth function of the velocity. For v = 2000 μm/s, the angle tends to go to 90° (see Fig. 4b), and $R\pi - R\sigma$ becomes zero for that angle (see Fig. 1). The apsect ratio in this case becomes 1. An additional effect is the decreasing penetration depth and, therefore, decreasing multiple reflections in the groove, which makes the aspect ratio 1 also.

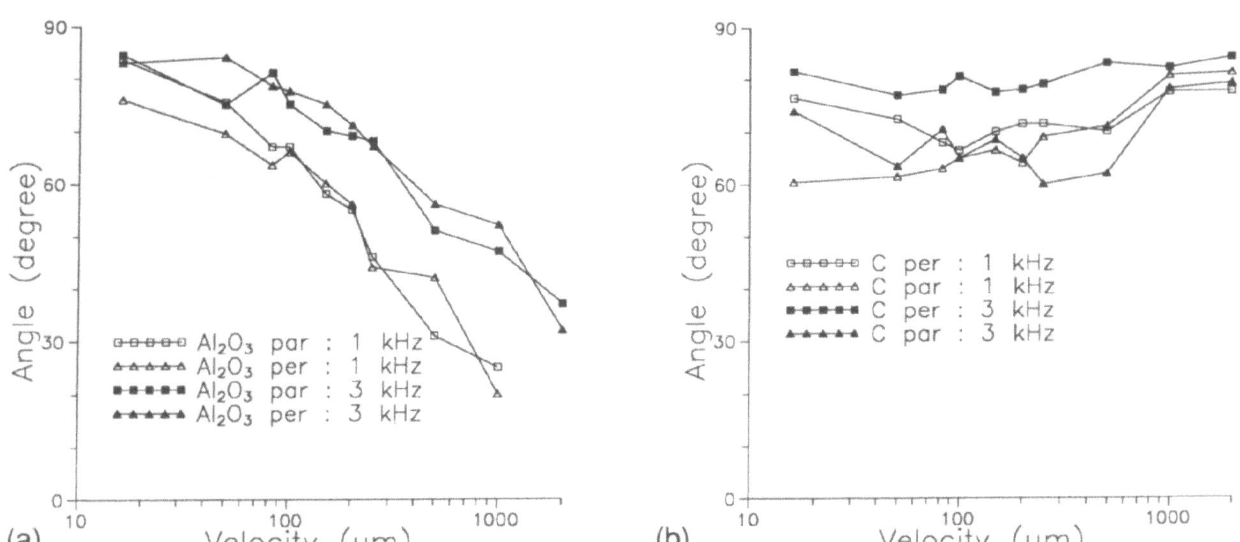

Figure 4. (a) Angle of incidence at the bottom of the cut for sapphire (par = $R\pi$ component, per = $R\sigma$ component). (b) Angle of incidence at the bottom of the cut for diamond (par = $R\pi$ component, per = $R\sigma$ component).

Photo 1. Circular cuts in diamond with linear polarized laser light seen through a microscope (100×). It is clear that an ellipse-like shape arises. The diameter of the circle is 300 μm.

The case of sapphire is totally different. For 1 kHz and a velocity of ~50 μm/s, the angle is 60° (see Fig. 4a) and the difference between Rπ and Rσ is large (see Fig. 1). In Fig. 5, a maximum aspect ratio is seen at ~50 μm/s. For higher and lower speeds, the aspect ratio tends to go to 1 because the difference between Rπ and Rσ becomes less.

In the case of P = 4.5 W, the reasoning is the same as above. Now it is seen that for v = 250 μm/s, an angle of 60° is reached (see Fig. 4a). For this velocity and power also, a maximum occurs (see Fig. 5). The first point on the curve has a high value, which may be caused by vibrations in the *xy* table for very low velocities.

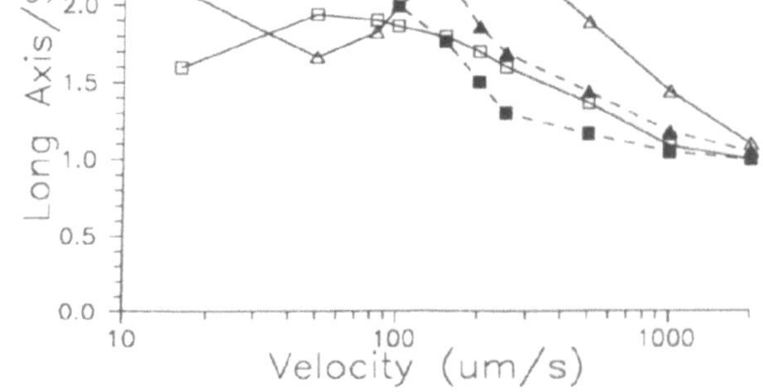

Figure 5. Aspect ratio as a function of velocity. For diamond it decreases smoothly. Sapphire shows maxima around 50 μm/s resp. 250 μm/s for low resp. high power.

Conclusions

For straight cuts, the penetration depth in diamond is less than in sapphire, mainly because of the larger heat loss in diamond. The angle of incidence is a function of the cutting speed and varies from 60°– 80° for diamond and 20°– 85° for sapphire. Circular cuts in diamond are elliptical in the range of 10–2000 μm/s because the difference in Rπ and Rσ is large.

For sapphire, aspect ratio of the ellipse has a maximum where the velocity gives an angle of incidence around the Brewster angle (60°). For lower speeds, the angle become large; for higher speeds, the angle becomes small. In both cases, the aspect ratio will be 1.

References

1. R.J. Wallace et al. "Curvature of laser-machined grooves in Si_3N_4," *J. Appl. Phys. 59* (10), 15 May.
2. W. Koechner. *Solid-state Laser Engineering*, second competely revised and updated ed., Springer Verlag (1988).
3. C.Y. Huang et al. "Time-resolved picosecond optical study of laser-excited graphite," *Proc. of Materials Research Society Symposia*, 51 (1986).
4. R. Wood. *Laser Damage in Optical Materials*. Adam Hilger (1986).
5. D.T. Swift-Hook and A.E.F. Gick. "Penetration welding with lasers," *Welding Research* supplement (Nov. 1973).
6. M.H.H. van Dijk. "Pulsed Nd:YAG laser cutting," *The Industrial Laser Annual Handbook*, D. Belforte and M. Levitt, eds., PennWell Books, Tulsa, OK (1987).
7. W.N. Reynolds. *Physical Properties of Graphite*. Elsevier Publishing Co. Ltd. (1968).

Drilling of aero-engine components: Experiences from the shop floor

Martien H.H. van Dijk
Eldim b.v., The Netherlands

Introduction

Although the efficiency of gas turbine engines, in terms of fuel consumption vs. thrust, has increased over the last 30 years, there is still constant pressure on engine manufacturers to increase the efficiency of the engine.

Performance can be improved by increasing the temperature of combustion; however, this leads to increased temperature in the combustion chamber and turbine. The temperature of the combustion gas can be as high as 2000°C, which is higher than the melting point of the superalloys used for combustion chamber and turbine components. Therefore, cooling air is introduced through thousands of very small holes in the walls of the combustion chamber. The turbine is also cooled by ducting air through very small holes in the blades and vanes of the turbine. These holes, often of a very special shape, are placed to provide films of insulating air or to make use of impingement cooling.

Development of new-heat resistant materials and the improvement of hole drilling in aero-engine components are necessary to further increase the efficiency of gas turbine engines. Improvement of drilling processes is also needed to reduce the production costs.

Drilling processes

Components of the combustion chamber and the turbine of aero-engines are made of superalloys (see Table 1), which are unsuitable for conventional drilling processes. Typical hole diameters are 0.3–2.5 mm. The number of holes per component may vary from 25 to 40,000 (see Table 2). Processes used for drilling these holes are electro-chemical drilling (ECD), electro-discharge machining (EDM), electro-chemical machining (ECM), and laser drilling. A comparison of these drilling techniques is given in Table 3. The actual results depend on material and processing parameters.

Laser-drilling processes

The basic premise of the laser-drilling process is heating of the material surface by a focused laser beam. Laser energy absorbed at the surface is converted to heat that melts and partly evaporates material. Due to the high pressure of the vapor, molten material will also be ejected. A gas jet, from a coaxial nozzle, may be used to improve the process of ejecting molten material. The diameter of the drilled hole depends on the intensity distribution in the focused laser beam. For the

Table 1. Materials for aero-engine components

Material Alloys	Component	Operating condition: Temp (deg. C)	Stress (MPa)	Kind of stress
Ni-based	Disc	400–700	700–900	Yield
	Blade	400–750	300–650	Yield
	Root			
Ni-based	Rotating blades	600–1000	70–250	Rupture
	Airfoils			
CO-based	Nozzle guide vane	800–1100	30–100	Rupture
	Airfoils			

Table 2. Typical hole dimensions

Component	Diameter (mm)	Wall thickness (mm)	Angle to surf. deg.	Number of holes
Blade	0.3–0.5	1.0–3.0	15	25–200
Vane	0.3–1.0	1.0–4.0	15	25–200
Afterburner	0.4	2.0–2.5	90	40,000
Baseplate	0.5–0.7	1.0	30–90	12,000
Seal ring	0.95–1.05	1.5	50	180
Cooling ring	0.78–0.84	4.0	79	4,200
Cooling ring	5.0	4.0	90	280

Table 3. Comparison of hole-drilling techniques for aero-engine components

	EDM	ECM	ECD	Laser
Minimum				
Hole diameter (mm)	0.3	0.75	0.5	0.1
Taper (mm)	0.0005	0.025	0.001	0.01
Recast layer (μm)	25	—	—	40
Angle to surface (degree)	20	—	15	15
Surface roughness (μm)	6	2	6	20
Maximum Depth/diameter ratio	25	20	250	100*
Complex cross section	yes	yes	no	yes
Simultaneous drilling	yes	yes	yes	no
Tooling complexity	high	high	high	low

*Depending on material thickness and laser peak power

same spot size, a higher pulse energy and thus a higher power density will result in a larger hole diameter (see Fig. 1).

The maximum depth of the hole also depends on the power density and the process of multiple reflection of the laser beam at the wall of the hole (see Fig. 2). Each time the laser beam is reflected at the wall, part of the beam energy is absorbed, which results in a decrease of the intensity of the laser beam at the bottom of the hole.

In single-shot drilling, thin material is penetrated with one laser pulse. This process can be used for drilling "on the fly," in which holes are drilled while the part is continuously moved relative to the laser beam. The proper combination of laser pulse frequency and velocity of the part will result in correct position of the holes.

In percussion drilling, several pulses are used to penetrate thicker materials. This may be done on the fly or with station-

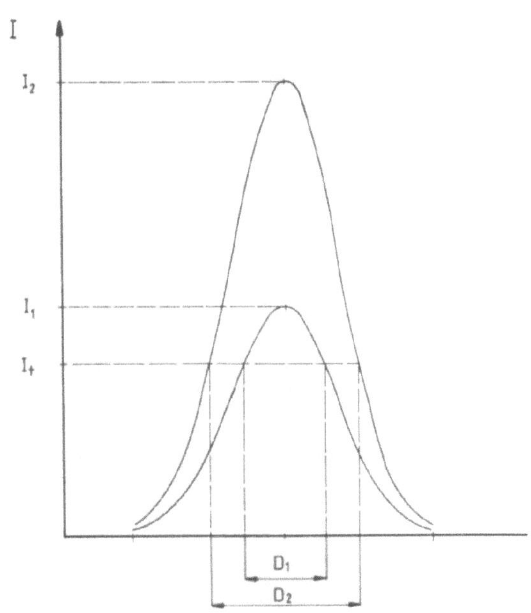

Figure 1. Percussion drilling, showing influence of pulse energy on hole diameter. I = power density, I_t = threshold power density, D = hole diameter. Pulse energy is proportional to power density.

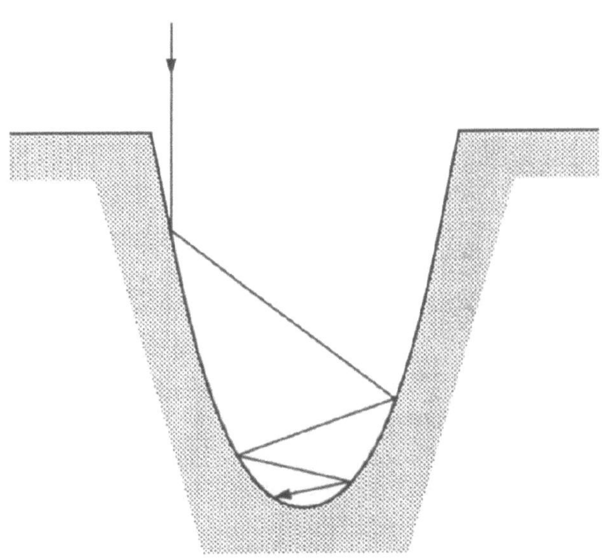

Figure 2. Multiple reflection of laser radiation at the wall of the hole.

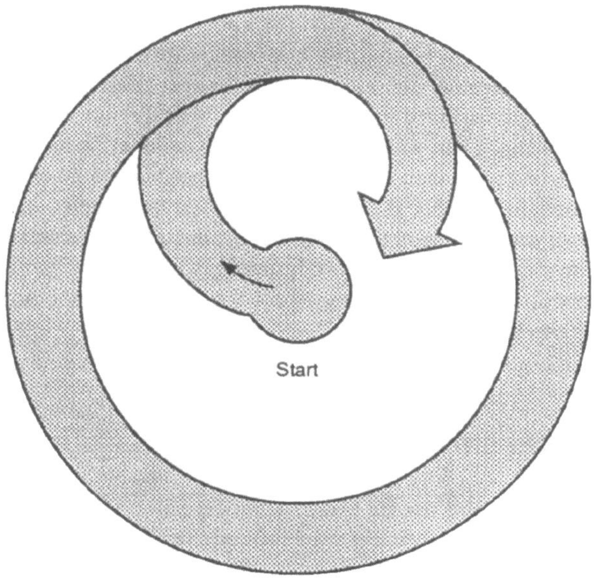

Figure 3. Contour in trepan drilling.

ary parts. Drilling on the fly is used for ring-shaped and circular cylindrical products. The advantage is a high processing speed/synchronization between the pulsing of the laser, and motion of the part is very critical. Percussion drilling with stationary parts is used for drilling blades, vanes, and cylindrical parts.

In trepan drilling, the beam is moved over a circular contour (see Fig. 3). The diameter of the hole can be controlled by changing the diameter of the contour. In general, trepanning results in better hole quality than percussion drilling. The main disadvantage of trepanning is lower drilling speed.

Traditionally, pulsed Nd:YAG lasers are used for laser drilling. Recently, F. Olsen developed the process of drilling on-the-fly with pulsed CO_2 lasers.[1]

Laser-drilling system

The main components of a system for laser drilling of aero-engine parts are:

- Laser source
- Beam guiding system
- Process gas system with coaxial gas nozzle
- Worktables with *x*, *y*, and *z* axes and two rotary axes
- Workstation enclosure
- CNC computer
- Trepanning head (only in systems for trepanning).

All of these parts are integrated into one system by systems suppliers, who fall into two groups: laser manufacturers and machine-tool manufacturers. Figure 4 shows a typical industrial laser drilling system; Photo 1 is a laser drilling system from Raycon Corp. (Ann Arbor, MI) that is used for aerospace applications. Technical specifications of a typical general laser-drilling system are given in Table 4. Dedicated systems for drilling blades and vanes will have shorter travel on the *x*, *y*, and *z* axes. The total mass of worktables for drilling blades, vanes and other small parts is in general 100 times higher than the mass of the part itself.

Although systems have become more user (operator) friendly and reliable over the last 10 years, there is need for further improvement. It is expected that equipment for laser-beam monitoring and process monitoring will be integrated into new systems. The higher costs of these systems will be compensated by reductions in downtime and improvements in process quality.

Monitoring flashlamp lifetime is very important when pulsed Nd:YAG lasers are used. To reduce the costs for consumption of flashlamps and the downtime for lamp change, one would prefer to use lamps as long as possible. However, lamps used too long may break, with the penalty of more time needed to replace the broken lamp and the risk of damage of protection tubes or even a broken laser rod.

In standard practice, a pulse counter is used to monitor lamp lifetime. However, the total electrical charge conducted through the lamp, not the number of pulses, determines the lifetime. Thus a pulsed counter is only adequate when a very limited range of pulse energies is used. A simple microprocessor that calculates the total transported charge can be used to monitor the flashlamp life time in cases where a wide range of pulsed energy is used.

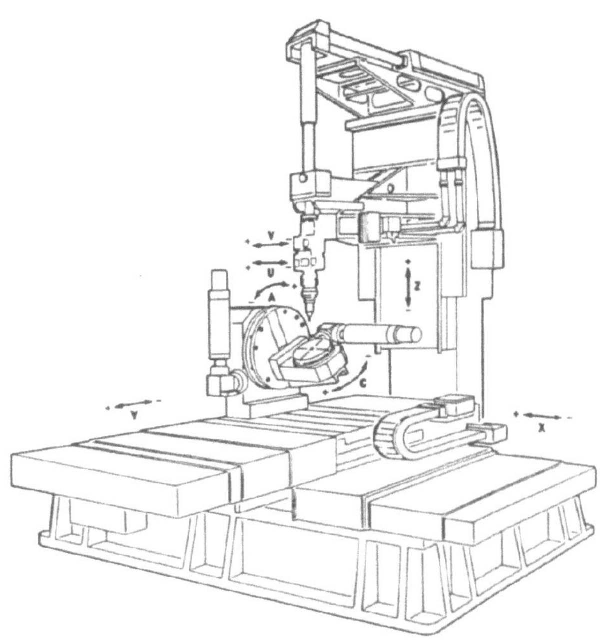

Figure 4. A typical industrial laser drilling system.

Photo 1. Laser drilling system from Raycon Corp. for aerospace applications.

Tooling

Parts are normally attached to a tooling that is mounted on the laser system's worktable. Typical weights of parts, weight of tooling, and the torque loading are given in Table 5. Values for original tooling and redesigned tooling are given. Standardization of tooling and tooling parts results in reduction of costs and lead time.

Shuttles may be used to reduce the time for mounting parts on the machine. The use of shuttles is cost-effective, especially for machined and sheet-metal parts. In the unrestrained condition, a part with a diameter of 500 mm may be up to 1 mm out of round. The tolerance on position of holes is normally less than 0.1 mm. So parts must be mounted in a restrained condition to restore the roundness. When shuttles are used, the next part can be restrained while the first part is machined. On a special test set-up, the roundness of the part is checked.

Cycle time

The cycle time needed to drill a part depends on the drilling speed, travel speed between hole positions, the time for mounting of the part or shuttle, the time for intermediate inspection, and the time to travel from the load/unload position to the process position. In many cases, it is necessary to prevent impingement of the laser beam on the surface behind the wall in which a hole is drilled. Normally, PTFE or metal strips or wax is used for this purpose. This protection can be applied before the part is mounted. Although this can be done outside the laser department, it is normally done by the laser operator to ensure that it is done properly. No extra machine

Table 4. Specifications of typical pulsed Nd:YAG laser drilling

	Travel	Accuracy	Repeatability	Feedrate
Work-handling system				
x axis	600 mm	0.02 mm	0.01 mm	0–5 m/min
y axis	900 mm	0.02 mm	0.01 mm	0–5 m/min
z axis	600 mm	0.02 mm	0.01 mm	0–5 m/min
a axis (rotary)	200°	0.01°	0.005°	0–600°
b axis (rotary)	360°	0.01°	0.005°	0–1200°
Trepanning head				
u axis (linear)	5 mm	0.005 mm	0.002 mm	0–0.5 m/min
v axis (linear)	5 mm	0.005 mm	0.002 mm	0–0.5 m/min
Laser source		Percussion driller		Trepanning driller
Average optical power (W)		150		250
Maximum optical peak power (kW)		6		50
Pulse length range (msec)		0.4–10		
Pulse length select. values (msec)				0,6: 1.0; 2.0
Pulse frequency (Hz)		0–200		0–60
Maximum pulse energy at:				
0.6 msec (J)		3		30
1.0 msec (J)		6		50
2.0 msec (J)		12		50

Table 5. Typical weight of parts and tooling

		Tooling (original)		Tooling (redesigned)	
Part Name	Mass (kg)	Mass (kg)	Torque (Nm)	Mass (kg)	Torque (Nm)
Ring segment	0.2	11	27	3	7
Blade	0.2	5.5	8		
Baseplate	1.3	33	58	10	17
Ring segment	1.43	28	116	3.6	38
Cooling ring	2.00	36	53	11	18

time is needed when there is a mixture of jobs with short and long cycle times.

Table 6 gives data on cycle time and set-up time for typical products. In drilling machined and sheet-metal parts, any reduction of cycle time must be found in the stop time for inspection reasons. Such reductions may be achieved by improving laser stability and/or in-process control of hole diameter.

In drilling of blades and vanes cost reduction can be realized by reduction of the time needed to go from one hole position to the next. This requires light and fast worktables.

The operator

One of the key factors in laser material processing is the operator. This is especially true for three-shift operation in which back-up from engineering and maintenance is available on short notice only one third of the time.

Table 6. Relative production times

	Blade	Sheet metal part
Parts per production run	200	14
Production		
Mounting of part	16.7%	0.0%
Set-up time	0.0%	2.6%
Travel to work position	9.2%	0.1%
Net drilling time	8.6%	57.0%
Positioning	34.6%	13.0%
Travel to load/unload position	9.3%	0.1%
Unloading	6.2%	%
Inspection during production	15.4%	25.7%
Total cycle time	100.0%	100.0%
Set-up time per part*	44.0%	2.0%
Part preparation time* (normally not on line)	4.0%	13.0%

*Time relative to cycle time

The operator's main tasks are:

Preparing the laser system for a new job:

- Mounting tooling on the worktable.
- Loading the CNC file in the computer.
- Checking reference point positions.
- Setting laser parameters.
- Tuning laser resonators.
- Aligning nozzle, setting nozzle pressure.
- Setting trepanning parameters.
- Drilling test plate.

Start of production:

- Drilling first product.
- Measuring hole dimensions and positions.
- Correcting settings when necessary.

Production:

- Operating the system.
- Mounting products on tooling/shuttle.
- Statistical process control to monitor quality.
- Correcting settings when necessary.
- Logging all relevant data.

Depending on the number of laser systems and operators, there will be one or more senior operators. The additional task of a senior operator involves preparing the production of new products and one-of-a-kind products. This task is performed in close cooperation with the engineering department and tool shop and includes:

- Defining operation sequence.
- Designing or selecting tooling.
- Checking CNC program on critical points.

Because new laser operators require in-house training, it is very important to have a proper training program on theoretical and practical aspects of laser materials processing. It is also important that an industrial engineer with at least a bachelor's of science is responsible for improving the processes. Problems that occur during production can result in a mess or in a significant improvement of the process; the outcome depends on the quality of the team: laser operator, senior operator, and industrial engineer. Laser materials processing generally asks for a higher level of experience from operators and support personnel than for normal CNC machining.

Job shops

Aero-engine manufacturers do have extensive facilities for nonconventional drilling of their components. They also use external job shops for extra capacity, experimental parts, and low-volume production. These job shops need to be flexible because production runs as short as eight hours are not unusual and production runs longer than 200 hours are rare.

Due to the relative short production runs and the need for flexibility, the job shop operators must be well trained. They should also be qualified for experimental work and machining critical parts.

Compared to other nonconventional drilling processes, laser drilling is faster, though a higher operator/machine ratio is usually needed. But because tooling for laser drilling is less complicated, the time to set up the production of a new part is much shorter. On the other hand, the penalty for inefficiency is much higher.

Statistical process control is needed for all processes to maintain a high level of quality.

The future

The market for laser drilling of aero-engine components will grow. Lasers with high average power will reduce drilling time in the trepanning process. On-the-fly drilling techniques will increase drilling speed in percussion drilling. New process development will result in deeper and more complex hole shapes. In-process control will improve quality and reduce time for inspection.

Although the future looks very promising, there are traps one may fall into. There is a tendency to provide laser drilling systems with advanced sensor systems to compensate for deviation in dimensions and shapes of castings, sheet metal, and machined parts. This makes laser drilling systems complex and expensive. Restraining sheet metal and machined parts on shuttles and improving casting technology are alternative solutions to this problem. In addition, in-process control is complicated because many variables are involved. Tests in different production situations and at different places are needed.

The first step on this promising road is to use process monitoring and to carefully analyze collected data.

Reference

1. D. Belforte and M. Levitt, "Industrial Laser Technology, Applications and Market Trends," *1991 Industrial Laser Review Buyers Guide*, pp. 11–27.

Laser hardening of boring tools

Vladimir S. Kovalenko
Kiev Polytechnical Institute, USSR

In bore hole surveying, the efficiency of the boring process depends on cutting speed. As the boring speed increases, wear resistance and reliability of the boring tools decreases. The weakest parts of the boring column are the internal threads of the tool joints and the bore pins. Laser hardening of the thread bottoms and the bore pin friction surfaces is a most effective way to increase wear resistance.[1] The tests conducted here prove bore reliability increases up to 30% and wear resistance of the tool joints doubles.

Conventional process

The boring tool joints are made from alloyed steel 40XH*, the bore pins from alloyed steel 14XH3MA*, subjected to full hardening. The element content of these alloys is shown in Table 1. The drawbacks of traditional heat treatment are component deformation, which creates the need for additional final mechanical machining, and low productivity of the conventional process.

Laser-hardening process

For laser hardening, a 1.5-kW CW laser was used. Beam-power density varied in the range of 3.0–8.0 kW/cm^2; scanning speed was in the range of 0.2–1.5 m/min. The working surface was oxidized to increase absorptivity. To irradiate the tool joints' internal threads, a special beam-delivery system and corresponding focusing device have been developed (see Fig. 1).

Experimental results

The material structural changes and hardened zone dimensions are dependent on power density and scanning speed. At high power density (8 kW/cm^2) and low scanning speed (0.2 m/min), a melted layer close to the surface was produced with a martensite-austenite structure of high dispersivity and microhardness. Below this was a layer of structureless martensite with a low amount of austenite, where the heating temperature did not exceed the melting point. On the border edge of the matrix material was a layer of partial hardening and then a transition layer.

At lower power density it was possible to avoid material melting. The hardened zone has a highly dispersive structure,

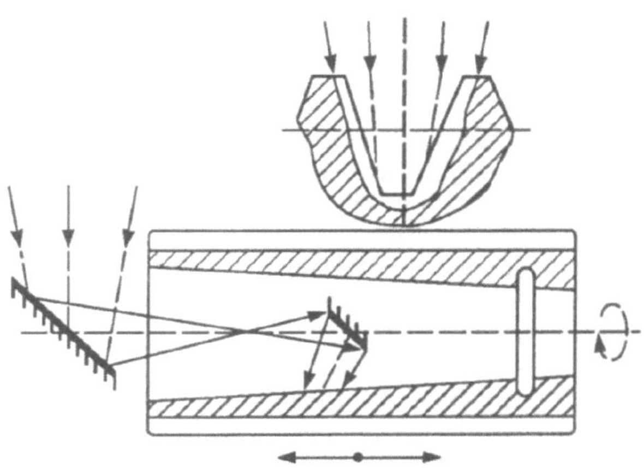

Figure 1. The beam-delivery scheme used to harden the internal threads.

Table 1. Elements of the steels used in boring tool process

	Elements content, %					
Steel	C	Si	Mn	Cr	Ni	Mo
40XH*	0.36–0.44	0.17–0.37	0.50–0.8	0.45–0.75	1.0–1.4	—
14XH3MA*	0.12–0.17	0.17–0.37	0.3–0.6	1.5–1.75	2.75–3.15	0.2–0.3

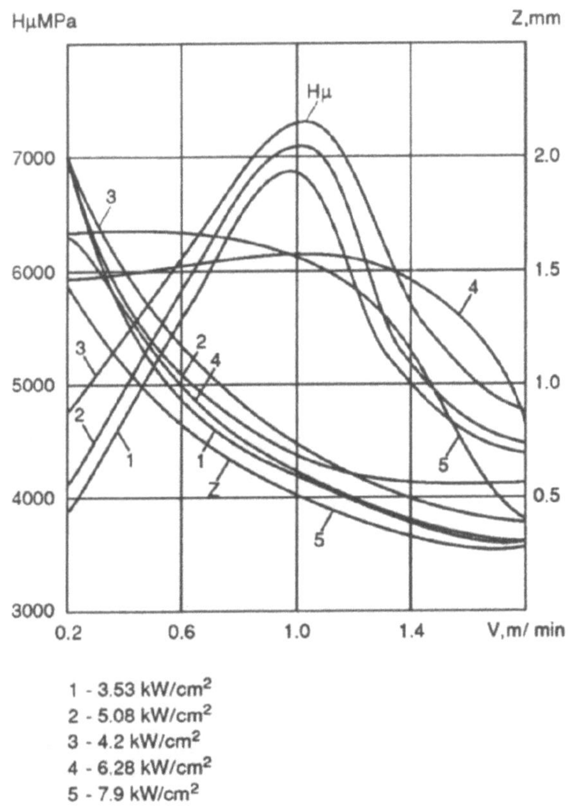

Figure 2. Heat-affected zone microhardness and depth vs. power density and scanning speed.

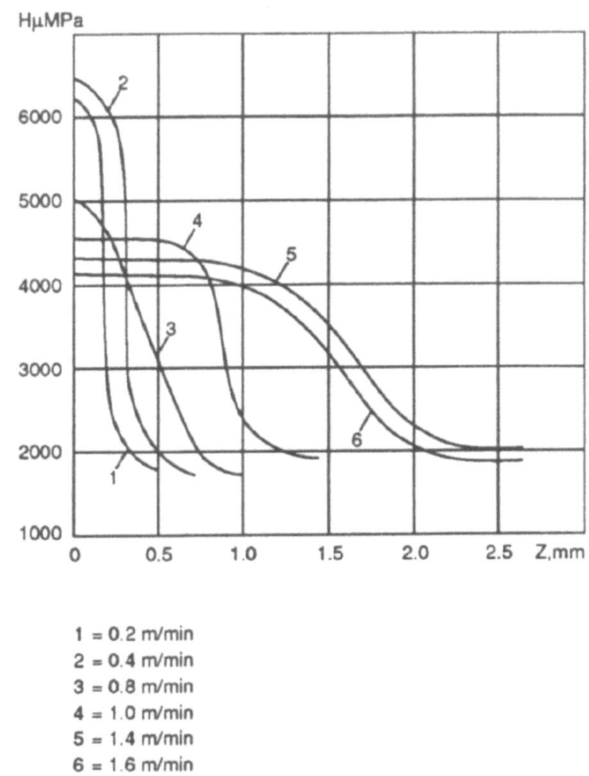

Figure 3. Microhardness changes vs. material depth for different scanning speeds.

and the microhardness is influenced more by scanning speed than by power density (see Fig. 2). The highest hardness was reached at a rate of 1 m/min for almost all power densities. The hardness zone depth was about 0.5–0.8 mm, and the surface did not differ from the initial roughness. The hardness depth is more sensitive to scanning speed changes than power density variations.

For 14XH3MA steel, the hardening effect for the given power density (5.3 kW/cm^2) and varied scanning speed (from 0.2 m/min up to 1.6 m/min) is shown in Fig. 3.

From experimental results, the optimum machining regimes have been chosen to achieve the highest productivity, microhardness, and appropriate microstructure. At the parameter regimes selected, the batch of tool joints and bore pins have been laser-hardened and then subjected to performance and field testing. They produced wear resistance and reliability increases up to 30% for bores and up to two times for tool joints.

Based on experimental results, a fully automated robotic production line for laser hardening has been developed and installed at the Kirovgeology plant, which significantly increased the productivity of the boring tools manufacturing process.

Conclusion

Laser hardening is a time-saving process, with productivity increasing up to 6 m/min and higher, depending on laser beam power. The thickness of the laser-hardened layer may reach 1 mm without changes in initial surface roughness. Laser hardening can be used to reach difficult-to-access locations on a component. The reliability and wear resistance of the boring tools increases 30%–100% after laser hardening.

Reference

1. L.F. Golovko, V.S. Kovalenko, L.K. Sinichenko, et al. "Surface hardening of boring tools by laser irradiation," *Technologia i organizacia proisvodtv, Kiev N2*, p. 4–6, in Russian (1989).

Laser hardening of chrome steels

Vladimir S. Kovalenko and Leonid F. Golovko
Kiev Polytechnical Institute, USSR

Chrome steels are now widely used for manufacturing components and tools working in heavy friction and/or high-corrosion conditions. The specific features of the laser-hardened chromic steels open new opportunities to form a variety of material structures with different characteristics.[1] A CW CO_2 laser was used in a study of the changes in the steel structure at different levels of irradiation. The wear mechanism and some tribology characteristics of the hardened steel are shown.

Experimental procedure and technique

For this experimental study, conventionally heat-treated samples made of chrome steel 20X13 (0.24% C; 0.36% Si; 0.43% Mn; 0.021% P; 0.018% S; 0.32% N; 12.54% Cr) were used. Chrome steel 20X13 is a martenistic stainless steel similar to 420. Laser hardening was performed using the power from a CW CO_2 laser focused with a 250-mm lens. The spot diameter was equal to 5 mm. Scanning speed varied from 0.2 to 1.5 m/min. To increase the absorptivity up to A=0.75, an absorption coating of ZnO was used. To produce a treated sample, overlapping paths of laser hardening were made in steps of 4 mm.

The samples were subjected to metallographic analysis and a friction surface microgeometry study. The wear-resistance tests were made on a friction machine using a "plane sample-roller" scheme with and without lubrication at a load of 50 N and slip speed of 2.5 m/s. The roller for this test was made from bearing steel and had a hardness of 55 RC.

Experimental results

Two specific structures are formed in the laser-hardened zone, depending on heating temperature. At temperatures exceeding the melting point, the hardened zone consists of two layers. One has a columnar structure with dendrites oriented in the direction of maximum heat transfer. The structure of this layer is a martensite-carbide eutectic with microhardness 6400–6800 MPa. Alongside this hard phase is a ferrite-perlite phase (x-ray analysis did not disclose the δ phase). Below the first layer is a structure of plate martensite and refined austenite with hardness of 5500–6000 MPa, which corresponds to the hardness after bulk heat treatment (see Fig. 1). At heating temperatures below the melting point, only one layer has the structure previously described.

The best antifriction characteristics result from the columnar structure. In this case, the grain-oriented layer is formed at the surface as a result of plastic deformation caused by friction forces (see Photo 1). The specific combination of hard and soft phases at the surface layer leads to a decrease in friction forces and excludes gripping.

In bulk heat-treated samples, the wear occurs as a result of brittle failure and the flaking of martensite and carbide particles. This causes lubrication-film destruction and, consequently, an increase in friction coefficient and wear. For normalized samples, the material hardness is lower, the friction coefficient increases up to 0.36, and wear is very high (see Fig. 2).

The described research data were used for laser hardening the internal working surfaces of the long slides on polymer-film-producing equipment (see Photo 2). Laser hardening, together with electroslag refining, reduced the mechanical machining (grinding after heat treatment) and thus reduced energy consumption.

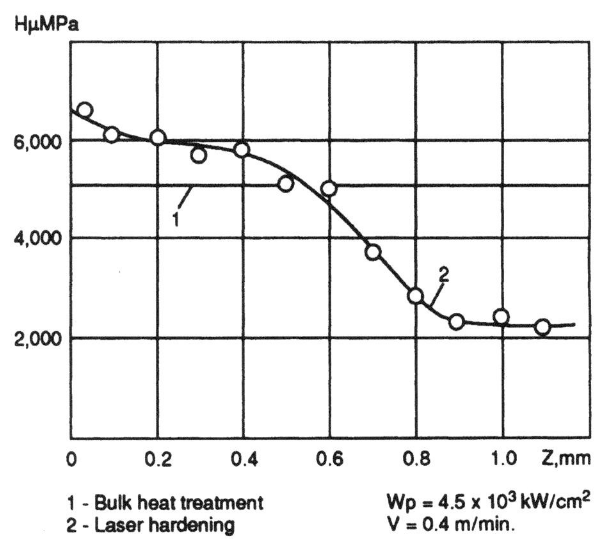

Figure 1. Microhardness changes with depth increase.

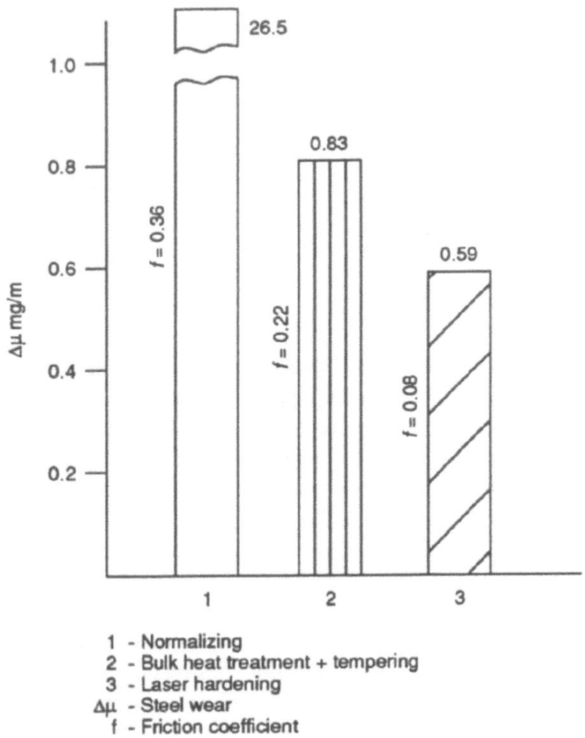

Figure 2. 20X13 steel wear for different types of heat treatment.

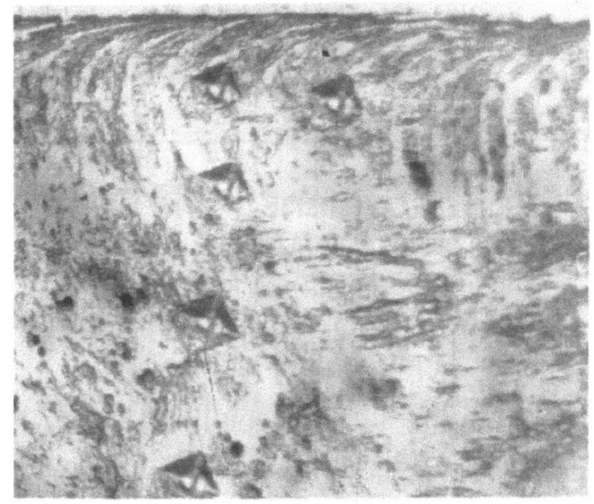

Photo 1. Surface layer microstructure after laser hardening and friction test.

Conclusion

Laser hardening of chrome steels leads to formations in hardened zones with specific structures consisting of hard and soft components. This improves antifriction characteristics of the working surfaces by better confinement of lubrication fluids, with increased impact load resistance and elimination of material flaking.

Photo 2. Laser hardening of the long slides.

Reference

1. V.S. Kovalenko, L.F. Golovko, V.A. Popchenko, et al. "Tribology characteristics of chromic steels, subjected to laser hardening," *Proc. of "The applications of lasers in science and technology,"* pp. 54–55, Omsk, Siberia (1988).

Laser treatment of lead alloys for battery grid application

Narendra B. Dahotre, Mary Helen McCay, T. Dwayne McCay, and C. Michael Sharp
Center for Laser Applications
The University of Tennessee Space Institute
Tullahoma, TN

Introduction

In the past several years, extensive research efforts have been conducted[1,2] to increase lead's mechanical strength, creep resistance, and corrosion resistance by conventional methods such as alloying and heat treatment. Out of these efforts, only a limited number of lead alloys were developed that are reasonably suitable for commercial application.

In an attempt to improve the strength of Pb-Sb-Cu-As-Se-Sn alloy, researchers at Asarco Inc.[2] developed a process that included mechanical working followed by heating rapidly to a desired temperature and then quenching. This sequence follows a path of unconventional solution treatment effects, resulting in strengthening of the alloy. Among the limitations of the process is the low percent of Sb- and As-Pb alloys. In addition, the mechanism behind this process is not understood. Thus, earlier techniques proposed for strengthening many Pb-alloys resulted in limited success.

In light of these developments, the present investigators conducted studies on continuous wave CO_2 laser processing of Pb-Ca, Pb-Sb, and Pb-Sb-Sn-As alloys for battery grid application.[3,4] The investigators have demonstrated the feasibility of the process for improved properties in these Pb-alloys. This unconventional treatment produces high cooling rates, which result in increased solid solubility in the alloy. The resulting microstructure was refined and contained unconventional and nonequilibrium phases. These changes in microstructure improved the mechanical and corrosion-resistance properties. To better understand the effects of laser processing on lead alloys, the study described herein was undertaken.

Experimental procedure

The experimental approach involved laser melting the Pb and Pb-alloys using a Rofin-Sinar 3000 laser, a commercially available fast-axial-flow, RF-excited (27.12 MHz) CO_2 laser (10.6 μm wavelength). For the present experiment, the laser was operated in continuous mode. The processing parameters are given in Table 1.

In this experiment, the surface of the sample was melted by laying adjacent tracks with the beam focused 25 mm above the surface of the workpiece. This provides a 2-mm-wide melt track on the surface and almost complete melting through the thickness of the sample. The sample was laser-treated on both sides to achieve uniform surface appearance. A CNC system was used in conjunction with a five-axis Aerotech workstation to provide constant linear translation of the workpiece.

During laser treatment, the sample was mounted on a workstation by sandwiching it between a water-cooled copper block substrate and another copper block with a window in it. This arrangement provided a high cooling rate in the sample and also prevented it from buckling due to thermal stresses developed during laser treatment. The compositions of the samples treated in this experiment are given in Table 2.

Tensile and creep tests

Tensile testing was performed on as-received and laser-processed alloys using an Instron testing machine. To enable the use of small specimens and avoid weight inertia effects due to the heavy grips on relatively soft Pb and Pb-alloy samples, a specially made microtensile sample holder was used (see Fig. 1). The tensile test parameters are given in Table 3.

Table 1. Laser-processing parameters

Laser mode	Continuous-wave CO_2 gas
Power delivered at workpiece	1650 W
Beam mode	TEM_{10}
Beam polarization	Circular
Traverse speed	100 mm/s
Focal position	25 mm above surface
Nozzle	Cross-flow nozzle for optics protection
Shielding gas	Argon, 4 L/min

Table 2. Compositions (weight percent) of materials

Alloy No.	Sb	Ca	Sn	As	Pb
1	—	—	—	—	100
2	1.4	—	—	—	balance
3	2.0	—	—	—	balance
4	2.5	—	—	—	balance
5	—	0.08	—	—	balance
6	1.4	—	0.2	0.2	balance
7	2.0	—	0.2	0.2	balance

Table 3. Tensile test parameters

Test type*	Uniaxial tensile test
Strain rate	3.3×10^{-4} s^{-1}
Temperature	Ambient
Gage length	25 mm
Sample thickness	1 mm

*Tensile axis was parallel to laser tracks on sample.

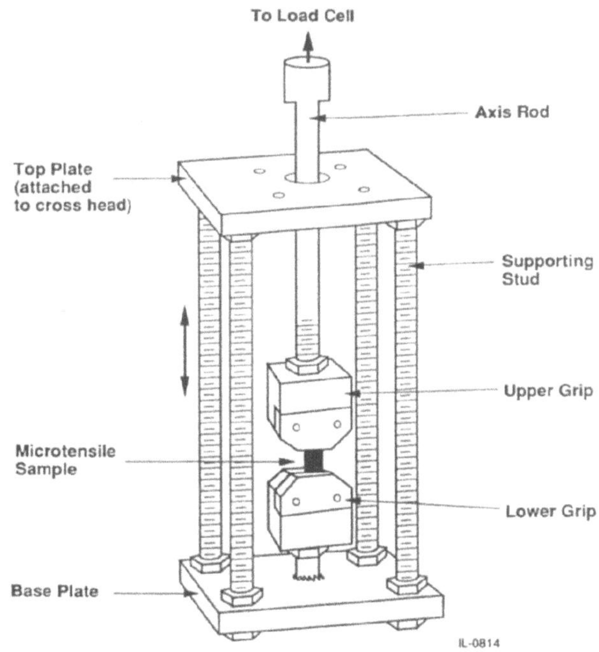

Figure 1. Schematic of microtensile sample holder.

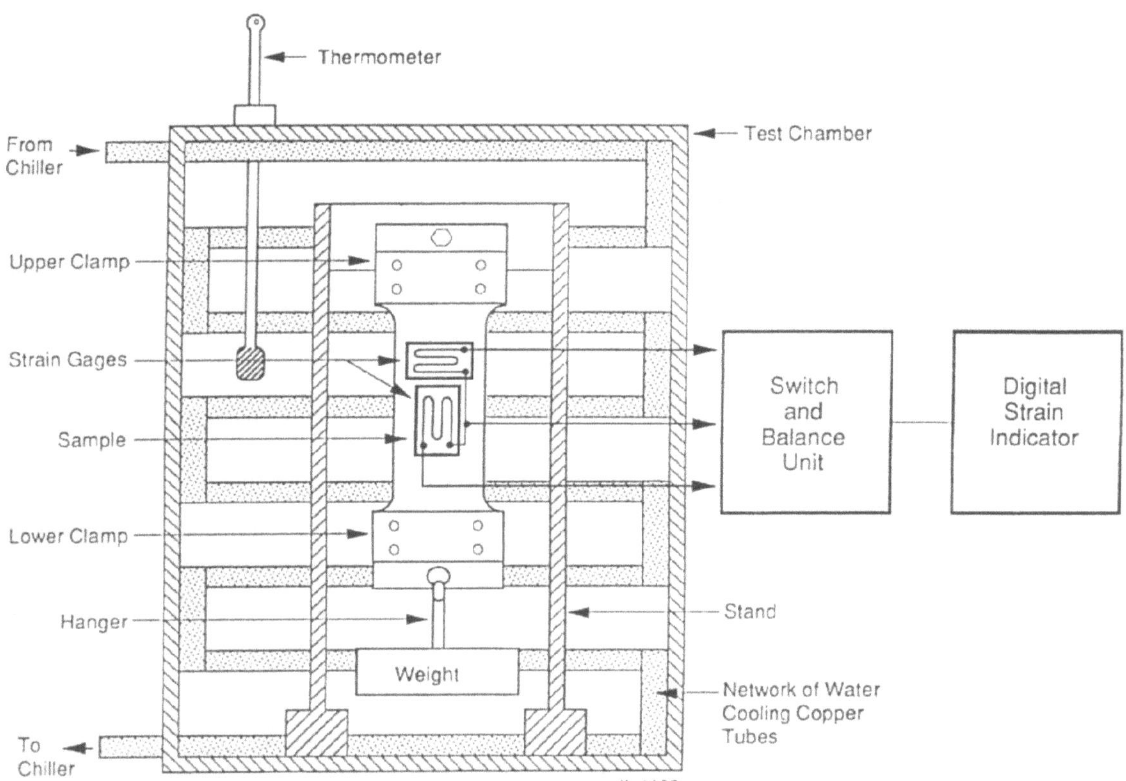

Figure 2. Schematic of creep test set-up.

Creep testing was conducted in a specially designed temperature-controlled chamber: a rectangular box ($90 \times 40 \times 36$ cm^3) used for multiple sample testing. The temperature of the chamber was controlled within a ±2°C range by running water from a chiller through copper coils mounted on the inside walls.

In this experiment, creep tests were conducted at 25°C. The creep test samples had the same geometry and dimensions as those used for tensile tests. The sample was uniaxially loaded with 1 kg weight under static conditions. The deformation was recorded as an output of a resistance strain gage (active tension gage) mounted in the gage section parallel to the tensile axis of the sample. In addition, an active compensation gage was mounted perpendicular to the tensile axis in the gage section to compensate for the response to changes in temperature and compressive forces. The tension and compensation gauges were configured into the half-bridge Wheatstone circuit on the digital strain indicator unit, which displayed the deformation as a microstrain (μin./μin.). The readings were initially recorded at eight-hour intervals, followed by 24-hour, 96-hour, and 360-hour intervals. Figure 2 shows a schematic of the creep test sample configuration for a single sample.

Corrosion tests

The potentiostatic tests on as-received and laser-processed Pb-alloys for corrosion properties were conducted at the Central Laboratory of Yuasa Battery Co. Ltd., Japan. The samples were rectangular coupons ($10 \times 2 \times 0.1$ cm^3). Figure 3 shows the schematic of the experimental set-up; Table 4 gives the experimental parameters. The samples were removed from the electrolyte after every 20 days for weight-loss, microstructural, and x-ray diffractometry analyses. The potentiostatic tests were conducted for a total of 60 days.

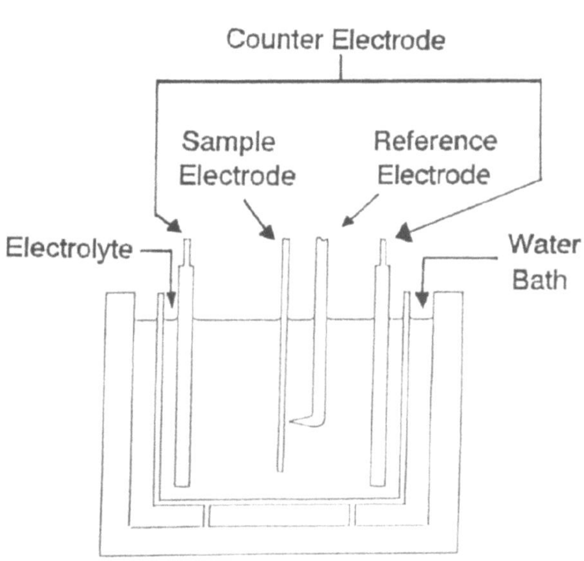

Figure 3. Schematic of corrosion cell assembly.

Table 4. Corrosion (potentiostatic) test parameters

Electrolyte	1.285 sp. gr. H_2SO_4
Reference electrode	Hg/Hg_2SO_4
Counter electrode	Commercial lead-acid battery grid
Temperature of bath	50°C
Test potential	1.35 V
Test duration	60 days

For microstructural observations, the samples were fixed in cold resin and then prepared for longitudinal and cross-sectional views. Each sample was etched with a mixture of glacial acetic acid (2 parts) and 30% H_2O_2 (1 part). The microhardness measurements were done with a 100-gm load and Vickers indentor.

A parallel set of experiments was conducted on as-received and laser-treated samples for x-ray diffractometry analysis. A Philips XRG3100 diffractometer with Cu tube target (K_α radiation, 1.54 Å wavelength) was used for this purpose. The samples were scanned from 20° to 100° for 2Θ at a rate of 1°/min. The results were plotted as intensity vs. 2Θ, and the peaks obtained on these graphs analyzed for existing phases.

Results and discussions

Microstructural evolution

Optical micrographs of as-received and laser-treated Pb + 0.08% Ca and Pb + 1.4% Sb are shown in Photos 1 and 2, respectively. At 300° C, about 0.06% Ca is soluble in lead, but at room temperature the solubility drops to about 0.01%, causing precipitation of Pb_3Ca. Photo 1a illustrates the occurrence of these Pb_3Ca precipitates (dark phase) in as-received Pb + 0.08% Ca alloy. After laser treatment of the Pb-Ca alloy, the original cubes/blocks of Pb_3Ca precipitates appear as extremely fine acicular precipitates uniformly distributed within the grains (Photo 1b). The grain structure changed from polygonal to irregular grains, and the relative size of the grains was reduced.

In hypoeutectic Pb-Sb alloys (< 11.2% Sb), α-phase of the eutectic—a mixture of antimony (β-phase) in a matrix of lead-rich (α-phase) solid solution—blends into the primary lead-rich phase, giving the eutectic a divorced appearance. Photo 2a shows such a divorced structure with α-dendrites (dark) and an interdendritic filling of eutectic (α + β). The white particles are β-crystals. The laser treatment of Pb-Sb produced a structure that was extremely fine dendritic with an interdendritic network of antimony in a matrix of lead-rich solid solution (see Photo 2b). Further, the higher content of

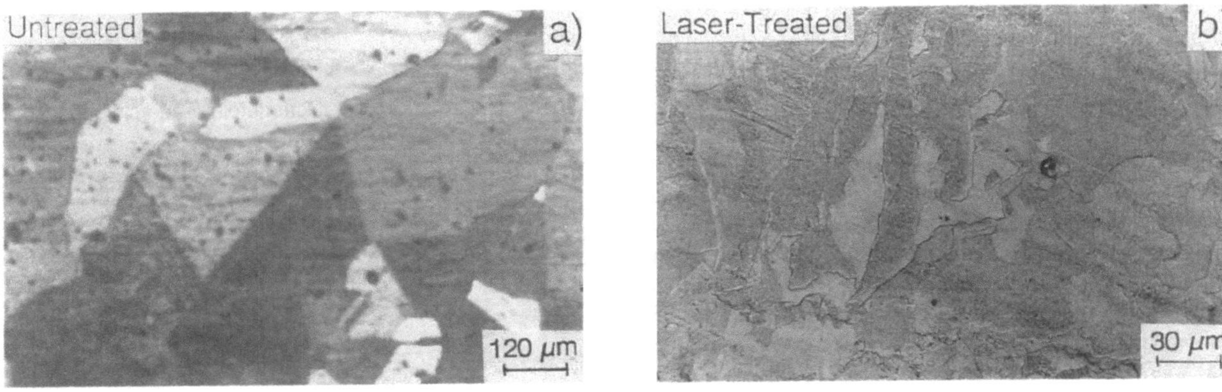

Photo 1. Microstructures of Pb + 0.08 wt% Ca alloy.

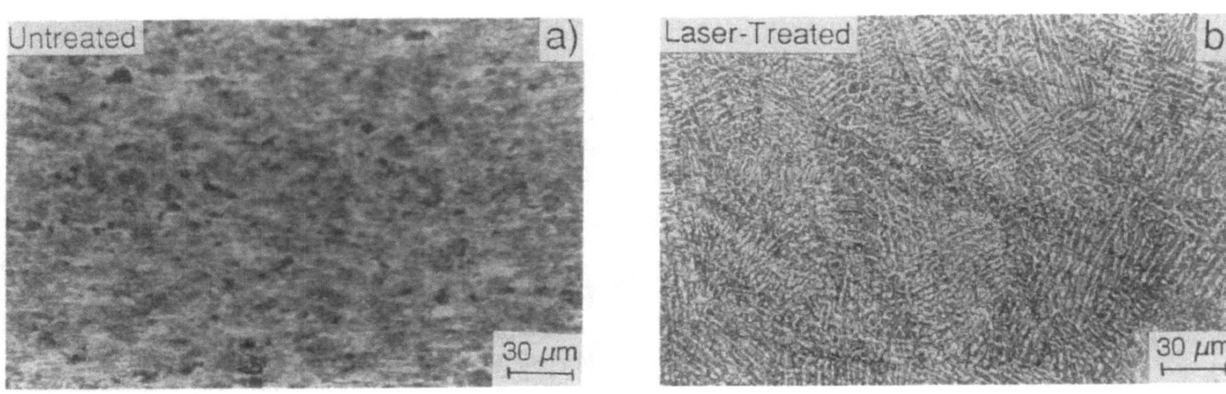

Photo 2. Microstructures of Pb + 1.4 wt% Sb alloy.

antimony changes the kinetics of solidification and offers a larger number of nucleating sites.[3,4] Similar effects were also observed in laser-treated Pb-Sb-Sn-As alloy.[3]

The microstructures evolved during laser processing Pb-alloys is the result of the higher cooling rates ($10^5 - 10^6$ °C/s) associated with the process. The extremely higher cooling rates produce large undercoolings below the liquidus temperature. Thermally, it causes the eutectic reactions in Pb-alloys to occur at a temperature several degrees below that shown in the equilibrium diagrams. Structurally, it causes a refinement of the particle size of the phase participating in the reaction. Thus, the dynamic conditions of rapid solidification induce heterogeneous microstructures made up of fine dendrites. Further investigation of formation of unconventional and nonequilibrium phases in Pb-alloys during laser treatment was performed using such techniques as x-ray diffractometry analysis.

X-ray diffractometry analysis

The x-ray diffractometry analysis of as-received and laser-treated Pb-alloys (see Fig. 4 and Fig. 5) illustrates the existence of several significant peaks that remained unidentified.[4] These peaks and corresponding d-spacing (interplanar distances) do not match with existing standard phases referenced in the available literature.[5] To characterize these unidentified, unconventional phases formed during laser treatment, it is essential to use a technique such as transmission electron microscopy. Also, the volumes of many phases formed during laser treatment of Pb-alloys are very small; therefore, it was not possible to detect them by x-ray diffractometry technique.

Though the surfaces of the samples were carefully cleaned for x-ray diffractometry studies, the presence of an oxide phase was noted in all samples in ranges from diffraction angle (2Θ) of 23° to 32°. This may be due to the surface

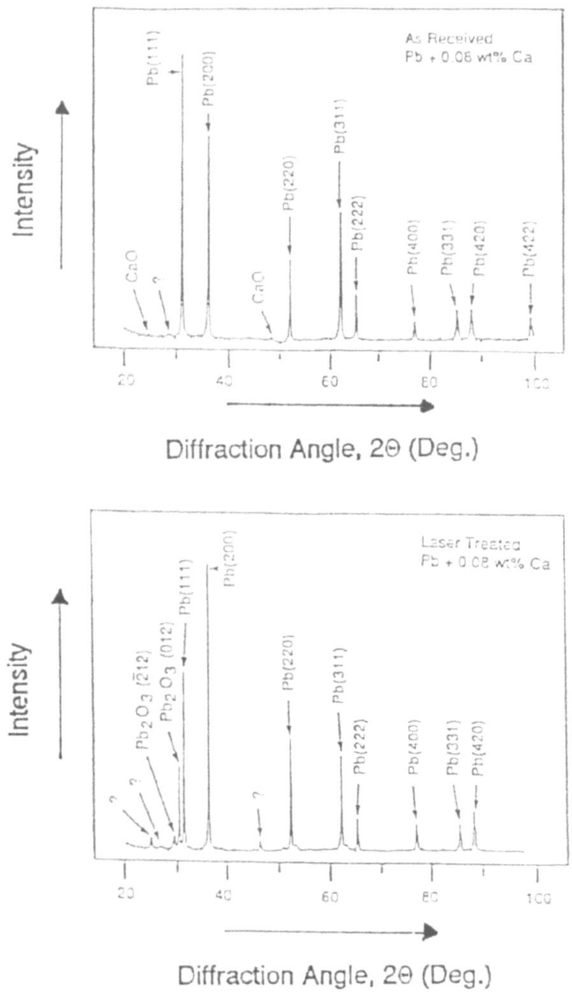

Figure 4. X-ray diffractometry data from Pb + 0.08 wt% Ca alloy.

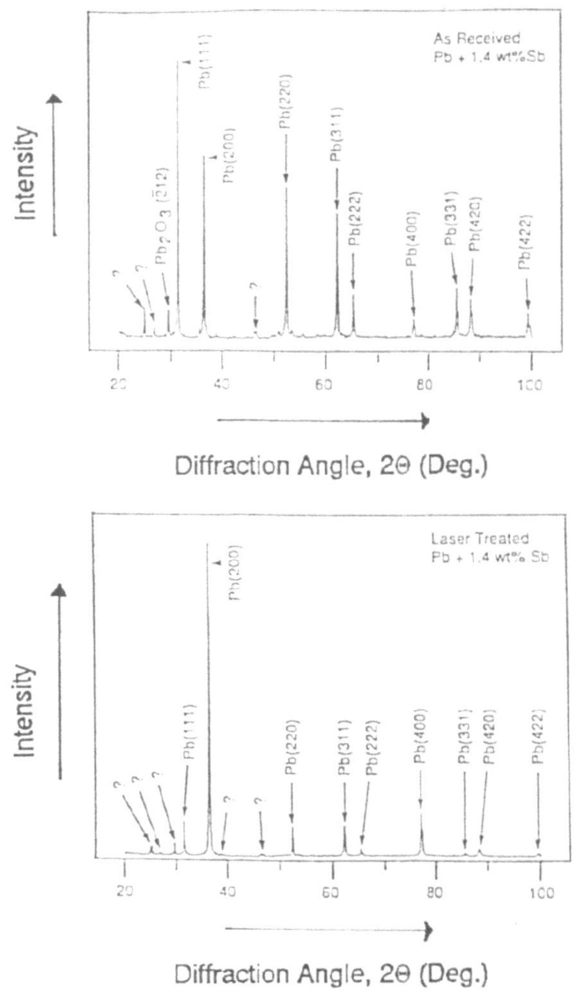

Figure 5. X-ray diffractometry data from Pb + 1.4 wt% Sb alloy.

oxidation or dispersion of oxide particles formed during earlier synthesis of the alloys. The variation in the intensity on peaks of Pb(111) and Pb(200) is attributed to the grain-orientation effect.

Mechanical properties

The extremely fine precipitates produced during laser processing provide the pinning locations for dislocations. In addition, the evolution of unconventional and nonequilibrium phases and their distribution in the soft lead-rich phase matrix strengthens the material. These mechanisms together enhance the mechanical properties of the Pb-alloys processed with a laser. The effects are more evident in Table 5, which presents the tensile tests and hardness data on as-received and laser-treated Pb-alloy samples. The values of ultimate tensile strength (UTS), yield strength (YS), and modulus of elasticity (E) are substantially higher for the laser-treated Pb-alloy samples than those of the untreated samples.

As a result of the increase in the values of these parameters, the elongation at fracture for the laser-treated samples decreased by almost 50% below that for untreated samples. In spite of this, however, it is still comparable to the value of elongation at fracture for pure lead. The hardness is higher for all laser-treated samples than for untreated samples, and it is about two to three times higher than for pure lead. These changes in the laser-treated Pb-alloy samples produce mechanically improved material for storage batteries.

Creep behavior

Figures 6 and 7 show creep curves for as-received and laser-treated Pb + 0.08% Ca and Pb + 1.4% Sb, respectively. The elongation occuring at the instant of loading, the instanta-

Table 5. Tensile test and hardness data on Pb and Pb-alloys

Material*	Treatment	UTS (MPa)	YS (MPa)	E (MPa)	Elongation at fracture (%)	Hardness Vickers (100 gm) kg/mm^2
Pb	as received	12.84	3.85	2898	33	88
Pb +1.4 Sb	as received	24.25	12.70	3174	51	125
	laser-treated	33.88	19.32	3795	20	207
Pb +2.5 Sb	as received	25.39	13.97	2415	49	227
	laser-treated	36.75	29.39	4485	25	237
Pb +0.08 Ca	as received	35.58	23.54	3243	36	110
	laser-treated	42.16	27.50	5382	25	150
Pb +1.4 Sb						
+0.2 Sn	as received	23.87	11.19	3174	47	148
+0.2 As	laser-treated	33.81	17.60	4347	25	247
Pb +2 Sb						
+0.2 Sn	as received	24.56	13.46	3519	49	135
+0.2 As	laser-treated	39.95	22.47	4761	38	237

*Compositions are in weight percent.

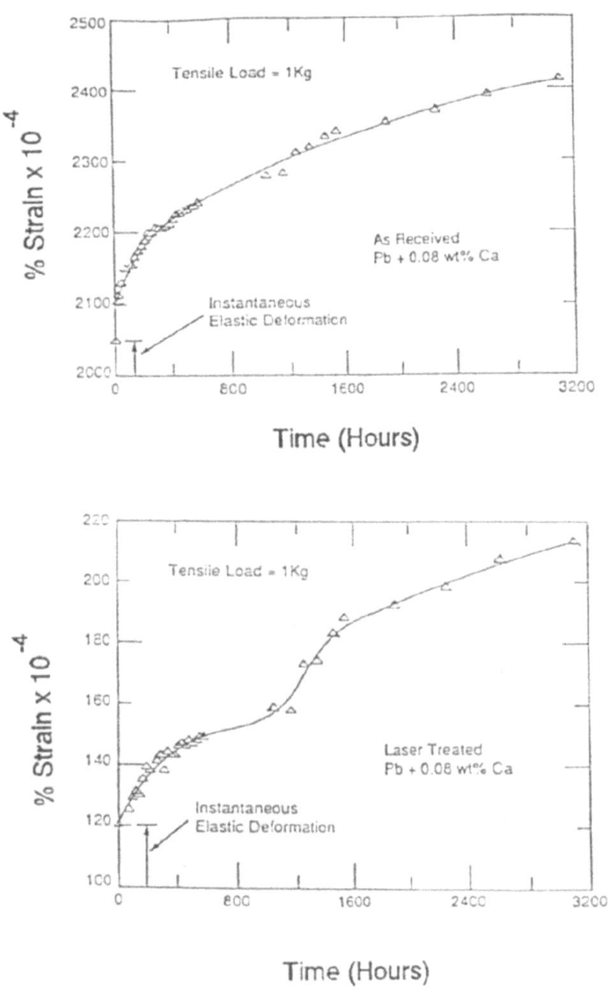

Figure 6. Creep curves for Pb + 0.08 wt% Ca alloy.

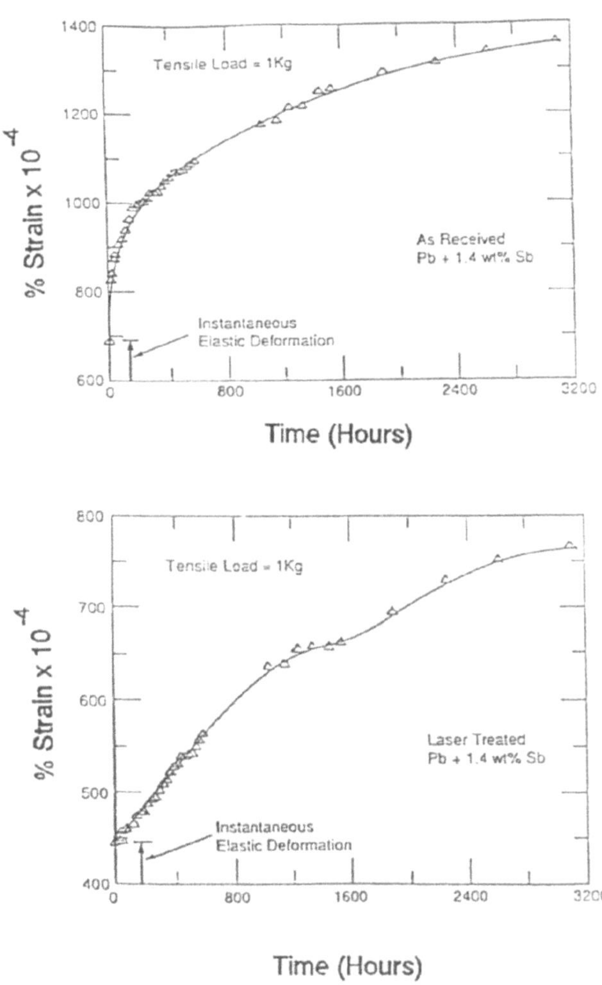

Figure 7. Creep curves for Pb + 1.4 wt% Sb alloy.

neous creep, consists mainly of elastic strain. The values of instantaneous creep for as-received Pb-alloys are lower than those for pure lead (2375×10^{-6}). The values of instantaneous creep for laser-treated Pb-alloys are about eight times lower than corresponding untreated alloys. The total primary creep (which follows instantaneous creep) values for laser-treated Pb-alloys are about eight to 12 times lower than those for untreated alloys. In some cases only primary creep was observed, indicating that the secondary creep was absent in these alloys.[4]

Alloying lead provides the strengthening mechanism via a combination of precipitation hardening (in Pb-Ca and Pb-Sb alloys) and solution strengthening (in Pb-Sb-Sn-As alloy). Further, as mentioned earlier and also observed in several laser materials interaction experiments,[6–8] the rapid solidification rate produces large undercoolings below liquidus temperature of alloys, and solidification occurs without rejection of solute (in present case: Ca, Sb, Sn, and As) in solvent (Pb). These solute atoms in solid solution are able to form atmospheres (Cottrell type atmosphere) around moving dislocations. Thus, the creep rate in laser-treated Pb-alloys decreases rapidly because of the restraining action of the dislocation-atmosphere interaction.

The creep resistance of precipitation alloys (such as Pb-Ca and Pb-Sb) are affected by particle size, their distribution, volume fraction of the second phase, and coherency stresses between particle and matrix.[9] Also, alloy systems and heat treatments that lead to formation of complex compounds are usually more stable than binary compounds.[9] The laser treatment of alloys results in formation of nonequilibrium metastable and complex phases, precipitates, and fine structures. This process also modifies the interface structure between matrix and particles.[8] Thus, based on earlier microstructural and x-ray diffractometry analysis, the same effects are assumed to occur in laser-treated Pb-alloys that resulted in a lower creep rate in the alloys.

Corrosion behavior

Figure 8 shows weight loss as a function of time in as-received and laser-treated Pb + 0.08% Ca and Pb + 2.5% Sb alloys. The topographical features of these alloys subjected to potentiostatic tests over a period of 60 days and after removal of the corrosion layers are illustrated in Photo 3.

A remarkable difference in weight-loss values between as-received and laser-treated Pb + 0.08% Ca was observed. The weight loss of the laser-treated sample was about 25% of that of the untreated sample during the test period (see Fig. 8a). X-ray diffractometry analysis of the corrosion layer revealed the presence of mainly α-PbO_2. After removing the corrosion layer from an untreated Pb-Ca alloy, several craters were observed (see Photo 3a), an indication of pitting corrosion. On the other hand, in the case of laser-treated Pb-Ca samples, the corrosion appeared to progress via formation of cracks along the grain boundary (see Photo 3b). This is an intergranular corrosion.

The weight loss in untreated Pb + 2.5% Sb was greater than that in the laser-treated alloy (see Fig. 8b) throughout the test period. The difference between the weight-loss values for untreated and laser-treated Pb + 2.5% Sb alloys was small compared to the weight-loss difference in Pb + 0.08% Ca alloys. In both untreated and laser-treated samples of Pb-Sb,

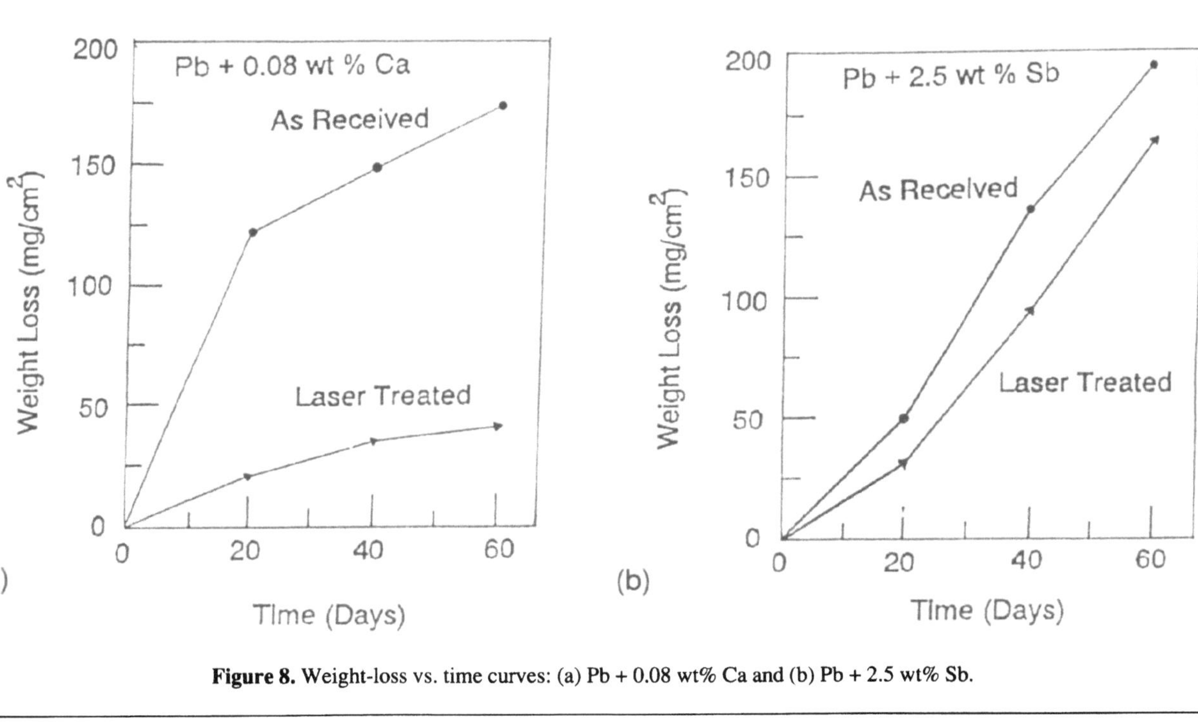

Figure 8. Weight-loss vs. time curves: (a) Pb + 0.08 wt% Ca and (b) Pb + 2.5 wt% Sb.

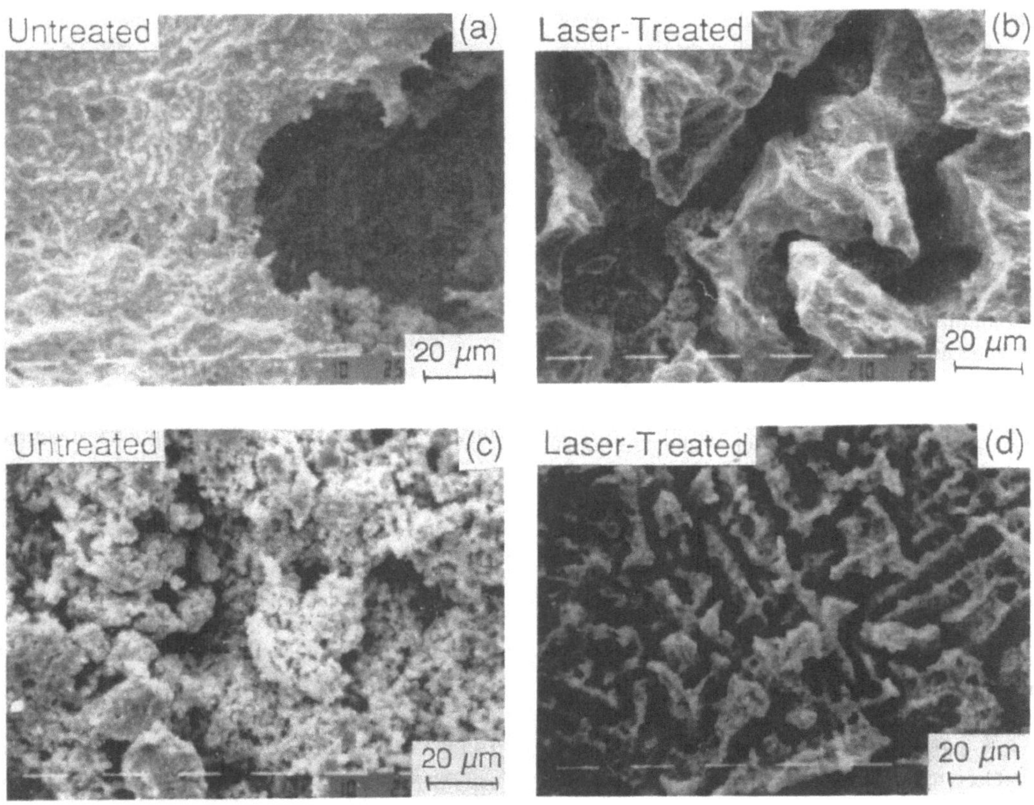

Photo 3. Scanning electron micrographs of Pb-alloys after removal of corrosion layer. (a) and (c) are for untreated Pb + 0.08 wt% Ca and Pb + 2.5 wt% Sb, respectively. (b) and (d) are for laser-treated Pb + 0.08 wt% Ca and Pb + 2.5 wt% Sb, respectively.

the corrosion layer was mainly made up of α-PbO_2. β-PbO_2 was also present in the corrosion layer of the untreated sample. Corrosion of the as-received Pb-Sb alloy produced pitting on the surface (see Photo 3c). The laser-treated Pb-Sb alloy showed a few large-diameter and shallow-depth craters as a result of the corrosion process (see Photo 3d). This effect was localized. Intergranular corrosion attack was also observed in laser-treated Pb-Sb alloys.

Conclusions

Laser treatment of conventional battery grid alloys (Pb-Ca, Pb-Sb, and Pb-Sb-Sn-As) has demonstrated its ability to produce improved battery grids compared to untreated alloys. The extremely high cooling rate and increased solid solubility associated with laser processing produced microstructures that have a heterogeneous distribution of unconventional and nonequilibrium phases. These microstructural changes produced enhanced mechanical properties (ultimate tensile strength, yield strength, and modulus of elasticity) and increased hardness of the Pb-alloys.

The laser process also improved the creep and corrosion resistance of the alloys. The values of instantaneous creep and primary creep for laser-treated Pb-alloys were much lower than those for corresponding as-received alloys. The corrosion test shows significant reduction in weight loss in laser-treated Pb-alloys compared to the untreated alloys. The corrosion mode in as-received alloys was pitting; in laser-treated alloys, it was intergranular. Based on these observations, laser processing of Pb-alloys holds great potential for battery-grid application.

Acknowledgments

The authors wish to express their gratitude for constant encouragement by Dr. Michael Kim at the International Lead Zinc Research Organization. This work was supported by the International Lead Zinc Research Organization Inc., under Contract No. ILZRO-LE-395. The authors also gratefully acknowledge the assistance of Santosh Gopinathan and William Stephens in conducting the experiments.

References

1. H.E. Howe, "The Creep Strengths of Lead and Lead Alloy," AIME Annual Meeting, Feb. 27–March 3, 1966, The Metallurgical Society of AIME, Reprint No. 2A-RF-5.7
2. M. Myers, "Process for Strengthening Lead-Antimony Alloys," US Patent No. 4,629,516, Dec. 1986.
3. N.B. Dahotre, M.H. McCay, and T.D. McCay, "Improved Lead-Alloy Grids by Laser Treatment," ILZRO Project LE-395, Progress Report No. 1, June 1990.
4. N.B. Dahotre, M.H. McCay, and T.D. McCay, "Improved Lead-Alloy Grids by Laser Treatment," ILZRO Project LE-395, Progress Report No. 2, December 1990.
5. Powder Diffraction File (Inorganic Section), Vol: 1-40, JCPDS-International Center for Diffraction Data, Pennsylvania, USA.
6. J. Singh and J. Mazumder, "Microstructure and Wear Properties of Laser Clad Fe-Cr-Mn-C Alloys," *Met. Trans 18A*, 313 (1987).
7. N.B. Dahotre, T.D. McCay, and M.H. McCay, "Laser Surface Modification of Zinc Base Composites," *J. of Metals 42*, 6 (1990).
8. N.B. Dahotre, M.H. McCay, T.D. McCay, S. Gopinathan, and L.F. Allard, "Pulse Laser Processing of SiC/Al-alloy Metal Matrix Composite," *J. Materials Research 6*, 3 (1991).
9. F. Garafalo, *Fundamentals of Creep and Creep Rupture in Metals*, The Macmillan Co., New York, 1966.

Soviet development of laser equipment for commercial applications

G.A. Baranov and V.V. Khukharev

Scientific Industrial Amalgamation Elektrofizika
Leningrad, USSR

Introduction

In the Soviet Union, development of industrial laser applications began in the mid-1960s. The earliest development efforts were directed toward the material-piercing properties of lasers; as a result, drilling applications were identified. Shortly thereafter, cutting applications evolved, and by the early 1970s surface-treatment and welding applications were being developed.

Since that early research period, many different types of industrial lasers have been developed in the USSR. Generalizations are risky, especially in a field this diverse. However, it is fair to say that the state of CO_2 laser development is similar in the US and USSR, though certainly there are differences in details; in some cases US industry may have a better solution to a technical problem, while in others Soviet industry may have a better solution.

Currently, several USSR manufacturers are producing a variety of industrial lasers. The applications for these lasers are similar to those in the US, although we believe that in the USSR there is much more emphasis on surface treatment and less on metal cutting than in the US.

The D.V. Efremov Scientific Research Institute of Electrophysical Apparatus (NIIEFA) was founded in 1945 in Leningrad to design, develop, and build electrophysical installations. These projects include charged particle accelerators and large installations to study the feasibility of controlled thermonuclear fusion reactions. In the USSR, NIIEFA is the main designer and producer of very-high-power gas lasers for industrial applications.

NIIEFA's staff of 3,500 includes 700 designers and technologists; 1,200 engineers, scientists, and laboratory technicians; 1,100 production workers, and 500 maintenance and service personnel. These people work in several design departments, an instrumentation department, and the production, maintenance, and service departments.

Numerous electrophysical installations now operating in the Soviet Union and other countries have been designed and manufactured by NIIEFA, including several types of accelerators, cyclotrons, superconducting coil electromagnets, and electromagnetic pumps. Specifically, the following major laser facilities have been developed by NIIEFA: Izhora, Slavyanka, Titan, and Maxim. Each of these is briefly described in the sections that follow.

Izhora

Developed several years ago, Izhora is a laboratory-type, self-sustained transverse discharge laser with output power up to 15 kW. It has been used regularly for small-scale industrial production as well as test projects. A general view of the machine is shown in Photo 1. The following description of Izhora is more detailed than the age of the machine might seem to justify, but it is useful in understanding the baseline from which the other NIIEFA lasers described here were developed.

Izhora's closed gas loop consists of two standard rotary vacuum pumps, two heat exchangers, the discharge chamber, and connecting ducts. The pumps and the optical system are installed on vibration isolation foundations to prevent pump vibrations from affecting the performance of the optics in delivering laser power to the workpiece. The vacuum pumps circulate the working gas mixture at specified rates through the laser discharge chamber.

Izhora's gas discharge chamber is designed for glow discharge in the laser working gas. Ballast resistor units ensure

Photo 1. Early version of a transverse-flow CO_2 laser with output to 15 kW.

the stability of discharge. The unstable resonator used in the Izhora installation is connected to the discharge zone in a design that guarantees a uniform distribution of radiation intensity around the circumference of the doughnut-shaped laser beam.

Stabilization of the power-supply current is accomplished by means of a feedback mechanism, which works by controlling a thyristor regulator through an electronic regulator. Output power can be controlled locally from the electronic stabilization regulator unit or remotely from the system control desk.

On-line control of the resonator shutter, supply current settings, application of high-voltage circuitry and oil switch is done from a system control desk, which is connected by cable to the control board. Except for initiating automatic shutdown, these operations cannot be controlled from the control board.

Conversion of the electric energy into laser radiation energy and removal of the latter out of the discharge region is accomplished by means of a three-pass unstable resonator. Such resonators have a number of advantages, allowing operations with large volumes of the active medium- and high-radiated power and, in this case, obtaining low divergence.

The laser radiation is directed by the output mirror into the gas-dynamic window, placed in its focal plane. The window simultaneously passes a laser beam through a hole in a diaphragm and assures that the pressure in the chamber is kept at a specified level. The gas make-up system is designed so that the fresh replenishment gas mixture is introduced through the orifice plates, while the depleted mixture is evacuated through the gas-dynamic window. Hence, it is necessary to maintain a stable flow of gas from the loop out through the gas seal. In turn, this requirement sizes, within limits, the diameter of the diaphragm hole and the capacity of the gas loop pumps.

The laser-beam delivery system is directed to a station for production work and technological testing or to a station with a rotary motion device or a coordinate transverse machine. Multi-axis capabilities have not been found necessary in this laboratory installation, although they could be incorporated readily if required.

Slavyanka

From the Izhora experience, the industrial laser system Slavyanka developed. This unit includes a 15-kW CO_2 laser, a beam-delivery and focusing subsystem, an automatic control subsystem, an electrical power supply subsystem for the discharge chamber and system actuators, and an optical manipulator based on a robot chassis.

Photo 2 shows a recent version of Slavyanka. This laser includes a gas-discharge chamber, closed gas loop with a pump and heat exchangers, ballast resistor unit, radiation forming device, and subsystems for gas injection, gas evacuation, and thermal stabilization of the mirrors.

Two alternative forms of the gas-discharge chamber have been developed: a metallic case chamber and a phenolic case chamber. Increased service life electrodes, producing no contamination of the laser working mixture, have been designed. The material of the electrode emitters will support up to a 150-mA current through the cathode element and is highly erosion-resistant.

Photo 2. Industrial version of a 15-kW transverse-flow CO_2 laser.

The construction of the gas loop and pump provide for excellent homogeneity of the gas-dynamic parameters of the working gas stream at the entrance to the gas discharge chamber, with a modest overall unit size.

A compact, water-cooled ballast resistor unit is mounted on the laser case. The metered gas-injection system is mounted on the gas loop of the laser and is quite easy to use. Laser radiation is emitted through a five-chamber, vacuum-type gas dynamic window, which was designed with the aid of numerical optimization techniques.

A Z-shaped, three-pass unstable telescopic resonator with a fourth amplification pass is used as a radiation-forming device. A resonator of this type is the most energy efficient and generates radiation with low divergence (<1.5 mrad), suitable for industrial applications. A special feature of this design is its versatility. It is possible to set up different resonators to form various configurations of output beam, and the laser can also be used as a radiation generator or amplifier.

The beam-delivery system includes accessory focusing lenses in the 25–100-cm range of focal lengths, provisions for beam splitting to support several work stations, beam alignment devices, and an in-line power meter.

A stabilized power supply system, using power semiconductor devices, has been developed for Slavyanka. The power supply provides for continuous and pulsed modes of operation. In the pulse mode, a wide range of pulse repetition rates (5–100 Hz) is available, in both manual and automatic control. Single-pulse operation is also available with continuous pulse duration adjustment of 0.005–0.1 s.

Control of each laser system is accomplished by a programmable controller independently or in accordance with commands from a central computer. The main functions of the automatic control system are automated start-up of the laser system, establishment of operation in a specified range, switching-off the laser system after finishing operations, and self-diagnostics and filing of faults during operation.

Photo 3. A 50-kW transverse-flow CO_2 laser with two gas-discharge channels.

Titan

A more recent follow-up to the Izhora/Slavyanka line has been developed and manufactured in NIIEFA: A 50-kW laboratory model laser named Titan (see Photo 3). At the time Titan was designed, the most developed concept for a continuous gas discharge CO_2 laser was one featuring a self-sustained transverse discharge in a chamber that incorporated a finely divided cathode structure with ballast resistors for controlling cathode currents. Although such a scheme has some disadvantages, it was chosen because it had been used with good results in a number of experimental, less-powerful installations.

In any case, we do not believe that it is best to increase laser-radiated power simply by increasing the dimensions of the device. Therefore, Titan is designed so its discharge volume consists of two identical gas-discharge chambers. The first contains a resonator generating laser radiation, while the second contains an amplifier.

Discharge in each of the chambers is initiated between partitioned knife cathodes and anodes. Voltage is fed to cathodes from the controllable thyristor converter through ballast resistors, which are housed in an oil-cooled unit.

The Titan radiation forming system consists of a resonator and amplifier, as indicated earlier, and a radiation extraction device. A three-pass unstable resonator—constructed in one of the discharge chambers and extended along the flow path —is formed by concave and convex mirrors. Flat mirrors are used for beam bending. Radiation, formed in the resonator, penetrates by means of the folding mirrors into the second chamber where, after several passes, it is amplified up to the desired level and then focused by means of mirrors into the gas-seal diaphragm.

Radiation, amplified in the amplifier, passes to a mirror and is focused in the diaphragm of the gas seal. That is, the output hole of the gas window is located in the focus of the last mirror. The structure of the seal includes a case, the first accessory diaphragm, the output diaphragm, attachment points, and a flange for connecting the pump-out pipeline. The seal is set directly on the resonator tank, and a diverging beam comes out of the gas window. An additional optical device is necessary to convert the diverging beam into a parallel beam that is suitable for further transmission.

The Titan laser design provides for the possibility of installing an independent resonator in each of the gas-discharge chambers. In this case, the device produces two independent beams that can be used simultaneously.

Maxim

Maxim—a $2 \times 2 \times 2$ m^3 electro-ionization (EI) CO_2 laser (see Fig. 1) designed in 1980 and now operating at NIIEFA—is the result of a different line of development.

Interest in the use of EI-CO_2 lasers arises from the possibility of obtaining high densities of laser radiation (on the order of 106 W/cm^2 without focusing) with good specific parameters (i.e., per unit of working-medium volume) and acceptable electro-optical efficiency, around 10%. Operating in the frequency-pulsed mode, the laser and gas mixture regeneration system can operate continuously for at least eight hours, at average powers up to 10 kW without replacement of the gas mixture. During this period, radiation divergence is kept at better than 1×10^{-3} radians. The use of a closed gas loop and a regeneration system minimizes the expenses related to working-gas mixture changes.

At the same time, the presence of complicated units and systems in the EI-CO_2 laser device (in particular, the electron accelerator with its beam-delivery device and associated biological shielding)—which results in the comparatively high cost of the laser as a whole—implies that the predominant use

Figure 1. This electron-beam ionizer-type laser can handle 10-kW power output.

of this type of laser will be in technical applications for which the peculiarities of the radiation produced by this type of laser are especially important. Among such peculiarities are high-radiation pulse densities (higher by three orders of magnitude than most pulsed lasers) and short pulse lengths (10–20 ms) in the frequency pulsed mode of operation.

Based on our experience, a modified EI-CO_2 laser, Maxim-2, has been designed and is currently being built by NIIEFA.

Applications

A number of practical industrial-material-processing applications have been accomplished by NIIEFA using the CO_2 lasers described in this article. The following exemplify the range of processing experiences.

Welding

Steels used in fabrications including stainless steel have been butt welded in thicknesses up to 12 mm. For example, using a 10-kW CO_2 laser with a beam focused to 0.45 mm, single-pass autogenous butt welds in 12 mm thick, 12X18H10T stainless steel was made at rates up to 15 mm/s. Joints with gap held to <0.1 mm were produced in 200-mm lengths.

Similarly, butt welds in titanium were produced with the same parameters except for the addition of a shielding gas to prevent oxidation. In these experiments, the heat-affected zone was held to less than 2 mm for the thickest section (12 mm). Successful butt welding of copper has been demonstrated in thicknesses up to 6 mm at rates up to 30 mm/s for piping used in the refrigeration industry.

NIIEFA has active development programs involving the use of filler metals in multipass welding of steel sections up to 100 mm thick. Also being studied is the welding of other dissimilar metal combinations: aluminum to steel, titanium to steel, aluminum to titanium, titanium to copper, and copper to aluminum. Of these, copper to stainless steel (see Photo 4) and aluminum to carbon steel have been successfully accomplished.

Cutting

High-power CO_2 laser cutting of thick-section ferrous and nonferrous metals has been successfully accomplished in the laboratory. For example, 60-mm thick steel was cut, using O_2 assist, at 7–12 mm/s using 18 kW of laser power. In these experiments, a 450-mm focal-length lens produced a 2–4-mm-wide kerf that exhibited striations of 1 mm depth. Although clinging dross was present, it was easily removed by mechanical means.

Copper up to 10 mm thick was cut with 18 kW of power at rates up to 20 mm/s. Using O_2 as the assist gas and a 450-mm focal-length lens, kerf widths of 1.0–1.5 mm and striations of 0.5 mm resulted. A thin oxide formation was removed by mechanical treatment.

Aluminum alloys in thicknesses up to 25 mm were also cut with 18 kW of power. Using N_2 as the assist gas, rates of 15–20 mm/s and a kerf width of 2 mm resulted, while the cut edges exhibited 0.5-mm-deep striations. Samples of marble up to 20 mm thick were cut using 6 kW of laser power with helium as the assist gas. There was no evidence of surface material spalling, and the cut edges were smooth.

As in the welding applications, some of the cutting work was experimental, while other work was performed for specific customer requests. NIIEFA is studying the cutting of steels up to 100 mm thick for the smelting industry and the use of high-power lasers to cut 400-mm-thick reinforced concrete.

Photo 4. Circumferential butt weld of copper to stainless steel at rates of 10–15 mm/sec.

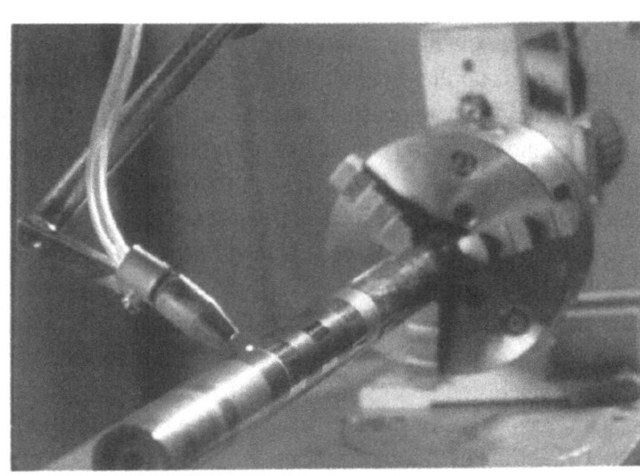

Photo 5. Surface treatment involving transformation hardening of carbon steel.

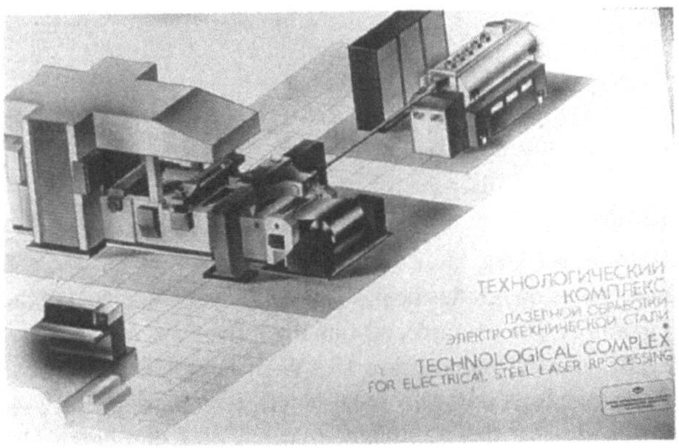

Figure 2. Two 10-kW CO_2 lasers arranged for grain orientation in strip steel.

Surface treatment

Transformation hardening of carbon and heat-resistant steels (see Photo 5) has also been investigated. Case depths of 0.01–0.5 mm at rates of 25–200 mm/s are typical. Using high-power scanned beams eliminates the need for absorbent coatings.

To harden steel saws using stone-cutting machine applications, a very deep case (about 6.0 mm) is required. While this depth can be achieved, surface melting is apparent. Other surface-hardening applications by NIIEFA include strengthening molds for hot-stamping turbine blades and hardening critical parts used in flame-cutting machines.

Other surface treatments that have been demonstrated include gas nitriding by flowing an alloy bearing gas over a laser-heated surface, and cladding wear-resistant layers on steel with dilution layers as shallow as 0.2–0.3 mm. In the latter operation, cladding material is fed as powder.

Continuing applications-development projects include further studies in wear-resistant and oxidation-proof coatings on ferrous and nonferrous metals, case hardening of steel, and copper cladding of steel surfaces. Also being studied are combination laser/plasma surface treatments.

One interesting piece of work being done with the EI-CO_2 laser is radiation treatment of anisotropic electric steels used in transformers. In the process of rolling these steels, large monocrystals (measuring 30 mm or more) are formed. To minimize operation energy losses due to remagnetization of these crystals, it is advisable to arrange them perpendicularly to the direction of magnetization. While there are several ways to do this, one of the most interesting uses laser radiation to produce thermodeformations by the localized heating of narrow-width (<0,3 mm) zones spaced at intervals of about 5 mm.

Similar technologies using solid-state lasers are being successfully applied now in the US and Japan. Germany used a CO_2 laser for this purpose in 1991.

Preliminary experiments at the Maxim facility demonstrated 6%–10% reductions in losses due to remagnetization of test samples. Estimates of the power requirements based on our experimental results suggest that two Maxim-2-type lasers will be sufficient for processing a 1-m-wide band, moving at up to 1 m/s (see Fig. 2).

Conclusion

In the course of developing and using these and other lasers, we have arrived independently at conclusions that are very similar to those reached by other investigators. For example, we share the view that laser cutting and welding are cost-effective in large-scale production processes. And our experience convinces us that higher power levels than are now being generally used in such processes will prove their worth in the next few years. Also, our experience suggests that the importance of surface treatment of metals—heat treating, alloying, cladding, etc.—will grow as very high-power lasers become more generally available.

We also expect a growing recognition of the utility of very high-power lasers in unique applications such as welding high-quality joints, including bimetallic joints, welding articles while preserving their magnetic properties, and cutting structures where radioactivity, noise, or other environmental limitations must be met.

Each application may be small, but collectively they can be a large market to be served by multipurpose complexes. Perhaps the growing number of laser job shops that has been remarked upon in recent industry meetings and in the trade press is a reflection of this.

Such a growing market convinced NIIEFA to build a new center for laser services in Leningrad to expand the services offered to European and American clients (see Fig. 3). The center (which was to be completed in mid-1991) will feature some multipurpose complexes centered on NIIEFA lasers described earlier. It will also include complexes centered on lower power lasers obtained from other manufacturers. Each multipurpose complex will include two or more stations at which workpieces are positioned so the laser beam can be switched from station to station to perform different functions on different workpieces as each is readied for treatment.

During the latter part of 1991 and into 1992, we expect our principal activities with American and European industries to be through such laser services. These services have been provided to Soviet clients for some years, giving considerable experience in tailoring basic laser capabilities to unique applications and then satisfying our clients' orders. The scale of the services will be great enough to serve a larger market when the new center is completed.

In summation, we believe that this prognosis for NIIEFA is a representative example of how Soviet industrial laser manufacturers will become increasingly involved in European, Asian, and American markets, by seeking opportunities

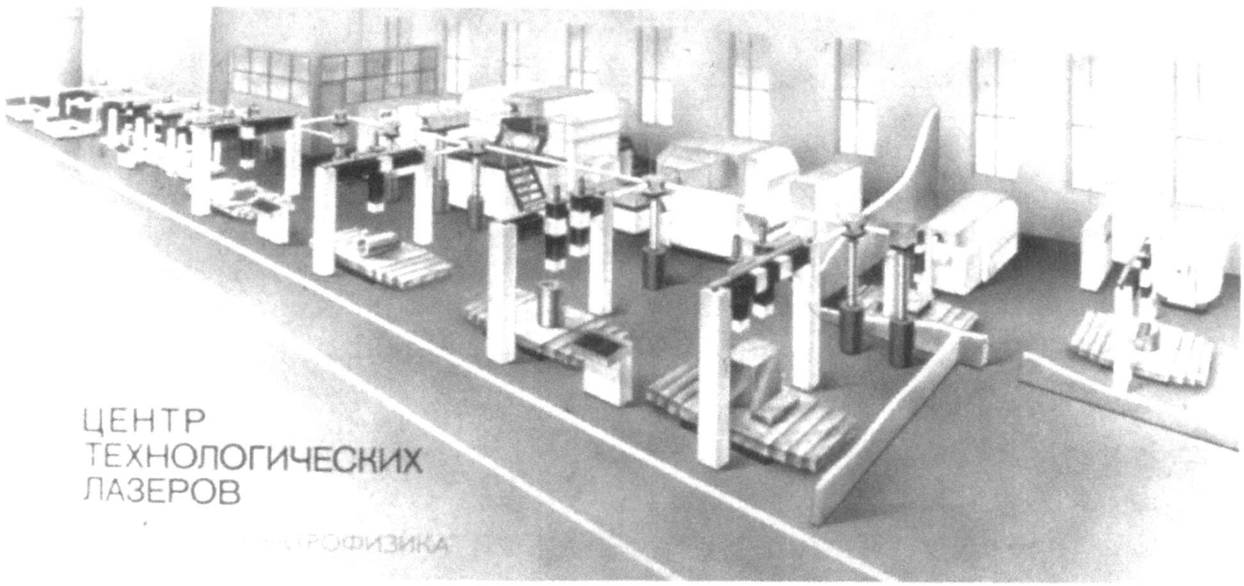

Figure 3. Rendering of NIIEFA's laser-processing center, which was scheduled to be completed in 1991.

to cooperate in the development, manufacture, and installation of highly effective, reliable, maintainable systems to serve the needs of worldwide users.

At the same time, we expect that Western laser manufacturers will find opportunities to become involved in Soviet markets. In particular, the demonstrated capabilities of Soviet organizations such as NIIEFA to produce laser components, lasers, and laser-based machine tools offer an important potential to manufacturing in the Soviet market and the global market.

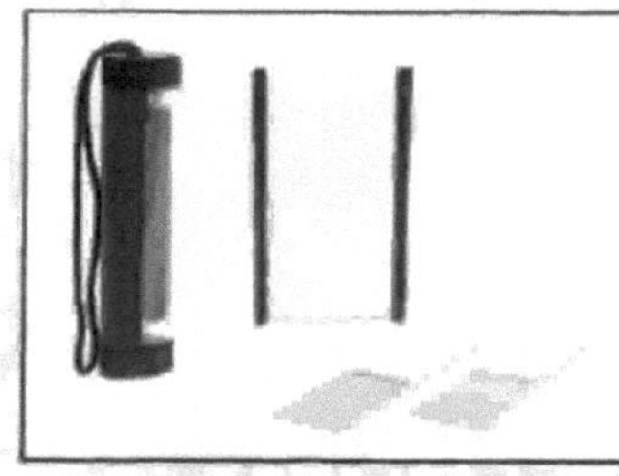
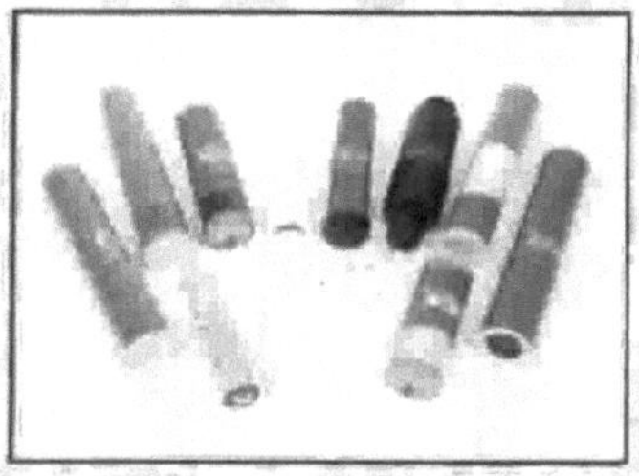

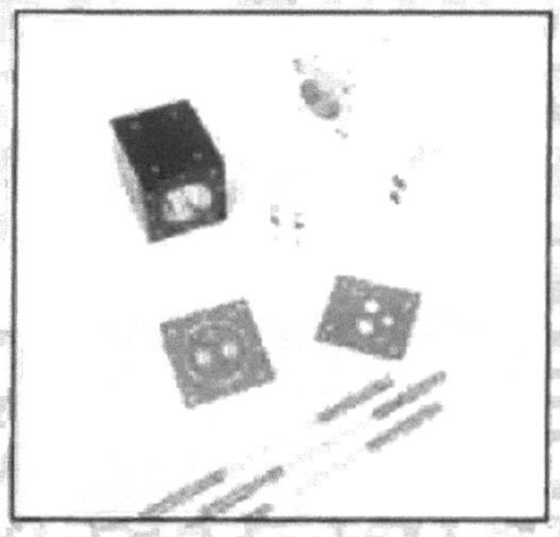

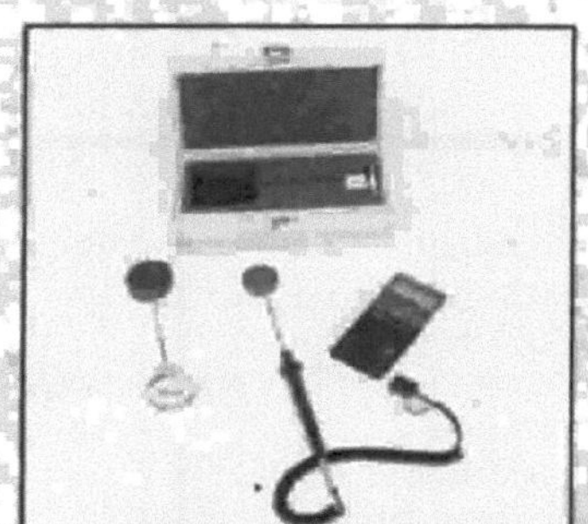

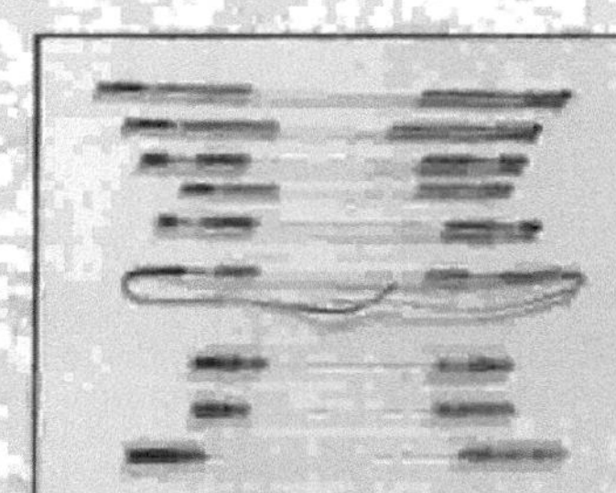

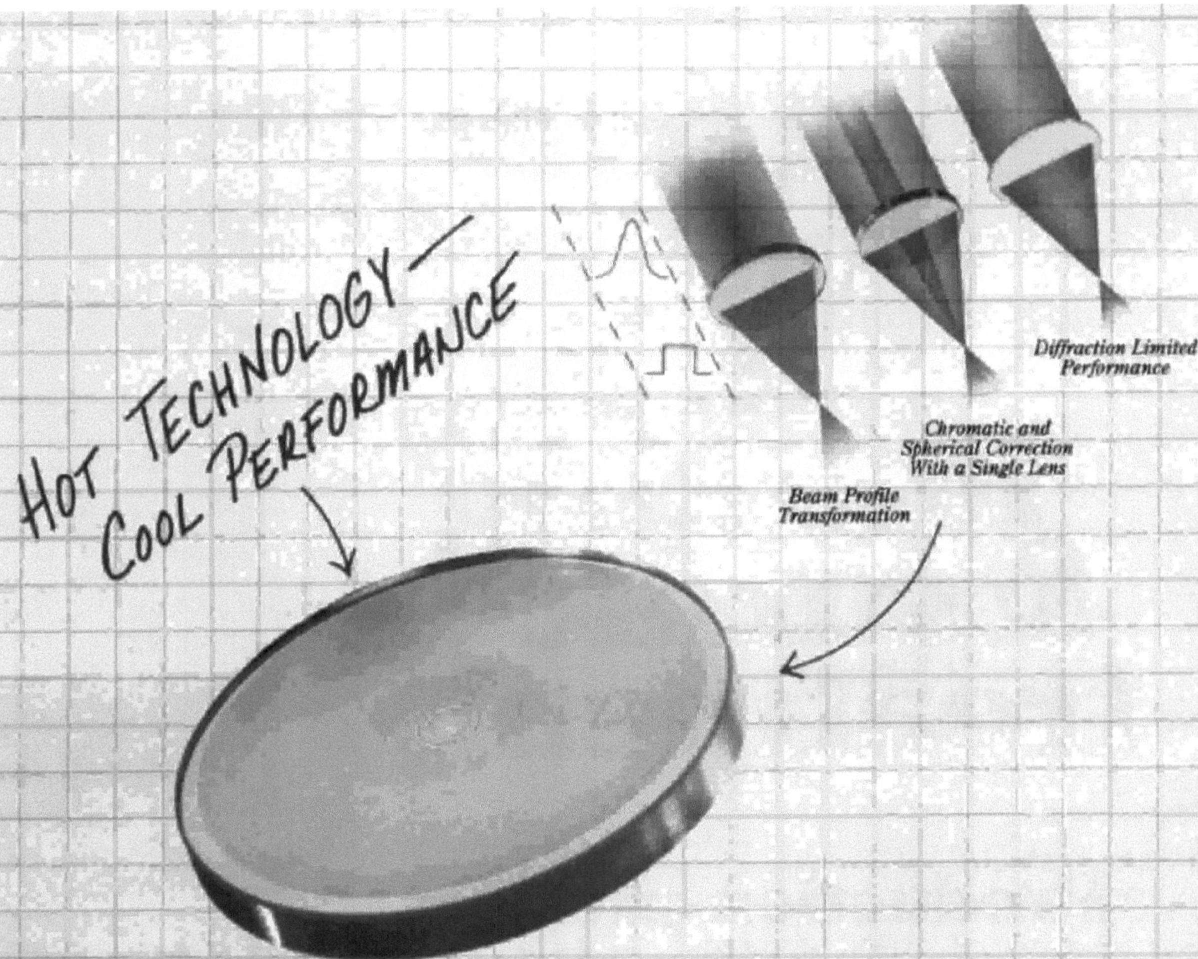
HOT TECHNOLOGY—
COOL PERFORMANCE
Diffraction Limited
Performance
Chromatic and
Spherical Correction
With a Single Lens
Beam Profile
Transformation

SECTION 3

Company & product directory

Questionnaires requesting company and product information were sent to more than 1600 companies, including more than 500 companies that provide lasers and systems. Responses received from about 30% of this segment included all of the major suppliers. Overall, a 27% questionnaire response was achieved. The 1992–1993 directory encompasses over 900 companies and organizations that sell products and services to the industrial market.

For completeness, the editors have chosen to repeat directory information on selected companies that appeared in the 1990 edition, even though revised information was not submitted. However, only those manufacturers that returned questionnaires for this edition have their equipment specifications listed. The manufacturers portion of the directory includes more than 300 company listings.

Once again, the editors have chosen to list U.S. and non-U.S. contract service (job shop) companies separately. Included in the listings are a number of contract shops known to the editors that did not return questionnaires. While far from complete, the directory represents the single most comprehensive listing of contract services available. This list includes a total of more than 650 shops.

Laser & system manufacturers

The following section lists companies that provide lasers and laser systems for industrial applications. Each company listed in the following specification chart has a more detailed listing in the company directory immediately following the chart.

For completeness, the editors have included listings for companies that did not return an updated questionnaire. An asterisk (*) after the address will identify these companies.

The editors strive to ensure that the information reported is current and correct. In an effort to present the most comprehensive directory of contract service companies, the editors have included information from some companies that did not provide updated information for this edition. These companies are noted by an asterisk (*) after their address.

MANUFACTURERS (CHART PAGE 1)

	Description		Lasers			Laser Systems to Perform												Motions	
	Laser Mfg.	System Mfg.	Solid-State	CO2	Excimer	Welding	Cutting (Metal)	Cutting (Nonmetal)	Surface Treatment	Drilling (Metal)	Drilling (Nonmetal)	Scribe/Engrave	Machining	Soldering	Marking	Resistor Trimming	Semiconductor	2-Axes	Multi-Axes
A-B Lasers Inc Acton MA 508-635-9100	•	•	•	•		•	•	•	•	•	•	•		•	•				
Acro Automation Systems Inc Milwaukee WI 414-352-4540		•				•							•						
Adapt Automation Inc Santa Ana CA 714-662-4454		•						•	•				•	•	•			•	•
Adlas GmbH & Co KG Luebeck Germany 0451-39 09300*	•		•																
Adron Sources S.A. Cestas France 33-56 78 81 81*	•		•	•		•	•	•		•	•	•			•				
Advanced Laser Systems Inc Waltham MA 617-891-8418	•	•	•	•		•	•	•		•	•			•	•				
Advanced Production Automation Veldhoven Netherlands 40-542445		•				•	•	•		•	•	•	•	•	•	•	•		
Advanced Technologies Inc Bay City MI 517-686-0676		•				•	•	•		•	•	•			•			•	•
Aerotech Inc Pittsburgh PA 412-963-7470		•																•	•
AGIE Laser AG Dietlikon Switzerland 01-835-9711		•					•	•		•	•		•					•	•
Aims Optronics S.A.-N.V. Kraainem Belgium (02) 731 04 40		•				•	•	•	•	•	•	•	•	•	•	•	•		
Alessi Inc Irvine CA 714-830-0660*		•														•	•		
Alltec GmbH & Co KG Luebeck Germany 49-451-3909350	•	•		•											•				
Amada Co Ltd Isehara Japan 463-96-3401	•	•		•			•												
Amchem Co Inc Woburn MA 617-938-0700		•				•	•			•									
Amoco Laser Co Naperville IL 708-961-8400	•		•																
Anorad Corp Hauppauge NY 516-231-1995		•				•	•	•		•		•	•		•	•	•		
Applied Laser Engineering Ltd Hersham, Surrey UK 0932 222260		•										•						•	•

MANUFACTURERS
(CHART PAGE 2)

	De-scrip-tion		Lasers			Laser Systems to Perform												Mo-tions	
	Laser Mfg.	System Mfg.	Solid-State	CO2	Excimer	Welding	Cutting (Metal)	Cutting (Nonmetal)	Surface Treatment	Drilling (Metal)	Drilling (Nonmetal)	Scribe/Engrave	Machining	Soldering	Marking	Resistor Trimming	Semiconductor	2-Axes	Multi-Axes
K.N. Aronson Inc Arcade NY 716-492-2400		•		•			•	•		•	•	•			•				
ASI Robotic Systems Clawson MI 313-288-5070		•					•	•											
Atlantic Machine Tools Inc Houston TX 713-445-3985		•					•	•										•	
Avimo Ltd UK 0823-331071	•	•		•		•	•	•		•	•	•						•	
Baasel Lasertech Starnberg Germany 08151/776-0	•	•	•	•		•	•	•	•	•	•	•	•	•	•	•	•		
C. Behrens AG Alfeld/Leine Germany 05181-75270*		•		•			•	•											
C. Behrens Machinery Co Inc Danvers MA 508-774-4200		•					•	•					•					•	•
Benchmark Industries Inc Goffstown NH 603-627-8484		•				•	•			•		•		•					•
Bihler of America Inc North Branch NJ		•		•		•	•												
BKLT Lasersystemtechnik GmbH Bad Aibling Germany 08061-6064		•				•	•	•					•					•	•
B.M. Industries Evry France 33-(1) 64 97 53 54	•	•	•			•	•			•			•						
Bridgeport Machines Inc Bridgeport CT 203-367-3651*		•	•	•		•	•			•			•						
Bystronic Inc Hauppauge NY 516-231-1212	•	•		•		•	•	•					•					•	•
Bystronic Laser Ltd Niederoenz Switzerland (41) 63 60 12 20	•	•		•		•	•	•	•				•					•	•
California Machine Tool Campbell CA 714-272-4687	•	•	•	•	•	•	•	•				•	•		•	•		•	•
Cheval Freres SA Ecole-Besancon France (33) 81 53 75 33	•	•	•			•	•	•		•	•	•	•	•	•				
Chicago Laser Systems Inc Chicago IL 312-282-2710		•	•													•		•	•
Chromatron Laser Systems GmbH Berlin 15 Germany 030-882 79 03	•	•	•			•	•	•		•	•	•		•	•	•			

MANUFACTURERS (CHART PAGE 3)

	Description		Lasers			Laser Systems to Perform												Motions	
	Laser Mfg.	System Mfg.	Solid-State	CO2	Excimer	Welding	Cutting (Metal)	Cutting (Nonmetal)	Surface Treatment	Drilling (Metal)	Drilling (Nonmetal)	Scribe/Engrave	Machining	Soldering	Marking	Resistor Trimming	Semiconductor	2-Axes	Multi-Axes
Cimcorp Precision Systems Inc St. Paul MN 612-484-7261		•																	•
Cincinnati Inc Cincinnati OH 513-367-7569*		•		•			•	•											
Coherent General Inc Sturbridge MA 508-347-2681	•	•	•	•		•	•	•	•	•	•	•		•			•		
Coherent Hull Ltd Hull UK 0482-831154	•	•		•				•			•	•			•				
Comau SpA Grugliasco Torino Italy (11) 3334-1*		•		•		•	•	•											
CONTEC GmbH Darmstadt 13 Germany 06151/5 40 77-79*		•				•	•	•	•	•	•	•	•	•	•		•		
Contek SpA Varallo, VC Italy 0163-54154		•				•	•	•	•						•				
Controllaser Orlando FL 407-438-2500	•	•	•			•	•	•		•	•	•	•	•	•	•	•	•	•
Control Micro Systems Inc Orlando FL 407-679-9716		•		•				•			•	•		•	•			•	•
Coteglade Photonics Ltd Swansea UK 0792 52931*		•	•				•	•		•	•	•							
CRILASER SA Barcelona Spain 345 74 40	•	•		•		•	•	•		•	•	•			•	•			
Cryophysics Ltd Witney, Oxford UK 993-773681	•	•		•											•				
Cutting Edge Inc Marblehead MA 617-631-1390		•		•				•										•	
Cymer Laser Tech San Diego CA 619-549-6760	•				•														
Daewoo Heavy Industries Ltd Changwon South Korea (0551) 80-4396		•		•		•	•	•	•	•		•	•		•				•
Daihen Corp Settsu City Osaka Japan 06-317-2521*	•	•		•		•	•	•	•	•	•	•							
Design Technologies Ltd Wellingborough UK 933-278871		•					•	•		•	•								
Directed Energy Inc Irvine CA 714-553-1225		•		•				•							•				

MANUFACTURERS (CHART PAGE 4)

MANUFACTURERS	Description		Lasers			Laser Systems to Perform												Motions	
	Laser Mfg.	System Mfg.	Solid-State	CO2	Excimer	Welding	Cutting (Metal)	Cutting (Nonmetal)	Surface Treatment	Drilling (Metal)	Drilling (Nonmetal)	Scribe/Engrave	Machining	Soldering	Marking	Resistor Trimming	Semiconductor	2-Axes	Multi-Axes
DKNE Design Services Inc Manchester CT 203-649-8068		•				•	•	•		•	•	•	•		•			•	•
Doerries Scharmann GmbH Mechernich 14 Germany 02484 12-0		•				•	•	•	•	•	•		•					•	•
Dorotek GmbH Berlin 42 Germany (030) 7850056	•	•	•					•			•				•	•		•	•
DTM Corp Austin TX 512-339-2922		•																	
DUAP Engineering Biel/Brugg Switzerland 032 53 29 26*		•				•	•	•					•						
Eberhard GmbH & Co KG Schierbach Germany 07021-7274-0						•													
Edinburgh Instruments Ltd Edinburgh UK 031 449 5844	•	•		•										•		•	•		
Electro Scientific Industries Inc (ESI) Portland OR 503-641-4141		•				•	•	•		•	•	•			•	•	•		
Electrox Inc Indianapolis IN 317-248-2632*	•	•	•	•		•	•	•	•	•	•	•	•		•	•			
Electrox Ltd Hitchin, Herts UK 0462 834848	•		•	•											•				
Elettronica Valseriana Fiorano Al Serio Italy 0039 035 712661*	•	•	•			•				•	•	•			•				
Emco Maier 5400 Hallein/Taxach Austria 06245-81551-260		•		•			•	•										•	
ESAB Group Heusenstamm Germany 06104-6431	•	•		•		•	•	•	•	•	•		•						•
Exatron San Jose CA 408-629-7600		•										•			•				
Fagor-Arrasate S. Coop. Mondragon (Guipuzcoa) Spain 34-43-792011*		•				•	•	•	•	•	•				•				
Fanuc Ltd Yamanashi Pref. Japan 0555-84-5555	•	•		•		•	•												
Florod Corp Gardena CA 213-532-2700	•	•	•		•	•	•			•	•	•	•	•	•	•	•		•
Fox Laboratories Danbury CT 203-791-1503	•	•	•					•			•	•	•		•	•	•		•

MANUFACTURERS
(CHART PAGE 5)

Manufacturer	Description		Lasers			Laser Systems to Perform												Motions						
	Laser Mfg.	System Mfg.	Solid-State	CO2	Excimer	Welding	Cutting (Metal)	Cutting (Nonmetal)	Surface Treatment	Drilling (Metal)	Drilling (Nonmetal)	Scribe/Engrave	Machining	Soldering	Marking	Resistor Trimming	Semiconductor	2-Axes	Multi-Axes					
Fuji Tool And Die Co Ltd Shizuoka Japan 0559-77-2300*		•		•			•																	
GE Fanuc Automation Charlottesville VA 804-978-5443	•			•																				
General Laser Inc Scottsdale AZ 602-991-0674	•	•		•				•			•	•			•									
General Scanning Inc Watertown MA 617-924-1010		•													•									
Gentec Inc St. Foy, Quebec Canada 418-651-8000*	•	•		•																				
GMFanuc Robotics Corp Auburn Hills MI 313-377-7000		•	•	•		•	•	•	•	•	•								•					
Golti S Corporation E. Elgoibar, Gulpuzcoa Spain 943-74-03-50		•		•			•											•	•					
D.A. Griffin Corp West Seneca NY 716-674-2300		•				•	•	•					•											
Group II Manufacturing Ltd Plainview NY 516-694-1334		•						•			•	•			•									
GSR Technologies Ltd Edmonton, Alberta Canada 403-451-9000*	•	•		•			•	•		•		•												
Haas Laser GmbH Schramberg Germany 07422-515-0	•	•	•			•	•	•	•	•	•	•		•	•									
Haco NV Rumbeke-Roeselare Belgium 051-220318		•				•	•	•	•	•	•	•	•	•	•				•					
Hahn & Kolb (GB) Ltd Rugby UK (0788) 577288		•				•	•	•	•	•	•	•	•		•									
Harland Simon Automation Systems Ltd Coventry UK 0203 473748		•				•	•	•		•	•	•	•	•	•	•		•	•					
Hauser Motion Control Inc Pittsburgh PA 412-835-9240		•				•	•	•		•		•			•			•	•					
HendrickSaw Inc Salem MA 508-745-5222		•										•												
Hoya Corp Shinjuku-ku, Tokyo Japan *	•		•																					
S.E. Huffman Corp Clover SC 803-222-4561		•				•	•	•	•	•	•		•											

MANUFACTURERS
(CHART PAGE 6)

	Description		Lasers			Laser Systems to Perform												Motions	
	Laser Mfg.	System Mfg.	Solid-State	CO2	Excimer	Welding	Cutting (Metal)	Cutting (Nonmetal)	Surface Treatment	Drilling (Metal)	Drilling (Nonmetal)	Scribe/Engrave	Machining	Soldering	Marking	Resistor Trimming	Semiconductor	2-Axes	Multi-Axes
Hughes Aircraft Co Carlsbad CA 619-931-3214	•			•															
IEF Werner GmbH Furtwangen Germany 07723-650-10														•	•			•	•
Image Micro Systems Inc Billerica MA 508-663-7070		•			•				•		•				•		•		•
Industrial Laser Technology Indianapolis IN 317-353-1333*		•																	
InTA Santa Clara CA 408-748-9955		•				•	•	•		•	•		•						
International Lasersmiths Compton CA 213-635-8536		•				•	•	•	•	•	•		•		•				
Isomet Corp Springfield VA 703-321-8301		•				•				•			•		•				
Jamieson Mfg Co Inc Torrington CT 203-482-6543		•				•	•			•			•						
JEC Lasers Inc Saddle Brook NJ 201-843-6600	•	•	•									•			•				
Jens Scheel Sondermaschinen GmbH Itzehoe Germany 4821 74062*		•	•	•				•											
Jipee Canada Technologies Inc Chambly, Quebec Canada 514-447-3473		•		•			•											•	
JRM International Inc Rockford IL 815-397-8515												•			•				
Jyoti Ltd, PO Gujarat India 82851-2-3*		•		•															
Karl Binder GmbH Reichertshofen Germany 8453-8015*		•		•		•	•	•											
Kitamura Machinery Co Ltd Toyama Japan 0766-63-1100*		•		•			•												
Komatsu Ltd Tokyo Japan 03-5561-2692		•		•			•												
Kuka Schweissanlagen & Roboter GmbH Augsburg 43 Germany (0821)797-0*		•	•	•		•	•												
Lambda Physik GmbH Goettingen Germany 49 551 69380	•				•				•						•				

MANUFACTURERS (CHART PAGE 7)

	De-scrip-tion		Lasers			Laser Systems to Perform												Mo-tions	
	Laser Mfg.	System Mfg.	Solid-State	CO2	Excimer	Welding	Cutting (Metal)	Cutting (Nonmetal)	Surface Treatment	Drilling (Metal)	Drilling (Nonmetal)	Scribe/Engrave	Machining	Soldering	Marking	Resistor Trimming	Semiconductor	2-Axes	Multi-Axes
Lambda Physik Inc Acton MA 508-263-1100	•	•			•					•	•	•			•		•		
Lasag Corp Arlington Heights IL 708-593-3021	•	•	•			•	•	•		•	•		•						
Lasag AG Thun Switzerland 033/224522	•	•	•			•	•	•		•	•		•						
Laser Application SA Baume-les Dames France 81.84.24.44*	•	•	•			•				•			•						
Lasercut Inc N. Branford CT 203-488-0031		•		•			•	•	•	•									
Lasercut SA Neuchatel Switzerland 038 25.97.86*	•	•	•	•			•	•											
Laser Dynamics Ltd Queensland Australia 075 396644	•	•	•	•			•	•				•							
Laserdyne Eden Prairie MN 612-941-9530		•				•	•	•	•	•	•	•	•	•	•			•	•
Laser Ecosse Ltd Dundee UK 0382 833377	•			•															
Laser Electronics Pty Ltd Queensland Australia (075)73-2066	•	•	•	•			•	•							•			•	•
Laser Lab Ltd Cheltenham, Victoria Australia 5849900*		•				•	•	•							•				
Laser Machining Inc Somerset WI 715-247-3285	•	•	•	•		•	•	•	•	•	•	•	•	•	•			•	•
Laserman Northridge CA 818-363-5485		•						•			•			•		•	•		•
Laser Material Processing Ltd (LMP Ltd) Nidau (Biel) Switzerland 032/51 67 77	•	•	•				•	•			•	•			•	•	•		
Laser Mechanisms Inc Farmington Hills MI 313-474-9480		•																	
Laser Nucleonics Inc W. Palm Beach FL 407-686-6867	•		•	•		•				•						•			
Laser Quanta S.A. Madrid Spain (1) 803.44.44	•	•		•		•	•	•		•	•	•	•		•	•	•	•	•
Lasers Industriels SA Nevers Cedex France (33) 86 59 01 44	•			•		•	•	•	•	•	•								

MANUFACTURERS (CHART PAGE 8)

	Description		Lasers			Laser Systems to Perform												Motions	
	Laser Mfg.	System Mfg.	Solid-State	CO2	Excimer	Welding	Cutting (Metal)	Cutting (Nonmetal)	Surface Treatment	Drilling (Metal)	Drilling (Nonmetal)	Scribe/Engrave	Machining	Soldering	Marking	Resistor Trimming	Semiconductor	2-Axes	Multi-Axes
Lasertechnics Inc Albuquerque NM 505-822-1123	•	•		•	•										•				
Laser Technology Inc So. Lyon MI 313-437-7625	•	•		•					•										
Laser Valfivre GmbH Furstenfeldbruck Germany 8141 24034	•	•	•	•		•	•		•										
Laser Valfivre Sorgenti e Sistemi SpA Florence Italy 055-439831*	•	•		•			•	•			•	•							
Laser-Work AG Pfungen Switzerland 052/31 17 21*		•				•	•	•		•			•						
Lectra Systemes SA Cestas/Bordeaux France 56 68 80 00							•	•		•	•								
Lectra Systems Inc Marietta GA		•		•				•											
Lee Laser Inc Orlando FL 407-422-2476	•		•																
Lillbackan Konepaja Kauhava Finland +358 644 830111		•					•												
Line Lite Laser Corp Mountain View CA 415-969-4900	•	•	•	•				•				•		•	•				
Lite AG Neuenhof Switzerland 056-86 20 04		•	•												•				
Lasercomb Weilheim/Teck Germany 7023-70090		•				•	•	•											
LPT Laser Physiktechnik GmbH Berndorf Austria 02672-3488-0		•				•	•	•		•	•	•	•		•	•	•		
LPT Laser Physiktechnik GmbH Sauerlach/Arget Germany 08104-1077		•				•	•	•		•	•	•	•		•	•	•		
Lumonics Corp Camarillo CA 805-987-2211	•														•				
Lumonics Corp Livonia MI 313-591-0101	•	•	•	•	•	•	•	•	•	•	•	•	•	•	•		•		
Lumonics Inc Kanata, Ontario Canada 613-592-1460	•	•		•	•			•			•		•		•		•		
Lumonics Ltd Rugby, Warwickshire UK (0788) 570321	•	•	•			•	•	•	•	•	•	•	•	•	•		•		

MANUFACTURERS (CHART PAGE 9)

	Description		Lasers			Laser Systems to Perform												Motions	
	Laser Mfg.	System Mfg.	Solid-State	CO2	Excimer	Welding	Cutting (Metal)	Cutting (Nonmetal)	Surface Treatment	Drilling (Metal)	Drilling (Nonmetal)	Scribe/Engrave	Machining	Soldering	Marking	Resistor Trimming	Semiconductor	2-Axes	Multi-Axes
LVD Company N.v. Wevelgem-gullegem Belgium 056/43 0511	•	•		•		•	•											•	•
Madrid Laser Madrid Spain (1) 332 4340						•	•	•	•	•	•	•			•				
Maho Pfronten Germany 8363-89-0		•					•	•		•		•	•						•
Manfred Fohrenbach GmbH Unadingen Germany 07707-159-0		•				•	•	•		•	•	•		•	•			•	•
Martek Lasers Livermore CA 415-294-8167	•		•			•	•	•	•										
Martin Pfaffen Laser Lengnau Switzerland (0041) (065) 529077		•				•	•	•		•	•	•	•		•	•		•	•
Maschinenfabrik GmbH Nordrach Germany 7838/84-0															•				
Maschinenfabrik GmbH & CO KG Halblech Germany 08368-18-0	•	•		•		•	•												•
Matsushita Ind Equip Co Ltd Toyanaka, Osaka Japan 06-866-8580*	•	•		•		•	•												
Mauser-Werke Oberndorf GmbH Oberndorf Germany 07423/700		•				•	•	•	•				•						•
Mazak Nissho Iwai Corp Schaumburg IL 708-882-8777		•		•		•	•	•				•							
MBB Ind Products Group Munchen Germany (089) 60731228	•	•	•	•		•	•	•	•					•				•	•
Meadex Technologies Springfield MA 413-567-3680		•				•	•	•		•	•	•	•		•	•			
Mechanical Technology Inc Latham NY 518-785-2211							•			•									
Meftech Inc Sunnyvale CA 408-745-1072		•														•			
Melco Industries Denver CO 303-457-1234*		•																	
Melles Griot Irvine CA 714-261-5600				•															
Melles Griot Carlsbad CA 619-438-2131	•			•															

MANUFACTURERS
(CHART PAGE 10)

	De-scrip-tion		Lasers			Laser Systems to Perform												Mo-tions	
	Laser Mfg.	System Mfg.	Solid-State	CO2	Excimer	Welding	Cutting (Metal)	Cutting (Nonmetal)	Surface Treatment	Drilling (Metal)	Drilling (Nonmetal)	Scribe/Engrave	Machining	Soldering	Marking	Resistor Trimming	Semiconductor	2-Axes	Multi-Axes
Meltec Laser AB Sandviken Sweden +46 26257401 *		•	•	•		•	•	•					•		•				
Merkle Schweissmachinenbau GmbH Kotz I Germany						•													
Messer Griesheim GmbH Puchheim Germany 089/80 92 0		•				•	•	•	•	•	•								•
Micro Controle Evry Cedex France 33-1-64 97 98 98 *	•		•																
Micromanipulator Co Inc Carson City NV 702-882-2400		•															•		•
Midwest Laser Systems Inc Findlay OH 419-424-0062 *		•				•	•	•		•	•								
Mitsubishi Electric Corp Tokyo Japan 03-218-3426 *	•	•		•		•	•	•	•	•					•				
Mitsubishi International Corp Wood Dale IL 708-860-4210 *	•	•		•		•	•	•	•	•	•		•						
Miyachi Technos Corp Chiba-pref Japan 0471-25-6177	•	•	•			•	•	•	•	•	•	•	•	•	•				
MLI Lasers Ltd Tel Aviv Israel (03) 492511	•	•		•		•	•	•	•	•	•				•				
MLS Laser Systems Inc Rockaway NJ 201-627-7787	•	•	•	•		•	•	•	•	•	•	•			•	•		•	•
mls munich laser systems gmbh Munchen 2 Germany 089/502 89 51	•	•	•			•	•	•	•	•	•	•	•		•	•			•
Modern Machine Tool Co Jackson MI 517-788-9120		•					•	•										•	•
Motoman Inc West Carrollton OH 513-847-6200*		•					•												
MPB Technologies Inc Dorval, Quebec Canada 514-683-1490	•	•		•		•											•		
M. Torres Disenos Industriales S.A. Torres de Elorz Spain 048-31-7811		•				•	•	•	•	•	•		•	•					
MTS Systems Corp Eden Prairie MN 612-937-4000		•		•		•	•	•											•
Nanavati Sales Pvt Ltd Bombay India 91-22-6147588/6146319				•		•	•	•	•	•		•	•			•			

MANUFACTURERS (CHART PAGE 11)

	Description		Lasers			Laser Systems to Perform												Motions	
	Laser Mfg.	System Mfg.	Solid-State	CO2	Excimer	Welding	Cutting (Metal)	Cutting (Nonmetal)	Surface Treatment	Drilling (Metal)	Drilling (Nonmetal)	Scribe/Engrave	Machining	Soldering	Marking	Resistor Trimming	Semiconductor	2-Axes	Multi-Axes
NEC Corp Kanagawa Japan 0427 73-1111 *	•	•	•			•	•			•									
NEC Electronics Inc Mountain View CA 415-965-6116	•	•	•			•	•	•				•			•	•	•		
Newall Manufacturing Technology Ltd W. Yorkshire UK 0535 667911		•	•	•		•	•			•									•
Newcor Bay City Bay City MI 517-893-9505 *		•																	
New England Laser Processing Danvers MA 508-774-4626																		•	•
Niigata Engineering Co Ltd Tokyo Japan 03-504-2151		•		•		•	•	•	•	•			•						•
Nippei Toyama Corp Minato-ku, Tokyo Japan (03)3434-8272		•				•	•	•							•				
Nisshinbo Industries Inc Okazaki, Aichi Japan 81-564-55-1101*		•		•			•	•											
NM Laser Products Inc Sunnyvale CA 408-733-1520		•																	
North China Research Institute of Electro-Optics Beijing PRC 4362761 Beijing	•	•	•			•	•	•		•		•		•		•		•	
Nothelfer GmbH Ravensburg Germany (0751) 886-247		•				•	•												
NTC/Marubeni America Corp Southfield MI 313-353-7060/800-3976427		•		•		•	•	•	•	•	•		•					•	•
NVL Balliu MTC Sint-Amandsberg Belgium 091/28.13.86						•	•	•	•				•		•			•	•
OPL (Oerlikon- PRC Laser SA) Gland Switzerland 011-41-22-64 19 35	•			•															
Optical Engineering Inc Santa Rosa CA 707-528-1080	•	•		•		•	•					•							
Optomic Lasers Ltd Migdal Ha'Emek Israel 972 6 545440	•			•															
Otto Borries KG Leinfelden Germany 0711 7 59 09 0		•													•				
Pacific Precision Labs Inc Chatsworth CA 818-700-8977		•				•	•	•		•	•	•	•	•			•		
Pacific Trinetics Corp San Marcos CA 619-471-2350		•						•			•							•	•

MANUFACTURERS
(CHART PAGE 12)

Manufacturer	Description: Laser Mfg.	Description: System Mfg.	Lasers: Solid-State	Lasers: CO2	Lasers: Excimer	Laser Systems to Perform: Welding	Cutting (Metal)	Cutting (Nonmetal)	Surface Treatment	Drilling (Metal)	Drilling (Nonmetal)	Scribe/Engrave	Machining	Soldering	Marking	Resistor Trimming	Semiconductor	Motions: 2-Axes	Multi-Axes
Panasonic Factory Automation Co Franklin Park IL 708-452-2500	•			•															
PermaNova Lasersystem AB Molndal Sweden +46 (0)31 870280		•	•	•	•	•	•			•		•		•	•				•
Permascand AB, Laser Div Ljungaverk Sweden 46-691-32940		•				•	•	•	•	•	•	•	•	•	•		•	•	•
Potomac Photonics Inc Lanham MD 301-459-3031	•				•														
PRC Corp Landing NJ 201-347-0100	•			•															
Precision Systems Ltd St. Ives, Cambridge UK 0480 67101		•				•													•
Prima Industrie Spa di Collegno-Torino Italy 011 410.31		•				•	•	•											•
Progressive Tool & Industries Co Southfield MI 313-353-8888		•				•	•												
PTR - Precision Technologies Inc Enfield CT 203-741-2281		•		•		•	•		•	•			•						
Pulse Systems Inc Los Alamos NM 505-672-1926	•	•		•				•	•						•				
Quanta System Srl Milan Italy 02 332 00239	•	•	•				•	•		•		•			•				
Quantel SA Les Ulis France 33 1 69 29 17 00	•	•	•	•		•	•	•	•	•	•								
Quantronix Corp Smithtown NY 516-273-6900	•	•	•			•	•			•		•	•	•	•		•		
Quantum Chromodynamics Inc Torrance CA 213-320-5717		•										•			•	•			
Quantum Laser Engineering Ltd Coventry UK 0203 680668		•				•	•	•							•				
Questek Inc Billerica MA 508-667-6790	•				•														
Raskin Machines SA Cheseaux Switzerland 21/731 94 00		•		•			•	•		•	•		•						•
Raskin USA Buford GA 404-271-7558	•	•		•		•	•	•	•			•	•					•	•

MANUFACTURERS (CHART PAGE 13)

	Description		Lasers			Laser Systems to Perform												Motions	
	Laser Mfg.	System Mfg.	Solid-State	CO2	Excimer	Welding	Cutting (Metal)	Cutting (Nonmetal)	Surface Treatment	Drilling (Metal)	Drilling (Nonmetal)	Scribe/Engrave	Machining	Soldering	Marking	Resistor Trimming	Semiconductor	2-Axes	Multi-Axes
Raycon Corp Ann Arbor MI 313-677-4911		•				•	•	•	•	•	•		•						•
Raytheon Co, Laser Products Quincy MA 617-479-5300	•	•	•	•		•	•	•		•		•	•	•	•			•	•
Reiner Kist Lasertechnik GmbH Eckental Germany 49-(0)-9126-8026		•										•							
Reis GmbH & Co Obernburg Germany 06022/503-0		•				•	•												•
Renault Automation Le Chesnay Cedex France (1)39.54.91.91*		•				•	•												
Resonetics Nashua NH 603-886-6772 *		•			•				•	•	•	•	•		•				
Robomatix International Inc Farmington Hills MI 313-471-6121		•				•	•	•							•				•
Rofin-Sinar Inc Plymouth MI 313-455-5400	•	•	•	•	•										•				
Rofin-Sinar Laser GmbH Hamburg 74 Germany 49 (0) 40 73 36 30	•	•	•	•	•										•				
Rouchaud Industrie Limoges Cedex France 55-341734 *		•		•		•	•	•	•				•						
RtMc Phoenix AZ 602-493-9343		•					•	•			•	•					•	•	
Safmatic Parthenay Cedex France 33-49943155		•	•	•		•	•	•	•										
Sala SpA Levico Italy 0461/701388		•					•	•											
Geo T. Schmidt Inc Niles IL 708-647-7117		•	•												•				
Sciaky Industries Cedex France 1-45 73 43 00		•	•	•		•	•												•
Sciaky Inc, Ferranti Chicago IL 708-594-3800		•		•		•	•	•	•	•	•	•			•				
Servo Robot Inc Boucherville, Quebec Canada 514-655-4223		•				•	•	•											
Shibuya Kogyo Co Ltd Kanazawa Japan 0762-62-1200	•	•		•	•	•	•	•		•	•	•			•		•	•	•

MANUFACTURERS (CHART PAGE 14)

MANUFACTURERS	Description: Laser Mfg.	Description: System Mfg.	Lasers: Solid-State	Lasers: CO2	Lasers: Excimer	Laser Systems to Perform: Welding	Cutting (Metal)	Cutting (Nonmetal)	Surface Treatment	Drilling (Metal)	Drilling (Nonmetal)	Scribe/Engrave	Machining	Soldering	Marking	Resistor Trimming	Semiconductor	Motions: 2-Axes	Multi-Axes
Spectra Gases Irvington NJ 201-372-2060	•																		
Spectron Laser Systems Rugby/Warcs UK (788) 544694	•		•																
Strippit Inc Akron NY 716-542-4511		•		•			•	•				•							
Synrad Inc Bothell WA 206-483-6100	•		•	•															
Tay Technology Inc Melville NY 516-351-1080		•													•	•		•	•
3D Systems Inc Valencia CA 805-295-5600		•																	
Thyssen Laser-Technik GmbH Aachen Germany 0241-8906-160		•				•	•	•	•	•	•	•			•				
Toeller Laserkraft GmbH Hagenburg Germany 05033-7786		•					•	•											
Toshiba Corp Mie-Pref Japan 0593-54-6047	•	•	•	•	•	•	•	•	•			•		•	•				
Toyama America (NTC) Arlington Heights IL 708-640-0940		•		•		•	•	•		•	•				•				
Tri Sigma Corp Tucson AZ 602-294-5300		•	•	•	•	•	•	•		•	•	•	•						
TRUMPF GmbH & Co Ditzingen Germany 07156-303-0	•	•		•		•	•	•	•				•					•	•
TRUMPF Inc Farmington CT 203-677-9741	•	•		•		•	•	•	•				•					•	•
Tungsram Laser Technology Inc Budapest Hungary 36 1 1696-619/1600-233*	•	•		•		•	•	•	•	•	•		•						
Utilase Blank Welding Technologies Detriot MI 313-521-2488		•				•	•	•											
United Technologies Corp So. Windsor CT 203-282-4200	•			•		•	•		•	•									
Universal Laser Systems Inc Scottsdale AZ 602-483-1214		•						•				•		•	•				
Universal Voltronics Mt. Kisco NY 914-241-1300	•			•															

MANUFACTURERS
(CHART PAGE 15)

	Description		Lasers			Laser Systems to Perform												Motions	
	Laser Mfg.	System Mfg.	Solid-State	CO2	Excimer	Welding	Cutting (Metal)	Cutting (Nonmetal)	Surface Treatment	Drilling (Metal)	Drilling (Nonmetal)	Scribe/Engrave	Machining	Soldering	Marking	Resistor Trimming	Semiconductor	2-Axes	Multi-Axes
Uranit GmbH Julich Germany 02461/65-388	•	•		•					•						•				
US Amada Ltd Buena Park CA 714-739-2111		•				•	•	•				•	•					•	•
US Laser Corp Wyckoff NJ 201-848-9200	•	•	•			•	•		•	•		•	•	•		•	•		
Utilase Engineering Inc Detroit MI 313-521-2488		•				•	•	•	•	•	•	•	•		•				
Utilase Systems Inc Detroit MI 313-521-2488		•				•	•	•	•	•	•	•	•		•				
Vanzetti Systems Inc Stoughton MA 617-828-4650		•												•					
Vero Precision Eng Ltd Southhampton UK*		•		•		•	•												
Waldrich Siegen Werkzeugmaschinen GmbH Burbach Germany 2736-400		•																	
W.A. Whitney Italia Torino Italy 011-470-27-02							•											•	
Weidmuller Sensor Systems McHenry IL 815-344-4141						•	•	•											
Wustefeld Maschinenbau GmbH Mehren/Eifel Germany 06592/556		•				•	•	•		•	•							•	•
WA Whitney Co Rockford IL 815-964-6771		•		•			•												
Wiedemann Division King of Prussia PA 215-265-2000 *		•		•			•	•											
XMR Inc Santa Clara CA 408-988-2426	•	•			•		•	•	•	•	•	•			•		•		
ZED Instruments Ltd Hersham, Surrey UK 0932-228977		•										•							

LASER & SYSTEM MANUFACTURERS DIRECTORY

A-B Lasers Inc, 4 Craig Rd, Acton, MA 01720; 508-635-9100, FAX 508-635-9199
ceo/pres, Dr. Charles B McGough; vp sls/mktg, O.J. Jones; vp tech, Patrick Schlather; vp ops, Al Hamelin; prod mgr-DPY, Gerhard Marcinkowski; 1985
Manufactures Nd:YAG lasers and CO_2 TEA lasers based on industrial marking systems, Nd:YAG laser soldering systems, micromachining systems, and CO_2 laser material processing systems.

ACEC Energie S.A., Rue Cahpelle Beaussart, 80, Charleroi, 6030, Belgium; 32.71.44.3287, FAX 32.71.43.35.46
plant mgr, A.M. Hannecart; 1988
Specializes in laminations for electric motors and generators.

Acro Automation Systems Inc, 2900 W. Green Tree Rd, Milwaukee, WI 53209; 414-352-4540, FAX 414-352-1609
pres/ceo, Cliff Loomis; vp/cfo, Mike Puhl; vp sls/mktg, Guy Castleberry; sr vp tech, Dino Giacomini; 1936
Provides custom designed, automated welding and assembly machinery systems for the automotive, applicance, electric motor, and electrical equipment industries. Specializing in semi-automatic machinery as well as high speed assembly applications. Single-source responsibility provided.

Adapt Automation Inc, 2020 S. Hathaway, Santa Ana, CA 92705; 714-662-4454, FAX 714-662-0838
pres, C.A. Van Mechelen; vp, T.B. Van Mechelen; mktg dir, Bruce Tilden; 1971
Offers design, development and fabrication of industrial systems and equipment.

Adlas GmbH & Co KG, Seelandstrasse 67, Luebeck, D-2400, Germany; 0451-39 09300*, FAX 0451-39 09 399
pres, Dr. B. Steyer; gen mgr, Dr. H.P. Kortz; sls mgr, J. Reingruber; emp 20, 1986
Diode-laser-pumped solid-state lasers.

Adron Sources S.A., Chemin de Marticot - B.P.M., Cestas, F-33610, France; 33-56 78 81 81*, FAX 33-56 78.14.82
indus mgr, Louis Martinez; r&d mgr, Herve Coutard; sls mgr, J. Pierre Lepez; emp 32, 1988
Develops, manufactures, and sells CO_2 and YAG lasers, and high pressure water jet systems. CO_2 lasers range from 50 W to 3000 W; YAG lasers from 130 W, 250 W, and 450 W.

Advanced Laser Systems Inc, PO Box 543, Waltham, MA 02254; 617-891-8418
pres, George Zahaykevich; emp 5, 1982
Manufactures precision welding and drilling laser systems, computer-controlled. Maintains an applications laboratory for research & development.

Advanced Production Automation, Industrieweg 6, PO Box 200, Veldhoven, NL 5500, Netherlands; 40-542445, FAX 40-545635
sls/mktg mgr, Gerard Gerritsen; acct mgr, Coen Nieuwenstein; 1981
Manufactures production equipment and systems for widely differing industries. Specialties are product development and design and assembly of production installations. Applied technologies are: mechanical/electronic control software; optics/laser technology; opto-mechanics; vision technology.

Advanced Technologies Inc, 2490 E. Midland Rd, Bay City, MI 48706; 517-686-0676, FAX 517-686-7560, telex 164310
pres, Thomas V. DeAgostino; ex vp/gen mgr/sls/mktg, Edward A. Cooke; eng mgr, John Wetzel; emp 75, 1967
Assembles and integrates automated laser welding systems. Laser welding systems manufactured by the company incorporate the latest in laser welding power supplies, YAG to high power CO_2. All of these systems can be designed to operate independently or as part of an integrated system.

AGIE Laser AG, (sub of AGIE Holding Ag), Neue Winterthurerstrasse 30, Dietlikon, CH-8305, Switzerland; 01-835-9711, FAX 01-833-5663
mgr, Giorgio Rigamonti; 1991
Manufactures AL2 laser machine. A precision cutting, drilling and welding system.

Aims Optronics S.A.-N.V., Rue F. Kinnenstraat 30, Kraainem, B-1950, Belgium; (02) 731 04 40, FAX (02) 731 89 18, telex 61501
gen mgr, L. DeSchutter; mktg mgr, G. Merckaert; tech mgr, M. Denis; admin mgr, H. Salah; 1973
Manufactures lasers and optical systems, all machine vision and smart sensors. Visual inspection systems: encoders, fiberscopes, fiberoptic interface systems.

Alessi Inc, div of Dukane Corp, 35 Parker, Irvine, CA 92718; 714-830-0660*, FAX 714-830-0953, telex 8304191
pres, Al Spriet; sls mgr, Don Miller; r&d dir, Gary Hunt; 1967
Manufactures a variety of laser cutting systems for integrated circuits. These lasers mount directly to the optics on analytical probing stations and are used in trimming, debugging, or failure analysis operations requiring localized passivation removal or circuit isolation.

Alltec GmbH & Co KG, Seelandstrase 67, Luebeck, D-2400, Germany; 49-451-3909350, FAX 49-451-3909399
pres, Dr. Bernhard Steyer; gen mgr, Dr. Ernst Albers; sls mgrs, Dr. Frank Marnitz, Dipl.-Vw. Peter Frahm; mfg mgr, Dipl.-Ing. Holger Tank; 1985
Manufactures pulsed CO_2 lasers and marking systems based on all types of pulsed lasers.

Amada Co Ltd, Laser Machine Sales Div, 200 Ishida, Isehara, 259-11, Japan; 463-96-3401, telex 3882311
1985
Manufactures laser sheet-metal cutting systems.

Amchem Co Inc, 155 N. New Boston St, Woburn, MA 01801; 617-938-0700, FAX 617-935-8395
vp oper, J.P.A. Haigh; west coast oper, G. LeGrove; 1968
Laser systems builder and integrator of equipment for drilling, cutting, and welding applications. Skills and expertise, demonstrated for many years on EDM and ECM equipment, are successfully applied to laser systems. Systems incorporate closed-loop control via various on machine-inspection capabilities wherever required.

Amoco Laser Co, (sub of Amoco Technology Co), 1251 Frontenac Rd, Naperville, IL 60563; 708-961-8400, FAX 708-369-4299
mfg dir, Chuck Crowder; mktg dir, Dennis Werth; r&d dir, John Clark; 1987
Manufactures small, efficient diode laser pumped, solid-state microlasers. Infrared and visible light products available in CW, Q-switch or single frequency. Contract development capabilities in these and related fields also available.

Anorad Corp

Anorad Corp, 110 Oser Ave, Hauppauge, NY 11788; 516-231-1995, FAX 516-435-1612
vp, mfg mgr, Stanley Squires; sls/mktg mgr, Richard Scherer; intn'l sls mgr, Amir Chitayat; dir eng, Mustan Faizullabhoy; emp 300, 1987
Manufactures precision linear, rotary, and tilting rotary positioning tables for the laser machine industry. These modular positioning elements feature closed loop dc servo drives that utilize a variety of position sensors, including laser interferometers. Anomatic series of contouring CNC' s provide high performance positioning and control of laser functions.

Anritsu Corp, 5-10-27 Minamiazabu, 5-chome, Minato-ku, Tokyo, 106, Japan
1987

Applied Laser Engineering Ltd, 328 Molesey Rd, Hersham, Surrey, KT 12 3PD, UK; 0932 222260, FAX 0932 222271
engrg dir, Ed Birch; 1987
Designs and manufactures systems and instrumentation utilizing both high and low power lasers, especially as associated with the printing industry. Also offers expertise in optical, mechanical, electronic, and software design and manufacture.

K.N. Aronson Inc, 635 Main St, Arcade, NY 14009; 716-492-2400, FAX 716-457-3517, telex 710-278-1491
1946
Offers a wide range of computer numerical controlled *x-y* coordinate plate cutting machines. Various models are offered with 4, 5, 6, 8, 10, and 12' cutting widths. Also offers a variety of cutting processes such as laser, plasma, oxy-fuel, waterjet, router, and knife cutting.

Asahi Seiki Manufacturing Co Ltd, 5050-1 Shindeno Asahimaecho, Owari-Asaki-Cho, Aichi-Ken, 488, Japan; 05615-3-311
1946

ASI Robotic Systems, 1250 Crooks Rd, Clawson, MI 48017; 313-288-5070, FAX 313-288-9330
pres, John Cargill; gen mgr, Mike Radeke; sls/mktg mgr, Dick LeBlanc; emp 1970, 1946
Manufacturer of gantry style systems for 6 axis of motion. X axis table style cell. Precision and standard accurate machines available. System is built with a constant tool point feature for the rotary axis.

Atlantic Machine Tools Inc, 11629 N. Houston Rosslyn Rd, Houston, TX 77086; 713-445-3985, FAX 713-445-3989
1976
Provides top quality machines utilizing American standard electrical, hydraulic, and tooling components such as Allen Bradley, Rexroth, Parker, and Telemecanique. Offers state-of-the-art CNC and DNC machining facilities producing machines in a highly productive and cost effective manner.

Avebury Circle Ltd, Deal Ave, Slough, Berks, SL1 4SH, UK
1976

Avimo Ltd, Photon Laser Technology Div, Lisieux Way, Taunton, TA1 2JZ, UK; 0823-331071, FAX 0823-274413, telex 46126
prod mgr, Dr. Stuart Howells; sls/mktg mgr, Jeffery J. Harris; 1937
Manufacturer of CW CO_2 lasers with output powers from 50 - 2000 W. Also manufactures integrated CNC laser dual-axis workstations for welding, cutting, drilling, and scribing processes. Consultancy services and feasibility studies for all laser and electro-optic applications.

Baasel Lasertech, Petersbrunner Stasse IB, Starnberg, D-8130, Germany; 08151/776-0, FAX 08151-776159, telex 526486 cblas d
pres/gen mgr, Carl F. Baasel; eng mgr, Dr. L. Langhans; mfg mgr, R. Schloss; sls/mktg mgr, F.G. Meyer; emp 300, 1975
Manufactures laser systems for marking, scribing, drilling, cutting, welding, soldering, engraving, and perforating. Also manufactures OEM-YAG-lasers, OEM-CO_2 lasers, YAG-lasers for scientific applications, and medical lasers. Offers development, construction, production, installation and service performances. Sale of laser components and spare parts. Provides laser academy.

C. Behrens AG, Postfach 1340, Alfeld/Leine, 322, Germany; 05181-75270*, FAX 05181-75300, TWX 8-11-92959
vp eng, Walter Tamaschke; 1858
Manufactures sheet-metal cutting systems; CNC laser cutting systems (fixed laser station, precise contouring); CNC laser/punch combinations (flexible cells and/or integrated in FMS with existing machinery).

C. Behrens Machinery Co Inc, (sub of C. Behrens AG), Danvers Industrial Park, Danvers, MA 01923; 508-774-4200, FAX 508-774-0532, telex 286804
pres, Richard Jacob; vp, Dr. Joachim Sahm; emp 13, 1974
Manufactures CNC turret punch presses, laser sheet-metal cutting systems, and flexible fabricating systems including automatic material storage and retrieving.

Benchmark Industries Inc, 215 St. Anselm's Dr, Goffstown, NH 03045; 603-627-8484, FAX 603-627-6788
pres, William L. Duschatko; chf eng, R.G. Linares; 1982
Manufactures turnkey hermetic welding systems providing laser, glove box, vacuum ovens, and motion control systems. All components are under dedicated computer control. Also manufactures custom designed fully automated systems including all parts handling.

Beveler Machines Deutschland GmbH, Robert Bosch Strasse 5, Koln 71, D-5000, Germany
1982

Bihler of America Inc, PO Box 5038, North Branch, NJ 08876
gen mgr, V.M. Schoen; 1982
Integrates lasers into metal-cutting and welding systems.

BKLT Lasersystemtechnik GmbH, Ludwig-Thoma Strasse 5, Bad Aibling, D-8202, Germany; 08061-6064, FAX 08061-30723
M. Krimpmann, A. Bernhard; 1982
Manufactures CO_2 lasers and laser systems.

B.M. Industries, Z.I. du Bois Chaland, 7, rue du Bois Chaland, Evry, F-91000, France; 33-(1) 64 97 53 54, FAX 33-(1) 64 97 52 03, telex 603671 F
pres, G. Brassart; sls mgr, G. Riboulet; mfg mgr, R. Garslian; tech mgr, J.P. Treton; 1984
Offers a complete range of equipment and support tailored to the needs of industrial manufacturers. CO_2 laser beam equipment: mirror mounts, laser beam switches, beam steering devices, CO_2 beam display systems, mirror focusing head, lens focusing heads, laser beam shaping optical systems, multi-stations optical systems, custom tailored optical systems. Industrial YAG lasers: YAG beam delivery optics, optical fibers systems.

Borries Marking Systems USA, 3135 S. State St #300, Ann Arbor, MI 48108; 313-761-9549, FAX 313-761-1171
1984

Bridgeport Machines Inc, 500 Lindley St, Bridgeport, CT 06606; 203-367-3651*, FAX 203-335-0454
vp sls/mktg, Michael S. LaMonica; dir mktg, Richard D. Jones; emp 1000, 1938
Manufactures a full range of vertical laser machining centers utilizing both solid-state and CO_2 laser technology.

BYSTRONIC

(sub of Bystronic Machine Ltd), 30 Commerce Dr, Hauppauge, NY 11788; 516-231-1212, FAX 516-231-1040
vp, U. Troesch; sls/mktg mgr, K. Barrette; mfg mgr, D. Pymm; 1975
Complete laser cutting systems including machine with automatic change table, loading system resonator, CNC control with normal interface, and programming station. Totally automatic laser systems including automatic part removal sheet storage and retrieval.

Bystronic Laser Ltd, Industriestrasse 21, Niederoenz, CH-3362, Switzerland; (41) 63 60 12 20, FAX (41) 63 61 67 37
pres, E. Zunstein; sls/mktg mgr, U. Singer; mfg mgr, A. Horisbeger; eng mgr, U. Hunziker; 1986
Manufactures laser cutting systems, including CO_2 high-power lasers with 1000, 1500 and 2300 W respectively. BYLAS for efficient cutting of flat material with exchange table system and flying optics. BYFLEX for versatile laser processing: cutting and welding of tubes (round or rectangular) and many other profiles. BYSMALL for small to medium size shops to laser process flat material and tubes.

California Machine Tool, Mitsubishi Laser Div, 1153 San Tomas Aquino Rd, Campbell, CA 95008; 714-272-4687, FAX 714-272-4688
pres, Glenn Kline; gen mgr, Steve Hackett; 1987
Distributes and sells Mitsubushi laser machines and Weidmuller sensor systems to manufactures and O.E.M.' s.

Carl Zeiss, Postfach 1380, Carl Zeiss Strasse, Oherkochen, D-7082, Germany
1987

Cheval Freres SA, Rue des Bosquets, Ecole-Besancon, F-25045, France; (33) 81 53 75 33, FAX (33) 81 53 72 39, telex 361162
gen mgr, Michel Bezin; eng mgr, Herve Picaud; mfg mgr, Joseph Simplot; sls/mktg mgr, Didier Cheval; exp mgr, Joachim Vogt; 1848
Manufactures Nd:YAG lasers, machines, automation systems incorporating the laser, special machines to customer specification, laser machining centers, laser machining centers with glove-box and controlled atmosphere. Lasers and machines are used for drilling, cutting, scribing, welding and engraving. Power: 10 - 900 W.

Chicago Laser Systems Inc, 4034 N. Nashville Ave, Chicago, IL 60634; 312-282-2710, FAX 312-282-8455, telex 6711269 Laser UW
1976
Standard and wide area, high-speed production laser trim systems used worldwide for precision active and passive laser trimming of hybrid circuits.

Chromatron Laser Systems GmbH, Olivaer Platz 16, Berlin 15, D-1000, Germany; 030-882 79 03, FAX 030-883 12 56
tech dir, A. Kuntze; mktg dir, T. Klesse; 1986
Specializes in the development, production, and selling of Nd:YAG lasers (pulsed or CW) and additional equipment. The company is mainly concerned with material processing, and major stronghold is laser marking with the moduler laser marking system LAS WRITE 3000. Products include a laser drilling system for drilling very small holes and air-cooled high-power YAG lasers. Services include applications and technical consulting.

Cimcorp Precision Systems Inc, 899 W. Hwy 96, St. Paul, MN 55126; 612-484-7261, FAX 612-483-2689
pres, Matti Korpinen; prod mgr, Al Sturm; eng mgr, Dan Kedrowski; acct mgr, Bo Billingham; 1961
Manufacturer of large gantry robots with controls.

Cincinnati Inc, 7420 Kilby Rd, PO Box 11111, Cincinnati, OH 45211; 513-367-7569*, FAX 513-367-7552
mgr prod/sls R.L. Kloczkowski; emp 600, 1898
Manufactures laser sheet-metal cutting systems.

Cloos Schweisstechnik GmbH, Industriestrasse, Haiger, D-6342, Germany
1898

COHERENT GENERAL

1 Picker Rd, Sturbridge, MA 01566; 508-347-2681, FAX 508-347-5134
pres/ceo, Albert D. Battista; vp sls/mktg, William H. Shiner; dir of mktg, James P. Rutt; eng mgr, Randy Thompson; 1966
Manufactures CO_2, Nd:YAG, and Nd:glass lasers for industrial applications. Manufactures lasers for OEMs and laser systems for end users. Laser applications development center available for applications R&D work.

Coherent Hull Ltd, (sub of Coherent), Gothenburg Way, Suttonfields Ind Estate, Hull, HU8 0YE, UK; 0482-831154, FAX 0482-839233
pres, Dr. K.S. Lipton; sls/mktg mgr, A.J. Chambers; eng mgr, K. Clark; mfg mgr, A. Stacey; 1966
Manufactures pulsed and CW CO_2 lasers and CO_2 laser marking systems.

Comau SpA, Via Rivalta 30, Grugliasco Torino, Italy; (11) 3334-1*, telex 221511
1966
Manufacturer of multi-axes laser processing systems.

CONTEC GmbH, Pfungstadter Str. 35-37, Darmstadt 13, D-6100, Germany; 06151/5 40 77-79*, FAX 06151/5 40 70, telex 6151817 contda
emp 10, 1983
System manufacturer, including CMD-3 laser dual-mode precision machine, CAD-systems; material processing and micromachining for electronics/mechanics with Nd:YAG-laser; welding, cutting, surface treatment, drilling, scribing/engraving and marking of metals and nonmetals; soldering; semiconductor processing; applications R&D and feasibility.

Contek SpA, Via Don Maio 40, Varallo, VC, I-13019, Italy; 0163-54154, FAX 0163-53630, telex 223162
mng dir, Daniele Conserva; tech mgr, Sisto Berta; 1983
Manufactures micropositioning systems - CO_2 laser optics (copper mirrors): plane, spherical and aspherical - CO_2 laser systems: welding, cutting and marking.

Controllaser

7503 Chancellor Dr, Orlando, FL 32809; 407-438-2500, FAX 407-851-2720, telex 514 215
pres, Roy Tanner; sls mgr, Steve Graham; mktg dir, Richard L. Stevenson; eng mgr, Angelo P. Chiodo; 1966
Manufacturer of CW Nd:YAG lasers and complete turnkey laser machining centers to customer specifications. Products include laser systems for marking, engraving, cutting, drilling, diamond processing, scribing, soldering, trimming and welding. Control laser supports industrial customers utilizing advanced applications development, systems engineering for custom turnkey systems and systems integration and multilevel training programs.

Control Micro Systems Inc

Coteglade Photonics Ltd, Brunel House, 1275,, Neath Rd, Swansea, SA1 2LB, UK; 0792 52931*, FAX 0792 52805
pres, P. McBride; sls mgr, G. Davies; eng mgr, G.R. Bennett; mfg mgr, P. Mason; emp 12, 1985
Designs and manufactures laser based systems for perforation and cutting of thin materials: solid state and semiconductor laser rangefinders: laser system consultancy.

CRILASER SA, Civdad de Asuncion, 46 Bis, Barcelona, 08030, Spain; 345 74 40, FAX 346 60 35
mng dir, Jaime Morros; 1985
Research and manufacturer of CO_2 lasers.

Cryophysics GmbH, Landwehrstrasse 48, Darmstadt, D-6100, Germany; 06151-86281, FAX 06151-84481
1985

Cryophysics Ltd, Unit 4 Station Lane, Ind'l Est, Avenue 2, Witney, Oxford, OX8 6YD, UK; 993-773681, FAX 993-705826, telex 837131
sls mgr, P. Martin; 1967
Manufactures CO_2 (TEA) lasers and laser marking systems.

Cutting Edge Inc, 21 Lime St, Marblehead, MA 01945; 617-631-1390, FAX 617-631-3983
pres, Marvin Duncan, Jr.; vps, George Prince III, James Herman, Jr.; 1989
Manufactures laser systems for cloth cutting. High accuracy. Four ft/sec speed. Digitizing. Vacuum tables as large as 48' x 72" . CO_2 sealed tube.

Cymer Laser Tech, 7887 Dunbrook Rd, Suite H, San Diego, CA 92126; 619-549-6760, FAX 619-549-6795
pres, Robert Akins; vp mktg, Uday Sengupta; vp adv res, Richard Sandstrom; 1986
Manufactures and develops excimer lasers for the semiconductor fabrication industry. Primary focus is the microlithography stepper market and it is an OEM supplier of spectrally narrowed and wavelength stabilized krypton fluoride excimer lasers to the major stepper manufacturers in the world.

Daewoo Heavy Industries Ltd, Machine Tool & Factory Automation Div, A-6 Block, Changwon Industrial Complex, Changwon, 641-090, South Korea; (0551) 80-4396, FAX (0551) 84-2496, telex K534465 DHILTD
pres, K.H. Lee; exec mng dir, J.S. Yang; emp 9000, 1937
Provides several kinds of laser processing machines including cutting machines, multipurpose machining center, welding, heat treating, drilling, and scribing as a system manufacturer.

Daihen Corp, Mechataronics Div, 5-1 Minami Senrioka, Settsu City Osaka, 566, Japan; 06-317-2521*, FAX 06-317-2582
pres, K. Kobayashi; sls dir, M. Shiramizu; mktg dir, M. Yagyu; eng dir, M. Nagasaka; mfg dir, E. Tazaki; emp 2000, 1919
Manufactures fast-axial-flow CO_2 lasers 700, 1000, 1500, and 2000 W and cutting, welding, surface treatment processing systems.

Design Technologies Ltd, (sub of GFM AG STEYER Austria), Bradfield Rd, Finedon Rd Industrial Estate, Wellingborough, NN8 4HB, UK; 933-278871, FAX 933 223231, telex 311375
mng dir, Ron Pistell; tech dir, Don Pilkington; 1974
Designers and builders of CNC machine tools and manufacturing systems using laser cutting techniques and other techniques. Any proprietary laser can be incorporated.

Directed Energy Inc, 2382 Morse Ave, Irvine, CA 92714; 714-553-1225, FAX 714-553-8049
pres/engrg dir, Leroy Sutter, Jr.; cfo, Michael Garner; vp mktg, Stanley J. Parnas; 1982
Offers CO_2 lasers for industrial and medical applications. All are sealed-off and RF excited. Output power ranges from 10 to 240 W. All models exhibit very high power to size ratio. The DigiMark is an "on-the-fly" digital laser marking system. It contains 7 low power lasers to produce a 7 dot high matrix.

DKNE Design Services Inc, 363 Spring St, Manchester, CT 06040; 203-649-8068, FAX 203-649-9665
sls eng, Richard J. Nelson; 1981
Manufacturer of 2-axis and multi-axis cutting and welding systems, utilizing both CO_2 and solid-state laser sources. Custom installations are available.

Doerries Scharmann GmbH, Laser Technique, Doerriesstrasse 2, Postfach 1108, Mechernich 14, D-5353, Germany; 02484 12-0, FAX 02484 14 00, telex 8 33 619 DOEWE D
Dr.-Ing. Ruediger Rothe, Dipl.-Ing. Ralf Louis, Dipl.-Ing. Adolf Starken, Dipl.-Ing. Joseph Lehle; emp 1100, 1981
Machine tool manufacturer utilizing lasers integrated with products and services: laser systems for material processing (PROLAS), complete industrial systems, job shop in laser material processing (welding, surface treatment, cutting), engineering R&D laser material processing.

Dorotek GmbH, Thuyring 50, Berlin 42, D-1000, Germany; (030) 7850056, FAX (030) 7862803, telex 786724
mng dir, Wolfgang Krebs; mktg, optics, Maren Muller; mktg, detectors, Hanno Schmidt; ch eng R&D, Lech Boruc; 1989
Manufactures laser systems for marking with Nd:YAG and CO_2 lasers; laser drilling machines for diamonds; laser components; Pockels cells and drivers; optical components; and IR detectors for far infrared.

DTM Corp, 1611 Headway Circle, Bldg 2, Austin, TX 78754; 512-339-2922, FAX 512-339-0634
1989

DUAP Engineering, (member of the AGIE Group), Mattenstrasse 6, Biel/Brugg, CH 2555, Switzerland; 032 53 29 26*, FAX 032 53 27 83
emp 70, 1989
The ALC-2000 provides high precision cutting, drilling, and welding system. Offers equipped laboratory for laser materials processing and material analysis.

Eberhard GmbH & Co KG, Auchterstrasse 35, Schierbach, D-7311, Germany; 07021-7274-0, FAX 07021-727423
pres, Manfred Eberhard; sls mgr, Josef Forkl; 1964

Edel Stanzmaschinen GmbH, Postfach 40 02 69, Stuttgart, D-7000, Germany
1964

Edinburgh Instruments Ltd, Riccarton Currie, Edinburgh, EH144AP, UK; 031 449 5844, FAX 031 449 5848, telex 72553 EDINST G
pres, Dr. Richard Dennis; sls/mktg mgr, Douglas Neilson; eng mgr, Dr. Brian Davis; mfg mgr, Ian Wilson; emp 40, 1971
Manufactures conventional, CW, TEA, and waveguide CO_2 laser systems. Has wide experience designing and manufacturing laser-based and optical systems for industrial and research applications, hybrid circuit resistor trimmers, and microsoldering systems for surface-mounted components. Also sells laser accessories and detectors.

Electro Scientific Industries Inc (ESI), 13900 NW Science Park Dr, Portland, OR 97229; 503-641-4141, FAX 503-643-4873
ceo/pres, Michael Ellsworth; sr vp, Ed Swenson; vp op, Chris Nawrocki; vp mktg, Robert McGeary; sls mgr, Dick Donaca; 1953
Designs and builds laser trimming and processing systems for hybrid and semiconductor manufacturers.

Electrox Inc, (sub of The 600 Group), 2701 Fortune Circle, Indianapolis, IN 46241; 317-248-2632*, FAX 317-240-5787
pres, Dr. Douglas Telford; mktg mgr, Brian Urban; eng mgr, Chris Hadley; vp oper, Don Bishop; 1976
Manufactures OEM CO_2 and Nd:YAG lasers, welding systems and laser engravers.

Electrox Ltd, (sub of 600 Group PLC), Fen End, Astwick Rd, Stotfold, Hitchin, Herts, SG5 4BA, UK; 0462 834848, FAX 0462 835488
mng dir, D.G. Telford, PhD; sls/mktg dir, S.E. Mitchell; prod dir, D. Price; 1981
Full range of industrial CO_2 lasers from (80 W to 2000 W) CW output, all with electronic pulsing and power control. High average power Nd:YAG lasers (up to 2400 W), Nd:YAG laser based marker systems (60 & 120 W), beam delivery, laser accessories, infrared optics. Work routinely with manufacturers of industrial laser systems.

Elettronica Valseriana, Evlaser Div, Via Bombardieri 27, Fiorano Al Serio, I-24020, Italy; 0039 035 712661*, FAX 035 714956
emp 18, 1978
Manufactures Nd:YAG lasers for medical and industrial applications, lasers for OEM, laser systems for end users, and pulsed and CW lasers.

Emco Maier, Emco Lasertec Div, Salzachtal Bundesstrasse Nord, 5400 Hallein/Taxach, A-5400, Austria; 06245-81551-260, FAX 06245/81551-11, telex 631052 emco a
Wolfgang Brandl; emp 1100, 1943
Manufactures LS CO_2 laser compact cutting systems, especially designed for prototyping, small and medium batches; metal sheet- up to 4mm, plastic- up to 20mm, wood- up to 20mm, compound material and sealing material. Laser system is fully encapsulated to protect personnel and environment. Full CAD/CAM support including nesting software.

ESAB Group, ESAB-HELD GmbH Div, Industriestrasse 26, Heusenstamm, D-6056, Germany; 06104-6431, FAX 06104-62221
pres, Siegfried Held; sls mgr, Jurgen Held; r&d mgr, Klaus Hansel; 1948
Worldwide manufacturer of turnkey, laser material processing systems; laser beam cutting, welding, hardening, heat treatment; up to 8-axis, numerical controlled flying optic systems for laser and/or waterjet cutting; laser job shop; process development; and large-area laser equipment.

Euro Laser Technology, Westerring 19, Oudenaarde, B-9700, Belgium; 32 55 333 960*, FAX 32 55 333 900
emp 25, 1978

Exatron, 2842 Aiello Dr, San Jose, CA 95111; 408-629-7600, FAX 408-629-2832
pres, Robert Howell; vp, Eric Hagquist; puch, Cecil Baumgartner; sls, Chris Barrett; 1974

Fagor-Arrasate S. Coop., Apartado 18 Bo. San Andres s/n, Mondragon (Guipuzcoa), Spain; 34-43-792011*, FAX 34-43-799672
r&d dept, Felix Remirez; lsr dept, Ricardo Oterino; emp 350, 1957
Complete laser systems: flat tables for cutting (2&3 axis); gantry system (5&6-axis); robolaser; fixed beam flat tables; special installations for cutting, welding, drilling, surface treatment, tailor-made; Pipe' s processing laser systems. Systems for manufacturing sheet metal, stamping systems, cutting lines, transfer lines; plasma cutting systems.

Fanuc Ltd, Oshino-mura, Minamitsuru-gun, Yamanashi Pref., 401-05, Japan; 0555-84-5555, FAX 0555-84-5512
pres, Dr. Seiuemon Inaba; emp 2000, 1972
Manufactures, sales, maintenance service, and training of CNC system, industrial robot, wire-cut electric discharge machine, plastic molding machine and high-power laser oscillators.

Finn Power International, 710 Remington Rd, Schaumberg, IL 60173
1972

Florod Corp, 17360 S. Gramercy Place, Gardena, CA 90247-5212; 213-532-2700, FAX 213-329-1015
pres, R. Waters; mktg mgr, B. Harwood; eng mgr, F. Pothoven; 1974
Manufactures laser systems for microelectronics applications.

Fox Laboratories, 7 Hitching Post Lane, Danbury, CT 06811; 203-791-1503, FAX 203-744-5095
pres, Richard L. Fox; 1977
Manufactures laser trim systems, laser marking systems, and lying probe substrate testers. Offers laser trim services, laser marking services and testing services.

Fuji Electric Co Ltd, Mecatronics Div, 1-12-1 Yuraku-cho, Chiyada-ku, Tokyo, 100, Japan
1977

Fuji Tool And Die Co Ltd, 20, Matoba, Shimizu-cho, Sunto-gun, Shizuoka, 411, Japan; 0559-77-2300*
1977

FANUC

PO Box 8106, Charlottesville, VA 22906; 804-978-5443, FAX 804-978-5207
1977
Offers high-power, RF-excited, fast-axial-flow CO_2 lasers for industrial applications. Rated powers of 1000, 15000, 2000, and 3000 W. Integrated CNC control for power, pulsation, sequencing, and laser diagnostics. State-of-the-art cutting performance. Applications development assistance available Complete, 1 or 2 year, unlimited hour, parts warranty, including optics.

General Laser Inc, 7950 E. Redfield Rd, Suite 150, Scottsdale, AZ 85260; 602-991-0674, FAX 602-991-1462
pres, Dick Norris; sls mgr, Bruce Greenwood; sls eng, Steve Allred; eng mgr, Don Taylor; 1983
Manufactures CO_2 and YAG laser marking systems; CO_2, CW and pulsed laser wire stripping systems; and low-power cutting systems. Job shop for marking and stripping.

GS GENERAL SCANNING INC.

500 Arsenal St, PO Box 307, Watertown, MA 02272; 617-924-1010, FAX 617-926-0708
pres, C.D. Winston; vp Mike Kampfe; vp optics, R. Zack Moseley; vp mfg, George Hrono; vp eng mgr, Kurt Pelsue; emp 325, 1968
Designs and manufactures optical scanning components and systems based on galvanometer and resonant scanner technologies to provide fast, precise placement of light. Applications include laser printing and marking, factory automation, vision, bar-code generation and reading, laser trimming and graphics, and three-dimensional scanning.

Gentec Inc, 2625 Dalton St, St. Foy, Quebec, G1P 3S9, Canada; 418-651-8000*, telex 051-31591
emp 75, 1979

GMFanuc Robotics

2000 South Adams Rd, Auburn Hills, MI 48326-2800; 313-377-7000, FAX 313-377-7366
pres, Eric Mittelstadt; dir prod & app, Scott Melton; mgr laser proc & app, Steve Wertenberger; lsr app engrs, Dr. G.C. Lim, Russ Bell, Bill Zanley; 1982
Manufactures multiaxis (robotic) laser systems for welding and cutting. Applications development, system engineering, training and stand-alone products available. Worldwide sales and service organizations.

GMP SA, 19 ave des Baumettes, Renens 1/Lausanne, CH 1020, Switzerland; (004121) 6348181, FAX (004121) 6353295
1982
Engineering office with exclusive distribution in Switzerland of electro-optical products, lasers and instruments for telecommunication, optical fibers/analytics, spectroscopy and electronics for medical use.

Goiti S Corporation, Apartado 80, E. Elgoibar, Gulpuzcoa, Spain; 943-74-03-50, FAX 943-74-31-38
1982

D.A. GRIFFIN CORP.

240 Westminster Rd, West Seneca, NY 14224; 716-674-2300, FAX 716-674-2309
pres/ceo, Robert Griffin; mktg mgr, George McClean; mfg mgr, James Downey; proj mgr, John Marriott; eng mgr, Frederick McGee; contlr, Chris Kennedy; emp 60, 1959
Offers special laser systems integration. Designs and builds for welding, cutting and cladding.

Group II Manufacturing Ltd, 27 Newtown Road, Plainview, NY 11803-4301; 516-694-1334, FAX 516-694-1336
pres, Herbert D. Gresser; vp, Sidney Hausthor; 1974
Manufactures industrial laser systems for engraving, drilling, sawing, and scribing diamonds; and computer-controlled laser systems for engraving other materials. Also chemical and electronic devices for cleaning diamonds.

GSR Technologies Ltd, Laser Div, 12345-121 St, Edmonton, Alberta, T5L 4Y7, Canada; 403-451-9000*, FAX 403-451-1795
ceo, Miles Palmer; pres, Jimmy Haythornthwaite; dir tech mktg, Dr. Roger Ball; mfg mgr, Ted Zscherpel; 1967
Designs and manufactures CW CO_2 lasers and CAD/CAM single and multi-headed lasercutters for industrial fabrics, tubular, roll goods and plastics. Systems feature touch screen controls, "cut-on-the-fly" capabilities, and automated parts removal. Aerospace certified machine shop and R&D facilities.

GTU, Im Lindenbosch 37, Baden-Baden 22, D-7570, Germany; (0049) 07223-58915, FAX (0049) 07223-58916
1986
Designs and manufactures HeNe lasers, solid-state lasers, Co_2 lasers, laser power supplies, scanning systems, and laboratory accessories and components.

Haas Laser GmbH, Product Management Div, PO Box 572, Aichhalder Strasse 39, Schramberg, D-7230, Germany; 07422-515-0, FAX 07422-515-39
mng dir, Paul Seiler; 1972
Manufacturer of Nd:Glass and Nd:YAG lasers, modular beam guidance systems, laser light cable systems, CNC laser centers, and process monitoring.

Haco NV, (sub of ELAS), Oekensestraat 120, Rumbeke-Roeselare, B-8810, Belgium; 051-220318, FAX 051-220979
1972
Offers 3 different types of laser cutting machines, with power ranges from 1000 to 2500 W.

Hahn & Kolb (GB) Ltd, Leicester Rd, Rugby, CV211NY, UK; (0788) 577288, FAX (0788) 561051
1972
Manufactures a range of cutting machines and laser marking and engraving machines.

Hamul Werkzeugfabrik, div of Th. Kirschbaum GmbH & Co KG (sub of ZARIAN-Bewegungs-Systeme GmbH), Postfach 127, Meusselsdorfer Str 27, Marktredwitz, D-8590, Germany; 09231-808-0, FAX 09231/62229, telex (17) 9231 813
1927
Manufactures special customer developed base frames X+Y axis complete; laser welding and cutting machines; high speed milling machines for cfk plastics; portals and gantry type machines for milling and waterjets; portal robots for handling up to 20 kg.

Hanson Systems Inc, 200 John L. Dietsch Blvd, Attleboro Falls, MA 02763; 508-699-7550, FAX 508-695-7864
sls/mgtg dir, Robert A. Ciampa; 1927
Offers laser welding assembly of small, precision parts.

Harland Simon Automation Systems Ltd, Torrington Ave, Coventry, CV4 9XQ, UK; 0203 473748, FAX 0203 474196, telex 312355
mng dir, D.J. Harbour; mktg mgr, C.R. Davis; 1965
Systems integrator for laser and other high technology automated manufacturing systems providing: system concept and design; applications engineering; system building and integration; project management; installation, commissioning and support.

Hauser Motion Control Inc, 1700 N. Highland Rd, Pittsburgh, PA 15241; 412-835-9240, FAX 412-835-9151
gen mgr, Richard W. Kessler; 1986
Manufacturer of electric linear drives for single and multi-axis applications. Also fully digital motor controllers, used for precise position control and speed regulation. Load capability to 2400 lbs., speeds to 46 ft./sec. and repeatability to +/- .004 in. Customised motion to accommodate almost any length.

Helling KG, Sylvesterallee 2, 2000 Hamburg-54, D-2000, Germany; 004940547180, FAX 004940542061, telex 213611+2165082
1986

HendrickSaw Inc, 36 Commercial St, Salem, MA 01970; 508-745-5222, FAX 508-741-4809
1986
Manufacturer of horizontal and vertical panel saws, chop saws, radial arm saws, gang saws, countertop saws, planers and edge finishers, CNC routers and laser engravers.

Hitachi Ltd, New Prod Eng Dept, 4-6, Kanda Surugadai, Chiyodaku, Tokyo, 101, Japan
1986

Hitachi Seiki Co Ltd, 1, Abiko, Abiko City, Chiba Pref, 270-11, Japan; 0471-84-1111, FAX 0471-84-1511
1986

Hoya Corp, 7-5 Naki Ochiai, 2-chome, Shinjuku-ku, Tokyo, 161, Japan
1986
Fabricates Nd:YAG laser rods and manufactures Nd:YAG lasers.

S.E. Huffman Corp, 1050 Huffman Way, Clover, SC 29710-1400; 803-222-4561, FAX 803-222-7599
pres, Allen Turk; vp sls, Michael O. Harrington; mktg mgr, Alan Mandell; vp r&d, Ernst Borchet; vp mfg, Jerry Creech; vp fin, Bob McBurney; emp 145, 1979
Designs and manufactures intelligent machine tool systems which offer fully integrated CNC multi-axis machine motion with either Nd:YAG or CO_2 lasers. The systems are rugged, state-of-the-art designs for industrial drilling, cladding, cutting and welding.

Hughes Aircraft Co, Industrial Products Div, 2051 Palomar Airport Rd, Carlsbad, CA 92009; 619-931-3214, FAX 619-931-5197
div mgr, J. Kolostyak; prod line mgr, P.F. Robusto; r&d mgr, R.A. Tilton; 1979
Carbon dioxide lasers for military and commercial applications. All lasers have hard seals resulting from the use of space-qualified materials and process technology. RF-excited; 3 W to 25 W lasers and small systems with frequency stability and isotopic options.

IEF Werner GmbH, Wendelhofstr. 6, Furtwangen, D-7743, Germany; 07723-650-10, FAX 07723-6501-48
1980

IHI Trumpf Technologies Ltd, 616-1 Hakusan-Cho, Midori-Ku Yokohama, 226, Japan; 045-939-7811, FAX 045-939-7817
1980

Image Micro Systems Inc, 900 Middlesex Trnpk, Bldg #6, Billerica, MA 01821; 508-663-7070, FAX 508-663-0148
pres, Berhard Piwczyk; sls/mktg mgr, James M. Morrison; 1984
Manufactures fully integrated R&D and production excimer laser micromachining systems. Complete computer control of all laser and processing parameters for 193 nm, 248 nm and 308 nm. Large field excimer laser ablation systems for high resolution projection patterning of polymers. Unique, high quality achromatic optics allow through-the-lens alignment and viewing operation.

Industrial Laser Source, 26 Pearl St Ste 204, PO Box 7, Bellingham, WA 02019; 509-966-3638*, FAX 509-966-2261
1984

Industrial Laser Technology, 2151 North Franklin Rd, Indianapolis, IN 46219; 317-353-1333*, FAX 317-351-7038
sls mgr, L. Gunseor; 1985
Offers laser measurement and alignment systems, vision systems, industrial laser systems, standard products and custom design.

InTA, 2281 Calle de Luna, Santa Clara, CA 95054; 408-748-9955, FAX 408-727-3027
pres, Dr. Paul Lovoi; mgr adv prds, Dr. Len Reed; vp admin, Marsha V. Lasky; mgr cus serv, Karen Cabral; mgr int robotics div, Phil Barone; 1979
Manufactures advanced laser robotic workcells, machine vision guided welding and cutting workcells, advanced workcell controllers utilizing nonproprietary modular construction and laser paint stripping systems.

International Lasersmiths, 1306-B Alameda St, Compton, CA 90221; 213-635-8536, FAX 213-635-2313
pres, Richard Crews; vp, John Adams; sec/treas, Jerry Crews; 1988
Laser job shop and manufacturer of laser systems. Specializing in large parts & systems. 5-axis system travels of: 20 *x*-axis, 8 *y*-axis, 3 *z*-axis, with *a* & *b* axis also. 800-W and 1600-W pulsed Nd:YAG lasers, 1.1 kW CO_2, 100-W Nd:YAG marking system. Provides water jet cutting services.

Isomet Corp, 5263 Port Royal Rd, Springfield, VA 22151; 703-321-8301, FAX 703-321-8546
pres, Henry Zenzie; exec vp oper, Leo Bademian; tech dir des/dev, Allister McNeish; 1956
Manufactures a programmable laser marking system, Nd:YAG based with no moving parts, which generates up to 7 alphanumeric characters with each laser pulse.

Jamieson Mfg Co Inc, PO Box 966, 2500 So. Main St, Torrington, CT 06790; 203-482-6543, FAX 203-482-4051
vp, Scott Jamieson; gen mgr, Wolfgang Kesselring; emp 30, 1965
Designs and manufactures automatic machinery for medium and high production, to laser weld, drill, cut and assemble small parts.

JEC Lasers Inc, 441 Market St, PO Box 933, Saddle Brook, NJ 07662; 201-843-6600, FAX 201-843-3469
pres, John H. Wasko; vp op, Tracy Forman; mfg mgr, Andrew Smith; app mgr, Robert Fischer; emp 12, 1977
Manufactures Nd:YAG lasers and Nd:YAG industrial laser marking systems. Provides repair and maintenance service and parts for a variety of solid-state laser systems. Operates a laser marking job shop to provide both small lot and volume laser marking and engraving services.

Jens Scheel Sondermaschinen GmbH, Gasstrasse 16, Itzehoe, D-2210, Germany; 4821 74062*, FAX 4821 74142, telex 28203 SCHEEL D
ch pres, Jens Scheel; tech mgr, Jurgen Andresen; 1960

Jipee Canada Technologies Inc, 8927 Boul Industriel, Chambly, Quebec, J3L 5G8, Canada; 514-447-3473, FAX 514-447-3473
pres, Jean-Pierre Serrano; sls mgr, Jeffrey Sparling; 1987
Produces compact, low-cost 2-axis laser cutting systems that provide high speed, high precision processing of sheet materials up to 8 ft wide by any length. These systems incorporate proprietary CNC technology that automatically generates cutting programs from CAD and that is accessed through a menu driven touch screen.

Jobs SpA, Via Marcolini 13, Piacenza, I-29100, Italy; 39 523 5496*, FAX 39-523-549750
1987

JRM International Inc

JRM International, Inc.

5633 E. State St, Rockford, IL 61108; 815-397-8515, FAX 815-397-7617
pres, James R. Mattox; 1987
Complete sales and service for Baublys CNC marking and engraving systems including digitizing and scanning for artwork and logos. 3-D engraving capability. 4-axis capability is a standard option.

Jyoti Ltd, PO, Chemical Industries, Baroda,, Gujarat, 390 003, India; 82851-2-3*, telex 0175-214
emp 5000, 1943
Manufactures industrial CO_2 lasers and laser systems for nonmetals and ceramics, processing and alignment laser systems, including start-up and after-sales services.

Karl Binder GmbH, Postfach 1164, Reichertshofen, D-8077, Germany; 8453-8015*, FAX 8453-8015
1943
Manufactures laser processing systems.

Karl Suss America Inc, PO Box 157, Waterbury Ctr, VT 05677; 802-244-5181*, FAX 802-244-5203
1943

Kitamura Machinery Co Ltd, 1870, Toidecho, Takaoka-shi, Toyama, 939-11, Japan; 0766-63-1100*
1943
Manufactures laser sheet-metal cutting systems.

Koke Sanso Kogyo Co Ltd, 3-4-8 Ohira Sumida-KU, Tokyo, 130, Japan; 03-624-3111*
1943

Komatsu Ltd, Plasma & Laser Div, 2-3-6 Akasaka, Minato-ku, Tokyo, 107, Japan; 03-5561-2692, FAX 03-3586-7053.
dep dir, Masaaki Ichimura; 1921
Manufactures the laser machine KLM series which features silent, sharp, precise, non-contact material cutting with the high energy concentrated laser beam and the most advanced CNC control system. A wide variety of materials can be worked including steel, aluminum, glass, ceramics, plastic, rubber, wood, etc.

Kuka Schweissanlagen & Roboter GmbH, (sub of IWKA AG), PO Box 431349, Augsburg 43, D-8900, Germany; (0821)797-0*, FAX (0821)797-1991, telex 53838-20 KUK D
Karl-Heinz Zinke, Werner Busch, Hans-Dieter Krebs, Stefan Muller, Ernst Zimmer; emp 1450, 1898
Designs, manufactures and installs turnkey manufacturing equipment primarily for the automotive industry and its component suppliers. Includes four product divisions: Transfer Lines, Assembly Systems, Industrial Robots, and Special Welding Machines. Each of these divisions supplies components, machines and systems either independently or in cooperation with the other divisions. Worldwide reputation in the field of engineering consultancy.

L'Air Liquide, 75, Quai D'Orsay, Paris Cedex 07, 75321, France; (33) 1 40625298*, FAX (33) 1 45556806
trade mgr, Remi Charachon; r&d, Bruno Marie; r&d, Vincent Poncon; r&d, Vincent Poncon; 1898
Manufactures and supplies lasing and process gases for materials processing applications under LASAL trademark available worldwide with name quality, precision, consistency, and performance. Also manufactures and supplies rare gases and mixtures for excimer and scientific lasers commercialized by ALPHAGAZ divisions. Manufactures and supplies gas handling equipment and installations.

Lambda Physik GmbH, (sub of Coherent Inc), Hans-Boeckler-Strasse 12, Goettingen, D-3400, Germany; 49 551 69380, FAX 49 551 68691
pres, Dr. Dirk Basting; dir sls, Dr. Klaus Pippert; dir mktg, Dr. Hans-Juergen Kahlert; mfg mgr, Dr. Gerd Steinfuehrer; dir US oper, Dr. Gerard Zaal; emp 150, 1971
Developes and manufactures industrial excimer lasers for marking, microelectronic packaging and microlithography applications. The Lambda laser series are devices with computer communications and optimized to be used on laser system assemblies as deep UV light source.

LAMBDA PHYSIK

(sub of Coherent Inc), 289 Great Rd, Acton, MA 01720-4739; 508-263-1100, FAX 508-263-4296
dir US oper, Gerard Zaal; nat Ind sls mgr, Robert Battis; emp 20, 1971
Excimer lasers for industrial applications: laser systems with computer control unit (IBM-PC compatible system), optional interfaces for master-slave communication.

LASAG
INDUSTRIAL-LASER

(sub of Lasag AG), 702 W. Algonquin Rd, Arlington Heights, IL 60005; 708-593-3021, FAX 708-593-5062
nat'l sls mgr, Art Spera; gen mgr, Fritz Mueller; mfg mgr, Adrian Fiechter; 1971
Manufactures solid-state, Nd:YAG and pulsed YAG laser sources and systems for end user and OEM markets for industrial, electronic, scientific and medical applications; CNC laser processing machining center with CO_2 or YAG lasers for welding, cutting, and drilling. Variety of attachables are available such as fiberoptic beam delivery, beam scanning, split beam delivery, or trepanning optics. Also operates a laser applications laboratory for R&D. Offers complete servicing.

LASAG
INDUSTRIAL-LASER

Mittlere Strasse 52, Thun, CH-3600, Switzerland; 033/224522, FAX 033/224173
pres, Dr. Salathe; strategic mgr, Dr. U. Durr; emp 100, 1974
Manufactures solid-state, ND:YAG and pulsed laser sources and systems for end user and OEM markets for industrial, electronic, scientific and medical applications; CNC laser processing machining centers, type VEGA, with Nd:YAG lasers for welding, cutting, and drilling. Offers a variety of accessories such as fiberoptic beam delivery systems, or trepanning optics. Also operates a laser applications laboratory for R&D. Offers complete service support.

Laser Application SA, I Rue des Bouvreuils, BP 79, Baume-les Dames, F-25110, France; 81.84.24.44*, telex 362895f
1981
Manufactures YAG lasers from 20 to 450 W, with or without optical fiber. Laser workstations for cutting, drilling, and surface treatment.

Lasercut Inc, 101 Fowler Rd, N. Branford, CT 06471; 203-488-0031, FAX 203-483-0463
emp 19, 1983
Manufactures and markets turnkey laser cutting systems.

Lasercut SA, Evole 19, Neuchatel, CH-2000, Switzerland; 038 25.97.86*, FAX 038 25.58.80
emp 12, 1982
Manufactures industrial CO_2 lasers and systems for cutting applications.

Laser Dynamics Ltd, 17 Production Ave, Ernest Junction, Queensland, 4214, Australia; 075 396644, FAX 075 971 545
mng dir, John Kavanagh; mktg mgr, Tim White; eng mgr, John K. Kavanagh; mfg mgr, Jeff Lucas; emp 58, 1979
Manufactures laser systems and accessories for surgery, industrial profile cutting, and scientific research. Military products: weapons simulators, laser rangefinders, and metal optics. Industrial systems include: CO_2 lasers up to 1000 W output with CNC profiling tables and CAD package; CO_2 or Nd:YAG lasers precision micromachining systems.

Laserdyne, 6690 Shady Oak Rd, Eden Prairie, MN 55344; 612-941-9530, FAX 612-941-7611
pres, Robert H. Schmidt; vp oper, LeRoy E. Gerlach; int'l sls mgr, Richard W. Owen; North Amer sls dir, Mark W. Barry; 1981
Multi-axis laser systems with up to 8-axis of motion for cutting, drilling, and welding of formed 3D, tubular and flat metal and non-metal parts. Systems include the direct drive BeamDirector™ for rotary and tilt axes of laser beam motion, software developed specifically for multiaxis laser processing, and either CO_2 or Nd:YAG lasers. Contract laser services include feasibility studies, programming, tooling design, and process development.

LASER ECOSSE

Kings Cross Rd, Dundee, DD2 3EL, UK; 0382 833377, FAX 0382 833788, telex 76391
pres, Dr. J.G. Freeman; sls/mktg mgr, I.E. Ross; eng mgr, C.G.H. Courtney; mfg mgr, A. Payne; 1990
Manufactures CO_2 lasers from 4 W to 10,000 W output for OEM uses in industrial, military, scientific, and medical applications. Product range includes sealed-off waveguide, slow-axial-flow, fast-axial-flow, and cross-flow lasers.

Laser Electronics Pty Ltd, CNR Gaven Way, Habana St, Gaven, Queensland, 4211, Australia; (075)73-2066, FAX (075)73-3090
pres, Noel H.F. Walden; mktg mgr, R. Craig Holberton; mfg mgr, Les Darcy; dom/intl sls coord, Cecily Chesterton; eng mgr, Bob Munnings; 1967
Designs, manufactures, and markets a range of energy efficient, slow and fast-axial-flow industrial lasers and laser systems, primarily for cutting metals and nonmetals.

Laser Lab Ltd, 367 Warrical Rd, PO Box 204, Cheltenham, Victoria, 3192, Australia; 5849900*, FAX 5844131
emp 63, 1967
Manufactures, sells and services a standard product range consisting of; 2, 3 and 5-axis laser processing systems. The standard processing area varies from 1.2m x l.2m to 3.0m x l.6m.

Laser Machining Inc, 500 Laser Drive, Somerset, WI 54025; 715-247-3285, FAX 715-247-5650
pres, Noel Biebl; sls mgr, David Plourde; tech dir, William Lawson; 1978
Manufacturer of XY, moving, fixed, hybrid, single or multiple beam workstations, low power CO_2 lasers, beam delivery components, proportional laser pulse to motion control and special automated systems. Integrator of CO_2 lasers to 6 kW. Applications: platic cutting, perforating, scoring, metal cutting, welding and heat treating, wood engraving, cutting, composite cutting and consolidation.

Laserman, 19186 Dunure Place, Northridge, CA 91326; 818-363-5485, FAX 818-368-6355
ceo/pres, Gary Firment; gm, Mark Kluczynski; vp mktg, Brad Cantos; 1989
Manufacturer and supplier of laser machine components for industry, including laser beam delivery, fiberoptic delivery systems, linear motion systems and contoller, machine frames, machine bases, related accessories, and technical assistance in machine construction and application development.

Laser Material Processing Ltd (LMP Ltd), Egliweg 10, PO Box 347, Nidau (Biel), CH-2560, Switzerland; 032/51 67 77, FAX 032 51 13 93, telex 934077 lmp ch
pres, H.R. Niederhauser; ch exec, W. Schmid; mgr, D. Canal; 1981
Manufactures diamond drilling (dies), cutting (raw diamonds), and functional trimming machines. The drilling machine is especially designed for shaped drilling or wire dies of finest diameters (a few microns). Cutting machines for raw diamonds and diamond-tools are designed for high productivity and precise cutting. The functional trimming-machine for electronic circuits is handy and small and especially designed for figuration in automatic production lines for thin-, and thick-film electronic-circuits.

Laser Mechanisms Inc, 24730 Crestview Ct, Farmington Hills, MI 48335; 313-474-9480, FAX 313-474-9277
pres, William Fredrick; sls eng, Donald Sprentall; purch mgr, Jon Davies; mfg eng, Glenn Golightly; 1979
Specializes in innovative laser-beam-delivery components for high power YAG, CO_2 and other wavelengths that suit all facets of industrial applications.

Laser Nucleonics Inc, 2480 Presidential Way, PO Box 403, W. Palm Beach, FL 33401; 407-686-6867, FAX 407-471-8611
pres, Harry E. Franks; 1968
Manufactures lasers, conducts R&D studies, and applications engineering with all lasers.

Laser Photonics, Mattenstrasse 6, Brugg, CH-2555, Switzerland
1968

Laser Quanta S.A., Fragua #7, 287060 Tres Cantos, Madrid, 28760, Spain; (1) 803.44.44, FAX (1) 803.12.39
pres, Miguel Garcia; prd mgr, Jose Garrote; mktg & sls mgr, Juan-Miguel Garcia; lsr dev mgr. Jayier Galindo; 1983
Manufactures CO_2 lasers up to 1700-W; laser systems for cutting, drilling, marking, scribing, welding and heat treating; and HENE lasers and laser systems. Operates material processing center.

Lasers Industriels SA, BP 300, 96 route de Marzy, Nevers Cedex, 58003, France; (33) 86 59 01 44, FAX (33) 86 59 01 66
chrm, Alain Diard; sls ing, Bruno Potts; emp 15, 1983
Manufactures new type of integrated turbo CO_2 laser that lies in the low losses in the gas transport between the roots or turbine blower and the laser head. The power supply can be completely separated from the laser head to provide unprecedented installation flexibility and easy servicing by module exchange.

Laser Systemes, 3 Rue Denis Papin, Beauchamp, F-95250, France
1983

Lasertechnics Inc

Lasertechnics

5500 Wilshire Ave, Albuquerque, NM 87113; 505-822-1123, FAX 505-821-2213
pres, Louis F. Bieck, Jr.; mktg mgr, Richard Macklin; eng mgr, Dr. Shiv Dass; emp 60, 1981
Manufacturer of Blazer 6000 laser marking system, a pulsed CO_2 TEA laser for use in industrial consumer product marking; DIR Gray Scale Printer that outputs photographic, true continuous tone images from electronic image sources; full line of Sony monochrome and color video graphic printers; laser wavelength meters for pulsed and CW lasers; pulsed supersonic molecular beam valves for use in laser spectroscopy.

Laser Technology Associates Inc, 25 Ozalid Rd, Bldg 2, Rm 226, Johnson City, NY 13790; 607-798-9064
ch dir tech, Dr. David Brown; vp, David P. Benfey; vp, John R. Crittenden; 1987

Laser Technology Inc, 10131 Colonial Industrial Dr, So. Lyon, MI 48178; 313-437-7625, FAX 313-437-4310
mktg mgr, Michael Way; 1987
Manufactures the pulsed laser stripping system (PLS System) comprised of a high power pulsed laser and all associated components including the beam delivery, effluent collection, controls and safety circuits. The PLS System is used for paint stripping and surface cleaning of various surface coatings from various substratums.

Laser Valfivre GmbH, Kurt Huber Ring 4, Furstenfeldbruck, D-8090, Germany; 8141 24034, FAX 8141 26503
gen/sls/mktg mgr, Dr. Heinz B. Puell; emp 10, 1989
Manufactures high-power CO_2 lasers and systems for welding and heat treatment applications.

Laser Valfivre Sorgenti e Sistemi SpA, Via Panciatichi,70, Florence, I-50127, Italy; 055-439831*, FAX 055-433256, telex 570044 VFIVRE I
mng dir, Dr Stefano Pulsoni; mktg mgr, Dr Rinaldo Calvi; ind tech dept mgr, Ivano Garuglieri; emp 25, 1988
High technology equipment to be used both in medical field (CO_2 flow or sealed-off surgical and therapeutic laser systems) and in the industrial field- CO_2 slow and fast flow laser systems for die-boards and plexiglass cutting, metal sheet welding and cutting, ceramic substrates (alumina) cutting and drilling (scribing/drilling) for micro-electronics applications and Nd:YAG laser systems for processing of precious metals.

Laser Walzen Center GmbH, Postfach 101005, Essener Strasse 259, Oberhausen 1, D-4200, Germany; (0208) 889293, FAX (0208) 889390
Dr.-Ing. Rolf Steinmetz; 1989
Offers texturing of rolls by means of a laser beam (Lasertex) and electro discharge (EDT, starting in June 1992) to improve drawability and paintability of steel and aluminum sheets. Additionally, hardchrome plating of rolls is done for longer roll life and better quality.

Laser-Work AG, (member of the AGIE Group), Daettlikonerstr 5, Pfungen, 8422, Switzerland; 052/31 17 21*, FAX 052/31'24'29, telex CH-896 424 lawo
eng mgr, N. Meier; 1975
Manufactures flexible laser cutting and welding machines. Basic machine with moving laser optic. Capacities up to 2000 x 3000 mm. Automated loading and removal of materials will be adapted to suit customer' s specific production program. Machines with rotary axis for round parts and 3-axis package for 3-dimensional laser processing.

Lectra Systemes SA, Chemin du Marticot, Cestas/Bordeaux, F-33610, France; 56 68 80 00, FAX 56 78 22 72, telex 540 122 F
pres, Jean Etcheparre; gen mgr, Bernard Etcheparre; mktg mgr, Patrick Mayette; eng mgr, Francois Peiffer; emp 1100, 1973
Manufactures, designs and sells CAD/CAM equipment for shape creation, profile modification and subsequent nesting and preparation for cutting with lasers, water jets, knives in sheetmetal. For the automobile, aerospace composite, leather, shoe, and furnishing apparel industries. Offers training and maintenance with 25 subsidiaries and 40 offices worldwide.

Lectra Systems Inc, 844 Livingston Court, Marietta, GA 30067
1973
Manufactures laser cutting systems for the fabric industry.

LEE LASER

3718 Vineland Rd, Orlando, FL 32811; 407-422-2476, FAX 407-839-0294
1984
Manufactures CW and Q-switched Nd:YAG lasers for industry and science. Special OEM configurations available. Also manufactures conventional and fiberoptic beam delivery and focusing systems. Supplies Q-switches, YAG rods, arc lamps, and laser mirrors. Provides laser engineering services.

Lillbackan Konepaja, div of Finn-Power, PO Box 38, Kauhava, SF62201, Finland; +358 644 830111, FAX +358 644 830313, telex 72188
pres, Jorma Lillbacka; vp, Risto Makitalo; mktg mgr, Pekka Siponen; pres Finn-Power, Mikko Lindstrom; 1969
Manufactures turret punching machines with loading/unloading, sheet storages and integrated right angle shear/laser punch combinations. Effect of laser equipment between 1000 -1750 W. Laser equipment manufactured by Wegmann-Basel.

LINE LITE LASER

430 Ferguson Drive, Building 4, Mountain View, CA 94043; 415-969-4900, FAX 415-969-5480
pres/eng mgr, Jack D. Foster; mktg mgr, J.P. Evans; 1982
Manufactures low- & medium-power CW/pulsed sealed CO_2 lasers (0-200 W) and low-, medium-, & high-power, CW/pulsed Nd:YAG lasers (0-200 W); power supplies for gas and solid-state lasers; glassware; custom laser tubes; solid-state laser cavities; and heat exchangers. Supplies A/O Q-switches, flashlamps, attenuators, and complete scientific/medical/industrial laser systems.

Lite AG, Hardstrasse 42, Neuenhof, CH-5432, Switzerland; 056-86 20 04, FAX 056-86 21 51
dir, Bruno M. Voellmin; dir, Urs P. Murbach; emp 4, 1986
Develops and manufactures industrial laser marking systems with Nd:YAG and CO_2 lasers. Consultant for industrial laser applications. In-house job-shop and laser material processing application lab.

Lasercomb, Austrasse 29, Weilheim/Teck, D-7315, Germany; 7023-70090, FAX 7023-700944, telex 1770 2314
ch exec, Dr. Walter Ulmer; mng brd, Rainer Holder; 1973
Manufactures laser systems for cutting metal and nonmetal materials with CNC; laser systems for cutting and welding of metals; laser welding machines. Offers CAD/CAM software.

LPT Laser Physiktechnik GmbH, Industriezentrum, Obj. 85, Leobersdorfer Str. 26, Berndorf, A-2560, Austria; 02672-3488-0, FAX 02672-5164, telex 14237
ch exec, G. Brandstetter; sls mgr/vice dir, F. Singer; int sls mgr H.P. Brandstetter; ch sci, G. Elmer; emp 10, 1983
Operates the Europe Laser Exchange, buys and sells used lasers and systems. Offers repair services, application development, laser training, consulting and engineering, equipment rental, and job shop services. Builds customized laser material processing systems.

LPT Laser Physiktechnik GmbH, Oberhamerstr 8, Sauerlach/Arget, D-8029, Germany; 08104-1077, FAX 08104-9336
ch exec, C. Rapp; int sls mgr, H.P. Brandstetter; 1983
Operates the Europe Laser Exchange, buys and sells used lasers and systems. Offers repair services, application development, laser training, consulting and engineering, equipment rental, and job shop services. Builds customized laser material processing systems.

Lumonics Corp, Marking Systems Div, 3629 Vista Mercado, Camarillo, CA 93012; 805-987-2211, FAX 805-484-7959, telex 140499
pres, Pat Austin; dir NA sls, John Derzy; 1970
Manufactures, sells and services a complete line of laser marking systems incorporating imaged masks and focused spot techniques. Laser types include pulsed and Q-switched YAG, plus CW and TEA CO_2. Systems sold at OEM, custom, and application-specific standard product levels.

Lumonics Corp, Industrial Products Div, 19776 Haggerty Rd, Livonia, MI 48152-1016; 313-591-0101, FAX 313-591-0045
mng dir, Nigel Jinks; vp sls/mktg, S.A. Llewellyn; sls dir Nd:YAG prods, L.M. Heglin; sls dir excimer prods, S.W. Seamans; 1967
Manufactures industrial gas CO_2 lasers, industrial solid-state Nd:YAG lasers, industrial excimer lasers, industrial laser processing systems for cutting, welding, sealing, drilling, heat-treating, cladding, ceramic scribing and machining. Designs and builds custom-engineered laser processing systems, provides application-laboratory services, field service and repair, and training courses.

Lumonics Inc, 105 Schneider Rd, Kanata, Ontario, K2K 1Y3, Canada; 613-592-1460, FAX 613-592-5706, telex 053-4503
emp 180, 1970
Manufactures TEA CO_2 lasers and beam-delivery systems for marking of semiconductor, electronic, electrical, and consumer products. Also manufactures industrial excimer lasers for materials processing and semiconductor processing.

Lumonics Ltd, (sub of Lumonics Inc), Cosford Lane, Swift Valley, Rugby, Warwickshire, CV211QN, UK; (0788) 570321, FAX (0788) 579824, telex 311540
mng dir, N.G. Jinks; dir new tech, C.L.M. Ireland; dir ind'l prods, K. Withnall; dir new tech, Dr. T.M. Weedon; 1972
Manufactures Nd:YAG lasers and systems for industrial applications.

LVD Company N.v., Nijverheidslaan 2, Wevelgem-gullegem, B-8560, Belgium; 056/43 0511, FAX 056/402464, telex 85 317-8
mng dirs, Robert Dewulf, Marc Vanneste, Jean Pierre Lefebvre; dir/personnel, Gilbert Seynaeve; quality, Frans Derycke; finances, Georges Porte; engrg, Willy Leperre; emp 700, 1951
Manufactures machines for sheet-metal work.

Madrid Laser, Centro Technologico, La Araboleda Crtra Valencia, Madrid, KM7-28031, Spain; (1) 332 4340, FAX (1) 3324280
gen mgr, F. Maldonado; sci dir, J.M. Orza; tech dir, F. Bellido; ch r&d dept, J.M. Montejo; 1988
Offers services including cutting, welding surface treatment, drilling, engraving and marking; R&D projects both nationally and European; Technical training and diffusion activities.

Maho, Tiroler Strasse 85, Pfronten, D-8962, Germany; 8363-89-0, FAX 8363-89-222
1988

Maho Machine Tool Corp, PO Box 639, Naugatuck, CT 06770; 203-723-7481
1988

Manfred Fohrenbach GmbH, (sub of Positioniersysteme), Lindenstrasse 34, Unadingen, D-7827, Germany; 07707-159-0, FAX 07707-159-80
1988

Martek Lasers, 322 Earhart Way, Livermore, CA 94550; 415-294-8167, FAX 415-9294-9128
pres, Gordon McFadden; eng, Jeff Broome; apps, Tim Webber; 1987
Manufactures Nd:YAG CW lasers at 600, 1200, 1800 W output, with fiberoptic delivery for use in cutting and welding ferrous and nonferrous metals; the cutting of non-metals, surface modification and the applications of coatings.

Martin Pfaffen Laser, Laser Optik Mechanik, Kupfgasse 10A, Lengnau, CH-2543, Switzerland; (0041) (065) 529077, FAX (0041) 53 07 81, telex 934771 plom cti
pres, M. Pfaffen, Ing ETH; 1984

Maschinenfabrik GmbH, Postfach 25, Nordrach, D-7618, Germany; 7838/84-0, FAX 7838/1003
1984
Manufactures flexible marking systems suitable for batch production, small, medium, or large. Utilizing the unique JUNKER Laser Marking Machine, type Jumarker 450, a fully automatic marking system for marking precision cutting tools or similar workpieces.

Maschinenfabrik GmbH & CO KG, Otto Bihler, Lechbrucker Strasse 15, Halblech, D-8959, Germany; 08368-18-0, FAX 08368 18-105
gen mgr, Otto Bihler; 1953
Develops and constructs CO_2 high-performance lasers for material processing. Offers complete solutions in the field of laser welding from one source, whether it is application development, after sales service or warranty work.

Matsushita Ind Equip Co Ltd, Production Systems Equipment Div, 1-1, 3-Chome, Inazu-cho, Toyanaka, Osaka, 561, Japan; 06-866-8580*, FAX 06-866-7954, telex 528-6121
1977
Manufactures CO_2 lasers and systems for industrial applications.

Mauser-Werke Oberndorf GmbH, Ein Unteruehmender Firmengruppe DIEHL, Postfach 13 49 + 13 60, Teckstrasse, Oberndorf, D-7238, Germany; 07423/700, FAX 07423/70238
1977

Mazak Nissho Iwai Corporation

55 E. Commerce Dr, Schaumburg, IL 60173; 708-882-8777, FAX 708-882-0191
pres, Karl Sekikawa; vp mktg, Glenn Berkhahn; eng/serv mgr, Charles Zeman; 1987
Manufactures laser systems for programmable 3 and 6-axis cutting of ferrous, nonferrous, and nonmetalic materials. Systems designed for both precision and general purpose fabricating in table sizes from 40 x 80 in. to 84 x 132 in.; thickness capacities to .500 in. mild steel.

MBB Ind Products Group

MBB Ind Products Group, PO Box 801180, Munchen, D-8000, Germany; (089) 60731228, FAX (089) 60731442
1984
Manufactures industrial laser processing systems. Laser welding and cutting systems from 1 - 15 kW for two and three dimensional operation including periphery and integral clamping technology. Customer specific developments are tested and jobbing orders are executed in laser job shop center.

Meadex Technologies, PO Box 2528, Springfield, MA 01101; 413-567-3680, FAX 413-567-9068
pres, Edward F. Watson; mktg mgr, Mary B. Hayes; 1984
Technical consulting, marketing, sales, design, and manufacturing of high technology automated manufacturing equipment using laser and electron beam systems. Manufacturers representative services. Purchase, sale, and service of used laser and electron beam equipment. Laser and electron beam job shop services.

Mechanical Technology Inc, 968 Albany-Shaker Rd, Latham, NY 12110; 518-785-2211, FAX 518-785-2127, telex 685-4572 MTILATMUW
chrm, Harry Apkarian; ceo/pres, Donald VanLuvenee; gm/tech & engrg, Edward Zorzi; 1961
ACCULASER™ balancing system - high precision, easy-to-use, computerized control at-speed balancing. Integrates laser technology with computer control to balance rotors, turbines, electric motor armatures, turbopump rotors. By replacing milling, drilling, and other conventional processes it can lower unbalance levels and increase throughput rates.

Meftech Inc, 404 Tasman Dr, Sunnyvale, CA 94089; 408-745-1072, FAX 408-745-0506, telex 311405 MEFTECH CPTO
gen mgr, Eddie Chen; 1985
Manufactures laser trimming system LTS-704 specially designed to trim chip resistor and resistor network with fast speed, dual step/repeat, dual substrate pick/place mechanism, and user friendly software. High performance without high price.

Meistergram, 3517 W. Wendover Ave, Greensboro, NC 27407; 919-854-6200, FAX 919-292-6863
vp,gen mgr, Stephen R. Gluskin; nat'l sls mgr, James C. Wallace; 1932
Laser engraving system to create durable, permanent industrial and mimic panels, labels, and equipment tags that withstand adverse environmental conditions. Powerful software for 100% conversion of AutoCAD® drawings. Software includes complete 59-font library. Engraving area of 17 in. x 11.5 in. with material thicknesses up to 6 in. Training available at your location.

Melco Industries, 1575 W. 124th Ave, Denver, CO 80234; 303-457-1234*, FAX 303-252-0508, telex 168131 MLCO UT
vp mktg, Larry Pearson; vp eng, Bob Angliss; sls mgr, Steve Nunn; mktg mgr, Lisa Megna; 1972
Manufactures a laser engraving system capable of engraving on all types of solid wood and most glass, acrylic, slate, engraver' s marble, and coated metals.

Melles Griot, Catalog Div, 1770 Kettering St, Irvine, CA 92714; 714-261-5600, FAX 714-261-7589
vp/gen mgr, Albert A. Kusch; 1969
Markets and services RF excited sealed CO_2 lasers, 20-250 W output. These lasers are designed for research, industrial and OEM applications and feature compact size and long life. Designs and manufactures helium-neon lasers, .5 - 20 mW for commercial industrial, research, medical, and OEM applications.

Melles Griot, Laser Div, 2251 Rutherford Rd, Carlsbad, CA 92008; 619-438-2131, FAX 619-438-5208
vp/gen mgr, Michael J. Dorich; 1979
Designs and manufactures He-Ne lasers; portable, plasma tubes; packaged lasers; and power supplies for commercial, industrial, and medical OEM applications. Offers modification of standard products and custom-manufacture of lasers and related equipment. Distributes CO_2 pulsed and CW for medical, industrial, and OEM applications.

Meltec Laser AB, Industrivagen 2, PO Box 23, Sandviken, S-81121, Sweden; +46 26257401 *, FAX +46 26273709, telex 47055 EPCO
emp 4, 1982
Manufactures laser systems for material processing. Laser service. Job shop CO_2 and Nd-lasers. New and used lasers sold.

Merkle Schweissmachinenbau GmbH, Industriestrasse 3, Kotz I, D-8871, Germany; FAX (0)8221-32596, telex 8221817
1965
Offers MIG/MAG welders (140-600A), pulse arc welders (400A at 100% duty cycle) with 32 programs, 1 button adjustment, TIG welder, plasma: 18A to 250A at 100%, with compressed air and technical gases as plastic carrier.

Messer Griesheim GmbH, Steigerwald Strahltechnik, Benzstr. 11, PF 1365, Puchheim, W-8039, Germany; 089/80 92 0, FAX 089/80 92 215, telex 5 21 722
lsr dept mgr, Dr. Lothar Bakowsky; 1963
Supplies laser systems for 3-D welding and cutting. Power range between 700 and 6000 W. Special systems for surface treatment.

Micro Controle, P.A. de St. Guenault, B.P. 144, Evry Cedex, F-91005, France; 33-1-64 97 98 98 *, FAX 33-1-60 79 45 61, telex 691 105
1964
Manufactures solid-state lasers.

Micromanipulator Co Inc, 2801 Arrowhead Dr, Carson City, NV 89701; 702-882-2400, FAX 702-882-7694
1964

Midwest Laser Systems Inc, (sub of RWC Inc), 1101 Commerce Parkway, Findlay, OH 45840; 419-424-0062 *
1964
Custom designs and builds automated laser welding, cutting, and drilling systems for automotive and appliance manufacturers, ranging from single station manually loaded and unloaded systems to fully automated multiple station systems. Included is a standardized laser tube cutting system with a fully automated tube handling, loading and feeding system.

Mitsubishi Electric Corp, Nagoya Works, 14-1-5 Yadaminami Higashiku, Nagoya, Japan
1964

Mitsubishi Electric Corp, Overseas Mktg Div, Heavy Electrical Machinery, 2-3, Marunouchi 2-chome, Chiyoda-ku, Tokyo, 100, Japan; 03-218-3426 *, FAX 03-213-7356
1921
Engaged in the manufacturing and sales of various electrical equipment covering small IC chips to large turbine generators. In the industrial machinery field, EDM (electrical discharge machine), NC and CO_2 laser processing machines are typical products which are highly reputed in Japan and in the world.

Mitsubishi International Corp, Laser Group, 1500 Michael Dr, Wood Dale, IL 60191; 708-860-4210 *, FAX 312-860-4231, telex 254-688
1921
Sells and services a complete line of laser processing systems. Table travels range from 2 x 3 ft to 5 x 10 ft. Power capabilities range from 750 W to 10 kW. All lasers utilize silent discharge excitation and all system components are manufactured by Mitsubishi.

Miyachi Technos Corp, 95-3 Futatsu-zuka, Noda-city, Chiba-pref, 278, Japan; 0471-25-6177, FAX 0471-25-6178
pres, T. Nishizawa; mng dir, T. Jochi; mfg dir, T. Suzuki; dev dir, Dr. Y. Fujii; sls dir, M. Fujisawa; 1987
Manufactures Nd:YAG laser systems for welding, cutting, drilling, marking, engraving, soldering, and surface treatment applications.

MLI Lasers Ltd, PO Box 13135, Tel Aviv, 61131, Israel; (03) 492511, FAX (03) 495969
pres/ceo, Dr. A. Shachrai; dir r&d, D. Katz; 1979
Manufactures high-power CO_2 lasers, laser systems for diamond processing, and optics for high-power CO_2 lasers. Provides material processing services (job shop) with high-power CO_2 lasers.

mls munich laser systems gmbh, Gollierstrasse 70, Munchen 2, D-8000, Germany; 089/502 89 51, FAX 089/519 99 40
pres, Karl Sauer; sls mgr, Franz Westermeir; mktg mgr, L. Ploner; eng mgr, Max Zopfl; mfg mgr, Gunter Josef Mazurok; 1985
Manufactures marking systems, resistor trimmers, Nd:YAG lasers, ceramic scribing and cutting systems, and slab lasers for cutting, drilling and welding.

Modern Machine Tool Co, 2005 Losey, Jackson, MI 49203; 517-788-9120, FAX 517-788-2668
sec/treas, Steven G. Walker; vp engrg, Gregory A. Walker; emp 32, 1916
Manufactures single-spindle, lathe type cut-off machines for the production cutting to length around tube, pipe, or bar stock. Machines can be equipped with a laser for cutting off tubing, and "C and Z" axis contouring of tubular parts.

Motoman Inc, 805 Liberty Ln, West Carrollton, OH 45449; 513-847-6200*, FAX 513-847-6277
1916
Manufactures robotic YAG laser cutting systems. The K10 Series articulated arms (6 & 7 axis) gives end users the ultimate flexibility for YAG laser applications. The use of the fiberoptic cable facilitates the use of the total available work envelope of the K10S Series robot. Robots can be mounted as floor mount, or inverted (ceiling mounted).

MPB Technologies Inc, Lasers & Electro-Optics, 1725 North Service Rd, Trans Canada Highway, Dorval, Quebec, H9P 1J1, Canada; 514-683-1490, FAX 514-683-1727
mgr, ind laser sys, Wes Jamroz; 1977
Manufactures sealed-off, refillable industrial CO_2 lasers up to 140 W. Also available a complete line of accessories including programmable beam attenuator, X-Y scanning mirror system, programmable 'zoom' system. Also, manufactures laser vacuum welding stations. Fully equipped materials processing laboratory provides applications testing and demonstration services.

M. Torres Disenos Industriales S.A., Carretera Huesca KM9, Navarra, Torres de Elorz, E-31119, Spain; 048-31-7811, FAX 048-31-7952, telex 37866 mt e
1977
Manufactures machinery of 2 and 5-axis for the following applications: cutting, welding and surface treatment by laser beam; welding by electron beam; drilling of metal and nonmetal materials.

MTS Systems Corp, 14000 Technology Dr, Eden Prairie, MN 55344; 612-937-4000, FAX 612-937-4515
ceo/pres, Donald M. Sullivan; vp adv technology, William Beduhn; emp 1400, 1966
Manufactures industrial laser systems.

Nanavati Sales Pvt Ltd, Laser Division, 2A Dayaldas Rd, Laxmi Prasad Bldg, Vile Parle, Bombay, 400 057, India; 91-22-6147588/6146319, FAX 91-22-6147385
chmn, Mahendra M. Nanavati; mng dir, Uday C. Nanavati; 1971
Represents interests of M. Pfaffen Laser Optik Mechanik (Switzerland) in India and worldwide. Specializes in lasers, laser material processing, optics, electronics and handlings, and intergrating the laser system into the production line. Designs and constructs the materials-handling systems that bring the workpiece to the machine, perform the laser process, and then discharge the workpiece, after inspection state. The following products are offered: CO_2 Laser from 20 to 200-W, He-Ne laser from .5 to 50 mW, materials processing machines based on CO_2 laser, cutting machines based on Nd:YAG laser, scriber "Light Writer", laser system for diamond drilling and cutting.

NEC Corp, International Electron Device Div, 1120 Shimokuzawa, Sagamihara, Kanagawa, 229, Japan; 0427 73-1111 *, FAX 0427-71 0875
1971
Manufactures Nd:YAG laser and systems.*

NEC Electronics Inc, Electron Components, Laser Systems (sub of NEC Corp), 401 Ellis St, PO Box 7241, Mountain View, CA 94039; 415-965-6116, FAX 415-965-6000
pres, K. Saito; mktg mgr, S. Watanabe; 1981
Manufactures up to 2 kW high power YAG laser systems and optical fiber delivery systems. M702A laser cutter is for flexible and 3 dimensional sheet metal cutting. Also manufactures YAG laser marking systems.

Uranit GmbH, Stetternicher Staatsforst, PO Box 1411, Julich, D-5170, Germany; 02461/65-388, FAX 02461/65-449
Jurgen Huftle; 1969
Maunfactures TEA-CO_2 lasers and laser systems for marking, materials processing and scientific applications, and high voltage power supplies.

Newall Manufacturing Technology Ltd, Aireworth Rd, Keighley, W. Yorkshire, BD21 4DP, UK; 0535 667911, FAX 0535 664418, telex 51266 KAGEE G
mng dir, G. Hemingway; sls dir, I.P. Laven; 1900
Nonconventional machine tools including laser systems linked with vision and airflow inspection, compressor and turbine blade grinding as well as landing gear grinding machines.

Newcor Bay City, Div of Newcor Inc, 1846 Trumbull Dr, Bay City, MI 48707; 517-893-9505 *, FAX 517-893-8961
emp 300, 1933
Manufactures welding and heat treating equipment. Special machine with emphasis on welding and heat treating equipment.

Nichimen Co Ltd, 1-6, Takata-cho, Chuo-ku, Tokyo, 144, Japan
1933

Nihon Welding Oko Ltd

Nihon Welding Oko Ltd, 3-19-9 Nishikamata, Ota-ky, Tokyo, Japan

Niigata Engineering Co Ltd, Machine Div, 1-4-1 Kasumigaseki Chiyodku, Tokyo, 100, Japan; 03-504-2151, FAX 03-3595-2648, telex 03-222-7111
emp 5400, 1895
Manufactures CO_2 laser processing systems, standard systems, special machine systems. Also provides research for laser application and laser systems including handling systems and optics.

Nippei Toyama Corp, Laser System Div, Onarimon Yusen Bldg, 3-23-5 Nishi-shinbashi, Minato-ku, Tokyo, 105, Japan; (03)3434-8272, FAX (03)3434-0384, telex 242-5397 (NTCO J)
div mgr, Roy Horii; int'l dept mgr, Tom Ichimaru; system sls mgr, M. Yamashita; emp 1138, 1945
Offers TLM series 3-dimensional, 5-axis laser cutting system and TLV series 3-axis cutting system.

Nippon Sanso Corporation, Industrial Equipment Div, 1-16-7 Nishishinbashi, Minato-ku, Tokyo, 105, Japan; FAX 03-3593-6237, telex J24228 NSANSO
emp 1800, 1910

Nisshinbo Industries Inc, Miai Mechatronics Plant, 30, Azukizaka, Miai-cho, Okazaki, Aichi, 444, Japan; 81-564-55-1101*, FAX 81-564-55-0311
dir, Suguru Miyazu; exp mgr, Tatsumi Sakamoto; desng mgr, Nobuo Kita; mfg mgr, Osamu Enya; 1947
Manufactures CNC-turret punching presses, laser cutting machines, CNC-press brakes, NC shears and special purpose machines for drilling, milling, deburring, polishing, and welding.

Nissin Boseki, Cresent Bldg. 71 72 Kyomachi, Ikuta-ku, Kobe, 650, Japan
1947

NM Laser Products Inc, 140 San Lazaro Ave, Sunnyvale, CA 94086; 408-733-1520, FAX 408-736-1152
pres, David C. Woodruff; 1987
Manufacturer of high-speed, high-power laser shutters for industrial processing. Models for CO_2, YAG, and excimer. Built-in beam dump/heat exchangers for modulation and safety. Custom development.

North China Research Institute of Electro-Optics, Research of Laser Processing Technique, Da Shen Zi, Chao Yang District, PO Box 8511, Beijing, 100015, PRC; 4362761 Beijing, FAX 01 43 63 226, telex 211169 NCRI CN
div dir, Song Weilian; 1956
Manufactures industrial lasers and systems.

Nothelfer GmbH, BleicherstraBe 7, PO Box 1960, Ravensburg, D-7980, Germany; (0751) 886-247, FAX (0751) 88 61 11, telex 732822
1921
Manufactures laser equipment for welding and cutting sHeets of different thickness, including specialty steel. The positioning of the sheets to the laser beam focus is electronic controlled. A further possibility of application of the laser welding process is the welding of coils to standard or optimized coils.

NTC/Marubeni America Corp, Laser Machine Group, 2000 Town Center, Suite 2150, Southfield, MI 48075; 313-353-7060/800-3976427, FAX 313-353-2298, telex 230172
sls mgr, Thomas Earl; serv mgr, Steven Glovak; 1921
Manufactures five-axis laser cutting machines. Sole distributor of NTC' s three-axis and five-axis laser cutting machines. Also offers training, service and spare parts. A dealer network is established throughout the United States and abroad.

NVL Balliu MTC, E Van Arenbergstraat 43, Sint-Amandsberg, 9040, Belgium; 091/28.13.86, FAX 091/28.86.37
1894
Experience in precision mechanics, robotics, laser technology and waterjet cutting. Offers a complete and versatile series of 2D and 3D machining centers.

Polaris Electronics Corp, 630 S. Rogers Rd, Olathe, KS 66062; 913-764-5210
1894

OPL (Oerlikon- PRC Laser SA), Route des Avouillons 16, Gland, CH-1196, Switzerland; 011-41-22-64 19 35, FAX 011-41-22-64 27 48
mng dir, Jurg Steffen; sls mgr, Klaus W. Schenzinger; 1990
Manufactures low to high power industrial CO_2 lasers and accessories. Produces special laser packages and provides all engineering required to meet special environmental or floor-space requirements. Offers applications engineering and accessories. Also sells lasers in knocked down 'kit' form for OEM integration with turnkey systems.

Optical Engineering Inc, PO Box 696, 3300 Coffey Lane, Santa Rosa, CA 95403; 707-528-1080, FAX 707-527-8514
pres, John A. Macken; 1968
Manufactures CO_2 lasers, laser power probes, spectrum analyzers, thermal image plates, and digital power probes.

optomic

(sub of Optomic Technologies Corp Ltd), Ramat Gabriel Industrial Park, PO Box 153, Migdal Ha'Emek, 10551, Israel; 972 6 545440, FAX 972 6 545382
ceo/pres, Dr. Oded Amichai; exec vp, B. Vered; vp eng, A. Zajdman; vp mktg/bus devl, Y.J. Gleitman; 1985
Develops, designs and manufactures CO_2 lasers and other optical systems for industrial and military applications. Offers a 750-W compact, rugged and portable CO_2 laser suitable for gantry and robot mounting in various industrial applications.

OTC America Inc, PO Box 217037, Charlotte, NC 28221; 704-597-8240, FAX 704-333-9790
1979
Manufactures CO_2 laser cutting and welding equipment.

Otto Borries KG, div of Borries Markier Systeme, Ernst-Mey-Strasse 4, Postfach 10 03 56, Leinfelden, D-7022, Germany; 0711 7 59 09 0, FAX 0711 59 09 20
1979

PPL

9207 Eton Ave, Chatsworth, CA 91311; 818-700-8977, FAX 818-700-8984
1979
Offers laser CVD process stations for R&D, pilot, and production lines, IR, VIS, UV, YAG, Argon Ion, HeCad, and Excimer sources. Provides laser micro machining systems, IR, VIS, UV, for R&D, pilot, and production units, laser scribing, drilling, cutting and welding systems, and photon backscatter systems.

Pacific Trinetics Corp, 1125 Linda Vista Dr, Suite #109, San Marcos, CA 92069; 619-471-2350, FAX 619-474-0698
pres, Gordon Zablotny; 1987
Manufactures CO_2 machining systems for cutting and via forming in unfired ceramic tape. Primarily for multilayer ceramic package manufacturing. Complete computer controlled turn-key systems.

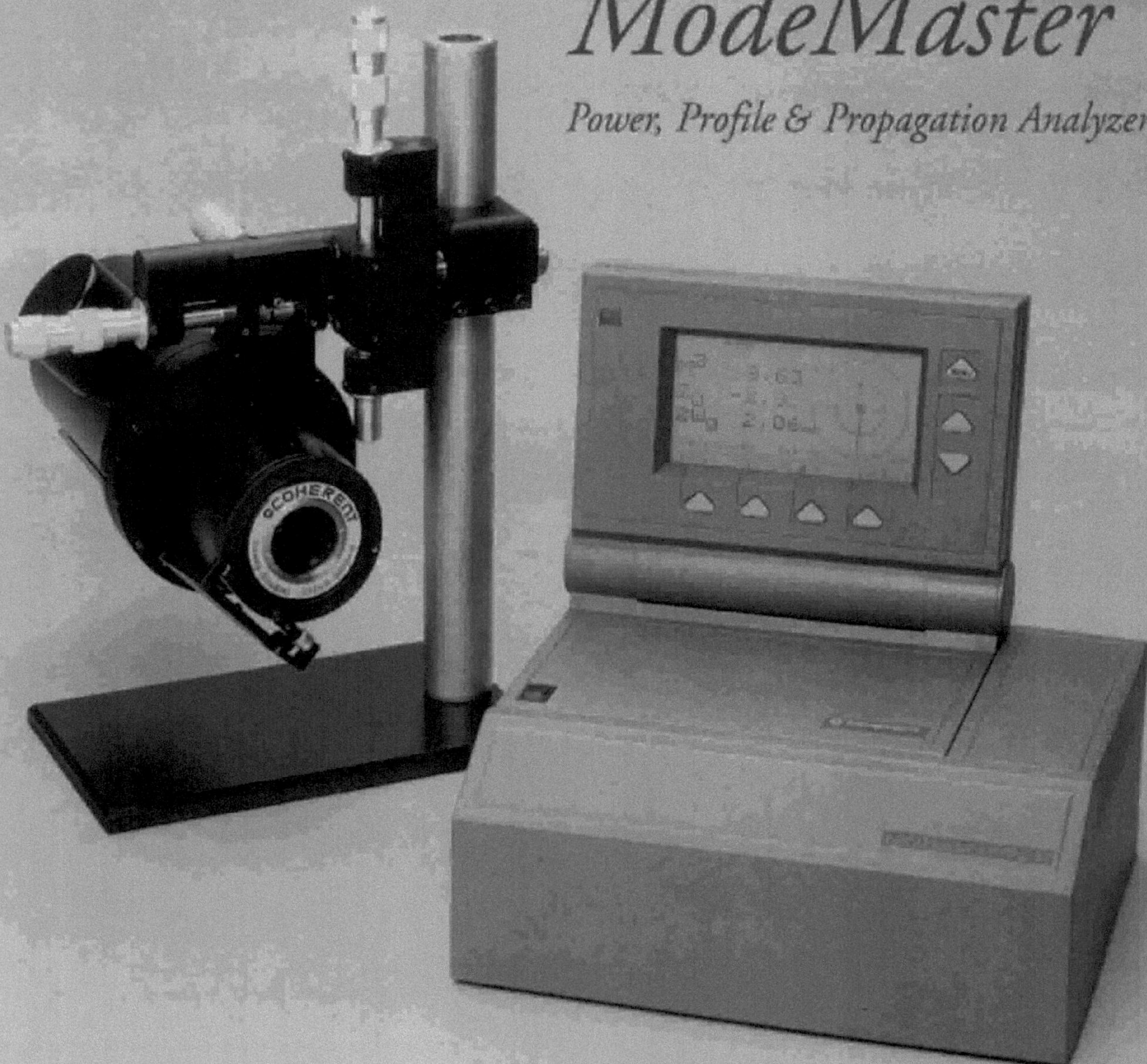

That's

Panasonic Factory Automation Co, Materials Processing Dept (sub of Matsushita Electric Corp of America), 9401 W. Grand Ave, Franklin Park, IL 60131; 708-452-2500, FAX 708-452-2575
pres, Y. Kobe; asst gen mgr, William R. Emery; sls mgr, Martin Weir; eng mgr, T. Yamamoto; 1987
The importation and resale on an OEM basis of fast axial flow carbon dioxide lasers. Units are available in 1000, 1500, and 2000 W. Units have become well known for their compact size, process repeatibility, ease of service, and excellent uptime. Engineering and service readily available.

PermaNova Lasersystem AB, Krokslatts Fabriker 30, Molndal, S-43137, Sweden; +46 (0)31 870280, FAX +46 (0)31 864612
gen mgr, Nils Ohrberg; mgr sys devlmt, Ulf Sandstom; sls lsr sys, Urban Widen; 1985
Laser processing with YAG, excimer and small CO_2 lasers. Turnkey systems for welding, cutting, drilling, marking, engraving, soldering and excimer processing. High power fiberoptics for YAG lasers. Specializes in glovebox welding, laser soldering of electronic components, surface or hole mounted.

Permascand AB, Laser Div, Box 42, Ljungaverk, S-84010, Sweden; 46-691-32940, FAX 46-691-33004, telex 71279
vp, Nils Ohrberg; 1979
System manufacturer and job shop. Job shop services for lasers include cutting, welding and heat treating. Applications R&D. Laser systems for material processing. Distributor.

Potomac Photonics Inc, 4720-E Boston Way, Lanham, MD 20706; 301-459-3031, FAX 301-459-3034
pres, C. Paul Christensen, PhD; mktg mgr, Sarah Cohn; 1982
Manufactures the UV waveguide laser, a small economical table-top Rf discharge excimer-type laser which is air cooled and operates on 110 current. Applications include small spot size direct-write photoablation, failure analysis, chemical analysis, dye pumping and wire stripping.

PRC® LASER

North Frontage Rd, Landing, NJ 07850; 201-347-0100, FAX 201-347-8932
pres, Carl Nilsen; exec vp, Benjamin Switzer; prod mgr, Ancel Thompson; vp eng svcs, John Kratsios; 1976
Manufactures low to high-power industrial CO_2 lasers and accessories. Produces special laser packages and provides all engineering required to meet special environmental or floor-space requirements. Also sells lasers in knocked-down kit form for OEM integration with turnkey systems.

Preciacier, BP10, Luzy, F-58170, France; 86.30.06.33 * FAX 86.30.21.75
1976

Precision Systems Ltd, Harding Way, Somersham Rd, St. Ives, Cambridge, PE174WR, UK; 0480 67101, FAX 0480 494462
mng dir, Simon Standley; mktg mgr, Dr. Michael Croker; 1981
Manufactures and supplies advanced precision welding systems based on the pulsed DC TIG and laser technologies. Specialists in automated and custom engineered machines for pressure sensors, capsules, bellows, electrical and gas turbine component fabrication and repair. In-house process development, consultancy and subcontract services available, with international maintenance support.

Prima Industrie Spa, Via Antonelli 32, Regina Margherita, di Collegno-Torino, 10097, Italy; 011 410.31, FAX 011-411-28-27, telex 222211 IMPRIV I
pres, F. Sartorio; sls/mktg mgr, A. Delle Piane; mfg mgr, E. Basso; tech mgr, F. Grassi; 1977
Manufactures multiaxis laser processing systems for industrial cutting and welding applications.

Prime Technology Inc, 4936 Kendrick S.E., Grand Rapids, MI 49512; 616-942-2104/800882-8196, FAX 616-942-2108
pres, Phil Pachulski; cfo, Lynn Calhoun; gen sls mgr, Mark Koetje; 1982
Specialists in precision metal cutting machine tools and services. Complete turnkey systems, documentation, and engineering.

Progressive Tool & Industries Co, Laser Process Development, 210000 Telegraph Rd, Southfield, MI 48034; 313-353-8888, FAX 313-353-5317
pres, L. Wisne; vp sls, J. Digiovanni; mgr lpd, H. Schlatter; app eng, P. Busuttil; emp 1800, 1940
Engineers and produces turnkey laser systems for cutting and welding for automotive, aircraft, and related industries. Job shop service for short run cutting and welding. Research and development center for CO_2 and YAG process application and feasibility development.

Prometec GmbH, Laser Products Div, Julicher Strasse 338, Aachen, D-5100, Germany; 49-241-166090, FAX 49-241-1660950, telex (17)2414027
J. van Gilse; 1985
Manufactures and sells diagnostic instruments for high power CO_2 lasers; beam analyzers for unfocused and focused laser beams up to 25 kW and 10 mW/cm^2; high accuracy laser beam power meters up to 25 kW and 25 kW/cm^2; measurement and calculation of beam quality M^2 or K; laser beam monitoring.

PTR - Precision Technologies Inc, 120 Post Road, Enfield, CT 06082-5699; 203-741-2281, FAX 203-745-7932
pres, Gottfried Kuesters; mfg dir, Bill Kennedy; sls mgr, Don Powers; 1989
Designs and builds a full line of electron beam welding and laser beam welding/cutting/drilling systems for industrial applications.

Pulse Systems Inc, 422 Connie Ave, Los Alamos, NM 87544; 505-672-1926, FAX 505-672-1934
pres, Edward J. McLellan; sls mgr, Will Arthur; mktg mgr, Linda L. McLellan; 1979
Manufactures state-of-the-art pulsed, flowing gas CO_2 lasers for scientific, industrial, marking, and government applications. Manufacturers laser marking systems for marking/coding of paper, glass, plastics, and anodized/painted metals in electronics, semiconductor, and consumer-product areas. Also manufactures laser detector preamplifiers for CO_2 lasers.

P&W Technic Corp, 4F33 Chung Cheng N. Rd, San Chung Taipei Hsien, Taiwan, ROC; (02) 982-1920, FAX (02) 984-9630
1979

Quanta System Srl, Industrial Laser Div, via Venezia Giulia 18, Milan, I-20157, Italy; 02 332 00239, FAX 02 35 75140
Dr. V. Fantini, Dr. A. Ferrario; 1984
Manufactures industrial solid-state Nd:YAG lasers for cutting, welding and drilling. Designs and builds custom laser processing systems. Also manufactures a complete line of laser marking systems for OEM applications.

Quantel SA, Avenue de l'Atlantique, BP 23, Les Ulis, F-91941, France; 33 1 69 29 17 00, FAX 33 1 69 29 17 29, telex 691329 F
chrmn, A. Orszag; mktg mgr, P. Aubourg; mfg mgr, A. Diard; 1970
Manufactures Nd:YAG lasers for industrial applications. YAG laser series IQL delivering from 300 to 2000 W of average power and CO_2 lasers for industrial applications delivering 1000, 1200, 3000 W. Specific laser systems upon request with the possibility of integration in specific machines.

Quantronix Corp, 49 Wireless Blvd, PO Box 9014, Smithtown, NY 11787-9014; 516-273-6900, FAX 516-273-6958
pres, Donald Mitchell; vp tech, Dr. Martin Cohen; prod mgr/ind lsrs, P. John Goodfellow; 1967
Manufacturers CW pumped Nd:YAG and Nd:YLF lasers for industrial, scientific, and medical applications. A-O-switched and harmonic versions at 1064 nm, 532 nm, 355 nm, 266 nm, and 1319 nm for Nd:YAG. Mode-locked and Q-switched versions at 1313 nm, 1053 nm, and 527 nm for Nd:YLF. Sync-pumped dye lasers and regenerative amplifiers for ultrafast market. Opaque/clear mask repair systems for semiconductor industry. General purpose Nd:YAG marking systems. Medical laser systems.

Quantum Chromodynamics Inc, Quantrad Trimmers, 2401 208th St, Unit 9, Torrance, CA 90501; 213-320-5717, FAX 213-320-5848
1967
Manufactures and sells laser systems for special applications such as cutting and scribing of substrates, resistor trimming and circuit tuning, and laser marking applications. Provides full service laser marking/engraving jobshop. Marking styles include: Bar Code 3 of 9, Helvetica, Gothic, OCR-A, Script, Old English, Roman, and custom logos.

Quantum Laser Engineering Ltd, Unit 1, Cross Rd Ind Estate, 31-41 Cross Rd, Coventry, CV6 5GR, UK; 0203 680668, FAX 0203 680924, telex 312309 QUALAS G
mng dir, Martin Rourke; tech dir, John Peter Hancocks; 1984
Manufactures standard CO_2 laser cutting systems up to 2500 W and 3500 x 2500 mm table size in 2- and 3-axis models. Manufactures special laser machining systems to customers requirements and Nd:YAG laser marking systems in hand, semi, and fully automatic mode.

Questek Inc, 44 Manning Rd, Billerica, MA 01821; 508-667-6790, FAX 508-667-9919
pres, Gary K. Klauminzer; sls/cus sprt mgr, Douglas H. Post; mfg/engng mgr, Howard Kaufman; 1982
Manufactures excimer lasers for scientific, medical and industrial applications. Custom designs for OEM customers.

Raskin Machines SA, Ch. Des Dragons 2, Lausanne-Vernand, Cheseaux, CH-1033, Switzerland; 21/731 94 00, FAX 21/731 94 09, telex 459 300
1947
Offers the largest range of equipment for cutting in the flat or in space by punching, laser, water jet and plasma in 3 to 5 axis, and loader/unloader, programming systems.

Raskin USA, 3150 Verona Ave, Buford, GA 30518; 404-271-7558, FAX 404-271-8260
1947
Manufacturer of laser machining centers, laser tube cutters, laser cutting/welding/heating systems and waterjet. Specializing in cutting of stainless steels and aluminium with ultra-high pressure assist gas, rotary table applications and grinding head.

Raycon Corp, A George Fischer Co, 2850 S. Industrial Hwy, Ann Arbor, MI 48104; 313-677-4911, FAX 313-677-2778
pres, Harry Moser; lsr prod mgr, John Ruselowski; 1965
Manufactures a line of standard CO_2 and YAG laser processing systems. These systems are typically three-to seven-axis systems with table movements of 60 in. or larger and full three-dimensional capability. Also, provides engineering and builds capability for specialized laser tools for dedicated manufacturing processes on the production floor.

Raytheon Co, Laser Products, 465 Centre St, Quincy, MA 02169; 617-479-5300, FAX 617-472-5084
mgr, Bernard Werschher; mfg/sys mgr, Dennis Breen; 1961
Manufactures industrial lasers and laser systems for welding, cutting and drilling. Oldest and most reliable manufacturer of industrial YAG and CO_2 lasers in US. Also manufactures fiberoptic laser machining centers, scribing systems and laser markers.

BERATUNG PROJEKTIERUNG DURCHFÜHRUNG KIST INGENIEUR BÜRO SERVICE ERSATZTEILLAGER LOHNBESCHRIFTUNG
LASERBESCHRIFTUNG

Sudetenstr 34, Postfach 16, Eckental, D-8501, Germany; 49-(0)-9126-8026, FAX 49-(0)-9126-7698, telex 626 916 kitec
1983
Provides all services associated with laser marking such as: engineering of complete laser lines, spare parts, consulting, start-up of laser marking lines, training, service, job shop, and applications.

Reis GmbH & Co, Im Weidig 1-4, Obernburg, D-8753, Germany; 06022/503-0, FAX 06022/503-110
W. Reis; 1957
Development, production, sales and service of robotic systems for cutting and welding with laser or plasma installations.

Renault Automation, Boite Postale 70 Centre Parly, Le Chesnay Cedex, F-78152, France; (1)39.54.91.91*, FAX (1)39.54.74.09, telex 695354 F
pres, Jacques Malavas; sls/mktg mgr, Michel Comes; emp 1600, 1957
Provides laser cutting and welding applications for the automotive industry. Standard product is a 3D laser gantry robot.

Resonetics, 4 Bud Way Bldg 21, Nashua, NH 03063; 603-886-6772 *, FAX 603-886-3655
1957
Excimer-laser materials-processing services, system manufacturer, and technology-development center. Full systems integration and custom-design capabilities. Independent consulting services provided. Specializes in industrial excimer-laser applications. Products include; excimer gas purifiers, air scrubbers, beam delivery and high-rep-rate direct-write materials-processing systems.

robomatix

23399 Commerce Dr, Suite B-10, Farmington Hills, MI 48335; 313-471-6121, FAX 313-471-3977
pres, Doron Laks; chrm, Zami Aberman; 1983
Manufactures multi-axis CO_2 laser cutting and welding systems for the body-in-white automotive production lines. Specializes in the delivery of turnkey projects for the automotive production lines of VW, Ford, Opel, GM and others. Products are: laser press-dedicated laser cutting systems; laser robot-flexible laser cutting systems; and MODULASER™ flexible laser cutting and welding systems.

ROFIN SINAR
INCORPORATED

(sub of Siemens Corp), 45701 Mast St, Plymouth, MI 48170; 313-455-5400, FAX 313-455-2741
pres, Peter Wirth; vp sls/mktg, Richard Walker; 1983
Manufacturers of CO_2 and Nd:YAG lasers and laser markers, beam-delivery modules, fiberoptics, and industrial laser systems. Process development and systems concepting is also provided.

ROFIN SINAR
LASER

(sub of Siemens AG), Berzeliusstrasse 87, PO Box 74 03 60, Hamburg 74, D-2000, Germany; 49 (0) 40 73 36 30, FAX 49 (0) 40 73 36 3100
pres, Samuel Simonsson; vp sls/mktg, Bernd Ladiges; 1983
Manufacturers of CO_2 and Nd:YAG lasers and laser markers, beam-delivery modules, fiberoptics, and industrial laser systems. Process development and systems concepting is also provided.

Rolf Wissner GmbH, Rudolf-Wissel Str. 14-16, Gottingen, D-3400, Germany; 05 51 63 12 30, FAX 05 51 63 12 30
1983
Production and development of CNC laser cutting machines and CNC engraving machines.

Rouchaud Industrie, Limoges Precision (sub of Smits/ Lieure), 36 Ave Saint Eloi, BP 565, Limoges Cedex, F-87012, France; 55-341734 *, FAX 55-320307
emp 115, 1987
Develops (under the trade mark Limoges Precision), laser cutting machines; flying optic systems, standard machines in 2 D and 3 D for cutting and welding applications (specially for job shops); machining centers (multiprocess); and special machines for welding and cutting production. All these machines can be integrated into production lines.

RtMc, 2414 E. Paradise Lane, Phoenix, AZ 85032; 602-493-9343, FAX 602-493-8003
pres, Richard T. Miller; vp, Roy J. Bruno; 1987
Laser wire stripping machines and services to remove all types of polymer insulations including teflon, kapton, fiberglass and other difficult to remove insulations as well as polyimide and polyurathane coatings. Customized laser machining systems using CO_2, YAG and Eximer lasers. OEM air cleaning and water cooling systems.

Rusch Wire Die Corp, 60 Brook St, Croton-on-Hudson, NY 10520; 914-271-3431
pres, E.W. Rusch; 1936
Drawing diamond and tungsen carbine dies, wire guided. Diamond and carbide wear resisting parts. Carbide extrusion diamond dies.

Safepack Masch GmbH, Postfach 1211, Mullheim, D-7840, Germany; 07631-18090, FAX 07631-180949, telex 763121-Weileng
1936
Manufacturer of laser welding machines (systems).

Safmatic, PO Box 009, Z.I. Rue Lavoisier, Parthenay Cedex, F-79201, France; 33-49943155, FAX 33-49943465
dir gen, Claude Obe; sls mgr, Paul Baronetti; dir tech, Jacques Feix; 1962
Offers a wide range of welding and cutting systems using many processes such as: Wig, Mag, Plasma, Submerged Arc, Electron Beam and Laser. For 20 years has manufactured special cutting and welding laser machines using YAG and CO_2 sources up to 5 kW power to offer the best welding/cutting "turnkey" solutions for efficiency, quality and payback.

Sala SpA, Via Per Barco, 11, Levico, I-38056, Italy; 0461/ 701388, FAX 0461/701410
mng dir, Marco Sala; lsr sys div dir, Maurizio Tabarelli; export dir, Giorgio Bellei; 1951
Manufactures four CO_2 laser powered systems. Three systems utilize automatic machines for cutting tubes up to 0-135 mm. One version has numeric control. The other system is for numerically controlled machines for cutting sheets up to 2000 x 4000 mm. An automatic pallet changing system and sheet loader is provided.

Geo T. Schmidt Inc, 6151 W. Howard St, Niles, IL 60648; 708-647-7117, FAX 708-647-7593
pres, Neal O'Connor; sls/mkting dir, Hal Durgin; eng dir, Ben Bergwerf; 1895
Specializes in equipment for permanent marking including YAG laser marking systems, hydraulic and pneumatic marking machines, stylus markers, numbering heads, engraved tools, and nameplate marking equipment.

Sciaky Industries, 119 Quai Jules Guesde, Vitry sur Seine, Cedex, F-94400, France; 1-45 73 43 00, FAX 1-46 825880, telex 265860
pres/gen dir, D.G. Akel; 1988
Manufactures precision laser equipment for the car industry. R&D laboratory for feasibility tests and technological development.

Sciaky Inc, Ferranti, (sub of Ferranti International), 4915 W. 67th St, Chicago, IL 60638; 708-594-3800, FAX 708-594-9213, telex 25 4099
pres, John A. Kasuba; natl sls mgr, James Stormont; 1940
Supplies complete laser processing systems, as well as electron beam, resistance welding, and fusion arc equipment. Maintains a multi-process equipped laboratory to provide sample work and R&D. In addition, service and spare parts departments provide international customer support.

Selto Laser Division, Sumgaitskaya Str 8, Cherkassy, 257336, USSR
1940

Servo Robot Inc, 1380 Graham Bell, Boucherville, Quebec, J4B 6H5, Canada; 514-655-4223, FAX 514-655-4963
pres, Jean-Paul Boillot; sls/serv mgr, Jean-Claude Fontaine; tech dir, Denis Villemure; prd mgr, Jean-Luc Cote; 1983
Manufactures, develops, and sells high-quality products and systems for industrial automation. Specializes in real-time robot guidance, particularly seam tracking and adaptive process control, offering hardware, software and peripherals as well as training and maintenance services to suit a large variety of applications. Company is active internationally and a number of industries world-wide can benefit from its expertise.

Shibuya Kogyo Co Ltd, Mameda-Honmachi, Kanazawa, 920, Japan; 0762-62-1200, FAX 0762-23-1795, telex 5122-428 SIBUYA J
sr mng dir, M. Maeda; mktg mgr, M. Ishikawa; eng mgr, K. Kishimoto; mfg mgr, Y. Kita; emp 1200, 1931
Manufactures CO_2 laser cutting machines (plane & 3-dimensional), CO_2 laser cutting robots (3-dimensional), CO_2 TEA lasers, laser marking systems, laser treatment systems for athlete' s foot and semiconductor processing systems.

Southeastern Die Co Inc, 5205 Snap Finger Woods Dr., Decatur, GA 30035
1931

Spectra Gases, 277 Coit St, Irvington, NJ 07111; 201-372-2060, FAX 201-372-8551
1980
Manufactures CO_2 and excimer laser gases and gas mixtures. Gas handling equipment such as regulators, gas panels, gas cabinets, cross purges, changeover manifolds, and other related equipment are also available.

Spectron Laser Systems

Spectron LASER·SYSTEMS

21 Paynes Lane, Rugby/Warcs, CV212UH, UK; (788) 544694, FAX (788) 575379, telex 317338 NDYAGG
mng dir, Dr. Paul Sarkies; 1983
Manufactures pulsed Nd:YAG lasers, CW and CW Q-switched Nd:YAG lasers suitable for OEM, industrial and R&D use. Flexible, modular design enhances special system capability.

Strippit Inc, (a unit of Idex), 12975 Clarence Center Rd, Akron, NY 14001; 716-542-4511, FAX 716-542-5957
1925
Manufacturer and leading supplier of fabrication equipment. Lasertool CNC holemaker, a patented laser cutting/punch press combination developed by Strippit, cuts a variety of production materials: aluminum, titanium, plastics, rubber, wood, paper products, leather, textiles up to 400 IPM in materials to .250" .

SYNRAD INC.

11816 Northcreek Pkwy N., Suite 103, Bothell, WA 98011; 206-483-6100, FAX 206-485-4882
pres, Peter Laakmann; chf sci, Dr. Stan Byron; 1984
Manufactures sealed "all metal" RF-excited CO_2, grating tuned CO_2, CO lasers. Power levels from 10-250 W. Manufacturer of high brightness CW, Q-Switched Nd:YAG lasers.

Tanaka Engineering Works Ltd, 11 Chikumasawa, Miyoshi-cho, Iruma-gun, Saitama, 354, Japan
1984

Tay Technology Inc, 900 Walt Whitman Rd, Melville, NY 11747; 516-351-1080, FAX 516-549-3944
Robert M. Schmidt; 1980

3D Systems Inc, (sub of 3-D Systems Inc), 26081 Avenue Hall, Valencia, CA 91355; 805-295-5600, FAX 805-295-0249
ceo, Robert Fletcher; pres, Charles Hull; cfo, Ed Kaftal; 1986
Stereolithography is used to rapidly create three-dimensional concept models, masters and patterns directly from CAD data without machining or tooling. Stereolithography is an enabling technology to concurrent engineering. Market time and product development costs can be drastically cut through this exciting technology.

Thyssen Laser-Technik GmbH, Steinbachstrasse 15, Aachen, D-5100, Germany; 0241-8906-160, FAX 0241-8906-121
1988
Application of laser technology in the Thyssen-Group and for related customers. Feasibility studies, partially at the Fraunhofer Institut fur Lasertechnik, system development, pilot plans, coordination and performance of test series and system integration.

Toeller Laserkraft GmbH, Toellerstrasse 1, Hagenburg, D-3055, Germany; 05033-7786, FAX 05033-7311
sls/design dept mgr, Manfred Toeller; 1989
Flying optics: mobile laser beam guidance by means of adjustable mirrors from 800 x 800 mm up to 2100 x 3200 mm. The workpiece rests on the cutting table during the processing operation. Advantages are in close tolerances at high feed rates. Special machines for the sign-making industry. Combinations with laser and milling head. Water jet cutting tables.

Toshiba America Electronic Comkponents Inc, One Parkway North, Suite 500, Deerfield, IL 60015; 708-945-1500
1989

Toshiba Corp, Mie Works Factory Automation Div, 2121 Nao Asahi-Cho, Mie-Gun, Mie-Pref, 512, Japan; 0593-54-6047, FAX 0593-77-2053
mgr, K. Tsuyuki; 1875
Manufactures 0.6 kW, 1.2 kW, 2.0 kW, 3.0 kW and 5.0 kW axial flow type CO_2 laser osillators and is supplying domestic and overseas markets. Can supply three standard CO_2 laser machining systems and special purpose CO_2 or YAG laser machining systems.

Toyama America (NTC), 657 E. Golf Rd, Suite 307, Arlington Heights, IL 60005; 708-640-0940, FAX 708-640-0948
mktg/tech mgr, Ken Miyata; 1984
The three-dimensional 5-axis controlled laser cutting system, TLM series, has been widely accepted. TLM series has been proven to drastically help minimize the production cost for the trimming process of press-formed products in trial stages.

Tri Sigma Corp, 3480 E. Britannia Drive, Tucson, AZ 85706; 602-294-5300, FAX 602-294-0099
pres, James W. Drain; exec vp, Paul Mioduski; sys mgr, Ron Mundt; 1979
Designs and manufactures turnkey laser systems: Model DDL4024 S Axis system, and Model 1248 Plastic Cutting CO_2 system.

TRUMPF GmbH & Co, Johann Maus Strasse 2, Ditzingen, D-7257, Germany; 07156-303-0, TWX 07156-303-309
pres, Berthold Leibinger; gen mgr laser sys div, Martin Benzinger; mng dir r&d, Hans Klingel; gen mgr laser div, Dr. Reinhard Wollermann; sls/mktg mgr laser div, Lebrecht von Trotha; emp 3000, 1923
Manufactures industrial CO_2 lasers and laser systems for cutting, welding, and surface modification. Lasers available in output powers from 750-6000 W as stand-alone units or integrated in 2-5 axis laser systems. Also manufactures laser/punch presses, punching and bending machines, and flexible manufacturing systems.

TRUMPF Inc, (sub of TRUMPF GmbH), Hyde Rd, Farmington Ind Pk, Farmington, CT 06032; 203-677-9741, FAX 203-678-1704
natl sls mgr/lsrs, Stanley J. Koczera; vp sls/mktg, Rainer Hundsdoerfer; 1967
Manufactures industrial CO_2 lasers for use in cutting, welding, and surface modification in output powers from 750-6000 W. Lasers available as stand-alone to OEM suppliers, system manufacturers, and end-users. Standard multi-axis laser systems are also available.

Tungsram Laser Technology Inc, (sub of Tungsram Co), IV. Megyeri ut 6, Budapest, H-1044, Hungary; 36 1 1696-619/1600-233*, FAX 36/1/1601-684
Dir, Dr. Tivadar Lippenyi; eng mgr, Dr. Otto Palya; proj mgr, Istvan Kreisz; 1988
Manufactures industrial CO_2 lasers and complete CNC controlled workstations, up to 1000 W power. Products include 60 W and 25 W surgical lasers as well as a mobile lidar system for environmental monitoring. Also service and maintain own products and foreign products in Hungary.

UTILASE, INC.
BLANK WELDING TECHNOLOGIES

20201 Hoover Rd, Detriot, MI 48205; 313-521-2488, FAX 313-521-4695
gen mgr, John Baysore; 1989
Joins flat sheet metal shapes of differing alloys, various thicknesses or different surface coatings into one-piece stamping blanks using laser butt-welding technology and conventional welding methods to produce custom stamping blanks having specific metallurgical properties and performance characteristics engineered in-place.

United Technologies Corp, div of United Technologies Industrial Lasers, 300 Pleasant Valley Rd, PO Box 981, So. Windsor, CT 06074-0981; 203-282-4200, FAX 203-282-4202
gen mgr, J.P. Carstens; mktg mgr, J.S. Foley; eng mgr, J. Timmins; chf sci, C.M. Banas; 1976
Manufactures and supplies multikilowatt, industrial CO_2 lasers at the 6,14,18, 25 and 45-kW power levels. All systems include I-year warranty, aerodynamic output window, I-week training course. Maintains and operates a complete applications testing laboratory for customer development.

UNIVERSAL
LASER SYSTEMS INC.

7707 East Acoma Dr, Scottsdale, AZ 85260; 602-483-1214, FAX 602-483-5620
pres, Yefim Sukhman; vp sls/mktg, Jennifer McDonald; 1988
Manufactures flexible, computer-controlled laser processing systems compatible with CAD and other HP-GL based software. Systems can be equipped with gas or solid-state lasers for processing a wide variety of materials and are ideal for cutting, marking, engraving, wire stripping and bar coding.

Universal Voltronics, (sub of TherMedics Inc/Thermo Electron Corp), 27 Radio Circle Dr, Mt. Kisco, NY 10549; 914-241-1300, FAX 914-241-3129
ceo/pres, Barry Ressler; sls mgr, Steve Meissel; eng mgr, Ray Black; 1960

US Amada Ltd, Laser Div, 7025 Firestone Blvd, Buena Park, CA 90621; 714-739-2111, FAX 714-670-1439, telex 183346
vp, Koji Fujita; div mgr, Donald Hoffman; mfg mgr, Jon Paul Edwards; emp 45, 1946
Manufactures, integrates, and markets industrial laser machines for metal cutting and welding with up to six axes of motion. Also operates a fully staffed service department. Training customers on specific aspects concerning the use and operations of the machines.

US Laser Corp, 825 Windham CT N, PO Box 609, Wyckoff, NJ 07481; 201-848-9200, FAX 201-848-9006
pres, Robert Regna; vp, Carl Miller; vp mktg, James Golden; gen mgr, Edward Krupp; emp 25, 1979
Manufactures standard and custom laser systems for industrial materials processing applications (including semiconductor processing); high power Nd:YAG lasers (CW and pulsed); complete line of spare parts and accessories for solid-state lasers. Maintains laser applications laboratory for feasibility studies. Provides technical support for laser systems form various manufacturers.

UTILASE ENGINEERING, INC.
LASER CONSULTING & TRAINING

20201 Hoover Rd, Detroit, MI 48205; 313-521-2488, FAX 313-521-4695
eng serv mgr, Randy K. Keller; 1990
Provides product design consultation, weld studies and process analysis services which examine the feasibility and comparative advantages of improved product quality, manufacturability and production efficiency afforded by the application of laser processing technology. Also conducts technical seminars and training activities promoting the practical application of laser technology.

UTILASE SYSTEMS, INC.
LASER SYSTEMS INTEGRATION

20201 Hoover Rd, Detroit, MI 48205; 313-521-2488, FAX 313-521-4695
vp mktg, Jeff Moss; gen mgr, Rand K. Keller; 1988
Develops, constructs and integrates custom laser-based manufacturing systems for the consumer and commercial product industries (automotive, aerospace, appliance, etc.). Distributor of Shibuya Kogyo multi-axis laser work stations and GMFanuc Robotics L-100 5- axis laser robots.

Vanzetti Systems Inc, 111 Island St, Stoughton, MA 02072; 617-828-4650, FAX 617-341-2084
chmn em, Dr. Riccardo Vanzetti; coo, Ashod S. Dostoomian; ceo, John P. Ward; 1968
Manufactures laser-based systems to automatically inspect the solder joints on printed circuit boards and the inner and outer lead bonds of TAB and fine pitch electronic assemblies; manufactures intelligent laser soldering systems that monitor in real-time the solder temperature and uses the information for feedback control of the process.

Vero Precision Eng Ltd, South Mill Road, Regents Park, Southhampton, S09 500, UK*
1968
Assembles laser cutting systems.

Waldrich Siegen Werkzeugmaschinen GmbH, Daimlerstrasse 24, Burbach, D-5909, Germany; 2736-400, FAX 2736-559, telex 87 28 49
pres, Helmut Belz; vp mktg, Dieter Feisel; vp eng/mfg, Theodor Petera; mktg dir, Manfred Kronauer; 1840
Manufactures laser texturing machines for rolls used in cold steel mills.

W.A. Whitney Italia, Esterlime Technologies, Strada DelFrancere 130-9, Torino, I-10156, Italy; 011-470-27-02, FAX 011-4702915
1969
Manufactures F.M.S. for plate processing, hydraulic punching machines combined with plasma cutting system, punching and nibbling machines, laser cutting system.

Weco Metal Products, 57 Commercial St, Webster, NY 14580; 716-872-3000, FAX 716-872-4266
pres, David Tierson; mfg eng, Rob Curtice; 1952
Offers precision sheet metal fabrication, laser cutting, CNC turret presses, CNC brake presses, electronic chassis, panels, and enclosures.

Weidmuller Sensor Systems

Weidmüller

div of Weidmuller Inc, 5400 West Elm St, McHenry, IL 60050; 815-344-4141, FAX 815-344-4152
pres, Wolfgang Schubl; sls/mktg mgr, Ernest J. Foldvari; chf eng, Ahmed Topkaya; 1952
Manufactures autofocus systems for maintaining optimal focal point and stand off distance during CO_2 and YAG laser cutting. Quick lens and nozzle change laser cutting heads for high pressure laser cutting. Laser power meters, laser welding monitors, weld seam tracking systems for closed loop power correction. Solid-state laser beam diagnostic systems for "on line" measurement.

Wesel Manufacturing Co, 1141 N. Washington Ave, Scranton, PA 18509; 717-311-7177 *
1952

Wustefeld Maschinenbau GmbH, Gewerbegebiet Mehren, Mehren/Eifel, D-5569, Germany; 06592/556, FAX 06592/4921
Wolfgang Wustefeld, Wolfgang Weisse; 1981
Cutting and welding problems can be solved by searching for the optimal (economical) solution along with the customer and then a special machine is built under the application of standardized modules.

WA Whitney Co, 650 Race St, PO Box 1206, Rockford, IL 61105; 815-964-6771, FAX 815-964-3175
vp mktg, Mike Donnelly; 1907
Manufactures fabricating centers for sheet-metal and plate, incorporating hydraulic punching and laser contour cutting. Capacity is 5/8 in. thick-mild steel, handling workpieces up to 1000 pounds. Positioning area is 60 x 90 in.; wider areas possible with repositioning in the *x*-axis. Machine has automatic tool changer with approximately 175 stations.

Wiedemann Division, (sub of Cross & Trecker Corp), 211 South Gulph Rd, King of Prussia, PA 19406; 215-265-2000 *, FAX 215-265-5768
emp 200, 1916
Manufactures sheet material fabrication equipment. Products include turret punch presses, combination turret punch/laser systems, right angle shears, laser contouring systems, and a panel press brake machine.

XENON

20 Commerce Way, Woburn, MA 01801; 617-938-3594, FAX 617-933-9804
pres, L.R. Panico; 1916
High performance flashlamps and krypton arc lamps of standard and custom design for OEM' s and end users of optically pumped lasers/ Also provided is the "incoherent" laser specially suited for customers who need a high intensity light in the nanosecond time region but do not require coherent light.

XMR Inc, (sub of Amoco Corp), 5403 Betsy Ross Dr, Santa Clara, CA 95054-1102; 408-988-2426, FAX 408-970-0742
emp 160, 1979
Manufactures industrial excimer lasers and laser-based systems for use in the microelectronics, semiconductor, and medical industries. Systems manufactured include an excimer laser-based micromachining system, a system for planarizing aluminum films in semiconductor processing, and other custom systems. Does extensive research and development using excimer lasers and laser-based systems to study the effects of laser processing on various materials. Also offers two complete applications laboratories to its customers.

ZED Instruments Ltd, 336 Molesey Rd, Hersham, Surrey, KT123PD, UK; 0932-228977, FAX 0932-243683
1973
Manufactures laser engraving machines for the production of flexographic printing plates and cylinders, anilox inking rollers and screens for rotary screen printing.

U.S. companies providing contract laser processing services

The following section includes information on U.S. companies that provide contract laser material processing services. These companies may be full-service laser job shops or sheet-metal shops with laser cutting capability. Each of these companies is prepared to offer services on a contractual basis for process R&D, application feasibility, prototype processing, or production.

The specification chart provides a sketch of each company. Addresses and details of each company can be found in the directory immediately following the chart.

The editors strive to ensure that the information reported is current and correct. In an effort to present the most comprehensive directory of contract service companies, the editors have included information from some companies that did not provide updated information for this edition. These companies are noted by an asterisk (*) after their address.

U.S. JOB SHOPS
(CHART PAGE 1)

	Lasers			Applications										Services								Motions	
	Solid-State	CO2	Excimer	Welding	Cutting (Metal)	Cutting (nonmetal)	Drilling (Metal)	Drilling (nonmetal)	Scribing	Heat-Treating	Other Surface Treat.	Marking	Engraving	Part Design	Prototype	Pilot Plant	Production	Spec. Sys. Supp.	Metallurgy & QC	Processing R&D	MIL Spec Qual	2-Axes	Multi-Axes
AccuDyne Co Inc Rochester NY 716-458-6564		•		•	•	•								•	•		•				•		
Accurate Laser International Burbank CA 818-843-8296		•			•	•								•			•					•	•
Accu-Tech Laser Processing San Marcos CA 619-744-6692		•		•	•	•	•	•	•						•	•	•				•		
Ace Laser Tek Inc Prospect Heights IL 708-537-4202	•	•							•			•	•		•	•	•	•	•	•	•	•	•
AcuCut Inc Bristol CT 203-582-2348		•			•	•	•	•	•					•	•		•			•			
Advanced Technology Co Pasadena CA 818-449-2696	•	•		•	•	•	•	•	•				•		•		•			•			•
Advance Laser Applications Inc Plymouth MI 313-451-0140		•			•	•									•		•						
A&E Manufacturing Co Inc Levittown PA 215-943-9460*		•			•										•		•						
Aero Precision Engineering Santa Monica CA 818-453-4515*		•			•										•		•						
AJ&L Tool & Manufacturing Co Inc Rochester NY 716-872-3226*		•			•									•	•					•			
Amber Precision Sheet Metal Riverside CA *		•			•										•		•						
American Fabrications Inc Nashville TN 615-834-8700		•		•	•	•	•	•							•		•			•		•	
Ameri-Die Inc Wadsworth OH 216-239-1484		•			•	•	•	•	•			•	•	•	•								
Amerson Precision Sheet Metal McMinnville OR 503-472-0659		•			•	•	•	•						•	•		•					•	
Andromeda Corp Huntsville AL 205-882-3974*		•		•	•	•			•	•		•	•	•	•			•		•			
Antique Auto Sheet Metal Brookville OH *		•			•										•		•						
Apertura Woburn MA 617-933-8088	•			•			•	•						•	•		•			•		•	•
Applied Fusion Inc San Leandro CA 415-351-8314	•	•		•	•	•	•	•	•	•		•	•	•	•		•	•	•	•	•		•

U.S. JOB SHOPS
(CHART PAGE 2)

	Lasers			Applications										Services								Motions	
	Solid-State	CO2	Excimer	Welding	Cutting (Metal)	Cutting (nonmetal)	Drilling (Metal)	Drilling (nonmetal)	Scribing	Heat-Treating	Other Surface Treat.	Marking	Engraving	Part Design	Prototype	Pilot Plant	Production	Spec. Sys. Supp.	Metallurgy & QC	Processing R&D	MIL Spec Qual	2-Axes	Multi-Axes
Applied Laser Technology Inc Beaverton OR 503-641-4400	●	●	●			●		●	●			●			●		●						
Aquarius Metal Elkgrove Village IL 312-956-6810*		●			●	●									●		●						
Artplak Studios Hauppauge NY *		●											●										
Assurance Mfg Inc Minneapolis MN 612-780-4252*		●		●	●	●	●	●	●	●	●	●	●	●	●	●	●	●	●	●	●		
Atlas Die Inc Elkhart IN 219-295-0050		●			●	●									●		●						
Atlas Tool & Die Works Inc Lyons IL 708-442-1661		●			●	●						●		●	●	●	●						●
Auto Metal Craft Inc Oak Park MI 313-398-2240		●		●	●	●	●	●	●		●	●	●	●	●	●						●	●
Badger Sheet Metal Works Green Bay WI 414-435-8881		●			●	●	●	●	●			●	●	●	●	●	●		●	●	●		
Ben-Mer Tool & Machine Inc Rochester NY 716-254-6090*		●			●	●			●			●	●	●	●		●			●	●		
Benteler Stamping Grand Rapids MI *		●			●										●	●	●						
Blackhawk Composites Rockford IL 815-968-7488		●		●	●	●	●	●							●	●	●				●		●
BME Engineering Inc Georgetown MA 617-352-8202*		●			●	●									●		●						
Bond Custom Manufacturing Arvada CO 303-423-8841*		●			●	●																	
Borba Mfg Inc So. San Francisco CA 415-761-1032*		●											●										
Boston Precision Parts Co Inc Hyde Park MA 617-361-1000		●		●	●	●	●	●			●			●	●	●	●			●	●		●
Boyd Machine & Repair Co Inc Kimmell IN 219-635-2195		●		●						●	●				●		●		●	●			
BuckBee Mears, St. Paul St. Paul MN 612-228-6400		●		●	●		●																
Byers Precision Fabricators/Laser Processing Hendersonville NC 704-693-4088		●			●										●		●					●	

U.S. JOB SHOPS
(CHART PAGE 3)

	Lasers			Applications										Services								Motions	
	Solid-State	CO2	Excimer	Welding	Cutting (Metal)	Cutting (nonmetal)	Drilling (Metal)	Drilling (nonmetal)	Scribing	Heat-Treating	Other Surface Treat.	Marking	Engraving	Part Design	Prototype	Pilot Plant	Production	Spec. Sys. Supp.	Metallurgy & QC	Processing R&D	MIL Spec Qual	2-Axes	Multi-Axes
California Saw & Knife Works San Francisco CA 415-861-0644		•			•										•	•	•						
Cantrell Ind Stafford TX 713-933-3051*		•			•																		
Carolina Laser Cutting Greensboro NC 919-292-1474		•			•	•									•		•					•	
Carolina Steel Rule Die Corp Greensboro NC *		•				•																	
Chromalloy Research & Technology Div Orangeburg NY 914-359-4700	•	•		•	•	•	•	•				•		•	•		•		•	•			•
Cincinnati Ventilating Co Florence KY 606-371-1320		•		•	•	•			•						•		•						•
Circle T&D Corp Woburn MA 617-935-7290	•	•		•	•	•	•	•	•	•		•	•	•	•		•						
Clark Metal Products Co Blairsville PA 412-459-7550		•			•	•									•		•						
Computer Fabrications Inc Pelham NH 603-635-2123*		•			•																		
Container Graphics Corp Cary NC 919-481-4200*		•			•	•									•	•	•						
Coors Ceramics Co Grand Junction CO 303-245-4000		•				•		•	•	•					•		•				•		
Corry Laser Technology Inc Corry PA 814-664-7212	•	•		•	•	•	•	•	•	•		•	•		•		•		•	•	•		
C.O.W. Industries Inc Columbus OH 614-239-1992		•			•	•								•	•	•	•						
C&R Specialty Co Celina OH 419-942-1421		•				•		•	•			•	•				•			•			
Custom House Engravers Inc Bohemia NY 516-567-3004		•											•										
Custom Laser Inc Lockport NY 716-434-8600*		•		•	•	•	•	•	•	•		•	•		•	•	•		•	•	•		
Custom Laser Services Corp Elkhart IN 219-293-0494*	•	•			•	•							•	•		•	•						
Cutting Dynamics Inc Westlake OH 216-871-4740		•			•	•	•	•							•		•				•		

U.S. JOB SHOPS
(CHART PAGE 4)

	Lasers			Applications										Services								Motions	
	Solid-State	CO2	Excimer	Welding	Cutting (Metal)	Cutting (nonmetal)	Drilling (Metal)	Drilling (nonmetal)	Scribing	Heat-Treating	Other Surface Treat.	Marking	Engraving	Part Design	Prototype	Pilot Plant	Production	Spec. Sys. Supp.	Metallurgy & QC	Processing R&D	MIL Spec Qual	2-Axes	Multi-Axes
D D Wire Company Temple City CA 818-442-0459*		•			•																		
Defiance Stamping Co Defiance OH *		•			•	•									•	•							
Dieknologist Inc Stratford CT 212-989-5959		•				•				•					•							•	
Die-Tronics Inc Northbrook IL 312-498-2110		•			•	•									•		•						
Digitron Tool Co Inc Miamisburg OH 513-435-5710*	•				•		•								•		•						
Directed Light Inc Sunnyvale CA 408-745-7300	•	•		•	•	•	•	•	•	•	•	•	•		•		•	•	•				•
Dolphin Laser Processing, Inc Versailles IN 812-689-7202		•			•	•								•	•	•	•			•			
Dubbeldee Harris Diamond Corp Mt. Arlington NJ 201-770-1420	•					•		•	•			•	•				•			•			
Eagle Laser Co Inc High Point NC 919-434-9400		•			•	•							•		•		•	•					
EBTEC Corp Agawam MA 413-786-0393	•	•		•	•	•	•	•		•		•	•	•	•		•		•	•	•		•
EBTEC West Huntington Beach CA 714-895-2725	•			•	•		•								•		•						
E&C Laser Houston TX 713-466-9556		•			•	•			•						•	•	•		•	•		•	
Edgar Barcus Co Inc Westville NJ 609-456-0204		•				•								•			•					•	
Edison Welding Institute Columbus OH 614-486-9400	•	•		•	•	•	•	•	•	•	•	•	•		•				•	•			
Elano Corp Xenia OH 513-426-0621		•		•	•		•										•						
Electro-Jet Tool & Mfg Co Inc Cincinnati OH 513-563-0800	•				•		•					•	•	•	•		•						
Electrolabs Inc Fraser MI 313-294-4150		•			•	•	•	•	•					•	•	•	•			•	•		
Erie Industrial Products Co Westchester IL 708-344-8200		•			•	•									•		•						

U.S. JOB SHOPS
(CHART PAGE 5)

	Lasers			Applications										Services								Motions	
	Solid-State	CO2	Excimer	Welding	Cutting (Metal)	Cutting (nonmetal)	Drilling (Metal)	Drilling (nonmetal)	Scribing	Heat-Treating	Other Surface Treat.	Marking	Engraving	Part Design	Prototype	Pilot Plant	Production	Spec. Sys. Supp.	Metallurgy & QC	Processing R&D	MIL Spec Qual	2-Axes	Multi-Axes
Ervite Corp Erie PA 814-838-1911		•			•										•		•						
Espe Mfg Co Inc Schiller Park IL 800-367-3773		•				•								•	•		•						
Estes Design & Mfg Co Indianapolis IN 317-899-2203		•		•	•	•	•	•	•			•	•	•	•		•				•		
Evans Industries Topsfield MA 508-887-8561*		•			•																		
Exotic Metal Forming Co Kent WA *	•				•		•																
F&B Mfg Co Phoenix AZ 602-272-3900		•		•	•	•	•	•	•			•					•			•	•		•
Ferco Tech Corp Franklin OH 513-746-6696	•	•			•	•	•	•						•	•		•				•	•	
Future Technologies Inc Linwood MI 517-697-5353																		•					
Gerard Metal Craftsmen Gardena CA *		•			•																		
Germantown Tool & Machine Works Inc Huntingdon Valley PA 215-322-4970		•			•	•			•						•		•						
Gladd Industries Detroit MI 313-537-2800		•		•	•	•											•	•				•	
Grayd-A Metal Fabricators Santa Fe Springs CA 213-944-8951*		•			•	•			•			•		•	•		•						
Great Lakes Laser Services Royal Oak MI 313-549-0450	•			•	•		•					•	•	•	•	•	•	•					
Greenfield Fabricating Canton OH 216-499-6778		•			•	•								•	•		•						
Hansman Industries Inc Stillwater MN 612-439-7202		•			•									•	•		•	•			•	•	
Harford Systems Inc Aberdeen MD 301-273-3400*		•			•										•	•	•						
HDE Systems Inc Sunnyvale CA 408-735-7272	•	•		•	•	•	•	•		•			•	•	•		•			•	•		•
Heiden Inc Manitowoc WI 414-682-6111				•	•		•							•	•	•					•	•	

U.S. JOB SHOPS
(CHART PAGE 6)

Company	Lasers			Applications										Services								Motions	
	Solid-State	CO2	Excimer	Welding	Cutting (Metal)	Cutting (nonmetal)	Drilling (Metal)	Drilling (nonmetal)	Scribing	Heat-Treating	Other Surface Treat.	Marking	Engraving	Part Design	Prototype	Pilot Plant	Production	Spec. Sys. Supp.	Metallurgy & QC	Processing R&D	MIL Spec Qual	2-Axes	Multi-Axes
Helgesen Industries Inc Hartford WI 414-673-4444		•			•	•								•	•		•				•	•	
Harold R. Henrich Inc Lakewood NJ 908-370-4455		•			•	•									•		•						
HI-TEK Manufacturing Mason OH 513-459-1094	•	•		•	•	•	•	•							•	•	•		•	•	•		
Hoechst CeramTec NA Inc Mansfield MA 508-339-1911		•				•		•	•	•	•			•	•		•				•	•	
HUI Kiel WI 800-877-8913		•		•	•										•		•						
Industrial Metal Specialties Inc Dallas TX 214-634-0211		•			•	•									•		•				•	•	
International Lasersmiths Compton CA 213-635-8536	•	•		•	•	•	•	•		•		•	•	•	•	•	•	•	•	•			
Iowa Laser Technology Inc Cedar Falls IA 800-397-3561	•	•		•	•	•	•	•		•	•				•		•		•	•	•		•
JAM Precision Copiague NY		•			•																		
Jensen Metal Products Inc Racine WI 414-637-5670*		•			•	•									•		•						
Jet Avion Corp Hollywood FL 305-987-6101*		•			•	•	•	•	•				•		•		•		•		•		
J&K Inc Waco TX 817-752-2557/800342-7297		•			•	•									•		•						
Kaiser Tool Company Inc Ft. Wayne IN 219-422-2408												•											
Klune Industries Inc No Hollywood CA 818-503-8100		•			•	•								•	•		•						
KMC Laserform Port Washington WI 414-284-3424		•		•	•	•	•	•	•	•	•	•		•	•	•	•			•	•		
LaFollette Machine & Tool Co LaFollette TN 615-562-5854		•		•	•	•	•	•	•			•	•	•	•		•					•	
Lane & Roderick Inc Santa Fe Springs CA 213-868-3465		•			•	•									•		•				•	•	
Laserage Technology Corp Waukegan IL 708-249-5900	•	•		•	•	•	•	•	•	•		•	•		•		•		•		•		

U.S. JOB SHOPS
(CHART PAGE 7)

	Lasers			Applications										Services								Motions	
	Solid-State	CO2	Excimer	Welding	Cutting (Metal)	Cutting (nonmetal)	Drilling (Metal)	Drilling (nonmetal)	Scribing	Heat-Treating	Other Surface Treat.	Marking	Engraving	Part Design	Prototype	Pilot Plant	Production	Spec. Sys. Supp.	Metallurgy & QC	Processing R&D	MIL Spec Qual	2-Axes	Multi-Axes
Laser Applications Inc Westminster MD 301-857-0770	•	•		•	•	•	•	•		•				•	•	•	•				•		•
Laser Automation Inc Chagrin Falls OH	•	•		•	•	•	•	•	•	•	•	•	•		•	•	•	•	•	•			
Laser Cutting Service Co York PA 717-767-6402*		•		•	•	•								•	•	•	•		•	•			
Laser Cutting Services Pawtucket RI 401-723-1910	•	•		•	•	•	•	•						•	•		•						
Laserdyne Eden Prairie MN 612-941-7694	•	•	•	•	•	•	•	•		•		•		•	•	•	•			•	•		•
Lasereliance Technologies Altamonte Springs FL 407-339-0737		•		•	•	•	•	•	•	•		•		•	•	•	•			•	•		
Laser Engineering & Fab Tulsa OK 918-437-7511		•		•	•	•	•	•	•	•				•	•	•	•				•		
Laserfab Inc of California Concord CA 415-676-2238	•	•		•	•	•	•	•					•	•	•		•		•	•	•		•
HGG Laser Fare Inc Smithfield RI 401-231-4400	•	•		•	•	•	•	•	•			•	•		•				•	•			•
Laser Fare Inc Smithfield RI 401-231-4400*	•	•		•	•	•	•	•	•		•	•	•	•	•	•	•	•	•	•	•		
Laser Impressions Inc Sunnyvale CA 408-734-2012	•								•			•	•		•	•	•				•		
Laser Industries Inc Lawrence MA 508-682-2460/800-6227040	•	•		•	•	•	•	•	•	•		•	•	•	•	•	•		•	•	•	•	•
Laser Industries Inc Orange CA 714-532-3271	•	•		•	•	•	•	•	•				•		•		•				•		•
Laser Innovations Woodruff WI 715-356-2422		•				•						•	•				•						
Laser Machining Inc Somerset WI 715-247-3285	•	•		•	•	•	•	•	•	•	•	•	•		•	•	•	•	•	•	•	•	•
Laser Magic Inc Hudson WI 715-386-2525		•										•	•		•		•						
Laser Manufacturing Inc Plymouth MI 313-471-5457*		•		•	•	•	•	•		•	•	•	•	•	•		•		•	•			
Laser Marking Inc Manchester CT 203-649-8068	•											•				•	•	•					

U.S. JOB SHOPS
(CHART PAGE 8)

	Lasers			Applications										Services								Motions	
	Solid-State	CO2	Excimer	Welding	Cutting (Metal)	Cutting (nonmetal)	Drilling (Metal)	Drilling (nonmetal)	Scribing	Heat-Treating	Other Surface Treat.	Marking	Engraving	Part Design	Prototype	Pilot Plant	Production	Spec. Sys. Supp.	Metallurgy & QC	Processing R&D	MIL Spec Qual	2-Axes	Multi-Axes
Laser Marking Services Inc Smithfield RI 401-232-2292	•								•			•	•								•		
Laser Mark-It Glendale CA 818-240-8122												•											
Lasermation Inc Philadelphia PA 215-228-7900		•				•		•	•			•	•										
Laseronics Inc Torrance CA 213-320-3700	•	•		•	•	•	•	•	•	•	•	•	•		•	•	•	•		•	•	•	•
Laser Precise Inc Knoxville TN 615-531-8016	•	•	•	•	•	•	•	•	•	•	•	•	•	•	•	•	•		•	•			
Laser Precision Cutting Inc Weaverville NC 704-658-0644		•			•	•									•		•						
Laser Services Inc Westford MA 508-692-6180	•	•		•	•	•	•	•	•			•	•	•	•	•	•		•	•	•		•
Laser Specialties Inc Oshkosh WI 414-233-6131*		•											•										
Laser Tech Inc Minnetonka MN 612-935-9277		•		•	•	•	•	•	•			•	•	•	•		•			•			
Laser Techniques Inc Duncanville TX 214-296-7934		•			•	•			•				•	•	•		•				•	•	
Laser Technologies Inc Wood Dale IL 312-766-0060*	•	•			•	•									•	•	•			•	•		
Lasertron Inc Sunrise FL 305-846-8600	•	•		•	•	•	•	•	•			•	•		•		•		•		•	•	•
Las Tec Hillsboro OR 503-640-2300		•			•	•	•	•	•	•		•			•	•	•			•			
Lazal Corporation Lititz PA 717-627-4242*		•			•	•									•		•						
Lenox Laser Phoenix MD 301-592-3106	•	•	•	•			•	•	•			•	•	•	•	•	•	•	•	•			
Light Touch Aurora CO 303-343-4773		•										•	•										
William Lo Dolce Machine Co Inc Malden on Hudson NY 914-246-7017*		•			•	•																	
Maloya Laser Inc Farmingdale NY 516-293-8333		•		•	•	•	•	•						•	•		•				•		•

U.S. JOB SHOPS
(CHART PAGE 9)

	Lasers			Applications										Services								Motions	
	Solid-State	CO2	Excimer	Welding	Cutting (Metal)	Cutting (nonmetal)	Drilling (Metal)	Drilling (nonmetal)	Scribing	Heat-Treating	Other Surface Treat.	Marking	Engraving	Part Design	Prototype	Pilot Plant	Production	Spec. Sys. Supp.	Metallurgy & QC	Processing R&D	MIL Spec Qual	2-Axes	Multi-Axes
Marking Methods Inc Alhambra CA 818-282-8823	●								●			●	●				●				●		●
Mathias Die Co So. St. Paul MN 612-451-0105		●			●	●								●	●		●				●	●	
McAfee Tool & Die Uniontown OH 216-896-9555		●			●	●	●	●						●	●		●		●	●	●	●	
McAlpin Industries Rochester NY 716-266-3060 *		●			●									●	●	●	●						
Meadex Technologies Springfield MA 413-567-3680	●	●		●			●					●											
Metalade Inc Rochester NY 716-424-3260				●	●	●	●	●						●	●		●						
Metal Equipment Fab Columbia SC 803-776-9250		●			●	●																	
Metko Inc New Holstein WI 414-898-4021		●			●	●								●	●		●			●			
Microcut Technologies Dallas TX 214-407-9094		●				●			●				●		●		●					●	
Midwest Laser Inc Lenexa KS 913-541-1813	●	●										●	●										
Midwest Machine Products Inc Golden CO 303-422-5388 *		●			●	●									●		●						
Midwest Trophy Del City OK 405-670-4545		●							●			●	●				●					●	
Mission Laser Works Inc El Monte CA 818-443-6904		●										●	●										
Modernistic St. Paul MN 612-291-7650	●	●				●		●	●						●		●						
Morrissey Inc Bloomington MN 800-798-6158		●			●	●						●	●		●		●						
Morton Metalcraft Co Morton IL 309-266-7176 *		●			●	●									●	●	●						
MSI Prototype & Engineering Warren MI 313-773-0800		●			●									●	●								
Myers Electric Products Inc Montebello CA 213-724-0450		●			●	●									●		●						

U.S. JOB SHOPS
(CHART PAGE 10)

	Lasers			Applications										Services								Mo-tions	
	Solid-State	CO2	Excimer	Welding	Cutting (Metal)	Cutting (nonmetal)	Drilling (Metal)	Drilling (nonmetal)	Scribing	Heat-Treating	Other Surface Treat.	Marking	Engraving	Part Design	Prototype	Pilot Plant	Production	Spec. Sys. Supp.	Metallurgy & QC	Processing R&D	MIL Spec Qual	2-Axes	Multi-Axes
National Laser Trimming Attleboro MA 617-226-3313 *	•														•		•				•		
National Metal Processing Inc Richmond KY 606-623-9291		•		•						•					•		•		•				•
Nelson Engineering Phoenix AZ 602-273-7114 *		•			•	•									•		•						
New England Laser Processing Danvers MA 508-774-4626	•	•		•	•	•	•	•				•	•		•	•	•			•		•	•
Northern Mfg Co Inc Oak Harbor OH 419-898-2821/800-9227339		•		•	•	•						•		•	•		•					•	
Northwoods Laser Wittenberg WI 715-253-2541		•										•	•	•	•		•						
Norton Metal Products Inc Fort Worth TX 817-232-0404		•			•	•	•		•			•	•	•	•		•					•	
Oakcraft Inc Coon Rapids MN 612-784-6305 *		•										•	•	•	•		•						
Oakley Industries Inc Mt. Clemens MI 313-792-1261		•			•									•	•								
Optimation Inc Manchester NH 603-623-2800	•						•					•		•	•								
Pacific Marine Sheet Metal Corp San Diego CA 619-234-6961		•			•	•	•	•							•		•				•		
Pacific-Mesa Engineering Corp Valencia CA 805-295-1330		•		•											•		•						
Parkway Metal Products Chicago IL 312-489-3144		•		•	•	•	•	•		•	•	•			•	•	•					•	
Peerless Laser Processors Inc Groveport OH 614-836-2073		•		•	•	•	•					•		•	•		•					•	•
Peterson Products Corp Schiller Park IL 312-678-0800 *		•			•	•											•						
Pickwick Co Cedar Rapids IA 319-393-7443/800397-9797		•			•	•	•		•			•	•	•	•	•	•			•	•		
P/M Industries Inc Portland OR 503-641-4646		•				•		•	•			•			•		•			•	•		
Precision Die Seattle WA 206-575-8217/800-7837773		•			•	•									•		•	•					

U.S. JOB SHOPS
(CHART PAGE 11)

	Lasers: Solid-State	Lasers: CO2	Lasers: Excimer	Applications: Welding	Applications: Cutting (Metal)	Applications: Cutting (nonmetal)	Applications: Drilling (Metal)	Applications: Drilling (nonmetal)	Applications: Scribing	Applications: Heat-Treating	Applications: Other Surface Treat.	Applications: Marking	Applications: Engraving	Services: Part Design	Services: Prototype	Services: Pilot Plant	Services: Production	Services: Spec. Sys. Supp.	Services: Metallurgy & QC	Services: Processing R&D	Services: MIL Spec Qual	Motions: 2-Axes	Motions: Multi-Axes
Precision Laser Services Inc Ft. Wayne IN 219-744-4375	•	•		•	•	•	•	•	•	•		•	•	•	•	•	•	•	•	•	•	•	•
Precision Machining & Fabricating Inc Milford CT 203-878-7414 *					•	•								•	•		•						
Process Equipment Co Tipp City OH 512-667-4451		•		•	•	•											•	•			•	•	
Proctor Products Co Inc Kirkland WA 206-822-9296		•		•	•	•	•	•							•		•						
Proto Dynamics Corp Faser MI 313-294-0500					•	•								•	•		•						•
Prototech Laser Inc Mt. Clemens MI 313-465-9944		•		•	•	•	•	•	•						•		•						•
Quality Laser Works Santa Clara CA 408-988-3377		•			•	•	•	•						•	•		•			•			•
Quantum Laser Corp Edison NJ 201-225-8686 *	•	•									•				•		•		•	•	•		
Quasar Industries Rochester Hills MI 313-852-0300 *		•		•	•	•	•	•	•	•	•	•	•	•	•	•	•	•	•	•	•		
Radtke & Sons Round Lake Park IL 708-546-3999	•			•	•		•		•			•	•		•		•			•			
Rapid-Line Inc Grand Rapids MI 616-530-0061		•		•	•	•	•	•				•	•	•	•	•	•			•			•
Red Mountain Laser Chandler AZ 602-786-3539		•			•	•	•	•	•	•		•	•		•		•						
Reid Industries Inc Roseville MI 313-776-2070 *		•			•	•	•	•	•						•	•	•						
Resonetics Nashua NH 603-886-6772			•				•	•	•		•	•	•	•	•	•	•	•	•	•			
CF Roark Welding & Engineering Co Inc Brownsburg IN 317-852-3163	•	•		•	•	•	•								•					•	•		
Roberts Precision Products Inc Ann Arbor MI 313-662-6569		•		•	•									•	•	•	•						
Rose Metal Products Springfield MO 417-865-1676	•				•	•								•	•		•				•		
Rotation Engineering & Mfg Co Plymouth MN 612-553-1090		•			•	•	•	•						•	•	•	•			•			

U.S. JOB SHOPS
(CHART PAGE 12)

	Lasers			Applications										Services								Motions	
	Solid-State	CO2	Excimer	Welding	Cutting (Metal)	Cutting (nonmetal)	Drilling (Metal)	Drilling (nonmetal)	Scribing	Heat-Treating	Other Surface Treat.	Marking	Engraving	Part Design	Prototype	Pilot Plant	Production	Spec. Sys. Supp.	Metallurgy & QC	Processing R&D	MIL Spec Qual	2-Axes	Multi-Axes
San Diego Sheet Metal Works San Diego CA 619-232-3153		•		•	•	•	•	•						•	•		•						
Seelye Plastics Inc Bloomington MN 612-881-2658		•			•	•	•	•	•					•	•		•						•
Simonds Industries Inc Fitchburg MA 508-343-3731		•			•																		
Skilcraft Sheetmetal Inc Burlington KY 606-371-0799		•		•	•	•	•	•						•	•		•	•					
Southern California Laser Products Inc Temecula CA 714-676-8083		•		•	•	•	•	•		•			•								•		
Southern Metalcraft Inc Lithonia GA 404-482-2923		•		•	•	•	•	•					•	•	•		•						•
Spectralytics Minneapolis MN 612-831-5511	•	•	•	•	•	•	•	•	•			•		•			•		•	•	•		•
Spectrum Manufacturing Wheeling IL 708-520-1553		•			•	•								•	•		•				•	•	
Stardyne Inc Johnstown PA 814-536-2000		•		•	•					•	•												
Stremel Manufacturing Co Minneapolis MN 612-339-8621		•		•	•	•	•	•	•			•		•	•	•	•					•	
Systi-Matic Kirkland WA 800-426-0000		•			•	•	•	•		•				•	•		•		•				
Texcel Inc Westfield MA 413-562-7593 vec	•	•		•	•	•	•	•	•			•	•	•	•	•	•			•			•
Tigart Laser System Inc Indianapolis IN 317-872-8113		•		•	•	•	•		•				•		•		•			•		•	•
Tru-Graphics Inc Centerville (Dayton) OH 800-776-4152		•										•	•				•						
Utilase Inc Detroit MI 313-521-2488	•	•		•	•	•	•	•	•	•	•			•	•	•	•		•	•	•		•
Viner Manufacturing Co Evergreen CO 303-674-0534	•	•		•	•	•	•	•	•	•		•	•		•		•	•			•		
Weber Knapp Jamestown NY 716-484-9135		•			•									•	•	•	•					•	
Webster Tool, Die & Mfg Co Webster NY 716-872-5650		•		•	•	•	•	•	•	•		•	•	•	•		•				•		•

U.S. JOB SHOPS (CHART PAGE 13)

	Lasers			Applications										Services								Motions	
	Solid-State	CO2	Excimer	Welding	Cutting (Metal)	Cutting (nonmetal)	Drilling (Metal)	Drilling (nonmetal)	Scribing	Heat-Treating	Other Surface Treat.	Marking	Engraving	Part Design	Prototype	Pilot Plant	Production	Spec. Sys. Supp.	Metallurgy & QC	Processing R&D	Mil. Spec Qual	2-Axes	Multi-Axes
Weiss Industries Inc Mansfield OH 419-526-2480		•		•	•	•	•	•		•	•			•	•	•	•		•		•		
Western Fab & Finish Inc Walla Walla WA 800-456-8818 *		•		•	•	•	•	•							•		•				•		
Westinghouse Electric Corp Pittsburgh PA 412-256-1785	•	•		•	•		•		•	•	•	•	•	•	•	•	•		•	•	•		
Wilke Enginuity Inc McSherrystown PA 717-632-5937		•			•	•			•					•	•	•	•			•			
Williams Machine Works Inc Memphis TN 901-774-5335		•			•	•								•	•		•					•	
Willie Laser Products Elk Grove Village IL 708-956-1344		•			•	•									•		•						
Wood Mizer New Point IN *		•			•										•		•						
W&W Metal Fab Portland OR 503-775-0834		•			•	•									•	•	•						
XL-Tool & Machine Corp Rochester NY 716-436-2250		•			•	•								•	•						•		
X-Mark Industries Washington PA 412-228-7373		•			•	•								•	•		•				•		

U.S. CONTRACT LASER PROCESSING SERVICES DIRECTORY

Alabama

Andromeda Corp, 8302 Whitesburg Dr, Suite B, Huntsville, AL 35802; 205-882-3974*
emp 50, 1986
A laser/optical system analysis/hardware small business that provides laser job shop services.*

Arizona

F&B Mfg Co, Metal Fabrication Div, 4316 No. 39th Ave, PO Box 14549, Phoenix, AZ 85063; 602-272-3900, FAX 602i-272-4117
gen mgr, Ron Affronti; eng mgr, John Stewart; sls dir, Rodger Reiland; 1923
Uses Cincinnati Hydroforms, 8" to 32" capacities, servicing the aerospace, medical and commercial industries. Complete secondary capabilities including 6 axis laser cutting and welding, fusion and resistance welding, CNC and conventional machining, aluminum and steel (vacuum) heat treating, with punch press capabilities up to 500 tons.

Laser Fab, 2017 S Pasio Lane, Mesa, AZ 85202
1923

Lindale Stainless Steel Services, 2618 Cypress, Phoenix, AZ 602-269-5070*
1923

Nelson Engineering, 4020 East Air Lane, Phoenix, AZ 85034; 602-273-7114 *
1923

Red Mountain Laser, PO Box 1348, Chandler, AZ 85244; 602-786-3539
vp admin, Jon Bliven; 1988
Laser job shop providing precision engraving, scribing, drilling, and cutting of ceramic, metals, plastic, and Kevlar. Specializing in material processing utilizing custom CAD driven systems.

California

Accurate Laser International, 3310 Vanowen St, Burbank, CA 91505; 818-843-8296
1988
Offers 3 & 5 axis laser cutting.

Accu-Tech Laser Processing, 1175 Linda Vista Dr, San Marcos, CA 92069; 619-744-6692, FAX 619-744-4963
pres, James Byrum; eng mgr Steven Slater; emp 350, 1976
Precision laser processing of ceramics, plastics, metals, and other materials.

Advanced Technology Co, (sub of Advanced Materials Joining), 2858 E. Walnut St, Pasadena, CA 91107; 818-449-2696, FAX 818-793-9442
pres, Jean L. DeSilvestri; sls mgr, Robert G. DeSilvestri; mktg mgr, David L. Velazquez; emp 360, 1971
Operates laser and electron beam job shop specializing in materials processing. Offers E.B. welding, laser welding, drilling, scribing, engraving, and cutting services. Equipment incudes CO_2 and Nd:YAG lasers with up to 8 axis CNC motion systems, TV systems, and 12 ft. of travel.

Aero Precision Engineering, 1707 Cloverfield Blvd, Santa Monica, CA 90404; 818-453-4515*
1971

Amber Precision Sheet Metal, 12155 Magnolia, Bldg 5, Riverside, CA 92503
1971

Applied Fusion Inc, 1915 Republic Ave, San Leandro, CA 94577; 415-351-8314, FAX 415-351-0692
1973
Laser machining and welding, electron beam and all conventional welding, CNC machine shops, vacuum brazing, conformance to all mil specs, hi-vac components and systems. Vapor deposition - flight hardware and nuclear components, surgical instruments and implants.

A.R. Peterson, 20478 Mission Blvd, PO Box 3940, Hayward, CA 94540
1973

Borba Mfg Inc, 206 Airport Blvd, So. San Francisco, CA 94080; 415-761-1032*
emp 300, 1949
Specialist in laser engraving of wood products.

California Saw & Knife Works, 721 Brannan St, San Francisco, CA 94103; 415-861-0644, FAX 415-861-0406
1949

Central Cal Metals, 4692 N. Brawley Ave, Fresno, CA 93772; 209-275-1391*
1949

D D Wire Company, 4335 Temple City Blvd, Temple City, CA 91780; 818-442-0459*
1949

Directed Light Inc, 1270 Lawrence Station Rd, Sunnyvale, CA 94089; 408-745-7300, FAX 408-745-7726
ceo, David Knox; eng mgr, Kevin Woolbright; 1983
Distributor of components for high-powered industrial lasers: optics, lamps, rods, cavities, filters, power supplies, RF drivers. Represents ILC Technology, Coherent Components Group, EG&G, Crystal Technology, and Noblelight. Offers equipment rebuilds of laser systems, laser service, maintenance contracts, and onsite/offsite training. Laser job shop with YAG and CO_2 lasers.

EBTEC West, Thermal Technology Group of TI Group, 5561 Engineer Dr, Huntington Beach, CA 92649; 714-895-2725, FAX 714-895-1659, telex 295644
pres, Roger O. Schultz; lsr sls, Marc P. Rubcich; emp 230, 1983
Laser cutting, drilling and welding in 3-D parts and flat sheet. Up to 5 axis laser capability. Other material processing services include electron beam welding, vacuum brazing, vacuum heat treating and TIG welding. Call for free consultation.

Gerard Metal Craftsmen, 151 W. Rosecrans Ave, Gardena, CA 90248
1983

Grayd-A Metal Fabricators, 13233 E. Florence, Santa Fe Springs, CA 90670; 213-944-8951*, FAX 213-944-2326
emp 300, 1964
Sheet-metal job shop.

HDE Systems Inc, 615 N. Mary Ave, Sunnyvale, CA 94086; 408-735-7272, FAX 408-739-1233
pres, Simon L. Engel; sls mgr, Dan J. Fritschen; 1976
Performs laser drilling and cutting of metals and non-metals. Also laser welding of all metals. Several multi-axis laser systems in house. The company specializes in serving the jet-engine, aerospace, medical device, and electronics industries. Advanced workshops are offered to laser users.

International Lasersmiths

International Lasersmiths, 1306 S. Alameda St, Compton, CA 90221; 213-635-8536, FAX 213-635-2313
pres, Richard Crews; vp/gm, John Adams; eng mgr, Jerry Crews; 1988
Laser job shop and manufacturer of laser systems. Specializing in large parts and systems. 5-axis system travels of: 20 ft *x*-axis, 8 ft *y*-axis, 3 ft *z*-axis, with *a* and *b* axis also. 800-W and 1600-W pulsed Nd:YAG lasers, 1.1-kW CO_2, 100-W Nd:YAG marking system. Provides water jet cutting services.

Klune Industries Inc, 7323 Coldwater Canyon Ave, No Hollywood, CA 91605; 818-503-8100, FAX 818-764-3199
emp 600, 1972
A highly versatile company with capabilities ranging from prototype sheet-metal fabrication and chassis designs to CNC laser cutting and machining production facilities.

Lane & Roderick Inc, 12640 Allard St, Santa Fe Springs, CA 90670; 213-868-3465, FAX 213-929-8791
vp, Daniel Bailey; 1955

Laserfab Inc of California, 940 Detroit Ave, Concord, CA 94518; 415-676-2238, FAX 415-676-4571
pres, Bill Furrow; gen mgr, Bob Pruett; 1984
Laser & electron beam welding, cutting, and drilling. Conventional CNC turning & milling, contract mechanical fabrication for medical hardware, scientific instrumentation, aerospace hardware, and vacuum components.

Laser Impressions Inc, 1203 Alderwood Ave, Sunnyvale, CA 94089; 408-734-2012, FAX 408-734-0235
pres, Stephen M. Irving; emp 80, 1980
Laser engraving of electronic, medical, ad-premium, and general industrial components. Short runs to production volumes. Service and spare parts of laser engraving systems.

Laser Industries Inc, 677 Hariton Street, Orange, CA 92668; 714-532-3271, FAX 714-639-6941
vp, John Butterly; sls mgr, Shareen Foster; 1986
CO_2 and YAG laser cutting and laser drilling. Four CO_2 lasers up to 1500-W of power, up to 5 axes of movement and 400-W YAG capability. Laser-cut and drill metal, plastics, composite materials, and laser weld most metals. 50 x 100 in. capabilities and 125 x 78 x 35 in. in 5 axes capabiality.

LASER MARK-IT

945 Air Way, Glendale, CA 91201; 818-240-8122, FAX 818-240-0227
pres, Phyllis Hannan; eng, Bob Kreycik; acct mgr/op, Brad Baumann; 1986
Provides laser marking and engraving services to industries requiring state-of-the-art identification. Process allows identification of most materials in inaccessible areas. Includes logos, alphanumeric bar and proprietary codes.

Laseronics Inc, 2808 Oregon Ct, Suite L-12, Torrance, CA 90503; 213-320-3700, FAX 213-320-0133
pres, Sy Hamadani; mktg mgr, Sharon Hobbs; eng mgr, Andy Petrossian; 1979
Offers five CO_2 and YAG lasers up to 1550 W for multiaxis cutting, welding, drilling and marking of metallic, ceramic, plastic and composite parts for aircraft, aerospace, military, medical and electronic industries. On line CAD/CAM capabilities. 4 ft. x 8 ft. table travel. Free consulting on tooling and fixture design and 48 hour rush order turn-around. Quality control is MIL-I-45208 approved.

Laser West Inc, 2655 S. Orange Ave, Santa Ana, CA 92707; 714-557-0340
1979

Marking Methods Inc, 301 So. Raymond Ave, Alhambra, CA 91803-1531; 818-282-8823, FAX 818-576-7564
sls/mktg dir, Ralph Gallella; 1954
Laser marking job shop service. Specializing in permanent identification of metals, plastics and silicons. Precise marking of logos, trademarks, part numbers, serial numbers, scales, bar codes and custom graphics. Prototype and full production runs. Identification also offered by Electro-Chemical Marking and impression stamping. Samples processed at no charge.

Mission Laser Works Inc, 10750 Lower Azuza Rd, El Monte, CA 91731; 818-443-6904, FAX 818-444-2118
pres, Peter Klein; eng mgr, James Clark; emp 200, 1980
Provides laser-engraved desk accessories and awards.

Myers Electric Products Inc, 1130 S. Vail Ave, Montebello, CA 90640; 213-724-0450, FAX 213-724-1650
sls mgr, Joe Long; sr programmer, Marco Pol; emp 125, 1947
Contract services for laser production runs, primarily metals from 20GA to .250 thickness and plastics to .250. Manufactures NEMA type sheet-metal enclosures with or without electrical circuit interiors. Also manufactures the Myers Scru-Tite hubs of zinc, aluminum, malleable iron, or stainless steel for the electrical industry.

Pacific Marine Sheet Metal Corp, Southwest Fabricators Div, 410 15th St, San Diego, CA 92101; 619-234-6961, FAX 619-234-6961
pres, Daren Weckerly; sls mgr, Jerry Weiss; mktg mgr, Chris Kingery; eng mgr, Tom Finch; prod suprvr, Dennis Richardson; emp 480, 1968
Full-service sheet metal fabricating facility offering CAD/CAM programmable laser and plasma cutting. CNC, shearing, punching, and forming through 1/4 in. plate. Certified welding in stainless, mild steel, and aluminum. Quality assurance approved to Mil-I-45208A.

Pacific-Mesa Engineering Corp, 28231 Ave Crocker, # 70, Valencia, CA 91355; 805-295-1330, FAX 805-295-1464
1968
Laser cutting, welding and heat-treating machinery, designed and fabricated, prototype or production.

Quality Laser Works, 2000 Wyatt Dr, Suite 6, Santa Clara, CA 95054; 408-988-3377, FAX 408-988-6027
dirs, Frank Bakonyi, Ferenc Ledniczky; 1983
Laser job shop. Provides laser cutting for semiconductor related materials- glass, ceramics, silicon, metals, and R&D.

Rayco-B Prod Inc, 1602 Raymond Ave, Monrovia, CA 91016
1983

San Diego Sheet Metal Works, 1440 J St, San Diego, CA 92101; 619-232-3153, FAX 619-234-6961
emp 400, 1905
Complete sheet metal facility. Offering CAD/CAM programmable laser and plasma cutting. CNC controlled shearing, punching, and forming through 1/4 in. plate. Certified welding in mild steel, aluminum, and stainless steel. Prototype and production.

Southern California Laser Products Inc, 42380 Rio Nedo, Temecula, CA 92590; 714-676-8083, FAX 714-676-1316
emp 100, 1985
Operates laser job shop using 1500-W CO_2 & YAG lasers with CNC, CAD/CAM system (.001 accuracy, 4 x 8' sheets) for cutting, welding, drilling, heat-treating, and wood engraving. Manufactures various sized CNC positioning tables up to 4 x 8'. Provides completely integrated 5-axis laser cutting & welding systems.

Trillwood Technologies, 11942 Woodlawn Ave, Santa Ana, CA 92705; 714-731-3663, FAX 714-838-8561
ceo, Richard Trillwood; 1989
Specialists in YAG laser systems for welding small components, both low and high volume production. Sophisticated tooling and workhandling systems. Twenty-five years experience in high energy beam welding (EBW and LBW).

Colorado

Bond Custom Manufacturing, PO Box 1086, Arvada, CO 80001; 303-423-8841*
1989
Laser job shop.

Coors Ceramics Co, Electronics Div, 2449 River Rd, Grand Junction, CO 81505; 303-245-4000, FAX 303-243-7031
sls mgr, Bill Czaplinski; 1977
Laser drilling, scribing, and machining of thick and thin film Alumina substrates, metals, plastics, glasses, and other materials for various electronic applications. Computer-assisted design engineering is available to provide superior quality at minimum cost. Fast service prototype quantities to large production volumes.

Light Touch, 684 Dearborn St, Aurora, CO 80011; 303-343-4773
1977

Midwest Machine Products Inc, 6255 Joyce Dr, Golden, CO 80403; 303-422-5388 *
emp 500, 1969
Complete precision sheet-metal fabrication, precision machining and laser processing. Includes Class "A" painting and sikscreening.*

Viner Manufacturing Co, PO Box 2871, Evergreen, CO 80439; 303-674-0534
pres, Ted Viner; emp 30, 1972
CO_2 and YAG job shop services.

Connecticut

Acme Steel Rule Die Corp, 5 Stevens St, Waterbury, CT 06704; 203-753-8944, FAX 203-753-9813
pres, Gerard DeHippolytis; 1972

AcuCut Inc, 71 Dolphin Rd, Bristol, CT 06010; 203-582-2348, FAX 203-582-0102
sls mgr, Scott Barmore; emp 500, 1978
Job shop specializing in precision laser cutting and E.D.M. Operates 21 machines 3 shifts a day and can meet rush deliveries.

Dieknologist Inc, 150 Long Beach Blvd, Stratford, CT 06497-7116; 212-989-5959, FAX 212-691-0692
pres, John A Passantino; 1978
Offers fancy steel rule dies, laser and jigged.

Laser Marking Inc, 363 Spring St, Manchester, CT 06046; 203-649-8068, FAX 203-649-9665
pres, Richard J. Nelson; 1988
Contract laser marking of aerospace products and R&D of aerospace applications. FAA approved source of laser marking.

Lee Co, 2 Pettipaug Rd, Westbrook, CT 06498; 203-399-6281*, FAX 203-399-2058
1988

Precision Machining & Fabricating Inc, 673 Naugatuck Ave, Milford, CT 06460; 203-878-7414 *
emp 500, 1973
Sheet-metal fabrications servicing the medical/electronic industries. CNC laser contours cut up to .250 in. steel, .125 in. stainless, .250 in. polycarbonate.*

Florida

ECC Int'l, 5882 S. Tampa Ave, Orlando, FL 32809; 407-859-7410, FAX 407-851-1871
exec off/gen mgr, George Frye; vp mfg, Spence Whitehead; 1973
R&D facility with complete manufacturing capability for building simulation devices for training and familiarization.

Jet Avion Corp, Jet Engine Componets Div (sub of Heico Corp), 3000 Taft St, Hollywood, FL 33021-4499; 305-987-6101*, FAX 305-966-2169
emp 200, 1964
Laser cutting services of most commerical materials in thickness up to 1/4 in. of stainless steel and in sheet dimensions of 40 x 80 in. Other manufacturing processes such as CNC turning and milling, GTAW welding, plasma spray brazing and heat treating can be integrated with laser cutting to provide complete in-house fabrication abilities in a variety of alloys including high temperature alloys. CAD/CAM programming provides a quick, accurate transition from engineering drawing to machine control.

Lasereliance Technologies, 774 S. North Lake Blvd, Suite 1016, Altamonte Springs, FL 32701-6732; 407-339-0737, FAX 407-339-7463
pres, Timothy H. Saunders; 1982
Custom CO_2 laser machining, scribing, drilling, welding, and heat treating.

Lasertron Inc, 14251 NW 4th St, Sunrise, FL 33325; 305-846-8600, FAX 305-846-8604
pres, Gary Geller; mktg mgr, Jack Dilts; 1979
Laser machining of ceramics, metals, and organic materials using CO_2 and YAG lasers. Capabilities include laser marking. FAA approved repair station.

Ultracut, 7655-4 Enterprise Dr, Riviera Beach, FL 33404; 407-863-5721 *
1979

Georgia

Southern Metalcraft Inc, 2675 Lithonia Ind Blvd, Lithonia, GA 30058; 404-482-2923, FAX 404-482-6292
pres, K.G. Williams; gen mgr, R.A. Little; engr, Greg Williams; 1974
Custom metal fabricated products using steel, stainless, aluminum, galvanized, brass and copper. Services include: engineering by fabricam and autocad; quality control, inspection, scanning and digitizing by fabrivision; laser and plasma cutting, shearing (1/2 steel), punching, forming, bending, rolling and MIG/TIG/electrode welding.

Illinois

Ace Laser Tek Inc, 65 East Palatine Rd, Suite 319, Prospect Heights, IL 60070; 708-537-4202, FAX 708-537-4701
pres, Greg Litcher; prod mgr, Paul Schroeder; 1988
Laser engraving/marking job shop. Offers contract R&D laser marking, laser machine design and sales for welding, drilling, cutting, and marking using ND:YAG and CO_2 lasers.

Aquarius Metal, 675 Greenleaf Ave, Elkgrove Village, IL 60007; 312-956-6810*
1988

Atlas Die Inc, Atlas Chicago Div, 2000 Bloomingdale Rd, Glendale Heights, IL 60139; 708-351-5140
mgr, Rick Storey; 1988

Atlas Tool & Die Works Inc, Accushim Inc, 4633 Lawndale Ave, PO Box 32, Lyons, IL 60534; 708-442-1661, FAX 708-442-0016
pres, Daniel J. Mottl; vp, Glen J. Mottl; 1918
Close tolerance cutting of steel, stainless and plastics, and short run and prototype fabricated assemblies.

Blackhawk Composites, 720 S. Wyman St, Rockford, IL 61101; 815-968-7488, TWX 815-968-8727
pres, Verne DeCourcy; vp, John Lundin; gen mgr, Rick Williams; 1918
Laser fabricating and job shop with CNC milling, laser cutting, sheet-metal fabricating, and 5-axis water jet cutting.

Danville Metal Stamping, 17 Oakwood Blvd, PO Box 856, Danville, IL 61832; 217-466-0647*
1918

Die-Tronics Inc, 3122 MacArthur Blvd, Northbrook, IL 60062; 312-498-2110
emp 870, 1974
A wire-EDM, laser cutting job shop.

Erie Industrial Products Co, 1234 Bristol Ave, Westchester, IL 60154; 708-344-8200, FAX 708-344-8205
pres, Peter Selleck; sls mgr, David Collins; lsr srvs engr, Frank Baker; 1945
Produces custom metal products of complex shapes and close tolerances. Provides full fabrication capabilities with a variety of materials. Utilizes state-of-the-art equipment including two lasers, a large CNC punching machine, and CNC brakes.

Espe Mfg Co Inc, 9222 Ivanhoe, Schiller Park, IL 60176; 800-367-3773, FAX 708-678-0253
1940
Laser cutting of nonmetallic materials such as fiber, nylon, phenolics, acrylic, and PVC. Also provides stamping, slitting, and drilling of these and other materials.

Laserage Technology Corp, 3021 Delany Rd, Waukegan, IL 60087; 708-249-5900, FAX 708-336-1103
emp 950, 1979
Provides industry with various laser capabilities. Using custom-designed CO_2 and Nd:YAG laser systems, can precisely cut, drill, mark, scribe, weld, and heat-treat a wide variety of materials to specifications.

Laser Technologies Inc, 766 N. Edgewood, Wood Dale, IL 60191; 312-766-0060*, FAX 312-766-1635
emp 90, 1985

Morton Metalcraft Co, 1031 W. Birchwood, Morton, IL 61550; 309-266-7176 *
1985

My-Lin Mfg Co, 820 N Russell Ave, Aurora, IL 60506
1985

Parkway Metal Products, 2301 W. Wabansia, Chicago, IL 60647; 312-489-3144, FAX 312-489-7144
pres, Ted Martin; vp, Daniel Brown; 1966
Precision metal stampings and CNC fabrications. CNC turret punching, CNC laser cutting and CNC brake forming. Stampings to 150 tons. Sub-assembly including all types of welding, riveting, staking and bonding.

Peterson Products Corp, 4848 N. River Rd, Schiller Park, IL 60176; 312-678-0800 *
1966

Radtke & Sons, Laser Works Div, 101 W. Main St, Round Lake Park, IL 60073; 708-546-3999, FAX 708-546-4008
gen mgr, John A. Radtke; 1989
Provides laser welding, cutting, drilling, scribing, and marking for aerospace, electronics, medical and other manufacturers. Division of precision CNC machining job shop.

Spectrum Manufacturing, 140 Hintz Rd, Wheeling, IL 60090; 708-520-1553, FAX 708-520-1639
1989
High precision laser cutting job shop.

Univ Illinois Dept Mach Eng, Center for Laser-Aided Mat'ls Proc, 1206 W Green St, Urbana, IL 61801; 217-333-1964, FAX 217-244-6534, telex 510-101-1969
dir, Prof. J. Mazumder; 1990
University research and education in the area of laser materials processing for quantitative understanding of processes required for successful industrial application and automation. Topics include: on-line optical diagnostics of mathematical modeling and characterization of physical and chemical properties associated with processes such as laser welding, drilling, cladding, alloying, chemical vapor deposition, ablation, heat treatment, etc.

WILLIE LASER PRODUCTS

div of Willie Washer Mfg Co, 2101 Greenleaf Ave, Elk Grove Village, IL 60007; 708-956-1344, FAX 708-956-7943
dir oper, Jim Neumann; sls mgr, Bob Urlakis; 1973
Precision laser job shop specializing in quick deliveries. CAD/CAM and digitizing capabilities. Able to produce one of a kind parts from prints or samples. 48 x 96 CNC capacity.

Indiana

Atlas Die Inc, 2000 Middlebury St, Elkhart, IN 46516; 219-295-0050, FAX 219-294-2793/800-2558786
pres, Herb Welsch; vp mktg, Kathleen Kauffman; vp sls, John Norgard; vp mfg-central, Mike Kush; vp mfg-east, Ken Roberts; vp metal prds, Alan Degenhart; vp fin, Jerry Hostetler; 1952
Laser produced steel rule dies for the folding carton, electronic, aerospace, gasket, label and other specialty industries. Also manufactures steel counterplates, bonded dies, Perma-Die®, embossing dies, layered stamping dies, combodies, tolerance gauges, transfer and injection molds for the rubber industry.

Boyd Machine & Repair Co Inc, 3794W - 505, Kimmell, IN 46760; 219-635-2195
pres, Larry Boyd; eng mgr, Mark Boyd; lsr eng, Bryan Biggs; emp 250, 1975
Service includes prototype, production, and research in laser cladding, heat-treating, alloying, and welding. Facility equipped with 5-Kw CO_2, three workstations and small metal laboratory with fabrication and machining departments.

Custom Laser Services Corp, 2120 W Franklin St, Elkhart, IN 46516; 219-293-0494*
emp 60, 1981
Custom cutting of wood, plastic, and metals. Etching wood and plastic.*

Decatur Industries Inc., PO Drawer 311, Decatur, IN 46733
1981

Die Tech Corp, 13130 S. U.S. Highway 27, Ft. Wayne, IN 46816
1981

Dolphin Laser Processing, Inc, PO BOX 688, Hwy 421 S, One Indt'l Park, Versailles, IN 47042; 812-689-7202, FAX 812-689-7203
pres, John Sheler; off mgr, Carolyn Sheler; 1989
Laser job shop service for cutting sheet metal and plate steel. Cuts wood, die boards, plastics, rubber, stainless steels, and other metals for prototypes and production. Other services include; forming, heat treating, and welding.

Estes Design & Mfg Co, 470 S. Mitthoeffer Rd, Indianapolis, IN 46229; 317-899-2203, FAX 317-898-2034
pres, Larry A. Estes; op dir, Ronald Estes; plant mgr, Larry Ruble; 1976
Offers custom metal fabrication, enclosures for medical and electronic applications, computerized production scheduling, CAD/CAM engineering, laser cutting and machining, CNC turret punching, forming with NC gauging, robotic welding, MIG and TIG welding, spot welding, CNC machining, tube bending, metal rolling, degreasing and painting capabilities, and silk screening.

Kaiser Tool Company Inc, Laser Images Div, 701 Sherman Blvd, Ft. Wayne, IN 46808; 219-422-2408, FAX 219-426-4479
ceo/vp, Lenore E. Perry; vp op, Douglas O. Perry; 1964
Provides permanent non-contact marking of most materials including metals, carbides, ceramics, silicones, plastics, etc. Available fonts include helvetica, block, OCR-A, script, and bar code-39.

Precision Laser Services Inc, 314 E. Wallace St, PO Box 10863, Ft. Wayne, IN 46803; 219-744-4375, FAX 219-744-5666
pres, Dale O. Ferrier; gen mgr, Ed Ferrier; emp 150, 1941
Provides precision laser machining for all types of materials. Services include laser cutting, welding, drilling, marking and engraving. Prototype and feasability studies to production processing.

CF Roark Welding & Engineering Co Inc, 136 N. Green St, Brownsburg, IN 46112; 317-852-3163, FAX 317-852-2738
pres, Charles T. Roark; sls mgr, William T. Golay; eng mgr, Bradley Beeson; 1949
Development and production programs for the aircraft engine and airframe industry in fabrication, stampings, laser cutting and welding, component overhaul and repair, and specialty welding.

Tigart Laser System Inc, 5630 W. 82nd St, Indianapolis, IN 46278-1300; 317-872-8113, FAX 317-872-8584
ceo/pres, H. Max Pugh; ops dir, David Archer; sls/off mgr, R. Smith; emp 120, 1985
Laser cutting corporation specializing in JIT production, engineering samples, and prototypes in steel to .5 in., stainless steel to .375 in., and aluminum to .500 in.

Wood Mizer, PO Box 128, New Point, IN 47263
1985

Iowa

Iowa Laser Technology Inc, 6122 Nordic Dr, Cedar Falls, IA 50613; 800-397-3561, FAX 319-266-8203
pres, Mark W. Baldwin; eng mgr, Donald G. Kobriger; 1978
Precision laser job shop performing laser cutting, welding, and heat treating and metal fabrication. Specializing in fast turn around, short lead-time manufacturing. Mil-I-45208 certified; helium leak detection; metallurgical analysis; applications R&D and process development available. 36 x 60" x 360° (*x-y*-theta) CNC capacity.

Pickwick Co, 1870 McCloud Pl NE, Cedar Rapids, IA 52402; 319-393-7443/800397-9797, FAX 319-393-7456
pres, Walter F. Corey; eng mgr/CAD appl, David Pearson; mfg mgr, Fred Mittan; sls/mktg mgr, Peter Ryder; 1939
Complete short-run sheet-metal fabrications, laser cutting up to 3/8" steel sheets, 5" x10", with tolerances of +/-.004. Also forming, welding, painting, and more. Manufacturing services including solids modeling, 1200 W full sheet (5'X10') laser with automated material handling, high speed punching, forming, welding, painting and machining. Primary materials: steel, aluminum, stainless and plexiglass.

Kansas

Midwest Laser Inc, 8281 Melrose, Lenexa, KS 66214; 913-541-1813, FAX 913-541-0704
1984
An industry leader in specialty advertising engraving. In addition, the company provides UL approved labels to many other manufacturers. The medical instrument industry is also supplied with quality engraving.

Kentucky

Cincinnati Ventilating Co, 7410 Industrial Rd, PO Box 217, Florence, KY 41042; 606-371-1320, FAX 606-371-2963
pres, R. Kevin Martin; 1952
Close-tolerance sheet-metal job shop working in carbon steels, stainless, and aluminum.

National Metal Processing Inc, 450 No. Estill Ave, PO Box 280, Richmond, KY 40475-0280; 606-623-9291, FAX 606-623-9291
pres, Norman S. Graves; sls mgr, Russ Bell; mktg mgr, Allen Graves; prod mgr, Boyd Parke; 1969
Laser job shop providing hardening and welding services on prototype or production basis. Four 650-W and 1200-W CO_2 units with CNC automation are utilized. Comprehensive in-house metallurgical lab and dimensional inspection center support processing. Conventional heat treating services also available.

Skilcraft

5184 Limaburg Rd, PO Box 896, Burlington, KY 41005; 606-371-0799, FAX 606-371-2627
ceo, Eric R. Thiemann; pres, Kenneth J. Anderson; vp, Richard Anderson; plant mgr, Bill Sears; 1965
Custom fabrication, sheet metal products. Specializing in precision stainless steel fabrication and laser cutting.

Maryland

Harford Systems Inc, PO Box 700, Aberdeen, MD 21001; 301-273-3400*, FAX 301-273-7892
1965
Precision custom fabrications.

Laser Applications Inc, 1110 Business Parkway South, Westminster, MD 21157; 301-857-0770, FAX 301-857-0774
pres, Bob Ulrich; sls eng, Jim Wollenweber; emp 550, 1979
Offers waterjet and laser processing, CNC machining, wire EDM, and light metal forming. Work stations include 6 axis contouring with travel limits up to 15' x 16'.

Lenox Laser

Lenox Laser, 1 Green Glade Ct, Phoenix, MD 21131; 301-592-3106, FAX 301-592-3362
pres, Joseph P. d'Entremont; sls mgr, Samuel W. Balzanna, Jr; mktg dir, Karen Janssen; ch eng, Michael Gardner; technologist, Gary Thornton; emp 60, 1981
Provides micro-drilling, micro-welding, and etching. Optical products include: high energy experimental apertures, precision apertures, aperture mounts, precision air slits, etc. Optical components include: optical benches and tables, solid aluminum breadboard, spatial filters, and honeycomb breadboard.

Massachusetts

Apertura, 14 Bruno Terrace, Woburn, MA 01801; 617-933-8088, FAX 617-938-7707
pres, Charles Hughes; 1976
Manufactures precision pinhole apertures and provides a service of micro-welding and micro-hole drilling. Lasers, EDM, abrasive and mechanical means are used to drill holes down to I um in diameter. Materials drilled vary from high carbon and stainless steel to high temperature alloys and precious stones and metals.

BME Engineering Inc, PO Box 849, Carleton Dr, Georgetown, MA 01833; 617-352-8202*
emp 750, 1981
Laser sheet metal cutting.

Boston Precision Parts Co Inc, 46 Sprague St, Hyde Park, MA 02136; 617-361-1000, FAX 617-364-0941
1958
Offers precision laser machining and precision sheet metal fabrication, and CNC machining.

Circle T&D Corp, 45 Commerce Way, Woburn, MA 01801; 617-935-7290, FAX 617-932-8936
pres, D. Hasty; sls mgr, F. Nickerson; 1972
Provides prototype and production work in the following areas: wire EDM, conventional EDM, lasers, metal stamping and die design and tooling, related fixtures.

EBTEC Corp, (sub of TI Ltd), 120 Shoemaker Lane, Agawam, MA 01001; 413-786-0393, FAX 413-789-2851
pres, David J. Maggs; vp sls/mktg, John Leveille; vp fin, Vincent A. Mammano; corp sec, Rita G. Ducharme; 1963
Contract laser and electron-beam job shop with complete machine shop, metallurgical lab, and quality-control department. Entire facility is FAA certified as a manufacturer' s service operation.

Evans Industries, PO Box 169, Topsfield, MA 01983; 508-887-8561*
1963

Hoechst CeramTec NA Inc, 171 Forbes Blvd, Mansfield, MA 02048-1148; 508-339-1911, FAX 508-339-5099
pres, James H. DiSorbo; exec vp, David B. Nicholson; elec prod mgr, Walt Dollman; controller, Helga Mueller; 1963
Custom laser services of technical ceramics, including cutting, machining, and drilling of alumina substrates for hybrid electronic applications. A large stock of substrates on site. Company has over 20 years of ceramic lasering experience. Prototype through volume capabilities.

Laser Industries Inc, 15 Union St, Lawrence, MA 01840; 508-682-2460/800-6227040, FAX 508-682-6774
pres, John R. Blutt; mfg mgr, Michael J. Catino; 1974
Laser processing job shop specializing in drilling, cutting, marking, and welding applications using CO_2, YAG, and ruby lasers. Also provides high pressure abrasive waterjet cutting services.

Laser Services Inc, 123 Oak Hill Rd, Westford, MA 01886; 508-692-6180, FAX 508-692-7271
pres, Bruce N. Beauchesne; vp & treas, June P. Beauchesne; oper mgr, Paul H. Wirzburger; 1979
Multi-shift laser job shop using 18 CO_2 and YAG lasers for cutting, drilling, welding, marking, substrate scribing, and resistor trimming. Materials include metal, plastic, wood, composites, rubber, etc. Positioning to 5 x 8 ft. Provides feasibility studies, prototype and high volume production.

Lighthouse Mfg Inc, 6 Fourth St, Peabody, MA 01960
1979

Meadex Technologies, PO Box 2528, Springfield, MA 01101; 413-567-3680, FAX 413-567-3680
pres, Edward F. Watson; mktg mgr, Mary B. Hayes; 1984
Technical consulting, marketing, sales, design, and manufacturing of high-technology automated manufacturing equipment using laser and electron-beam systems. Manufacturer' s representative services. Purchase, sale, and service of used laser and electron-beam equipment. Laser and electron-beam job shop services.

Micro Precision Technology, 530 Turnpike St, N. Andover, MA 01845; 508-686-6620, FAX 508-689-2261
1984
Resistor trimming for thick and thin films in semiconductor and hybrid industries.

National Laser Trimming, Div of Barry Industries, 67 Mechanic St, Attleboro, MA 02703; 617-226-3313 *, FAX 617-226-3317
1983
Contract laser trimming services of resistors/circuits. Low OHM, tight tolerance, mil-spec and prototype capabilities. Custom probe cards built in-house.

New England Laser Processing, Div of C. Behrens Machinery Co Inc, Danvers Industrial Park, Danvers, MA 01923; 508-774-4626, FAX 508-774-0532, telex 286804
pres Richard Jacob; gen mgr, William J. Forrestall; 1987
Laser job shop and applications lab for cutting, marking, engraving, and welding of metals and nonmetals. Also provides process development service, and prototype and pilot production.

Simonds Industries Inc, Intervale Rd, PO Box 500, Fitchburg, MA 01420; 508-343-3731, FAX 800-541-6224, telex 6711015
ceo, Charles Doulton; eng mgr, Ray Edson; bus unit mgr, Herb Kartanos; mktg mgr, Dale Petts; emp 480, 1832
Contract cutting of mild, alloy and tool steel flat products from sheets or round blocks to .4 in. thick and 84 in. diameter. Programming and laser cutting services utilizing DNC controlled 1500-W CO_2 laser on 2-axis custom table.

Stellar Industries Corp, 225 Viscoloid Ave, Leominster, MA 01453; 508-840-1884, FAX 508-840-4384
1832

Texcel Inc, 8 Elise St, Westfield, MA 01085; 413-562-7593 vec, FAX 413-568-9469
ceo, R. Lalli; pres, Laurence S. Derose; vp/gen mgr, J. Lovotti; 1987
Offers laser-based process development studies, contract R&D and feasibility studies. Special emphasis on hermetic sealing of high reliability products for aerospace electronics, biomedical, and instrumentation industries. CO_2 and YAG laser application labs. Laser hermetic sealing facility complete with glovebox laser system, leak checking, and metalurgical support.

Michigan

Advance Laser Applications Inc, 47808 Galleon Dr, Plymouth, MI 48170; 313-451-0140, FAX 313-451-0457
pres, Steven F. Fedor; emp 50, 1980
Manufacturer of templates for auto industry. Started laser job shop activity in 1986.

Alloy Tek Inc, 2900 Wilson SW, Grandville, MI 49418; 616-534-1000
1980

Auto Metal Craft Inc, Auto Chem Craft Div, 12741 Capital Ave, Oak Park, MI 48237; 313-398-2240, FAX 313-398-3411
ceo, Patrick N. Woody; exec vp, Donn Woody; vp, Kevin C. Woody; sls mgr, Kim Woody; 1950
Offers prototype sheet metal forming and fabrication, CAD/CAM, 3-axis and 5-axis laser cutting, and orbital EDM die spotting.

Benteler Stamping, PO Box 3124, Grand Rapids, MI 49501
1950

Cedar CNC Machining Inc, 104 Beech St, Cedar Springs, MI 49319; 616-696-2750, FAX 616-696-1436
pres, Ronald Alderton; 1980
Aerospace machining utilizing CNC turning, CNC milling, CNC grinding, CNC abrasive waterjet, EDM, wire EDM, laser cutting scheduled for fall 1991. Aerospace and government approved quality assurance program to MIL-STD 45208-A.

Electrolabs Inc, 18503 14 Mile Rd, Fraser, MI 48026; 313-294-4150, FAX 313-294-6090
pres, Dennis Suddon; eng mgr, Scott Suddon; mfg mgr, Al Suddon; sls/mktg mgr, Denise Dawson; 1963
Precision sheet-metal fabricator supplying mainly to the computer and electronics industries, but not limited to any field(s) in particular. Provides engineering, laser, turret punch press, shearing, forming, welding, spot welding, paint, and silkscreen services.

Future Technologies Inc, 810 S. Huron Ave, Linwood, MI 48634; 517-697-5353, FAX 517-697-3125
pres, Gerald Bechanko; sls mgr, Diego Calvo; 1989
Specializes in the design and construction of special laser welding machines.

Gladd Industries, 15450 Dale, Detroit, MI 48223; 313-537-2800, FAX 313-537-1523
1963
Laser job shop, sheet metal shop.

Great Lakes Laser Services, 2608 N. Woodward, PO Box 868, Royal Oak, MI 48068; 313-549-0450, FAX 313-549-0650
pres, Carl R. Hildebrand; eng mgr, L.G. LaMay; sec, Terri Swindle; 1981
Offers contract service and product development utilizing lasers. Principal services offered are laser engraving with computer-controlled YAG laser markers and laser cutting, drilling, and welding with a 200-W YAG laser.

Hillsdale Tool & Mfg, PO Box 286, Hillsdale, MI 49242
1981

Laser Dynamics Inc, 1064 Commerce, Birmingham, MI 48009; 313-646-1916, FAX 313-646-6720
1981
Offers laser and conventional computerized engraving services with custom CAD/CAM graphics. Marks industrial parts, precision medical instruments, advertising specialty items, etc. Manufactures specialized machine panels, barcode strips and signage.

Laser Manufacturing Inc, PO Box 344, Plymouth, MI 48170-0344; 313-471-5457*, FAX 313-471-1106
emp 120, 1985
Laser cutting, heat treating, and welding of complex parts various materials and thicknesses. Prototype and process development capability.

Milford Fabricating, 19200 Glendale, Detroit, MI 48223
1985

Modern Prototype Co, 2927 Elliott, PO Box 4540, Troy, MI 48099-4540; 313-585-0120, FAX 313-585-2466
pres, Dennis Cedar; 1975
High tech, full service prototype shop equipped with latest metal forming/metal molding technology, including computer operated, 5- axis laser cutting machines, nc (tarus) metal and clay mills, CMM inspection machines and updated foundry. Manufacturing prototype kirksite and expoxy composite dies, sheet metal stampings, and welding assemblies primarily for automobile industry.

MSI Prototype & Engineering, div of MSI Corp, 26269 Groesbeck, Warren, MI 48089; 313-773-0800, FAX 313-773-0054
gen mgr, Ron Nadrowski; purch, Leon F. Mikolaczyk; 1965
Automotive prototype.

Oakley Industries Inc, 35166 Automation Dr, Mt. Clemens, MI 48043; 313-792-1261, FAX 313-792-1332
pres, Ronald Oakley; vp, Michael Oakley; 1980
Provides automotive prototype stamping and assemblies. 3-axis and 5-axis laser cutting. Die tryout and binder development.

Proto Dynamics Corp, 34197 Doreka St, Faser, MI 48026; 313-294-0500, FAX 313-294-1980
pres, K.G. Murski; mfg eng, Ryan Burns; 1973

Prototech Laser Inc, 44444 Reynolds, Mt. Clemens, MI 48043; 313-465-9944, FAX 313-465-0764
pres, David E. Prue; vp, Fred French; sls mgr, Steven A. Prue; 1990
Offers high quality 3 & 5-axis laser cutting and welding services. Able to manufacture prototype as well as production orders.

Quasar Industries, 2687 Commerce Dr, Rochester Hills, MI 48309; 313-852-0300 *, FAX 313-852-0442
emp 750, 1965
Metal prototype laser process job shop specializing in high-tech prototype manufacturing. Features full CAD/CAM 3-D modeling capabilities, a 72 x 196 in. laser gantry, a 10-axis robot, wire feed weld systems for laser welding and metal cladding.

Rapid-Line Inc, 4900 Clyde Park Ave SW, Grand Rapids, MI 49509; 616-530-0061, FAX 616-530-0819
pres, Mark Lindquist; vp engrg, George Ignatiev; vp sls, Joe Jacques; 1989
Specializes in light guage sheet metal fabrications. With a Mitsubishi laser and other CNC equipment produces a wide variety of custom and production sheet metal products. Using CAD/CAM technology and vertical integration, can meet both your technical as well as delivery requirements.

Reid Industries Inc, 28440 Groesbeck, Roseville, MI 48066; 313-776-2070 *, FAX 313-776-0430
1976
Wire E.D.M and laser cutting service.

Roberts Precision Products Inc, 170 Enterprise Dr, Ann Arbor, MI 48103; 313-662-6569, FAX 313-662-9230
pres, Gordon McNutt; 1970
Fabricates precision sheet metal parts for electronics including enclosures, housings, front panels, NEMA type enclosures, brackets and weldments.

Troy Design & Mfg

Troy Design & Mfg, 12675 Berwin, Redford, MI 48239; 313-537-3880
1970

UTILASE, INC.
LASER PROCESSING JOB SHOP

20201 Hoover Rd, Detroit, MI 48205; 313-521-2488, FAX 313-521-4695
vp/mktg, Jeff Moss; vp/ops, Tim Metko; 1986
A contract laser processing job shop service center with more than a dozen CO_2 and YAG laser work stations, performing laser welding, laser cutting, laser heat treating, laser drilling and laser cladding to meet low-volume prototype and high-volume production volume requirements of manufacturers in the automotive, aerospace, appliance and office equipment industries.

Minnesota

Albers Sheet Metal, 200 W Plato Blvd, Saint Paul, MN 55107; 612-224-5428
1986

Assurance Mfg Inc, 9010 Evergreen Blvd, Minneapolis, MN 55433; 612-780-4252*, FAX 612-780-8847
emp 950, 1964
Laser cut prototyping, design services, tool room facilities, precision short and medium run stampings, SPC, CAD/CAM, electrical and mechanical assembly, secondary operations, traveling wire EDM, continuous quality control, fully computerized manufacturing from SPC to CAD/CAM.

BuckBee Mears, St. Paul, 245 E. 6th St, St. Paul, MN 55101; 612-228-6400, FAX 612-228-6572, telex 29-7080
sls/mktg dir, Fred Grimm; gen mgr, Tom Stevens; qa mgr, Jerry Miller; engr mgr, Tom Myers; 1907
Photo-etch job shop with laser machining capabilities. Offers 98 x 78 in. Laser Labs system with Rofin Sinar 1200-W CO power supply. Also offers two YAG lasers for seamless welding. Specializes in custom perforated cylinders.

Hansman Industries Inc, 10860 No. 60th St, PO Box 2058, Stillwater, MN 55082; 612-439-7202, FAX 612-439-1668
1946
A full service prototype, preproduction and production precision sheet metal job shop. Offers design, developement, fabrication and electro-mechanical assembly.

LASERDYNE PRECISION SERVICES

6690 Shady Oak Rd, Eden Prairie, MN 55344; 612-941-7694, FAX 612-941-7611
dir, Ron Sanders; sls mgr, Rick Schuette; plant mgr, Richard Jackson; 1981
Full service, multi-axis laser job shop specializing in machining, welding, heat treating, and marking of flat and 3-dimensional metal and non-metal parts.

Laser Tech Inc, 14860 DeVeau Place, Minnetonka, MN 55345; 612-935-9277, FAX 612-935-0433
ch exec/sls mgr, Jerry Peterson; qual mgr, Larry Whitman; emp 180, 1977
Laser process and custom machine ceramics and products primarily within the micro-electronics industry. Complete processing and metalization services in thick-film areas. Machining heat sinks, adhesives, and beta-stage for the PC board industry.

Mathias Die Co, 1200 N. Concord St, So. St. Paul, MN 55075; 612-451-0105, FAX 612-451-1943
pres/owner, Thomas Mathias; sls mgr, Joseph Sampair; plant mgr, Larry Monhaut; 1965
Manufactures laser steel rule dies used in stamping 2 axis, typically flat sheet substrates such as plastic, paper, foams, rubber, thin foils, etc. In addition, provides specialty die cutting & laser cutting services for all above stated applications.

Modernistic, 169 E. Jenks Ave, St. Paul, MN 55117; 612-291-7650, FAX 612-291-2571
vp, Scott Schulte; plant mgr, Garry Novak; laser dept mgr, Ron Vitali; 1939
Laser cutting and die cutting of wood and plastics up to 1/2 in. thick. CAD/CAM, digitizing, and scanning capabilities.

Morrissey Inc, 9305 Bryant Ave, South, Bloomington, MN 55420; 800-798-6158, FAX 800-798-7315
1939
Offers custom metal stamping with prototype, short-run and long-run stamping. Laser is used as a means to produce parts without the added cost of tooling.

NYBO Welding, 9649 Humboldt Ave S., Bloomington, MN 55431; 612-884-4379, FAX 612-884-6115
1939

Oakcraft Inc, Laser Carving Div, 9521 Foley Blvd NW, Coon Rapids, MN 55433; 612-784-6305 *
emp 90, 1984
Custom layout and design of production and prototype products to be laser engraved.

Rotation Engineering & Mfg Co, 14940 28th Ave North, Plymouth, MN 55447; 612-553-1090, FAX 612-553-9191
pres, James A. Lorence; plant mgr, Lloyd Larson; emp 280, 1975
Wire EDM work, dies, laser work and stampings.

Seelye Plastics Inc, 9700 Newton Ave South, Bloomington, MN 55431; 612-881-2658, FAX 612-881-6203
emp 750, 1967
Job shop utilizing CNC lasers, mills, lathe, and router to manufacture machined parts from plastic sheet and rod stock prototype to production runs.

SPECTRAlytics

7393 Bush Lake Rd, Minneapolis, MN 55439; 612-831-5511, FAX 612-831-9393
pres, Gary Oberg; sls/mktg mgr, William Shnowske; 1990
Laser job shop offering welding, cutting, drilling, skiving, and marking services. Full range of YAG, CO_2, and excimer lasers. Full 5-axis capability. R&D to production. Full CAD/CAM, clean booth, ESD, leak testing, and QC testing.

Stremel Manufacturing Co, 260 Plymouth Ave, Minneapolis, MN 55411-3465; 612-339-8621, FAX 612-339-2661
gen mgr, Dave Curless; sls mgr, Jack Garberg; 1896

Missouri

Rose Metal Products, PO Box 3238, Springfield, MO 65808; 417-865-1676, FAX 417-865-7673
pres, Ron Buchanan; eng dir, Larry Little; sr prog cnc op, Bruce Durnell; emp 560, 1960
Custom metal fabricators, specializing in laser cutting, punching, shearing, braking, welding and finishing metal components. Operations are CAD and CNC driven with close tolerances held accordingly.

New Hampshire

Computer Fabrications Inc, PO Box 418, Pulpit Rock Rd, Pelham, NH 03076; 603-635-2123*
emp 450, 1977
Precision sheet-metal fabrications of electro-mechanical assemblies.

Optimation Inc, Burr-Free Microhole Div, 8025 S. Willow (1-101), Manchester, NH 03083; 603-623-2800, FAX 603-623-3020
pres, J. Dean Jorgensen; 1971
Specializes in supplying microhole hole drilling using laser, micro EDM, and punch and die. Products include spatial filter instruments, and *xy* pinhole positioners. Micromotion motor micrometers for 0-1/4 in. travel. Laser consultant for industrial applications. Brokers for used laser systems specializing in laser marking systems.

Resonetics, 4 Bud Way Bldg 21, Nashua, NH 03063; 603-886-6772, FAX 603-886-3655
pres, Jeffrey Sercel; vp sls/mktg, Michael Scaggs; 1988
Excimer-laser materials-processing services, system manufacturer, and technology-development center. Full systems integration and custom-design capabilities. Independent consulting services provided. Specializes in industrial excimer-laser applications. Products include; excimer gas purifiers, air scrubbers, beam delivery and high-rep-rate direct-write materials-processing systems.

New Jersey

Atlantic Metal Products, 21 Fadem Rd, Springfield, NJ 07081; 201-379-6200*, FAX 201-379-6947
1988

Dubbeldee Harris Diamond Corp, 100 Stierli Ct, Suite 106, Mt. Arlington, NJ 07856; 201-770-1420, FAX 201-770-1549, telex 236064
1961
Provides precision diamond components, job shop laser machining, and other scientific services. Computer-controlled Nd:YAG Q-switched laser system to machine various materials including: diamonds, polycrystalline materials, sapphire, aluminum oxide, rubber (silicon R), titanium boride, and tungsten carbide.

Edgar Barcus Co Inc, Rte 45 & Park Ave, Westville, NJ 08093; 609-456-0204, FAX 609-456-5970
1961
Provides steel rule cutting dies, strippers, mylars. Laser burned die boards. Dies to cut cardboard, paper, rubber, leather, plastic, cork, etc. Dies for folding cartons, corrogated cartons, membrane switch, flexible circuits, set-up boxes, gaskets, etc. CAD/CAM system with modem.

Harold R. Henrich Inc, 300 Syracuse Ct, Lakewood, NJ 08701; 908-370-4455
emp 500, 1927
Laser metal cutting.

Laser Dies Inc, 32 Industrial Ave, Little Ferry, NJ 07043; 201-440-7600, FAX 201-440-8846
1927

Quantum Laser Corp, 300 Columbus Circle, Edison, NJ 08837; 201-225-8686 *, FAX 201-225-9328
emp 400, 1982
A service company based on DPF laser coating technology. Specialists in providing restorative welding buildups, hardfacing, and machining services for the gas turbine industry and end-user markets.

New York

AccuDyne Co Inc, 700 Colfax St, Rochester, NY 14606; 716-458-6564, FAX 716-458-6664
pres/owner, Frank Visca; vp ops, Lynn L. Livermore; 1974
Contract manufacturing and engineering services for prototype, shortrun, and volume production of sheetmetal components and value added assemblies. Quality control program to MIL-45208 and SPC capabilities to 5 SIGMA. Typical products medical, aerospace, office product, computer and commercial.

AJ&L Tool & Manufacturing Co Inc, 233 LaGrange Ave, Rochester, NY 14613; 716-872-3226*, telex 716-458-6400
emp 102, 1969
Custom sheet-metal fabricator-prototype through production. Robotic welding and assembly capabilities. Full engineering services available.

Artplak Studios, 175 Commerce Dr, Hauppauge, NY 11788
1969

Ascension Metals Inc, 1254 Erie Ave, N. Tonawanda, NY 14120; 716-693-9381
1969

Ben-Mer Tool & Machine Inc, 1255 Emerson St, Rochester, NY 14606; 716-254-6090*, FAX 716-647-8274
emp 135, 1971
Precision sheet metal/design/components/assemblies. Laser services include high-precision cutting to tolerances of +/- .00022 in., clean-cut operations, pulse beam with chopper, multilevel contour cutting, auto-following head with controlled beam spot size, capability to perform offload EDM work, short lead time, fully supported by CAD/CAM system with laser-specific software.

Blackstone Corp, 1111 Allen St, Jamestown, NY 14701; 718-665-2620*
1971

Chromalloy Research & Technology Div, (sub of Sequa), Blaisdell Rd, Orangeburg, NY 10962; 914-359-4700, FAX 914-359-4409, telex 137393 RESCH TECH ORAN
pres, Dr. Harry Brill-Edwardas; vp eng, Keith Chessum; mgr, Dr. Reza K. Mosavi; emp 1100, 1965
Job shop for CO_2, glass, and YAG laser drilling, cutting, welding, and marking. All lasers fully CNC-controlled, 5 axis. Full engineering programming support, development through production pieces.

Custom House Engravers Inc, 104 Keyland Court, Bohemia, NY 11716; 516-567-3004, FAX 516-567-3019
vp/gm, Terry McLean; sls mgr, Brian McLean; 1974
Laser engraving from camera-ready artwork on a variety of materials: wood, acrylic, stone, glass, etc. Can supply complete product or will engrave on supplied merchandise.

Custom Laser Inc, 410 Ohio St, Lockport, NY 14094; 716-434-8600*, FAX 716-434-1757
emp 100, 1987
Laser job shop service offering laser welding, cutting, engraving, drilling, and scribing.

JAM Precision, 182 N. Oak St, Copiague, NY 11726
1987
Sheet metal cutting.

William Lo Dolce Machine Co Inc, Malden Turnpike, Malden on Hudson, NY 12453; 914-246-7017*
1987

Maloya Laser Inc

Maloya Laser Inc, 48 Toledo St, Farmingdale, NY 11735; 516-293-8333, FAX 516-293-8373
pres, Paul Hug; vp, Rebo Hug; 1954
Precision fabrication of sheet metal, plate and rectangular tubing - shapes, rings, gaskets, component and miscellaneous hardware. Materials; metals, non-metals and inorganics. Maximum workpiece size; 5' x10' x.5" thickness. Secondary ops. available. Short and long run production facilitated.

McAlpin Industries, 255 Hollenbeck St, Rochester, NY 14621; 716-266-3060 *, FAX 716-266-3060 ext 61
emp 200, 1962
Offers CAD/CAD, sheet metal prototype, sheet metal laser cutting, sheet metal fabrication, short run stamping, production stamping, metal finishing (plating), and CNC turning and machining.

Metalade Inc, Fabricating Div, PO Box 9885, 2025 Brighton Henrietta Tn Rd, Rochester, NY 14623; 716-424-3260, FAX 716-424-3258
pres, Richard Groth; oper mgr, Richard McKay; engrg mgr, Ross Mazzola; emp 175, 1968
Company is a prototype, fabricating, finishing and assembly facility. CNC lasers and turrets are used in the initial phases of production to produce blanks. "World Class Job Shop" supplying parts, finishes and complete assemblies to mostly "blue chip" companies both in the commercial and military markets.

Moloya Corp, 48 Toledo St, Farmingdale, NY 11735
1968

Weber Knapp, 441 Chandler St, Jamestown, NY 14701; 716-484-9135, FAX 716-484-9142
pres, Carl H. Little; sls mgr, Tom E. Coffin; prd engr mgr, Donald R. Pangborn; 1909
Manufactures specialty hardware. Full capability of product design, tooling, fabrication, and finishing to meet customer requirements. Some products include hinges, counterbalances, and mechanisms for ergonomic furniture.

Webster Tool, Die & Mfg Co, 164 Orchard St, Webster, NY 14580; 716-872-5650, FAX 716-872-2839
1966
Provides sheet-metal and plastic fabrication, production assemblies, machine building, laser machining, CNC machining, production stamping, and prototypes. Prototype through production vertically integrated electro-mechanical assemblies. Four-axis CNC laser machining. Precision sheet metal fabrications. CNC and general machining. Plastic fabrication. Production stampings. Assembly rework, CNC and conventional EDM.

XL-Tool & Machine Corp, 950 Exchange St, Rochester, NY 14608; 716-436-2250, FAX 716-235-5260
pres, George Schott; vp qa, Christopher Schott; vp sls/mktg, Peter Schott; 1967
Full service prototype job shop providing design, engineering, machining, and sheet metal fabrication including laser cutting services. Manufactures assemblies and pre-production machine parts for the office equipment industry.

North Carolina

Ameritek Lasercut Dies, PO Box 21361, 122 S. Walnut Circle, Greensboro, NC 27420; 919-292-1165, FAX 919-292-1174
1967

Beyers Sheet Metal, PO Box 5127, Hendersonville, NC 28793; 704-693-4088
1967

Carolina Laser Cutting, (sub of GLC Enterprises Inc), 4536 Drummond Rd, Greensboro, NC 27406; 919-292-1474, FAX 919-852-1482
pres, Mack L. Gordy; sec treas, Harold Gordy; emp 60, 1985
Job shop specializing in cutting metal, plastic, wood, composite, and more. Capacity: 60 x 120", tolerance: +/- .002", prototype, production run, CAD/CAM programming, 24-hour FAX line, steel rule dies, blanking grids, and counter plates.

Carolina Steel Rule Die Corp, 3019 Pacific Ave, Greensboro, NC 27420
1985

Container Graphics Corp, PO Box 5489, Cary, NC 27511; 919-481-4200*, FAX 919-481-4200 x228
pres, Phil Saunders; vp, Donald Moore; vp eng, James Smithwick; emp 425, 1961
Converts steels, plastics, wood and cloth on a prototype and short run production basis. Table size up to 63 x 96 in. and power rating of 800 CW with pulse capability.

Eagle Laser Co Inc, PO Box 7008, High Point, NC 27264; 919-434-9400
emp 60, 1983
Laser job shop.

Laser Precision Cutting Inc, 181 Reems Creek Rd, PO Box 1654, Weaverville, NC 28787; 704-658-0644, FAX 704-645-3230
pres, Bruce Hafer; vp, Joseph Karpen; 1983
Laser cutting of all materials, especially mild steel and stainless steel. Prototype and production work.

Ohio

Advanced Laser Services, PO Box 99, Grove City, OH 43123
1983

Ameri-Die Inc, 5051 Ridge Rd, Wadsworth, OH 44281; 216-239-1484, FAX 216-239-2474
pres, David Young; 1981
Fully equipped steel rule die shop. Manufactures flat rotary dies and stripping units. Computer services include CAD/CAM autocad, graphic programming, 44 X 60 digitizing pad, multi-cavity patterning and phone modem service. Laser services include unknifed laser cut die boards, laser cut samples, drilling, boring holes, laser etching.

Antique Auto Sheet Metal, 718 Albert Rd, Brookville, OH 45309; FAX 513-833-4785
pres, Ray Gollahon; 1970
Manufactures antique auto parts.

CJ Laser Corporation, 3035 Dryden Rd, Dayton, OH 45439; 513-296-0513, FAX 513-296-1855
pres, M. Cem Gokay; vp, Nilesen Gokay; 1981
Manufacturing, consulting, and engineering, for metal vapor and HeNe lasers, laser systems, laser accessories, specialized electro-optical equipment/hardware, barcode readers, specialized measurement systems and electro-optical instrumentation and packaging.

C.O.W. Industries Inc, div of T.N. Cook Inc, 3520 E. Fulton St, Columbus, OH 43227; 614-239-1992, FAX 614-239-1769
1960
Job shop, light gauge metal fabricator serving the medical, computer, and process industries that do not fab in-house or send out overflow work.

C&R Specialty Co, 648 Skeels Rd, Celina, OH 45822; 419-942-1421, FAX 419-942-1573
owner, Ron Keller; 1985
Laser job shop with laser engraving services.

Cutting Dynamics Inc, 30103 Clemens Rd, Westlake, OH 44145; 216-871-4740, FAX 216-871-4742
pres, William V. Carson Jr; vp, Wilbur S. Kohring; sls mgr, Bill McCale; 1985
Provides laser and wire EDM job shop services for the aerospace, automotive, and related industries. Short run forming capability.

Defiance Stamping Co, 1090 Perry St, Defiance, OH 43512
1985

Digitron Tool Co Inc, 8641 Washington Church Rd, Miamisburg, OH 45342; 513-435-5710*
emp 200, 1976
Applications R&D consultant. Automotive engine parts, manifold flywheels, etc. Aircraft flight hardware, brackets, tubes, etc., machine tool building.

EWI

1100 Kinnear Rd, Columbus, OH 43212; 614-486-9400, FAX 614-486-9528
exec dir, Karl F. Graff; fwa dept mgr, Dave Edmonds; sr resch eng, Gene J. White; resch eng, James P. Hurley; emp 850, 1985
An applied engineering center providing problem solving, educational and research services in welding and related joining technologies. The laser facilities include both CO_2 and Nd:YAG lasers, allowing work to be performed in all areas of laser material processing.

Elano Corp, 2455 Dayton Xenia Rd., Xenia, OH 45385; 513-426-0621, FAX 513-427-0288
pres/ceo, Robert Hessell; dir mktg/sls, Robert G. Graham; 1956
Manufactures tubes, ducts, and manifolds for use on jet engines and within commercial aircraft. These products are made of thin-walled, high-temperature-capability metals such as stainless steel. Products also include general aviation engine exhaust systems and locomotive engine exhaust manifolds.

Electro-Jet Tool & Mfg Co Inc, 10400 Evendale Dr, Cincinnati, OH 45241; 513-563-0800, FAX 513-563-0935
emp 125, 1956
Specialist in aerospace manufacturing and difficult to machine metals. General, tooling and parts of all kinds. CNC milling, drilling, turning, and laser drilling and cutting. EDM, wirecutting, and conventional machining also available. Laser emphasis on hole drilling.

E & W Services Inc, 7876 Enterprise Dr, Willoughby, OH 44094; 216-951-1500
1956

Ferco Tech Corp, 291 Conover Dr, PO Box 607, Franklin, OH 45005-1944; 513-746-6696, FAX 513-746-9656
ceo, G.G. Fernandez; pres/eng mgr, G.A. Fernandez; mktg dir, R.W. Lucius; plant mgr, R.W. Angell; 1984
Aircraft engine brackets and aircraft hinges; sheet metal fabrications, light fabrications; machined parts such as castings, forgings, fabrications; tooling such as die details, fixtures, gages, templates, jigs; laser cutting and drilling of metals and other materials.

Greenfield Fabricating, PO Box 8859, Canton, OH 44711-8859; 216-499-6778, FAX 216-499-0881
pres, Neil R. Tyburk; sls/gen mgr, John J. Stalica; emp 150, 1977
High tech sheet metal fabricator job shop. CNC punching/laser, CAD/CAM computer support available specializing in one price or production quantities.

6050 HI-TEK Court, Mason, OH 45040; 513-459-1094, FAX 513-459-9882
pres/acting mktg mgr, Cletis M. Jackson; eng mgr, Gary W. Griessmann; 1979
Laser job shop with approvals from major aerospace manufacturers. All facilities are full 5-axis numeric controlled. Extensive CAD/CAM facilities enhance the capabilities of the equipment. In addition, CNC machining, wire EDM and conventional EDM processes are performed.

Laser Automation Inc, 16771 Hilltop Pk Place, Chagrin Falls, OH 44022
emp 140, 1974
Laser job shop for R&D applications and production work. Also buys, sells, and rebuilds used lasers and systems.

McAfee Tool & Die, 1717 Boettler Rd, Uniontown, OH 44685; 216-896-9555, FAX 216-896-9549
pres, Gary McAfee; plant mgr, Mike Francek; 1977
Offers tool & die making, CNC machining, wire EDM metal stamping, plastic injection molding, and CAD/CAM design.

Northern Mfg Co Inc, 132 N. Railroad St, Oak Harbor, OH 43499; 419-898-2821/800-9227339, FAX 419-898-4470
vps, Quintin Smith, Kellen Smith; sls mgr, Bob Gorski; 1951
Specializes in laser cutting, CNC punching, plasma burning, bending, and custom precision sheet metal fabrication

OSU Dept of Weld Eng, 190 W 19th Ave, Columbus, OH 43210
1951

Peerless Laser Processors Inc, (sub of Peerless Saw Co), 4353 Directors Blvd, Groveport, OH 43125; 614-836-2073, FAX 614-836-5824
pres, Con Wittkopp; operations mgr, Gregg Simpson; eng mgr, Al Walton; emp 420, 1931
Contract laser cutting, welding, drilling and marking on four large multiaxis motion systems. Metal cutting in aluminum to .312", clean cut stainless steel to .375", and alloy steel to .750" thick. Laser power to 3000-W CW.

Process Equipment Co, 6555 So. State Rt 202, Tipp City, OH 45371; 512-667-4451, FAX 513-667-4798
pres, L.D. Ewald; vp mfg group, D.E. Vandeveer; vp eng/design/assm grp, A.B. Hauberg; emp 170, 1946
Contract laser cutting up to 60 x 120". Special laser welding machines.

Tru-Graphics Inc, 95 Compark Rd, PO Box 708, Centerville (Dayton), OH 45459; 800-776-4152, FAX 513-438-9666
pres, Wm. H. Kirkeiner; 1989
Designers and manufactures of custom signage, gifts and awards, with laser engraving and sandblast etching capabilities.

Weiss Industries Inc, Metal Forming Div, PO Drawer 157, 2480 N. Main St Rd, Mansfield, OH 44901; 419-526-2480, FAX 419-526-1158
pres, Rudy Weiss; vp, Bob Bobst; sls/mktg mgr, Mick O'Donnell; 1949
Contract producer of light-gage, medium to high volume metal shapes and assemblies. Capabilities include stamping, forming, rolling, joining, assembly, and welding. Specialties include cylindrical, oval and other shaped tube type assemblies. Engineering and technical assistance from "black box" /prototype, tooling, debugging, and sample stages through production.

Oklahoma

Laser Engineering & Fab, 10815 E. Marshall St, Suite 105, Tulsa, OK 74116; 918-437-7511, FAX 918-437-8228
pres, Ron LaPelle; vp, Mike Van Schoyck; sls mgr, Jim Aitkenhead; 1989
Specializing in laser processing and metal forming.

Midwest Trophy, 3501 S.E. 29th St, Del City, OK 73115; 405-670-4545, FAX 405-672-0964
1989
Laser engraving on walnut, oak, clear acrylic, avonite, marble, etc.

Oregon

Amerson Precision Sheet Metal, 1915 Colvin Ct, McMinnville, OR 97128; 503-472-0659, FAX 503-472-1072
pres, Alan Amerson; emp 130, 1979
Complete precision sheet metal shop.

Applied Laser Technology Inc, 14155 S.W. Brigadoon Ct-B, Beaverton, OR 97005; 503-641-4400, FAX 503-641-6696
pres, Hossein Karamooz; sls mgr, R.S. Sig Jensen; qual assur mgr, G. Tankersley; 1987
Provides nonmetallic cutting and machining services, and resistor trimming for hybrid microcircuits. Provides excimer laser skiving and CO_2 laser cutting for hybrid microcircuits, flexible circuits, and synthetic electric/electronic subcomponent and material processing.

Gage Industries Inc, PO Box 1318, Lake Oswego, OR 97035; 503-639-2177*
1987

Las Tec, 5215 NE Elam Young Pkwy, Hillsboro, OR 97124; 503-640-2300, FAX 503-640-1350
emp 910, 1979
Precision laser trimming, scribing, drilling, and machining of all alumina based hard substrates as well as soft substrate materials such as Kapton and Teflon. Precision drilling and machining of specialty materials such as steel and plastics. Precision resistor trimming of thick and thin film resistors.

P/M Industries Inc, 14320 NW Science Park Dr, Portland, OR 97229; 503-641-4646, FAX 503-643-6983
pres, Paul Parks; sls mgr, Lee Timiney; mktg mgr, Paul Parks Jr.; emp 500, 1971
Contract laser trim thick film, thin film polymer, magnetic media, GaSa and other applied coatings. Laser drill and scribe alumina, beryllia, aluminum nitride, polyimide, pc board, metal, duriod, quartz, and other materials.

W&W Metal Fab, 6521 SE Crosswhite Way, Portland, OR 97206; 503-775-0834, FAX 503-777-9068
1982
Provides precision sheet-metal fabrication, CNC equipment and laser cutting. Full welding capabilities, including robotic welding. Cabinet and enclosure fabrication and assembly. Painting, powder coating, phosphating, and chromate capabilities.

Pennsylvania

A&E Manufacturing Co Inc, 2210 Hartell St, Levittown, PA 19057; 215-943-9460*
1982

17 Maple Ave, PO Box 86, Blairsville, PA 15717; 412-459-7550, FAX 412-459-0207
pres, J.W. Clark, Jr.; plant mgr, Ed Hendricks; sls mgr, Glenn Sheffler; inside sls, Jan Knechtel; 1954
Job shop precision sheet metal fabricator serving a wide variety of industries. Principal products are custom enclosures and cabinets for the electronic and electrical industries. Capabilities incude CNC punching and bending, laser cutting, welding, assembly and finishing. Offers quality work at competitive prices.

Corry Laser Technology Inc, 414 W. Main St, PO Box 18, Corry, PA 16407; 814-664-7212, FAX 814-664-4961
vp/gen mgr, Madi Rathinavelu; sls eng, Scott Brady; 1987
Laser cutting, drilling, and welding of metals, plastics, ceramics, and other materials using computer-controlled 7-axis system.

Ervite Corp, 4000 W. Ridge Rd, PO Box 8287, Erie, PA 16505; 814-838-1911, FAX 814-838-1031
pres, Charles Vicary; vp, Lucille Pluta; eng mgr, Bruce Ritts; 1945
Offers insulated and painted sheet metal frames, exterior covers, assemblies, and related products.

Gamlet Inc, 190 Carlisle Ave, York, PA 17104; 717-845-1651
1945

Germantown Tool & Machine Works Inc, 1681 Republic Dr, Huntingdon Valley, PA 19006; 215-322-4970, FAX 215-357-8024
pres, Paul F. Wolf; sls rep, Tina K. Wolf; 1947
Offers laser processing, metal stamping, sheet metal fabrication, tooling, CNC Centrum 1000 Weidemann, robotic welding, and assembly. Short runs, medium to long runs. Complete manufacturing facility for all tooling and fabrication needs from design through packaging.

Heintz Corp, 11000 Roosevelt Blvd, Philadelphia, PA 19116; 215-677-3600*, FAX 215-464-5390
1947

Laser Cutting Service Co, York Saw Co, PO Box 733, York, PA 17405; 717-767-6402*, FAX 717-764-2768
emp 600, 1971
Offers contract cutting. At present uses an 820 1500-W CO laser, 8x10 ft table, and an 810 600-W CO_2 laser, 4x4 ft table.*

Lasermation Inc, 2703 N. Broad St, 2nd Floor, Philadelphia, PA 19132; 215-228-7900, FAX 215-225-1593
pres, Joe Molines; emp 550, 1969
Laser-engraved awards. Drilling and cutting of nonmetals. CO_2 process and feasibility development. Intricate brass stencils for marking. Ceramic substrates for microelectronics-alumina and beryllia.

Lazal Corporation, Kleine & Warwick Sts, Lititz, PA 17534; 717-627-4242*
emp 50, 1985
Laser cutting job shop-materials include all metals (i.e. aluminum, stainless steel, brass, copper, monel, inconel, etc.), wood, plastics, rubbers. Thicknesses up to 1/4 in. and cutting speeds up to 300 in./min. Accuracies of contour cutting to +/- 0.001 in. *

Stardyne Inc, 575 Central Ave, Johnstown, PA 15902; 814-536-2000, FAX 814-539-3151
cob & ceo, Edward J. Sheehan, Sr.; pres, John I. Nurminen, PhD,PE; exec vp, Edward J. Sheehan, Jr.; vp mfg/ops, James E. Smith; 1989
Offers 1-kW to 25-kW laser output through the use of our two 25-kW CO_2 UTIL lasers. Welding, cladding and cutting services for products such as rolls, shafts, cylinders, sheet, plate and complex configurations. Laser metalworking offers enhanced quality and reduced overall product cost.

Westinghouse Electric Corp, Science and Technology Center, 1310 Beulah Rd, Pittsburgh, PA 15235; 412-256-1785, FAX 412-256-1348, telex 4909989013
vp, I.R. Barpal; gne mgr sptd, S.D. Harkness; mgr adv proc, A.T. Male; mgr lsr mtlwrk, G.J. Bruck; emp 1200, 1989
Performs laser metalworking research, process development, procedure qualification, prototype demonstration, and pilot production. Production processing also available. Unique laser beam manipulation, hot wire feeding and powder delivery equipment complement the high-power capability offered to corporate, industry, and military customers.

Wilke Enginuity Inc, 401 Ridge Ave, McSherrystown, PA 17344; 717-632-5937, FAX 717-632-6677
pres, J. Wilke; sec/treas, F. Wilke; mgr, E. Wolfgang; 1982
Job shop performs laser cutting of steel to 1/4" thick, wood to 2" thick and acrylic to 1" thick. CAD/CAM capabilities available with complete in-house machine shop support.

X-Mark Industries, 2001 N. Main St, Washington, PA 15301; 412-228-7373, FAX 412-228-2122
pres, Robert F. Kastelic; sls & mktg mgr, Steve Gibson; vp mfg, Carl Bruckner; emp 900, 1967
Precision metal fabricating company, employing state-of-the-art technology to meet the needs of customers. The primary industries served are: electronics, computer, communications, food service, mining, material handling, compressor, medical and high tech. A complete package, from prototype to production, including finishing.

Rhode Island

Laser Cutting Services, 485 Naragansett Park Dr, Pawtucket, RI 02861; 401-723-1910, FAX 401-728-0980
sls mgr, Dick Nichols; 1967
Contract job shop specializing in cutting, welding, and drilling. Supplies prototypes and production quantities. Materials processed include aerospace and marine alloys, hastelloy, monel, carbon steels, plastics and wood.

[HGG LASER FARE INC.]

One Industrial Drive South, Smithfield, RI 02917; 401-231-4400, FAX 401-231-4674
pres, Terry Feeley; gen mgr, Thomas McDonald; 1977
Laser job shop with multi-axis capabilities in cutting, welding, drilling, marking and R&D. Materials processed include composites, plastics, ceramics and metals to the superalloys.

Laser Fare Inc, Industrial Dr South, Smithfield, RI 02917; 401-231-4400*, FAX 401-231-4674
pres, Terry Feeley; vp/gen mgr, Thomas J. McDonald; 1980
Laser machining services including cutting, welding, drilling, and marking. Full quality assurance program SPC. Contract R&D in the areas of rapid prototyping and non linear optics, certified to many customers in the aerospace and medical industries.

Laser Marking Services Inc, 169 Douglas Pike, Smithfield, RI 02917; 401-232-2292, FAX 401-232-0510
pres, Jack Kebarian; vp, James Kaminski; sls/mktg mgr, John Kaminski; 1986
Contract job shop specializing in precision industrial and decorative marking and engraving on metal components. MIL-I qualified. Multiple and YAG laser marking systems and three shifts to ensure consistent on-time delivery.

South Carolina

Metal Equipment Fab, 141 Metal Park Dr, Columbia, SC 29290; 803-776-9250, FAX 803-776-9610
pres, Curtis P. Smoak Jr.; vp, Frances S. Smoak; 1977
Sheet metal shop.

Rieter Corp, PO Box 2378, Aiken, SC 29802-2378; 803-649-1371
1977

Tennessee

American Fabrications Inc, 711 Space Park So. Drive, Nashville, TN 37211; 615-834-8700, FAX 615-834-5859
pres, Milton Grief; vp, Clarence Bain; mfg mgr, Fred Hobb; sls mgr, Paul Sutter; eng, Eddie Barnett; 1984
O.E.M. precision sheet metal fabrication equipment includes laser, shears, turret punches, press brakes, C.M.M. inspection service, welding, cleanup and assembly department.

LaFollette Machine & Tool Co, PO Box 431, 218 First Street, LaFollette, TN 37766; 615-562-5854, FAX 615-562-8437
pres, Jerry Sharp; vp, Jarrett Sharp; 1936
Provides laser cutting from prototype to production. Using 1500-W, 4' x8' table and CAD/CAM programming.
Maximum material thickness: 1/2" steet, 1/4" stainless, 1/8" aluminum, 3/16" titanium, and 3/32" inconel.

LP LASER PRECISE, INC.

127 Perimiter Park Rd, Suite D, Knoxville, TN 37922; 615-531-8016, FAX 615-531-8017
pres, Kenneth R. Barks; sec, Carol Barks; 1990
Laser process service center specializing in tough tasks with fast, accurate turn-around. Offers cutting, drilling, welding, marking, etching and scribing of ceramic, metal, and organics like rubber or plastic.

Univ of Tennessee Space Institute

Univ of Tennessee Space Institute, Center for Laser Applications (sub of Laser Material Processing Group), B.H. Goethert Pkwy, Tullahoma, TN 37388; 615-455-0631, FAX 615-454-2271
prof engr sci/mech, T. Dwayne McCay; assoc prof eng sci/mech, Mary Helen McCay; post-doc res assist, Rob Mueller; res assist prof, Narendra Dahotre; eng, C. Michael Sharp; 1990
Process R&D laboratory specializing in laser welding, weld monitoring, weld plasma diagnostics, laser beam diagnostics, and mathematical modeling. Facilities include RF Excited 3 kW Co_2, 400 W YAG, 5 axis workstation and a comprehensive metallurgical facility. Activities range from fundamental research to the development of specific industrial applications.

Williams Machine Works Inc, 2270 Channel Ave, PO Box 13307, Memphis, TN 38113; 901-774-5335, FAX 901-774-5422
pres, Logan D. Williams; vp, L.D. Williams Jr.; vp/sec, Kathleen W. Peck; 1951
Small to medium job shop. Production turning, milling and laser cutting. General shop. Blanchard grinding to 120 in. diameter, turning to 13 ft.

Texas

Cantrell Ind, 13006 Mula Lane, Stafford, TX 77477; 713-933-3051*, FAX 713-933-0100
emp 420, 1960
Laser profile cutting, CAD systems, signage systems, solar power packs, and solar lights.*

E&C Laser, 6061 Thomas Rd, Houston, TX 77041; 713-466-9556, FAX 713-466-9902
pres, Nelson Karro; 1991
Laser cutting job shop. Precision metal fabrication facility offering CNC turret presses, CNC brakes, welding, painting, and plating.

G&H Diversified, 5112 Steadmont Dr, Houston, TX 77040
1991

Industrial Metal Specialties Inc, 4023 Singleton Blvd, Dallas, TX 75212; 214-634-0211, FAX 214-631-0917
pres, Joe O. Neuhoff, Jr.; mktg mgr, Stan McKay; eng mgr, Fred Ott; emp 200, 1973
Job shop CNC-CO_2 laser cutting from protype to production runs. Materials include: highly reflective aluminum and copper, stainless, Inconel, Monel, titianium, plastics, composites, rubber, wood, etc. 2500-W machine with 25 x 33 in. table--close tolerance for high-tech industry. Also plasma cutting.

J&K Inc, Laser Div, 405 S. 15th St, PO Box 504, Waco, TX 76703; 817-752-2557/800342-7297
gen mgr, Darrell Wilbanks; prod mgr, Todd King; op mgr, Bill Trammil; 1975
Laser cutting job shop. Specializing in cutting plastic, cold roll, hot roll, stainless steel, high carbon steel, and non-ferrous materials.

Laser Techniques Inc, 803 Gemini, Duncanville, TX 75137; 214-296-7934, FAX 214-780-7861
pres, Clifton Thrailkill; sls mgr, William Hunter; 1987
CO_2 laser job shop offering part design, prototype, production, cutting of metals and nonmetals.

Microcut Technologies, 4400 Sunbelt Dr, Dallas, TX 75248; 214-407-9094
1987
Laser job shop offering part design, prototype, and production. No minimum quantities. Cutting a wide variety of materials per customer specifications. Fast turnaround time.

Norton Metal Products Inc, 1350 Lawson Rd, Fort Worth, TX 76131; 817-232-0404, FAX 817-577-1634
ceo, J.M. Norton; pres, Larry Dunlap; vp, Larry Yates; 1947
Performs fabrication functions: shear 20', press brake 20' 1000 ton, punching, laser operations-plasma, oxyfuel CNC cutting; slitting: 54" -30000#, roll forming operations, 7 lines commercial building products. Stocking steel distribution & processing.

Precision Laser Cuts, 1406 Smith Rd, Unite F, Austin, TX 78721; 512-385-1911
1947

Virginia

Atles Die Inc, Atlas Richmond Div, 450 S. Lake Blvd, Richmond, VA 23236; 804-379-0901
mgr, Katie McDonnough; 1947

Grapha Mfg, 11850 Jefferson Ave, PO Box 6630, Newport News, VA 23606; 804-873-1234, FAX 804-873-1718
1947

Washington

Exotic Metal Forming Co, 5411 S. 226th St, Kent, WA 98032
1947
Laser job shop.

Precision Die, 984 Industry Dr Bldg 28, Seattle, WA 98188; 206-575-8217/800-7837773, FAX 206-575-8245
Owner, Kim Vo; 1983
Manufactures cutting dies. Specializes in laser cutting and die cutting of wood, metal, and plastics.

Proctor Products Co Inc, PO Box 697, 210 8th St South, Kirkland, WA 98033; 206-822-9296, FAX 206-634-2396
vp and gm, Harold Rathman; plant sup't, Randy Keen; eng, Fred Riehl; 1954
Job shop offering laser cutting, high-speed punching, steel, aluminum, and stainless steel fabrication, sheet metal work, and machining.

Systi-Matic, Precision Laser Cutting Div, 12530 135th Ave N.E., Kirkland, WA 98034; 800-426-0000, FAX 206-821-0804
pres/mktg, Rich Budke; sls/eng mgr, Rob Jones; 1956
Laser sheet-metal cutting. Ferrous, nonferrous, plastic, and wood cutting.

Western Fab & Finish Inc, 3301 E. Isaacs, Walla Walla, WA 99362; 800-456-8818 *, FAX 509-529-8213
emp 125, 1979
Full capability contract manufacturing in sheet metal. Laser cutting up to 3/8 in. steel with strippit 1350-W CO_2 laser center. Quality assurance program to NQA-1. Other capabilities include CNC punching, stamping up to 400 ton, welding, and powder or porcelain finishes.

Wisconsin

Biersach & Niedermeyer Co, 10245 N Enterprise Dr. 66th, Megwon, WI 53092
1923

Heiden Inc, 3403 Menasha Ave, PO Box 1477, Manitowoc, WI 54220; 414-682-6111, FAX 414-682-3174
pres, James G. Morrow, Sr.; vp/gen mgr, Bradd E. Steckmesser; mktg mgr, Charles C. Weier; 1959
Metal fabricators - specializing in stainless steel, aluminum, and high-grade carbon steel. Welting specialty in all types and metals plus certifications with AWS and MIL standards. Propriatory line of regular and hydraulically operated truck crane accessories.

Helgesen Industries Inc, 7261 Highway 60 West, Hartford, WI 53027; 414-673-4444, FAX 414-673-4309
pres, Ronald Marshall; vp, Carl Duveneck; sls mgr, Bryndon O'Hara; 1977
Suppplies custom multi-run sheet/plate metal fabrications ranging from simple components to large complex weldments. These parts are processed on the latest of machine technology including 1500-W multiple table lazer, plasma, NC punching, robotic welding and a five stage wash, electrostatic paint and powder system.

HUI, 10 East Park Ave, Kiel, WI 53042; 800-877-8913, FAX 414-894-7268
pres, Charles Deibele; gen mgr, Steve Deibele; mkt mgr, Kathy Vogel; sls/est, Joe Isely, John Moritz; 1977
Precision cutting with laser technology. CNC punch lasers combination as well as gantry style 60 x 120 in. bed with clean cut capability. Specializing in the fabrication of metal, plastics, and various composite materials functioning with CAD/CAM design, forming, heli arc and robotic welding, powder coating. Full job shop capabilities.

Jensen Metal Products Inc, 1222 18th St, Racine, WI 53126; 414-637-5670*, FAX 414-637-1247
emp 470, 1922
Family operated job shop, metal fabricator. Close tolerances, fast turnaround with three 1500-W lasers that have the flexibility to work with gasket material, various alloys, and plastics.

KMC Laserform, div of Kickhaefer Mfg Co, 1221 S. Park St, Port Washington, WI 53074; 414-284-3424, FAX 414-284-9774
pres/ceo, R.E. Davis; dir oper, Brad Gador; dir sls/mktg, S. Endicott; 1908
Trumpf 180 LW 1500-W CO_2 laser turret press is utilized to produce prototype and production quantities. Part size: carbon steel to .250", stainless steel to .125" . Superior edge quality, minimum production leadtime, documented quality plan.

Laser Innovations, 8822 Woodruff Rd, Woodruff, WI 54568; 715-356-2422, FAX 715-356-2422
owner, Steven J. Congdon; 1987
Quality laser engraving or light cutting of most nonmetals with very fine detail from any black and white artwork. Will engrave your products or ours. Specializing in both production and custom work. Two-week delivery on most orders. No minimum quantities or set-up charges. Rush service available.

Laser Machining Inc, 500 Laser Drive, Somerset, WI 54025; 715-247-3285, FAX 715-247-5650
pres, Noel Biebl; js mgr, Dennis Berke; lser process tech, Kurt Hatella; sls eng, Tom Daul; 1978
Precision job shop from prototype to production runs. Three shift operation. High pressure, rotary, roll handling, 2 x 2 ft. to 5 x 10 ft., multi-axis, CAD/CAM, automated and robotic applications. CMM, engineering support and development.

Laser Magic Inc, 1540 Livingstone Rd, Hudson, WI 54016; 715-386-2525, FAX 715-386-1557
owner, Ted Bauer; 1985
Full service contract CO_2 and YAG laser job shop. Capabilities include: engraving, scribing serializing, metal marking, bar coding, product imprinting, personalization, batch coding, parts indentification, etching, alphanumerics, logos, micro marking. Support services include: art layout, graphics, assembly, packaging, shipping, R&D, and prototyping. All materials, all quantities.

Laser Maintenance Inc, div of Laser Artistry Inc, 6420B S. Howell Ave, PO Box 220, Oak Creek, WI 53154-0220; 414-764-4002, FAX 414-764-6826
pres, Terry Michaels; 1990
Provides service and repair of industrial and medical laser systems. Maintenance programs available to insure laser systems continue operating at best possible performance. Ion laser leasing also available.

Laser Specialties Inc, PO Box 3063, 522 W 17th Ave, Oshkosh, WI 54903; 414-233-6131*, FAX 414-233-0377
emp 210, 1983
Job shop for customer-supplied items. Engraving available basically on wood, acrylic, and marble. Offer items including awards, office accessories, plaques, executive items, gifts, games, knives and more. Sell to the advertising specialty market.

Mayville Engineering, 715 South St, Mayville, WI 53050; 413-387-4500 *
1983

Metko Inc, 1251 Milwaukee Drive, New Holstein, WI 53061-1499; 414-898-4021, FAX 414-898-5293
prod mgr, Michael McCarthy; 1971

Northwoods Laser, Route 2, Box 202A, Wittenberg, WI 54499; 715-253-2541, FAX 715-253-3566
pres/eng mgr, Mark Onesti; sls/mktg mgrs, Pam & Mark Onesti; 1986
Complete custom laser engraving services from design to finished product. Dedicated to supplying only top-quality engraving with quick turn-around time. Also experienced laser repair of Apollo and Coherent brand CO_2 lasers.

International companies providing contract laser processing services

The following section includes information on international companies that provide contract laser material processing services. These companies may be full-service laser job shops or sheet-metal shops with laser cutting capability. Each of these companies is prepared to offer services on a contractual basis for process R&D, application feasibility, prototype processing, or production.

The specification chart provides a sketch of each company. Addresses and details of each company can be found in the directory immediately following the chart.

The editors strive to ensure that the information reported is current and correct. In an effort to present the most comprehensive directory of contract service companies, the editors have included information from some companies that did not provide updated information for this edition. These companies are noted by an asterisk (*) after their address.

INTERNATIONAL JOB SHOPS (CHART PAGE 1)

	Lasers			Applications										Services							Motions	
	Solid-State	CO2	Excimer	Welding	Cutting (Metal)	Cutting (nonmetal)	Drilling (Metal)	Drilling (nonmetal)	Scribing	Heat-Treating	Other Surface Treat.	Marking	Engraving	Part Design	Prototype	Pilot Plant	Production	Spec. Sys. Supp.	Metallurgy & QC	Processing R&D	MIL Spec Qual	2-Axes	Multi-Axes
Action Laser Pty Ltd North Ryde Australia 61-2-8878290	•			•	•		•																
ADB Mazzano (Brescia) Italy 030/2591831		•			•																		
Alex Neher AG Ebnat-Kappel Switzerland 74 314 14*		•			•	•											•						
Alfa Lasersystem GmbH 2000 Hamburg 26 Germany 49 (0)40 789 83 87	•	•		•	•	•	•	•	•			•	•	•	•	•	•	•				•	•
Alfa Lasertechnik GmbH Braunschweig Germany 49 (0)531-89 51 67		•	•	•	•	•	•	•		•				•	•	•	•	•	•	•		•	•
ALL Applikationslabor fur Lasertechnik Munchen 2 Germany 089-716033*	•	•		•	•	•	•	•	•	•		•	•	•	•	•	•	•					
Alpha Laser Pty Ltd Victoria Australia +613-870-2345*		•			•	•	•	•							•		•						
Alpha Laser Colombey Les Deuxeglis France 25 01 51 94	•											•											
IB Andresen Industri A/S Langeskov Denmark +45 65 38 12 34		•		•	•										•	•	•			•			
Applied Cutting Technology Essex UK 0206-395858		•		•	•	•	•	•	•			•	•	•	•	•	•						
Applikationslabor Fur Lasertechnik Munich 2 Germany (49) (89) 507080		•		•	•	•	•	•	•	•					•	•	•	•				•	•
Arbucias Industrial SA Arbucies (Girona) Spain *		•			•	•									•	•	•						
ARES St. Lev La Foret France 030 40 88 32	•	•		•	•	•	•	•		•	•			•	•	•	•	•	•	•	•	•	•
Armor Ploemeur France 97 83 24 61		•			•					•		•											
Aspa Oy Jarvenpaa Finland (90) 280 122*		•			•																		
Ateca Montauban France (33) 6363 7897	•			•	•	•	•	•						•	•	•	•			•			•
Atim Courtry France (1) 60 08 22 54		•			•																		
Atom Charleville Mezieres France 24 37 66 77		•			•																		

INTERNATIONAL JOB SHOPS
(CHART PAGE 2)

	Lasers			Applications										Services								Motions	
	Solid-State	CO2	Excimer	Welding	Cutting (Metal)	Cutting (nonmetal)	Drilling (Metal)	Drilling (nonmetal)	Scribing	Heat-Treating	Other Surface Treat.	Marking	Engraving	Part Design	Prototype	Pilot Plant	Production	Spec. Sys. Supp.	Metallurgy & QC	Processing R&D	MIL Spec Qual	2-Axes	Multi-Axes
ATS Cruissy France (1) 60 17 80 23		•			•																		
ATS Saronno (Varese) 02/9609465		•			•																		
Autz & Herrmann Heidelberg 1 Germany 62-21-50-60*		•				•																	
Azuma Co Ltd Chiyoda-ku, Tokyo Japan 3-834-4541*		•			•	•									•		•						
Bal Laser AG Schinznach Dorf Switzerland 56 43 28 21*		•			•	•									•		•						
B&B Metall und Kunststoffverarbertung GmbH Wuppertal 2 Germany 0202 1662988*		•			•	•						•	•		•		•						
Bellion Chollet France 41 71 90 90		•			•																		
Bernhard Lasertechnik Bad Aibling Germany 08061/6064*		•		•	•	•								•	•	•	•						
Berto Lamet Torino Italy 011/30021		•			•										•								
BG Industrial Laser Group Plc Huntingdon, Camb UK 0480-455441		•		•	•	•	•	•	•			•	•		•			•					
BIAS Forschungs und Entwicklungs Labor Bremen 33 Germany (49)421/21801	•	•	•	•	•	•	•	•	•	•	•			•	•	•			•	•		•	•
Bonomi Vergiate (Varese) Italy 0331/946 251		•			•																		
Bordogna Palazzolo Sull'Oglio Italy 030/7300261		•			•																		
Brooks-Koochew Pty Ltd Victoria Australia 458-2611		•			•	•								•	•		•						
Bruscoli Marcello Serrungarina (Pesaro) Italy		•				•																	
Bussetti Officine Moncalieri (TO) Italy		•			•																		
Cacir Nicco Lagny Sur Marne France (1) 64 30 43 64		•						•															
Calfetmat Villeurbanne France 72-43-83-04	•	•		•	•	•	•	•		•	•								•	•			

INTERNATIONAL JOB SHOPS
(CHART PAGE 3)

	Lasers			Applications										Services							Motions	
	Solid-State	CO2	Excimer	Welding	Cutting (Metal)	Cutting (nonmetal)	Drilling (Metal)	Drilling (nonmetal)	Scribing	Heat-Treating	Other Surface Treat.	Marking	Engraving	Part Design	Prototype	Pilot Plant	Production	Spec. Sys. Supp.	Metallurgy & QC	Processing R&D	MIL Spec Qual	2-Axes	Multi-Axes
Capital Lasers London UK 071-928-6235	•			•	•		•		•			•	•		•		•			•			•
Carlton Laser Service Ltd Leicester UK 533 761 177*		•			•	•									•		•						
Car Mec Malnate (Varese) Italy		•			•																		
Cassetto Lugnacco (TO) Italy 0125/789088		•			•																		
CBI Italia SpA Monza Italy 039 741741		•			•		•										•						
CBS Industrie SA Nemours France (1) 64 28 02 34		•			•																		
CEM Magnago (Milano) Italy		•			•																		
Centro Laser-Soc Cons a r.l. Valenzano (Bari) Italy 080-8774314*	•	•	•	•	•	•	•	•	•	•	•	•	•	•	•					•			
CERCA Creteil Cedex France 43.77.12.63*	•			•	•		•			•				•	•		•						
CETENASA Noain (Navarra) Spain (48) 238258		•		•	•					•	•				•				•	•			
CGM Igny France (1) 69 41 13 01		•			•																		
Cheval Freres SA Besancon Cedex France 81.53.75.33*	•			•	•		•		•			•	•										
Chugoku Giko Co Ltd Hiroshima Japan 82-249-4450*		•			•										•		•						
CIM Vignate (Milano) Italy 02/9566266		•			•																		
Cipiemme Carygo (Como) Italy 031/749455		•			•																		
CISE Milano Italy (02) 2167.1	•	•		•						•	•				•	•		•		•			
Claux Brive La Gaillarde France 55 86 90 07		•			•																		
CMB Srl Barone Canauese (TO) Italy 011/9898095-62		•			•																		

INTERNATIONAL JOB SHOPS
(CHART PAGE 4)

	Lasers			Applications										Services								Motions	
	Solid-State	CO2	Excimer	Welding	Cutting (Metal)	Cutting (nonmetal)	Drilling (Metal)	Drilling (nonmetal)	Scribing	Heat-Treating	Other Surface Treat.	Marking	Engraving	Part Design	Prototype	Pilot Plant	Production	Spec. Sys. Supp.	Metallurgy & QC	Processing R&D	MIL Spec Qual	2-Axes	Multi-Axes
CMS SpA Fisciano (Salerno) Italy 089/879233		•			•																		
Cobra Signs Brisbane, Queensland Australia 208 0722		•			•	•						•			•		•						
Collot Technologies Laxou Cedex France 83 97 20 10		•			•																		
Cols Cutting Formes P/L Bexley, NSW Australia 588 5999*		•				•											•						
Comela L'Arbresle France 74 26 41 55		•			•																		
Compel Cornate D'Adda Italy 039/6926555	•	•							•														
Control Laser PTI Ludvika Sweden +46 240 13660*	•	•		•	•	•				•		•											
Cosmetal Recanati (Marcerta) Italy 071/7570426		•			•																		
CREAS Nichelino (TO) Italy 011/621959		•			•																		
C&R Equipment Ltd Christchurch New Zealand 0011 64 3 3843154		•			•	•																	
CTM Igny France (1) 69 41 13 01		•			•	•																	
Delcros St. Marie Aux Mines France 89 58 75 49		•			•																		
Derby Carauaggio (Bergano) Italy 0363/52003		•				•											•						
Deville Sodery Charleville Mezieres France *		•			•	•									•		•						
DFVLR Stuttgart 80 Germany 0711-6862-770	•	•	•	•	•	•	•	•		•	•				•				•	•		•	•
Doll Sasbach Germany *		•			•										•		•						
Dorries Scharmann GmbH Mechernich Germany 02484 12 123		•		•	•	•	•	•		•	•			•	•	•	•	•		•			•
DRU Industrial Products & Services Ulft Netherlands 08356-89911		•			•	•					•			•	•		•			•			•

CONTACT SURFACE FOLLOWER WITH MAGNETIC CRASH BREAKAWAY

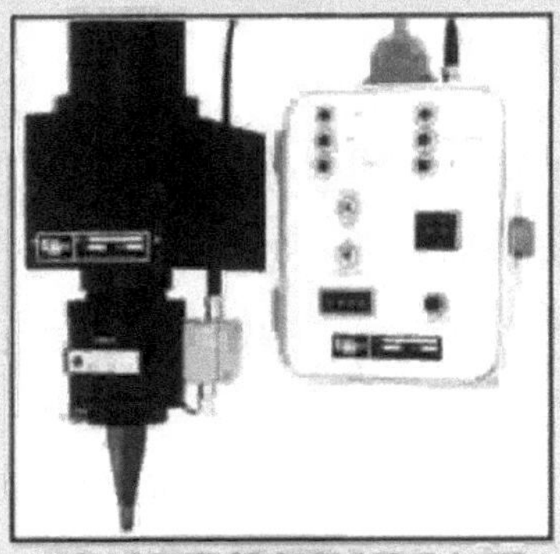

NON-CONTACT CAPACITIVE AUTO FOCUS WITH COMPACT Z SLIDE

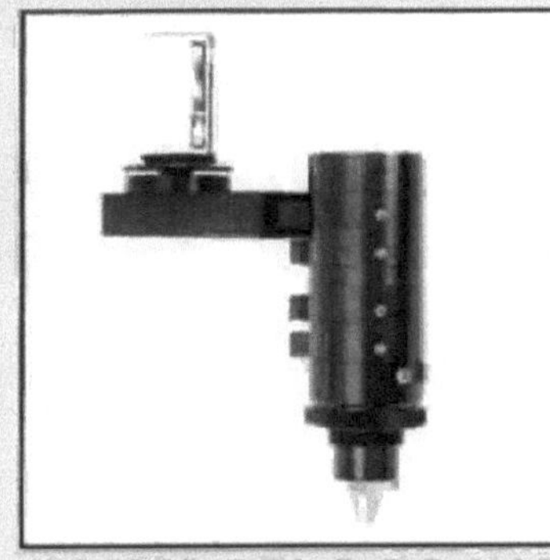

YAG OR CO_2 THROUGH THE LENS TV VIEWER

PROGRAMMABLE HIGH SPEED BORING HEAD

LIGHTWEIGHT PROGRAMMABLE X-Y END EFFECTOR

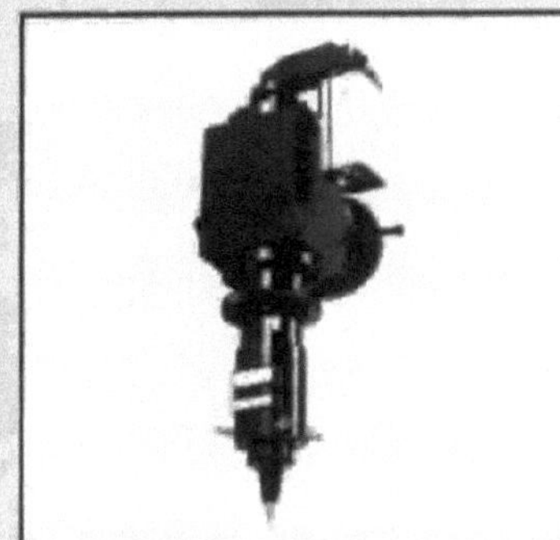

YAG DELIVERY HEAD WITH TV VIEW AND SWIVEL MOUNT

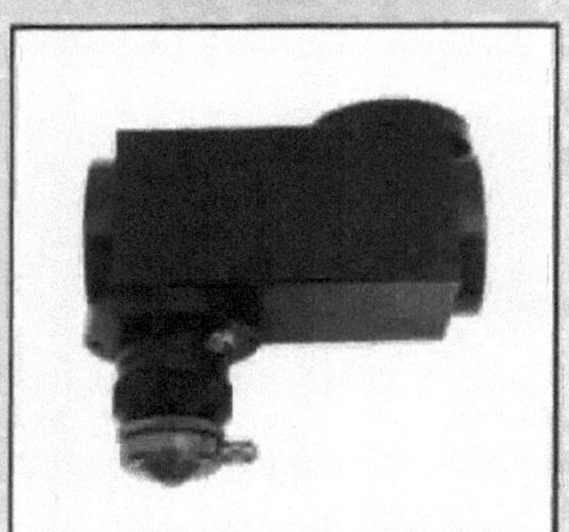

REFLECTIVE PARABOLA FOCUSSING HEAD WITH GAS JET

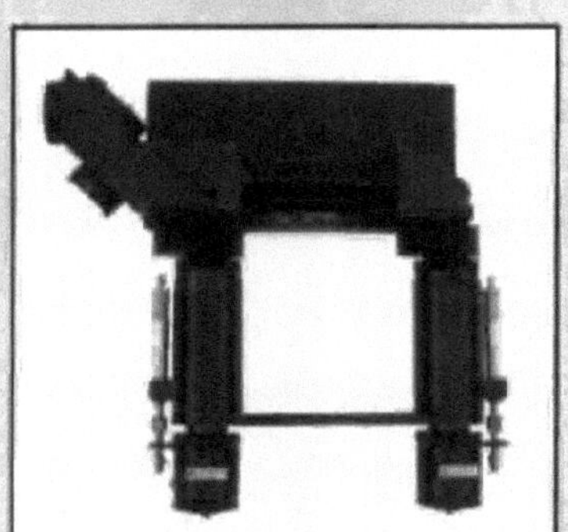

BEAM SPLITTER SYSTEM FOR CERAMIC PROCESSING

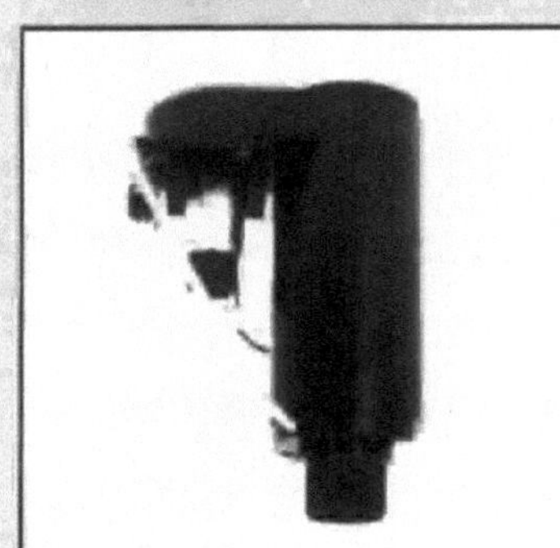

MOTORIZED Z SLIDE WITH INTEGRAL PATH ENCLOSURE

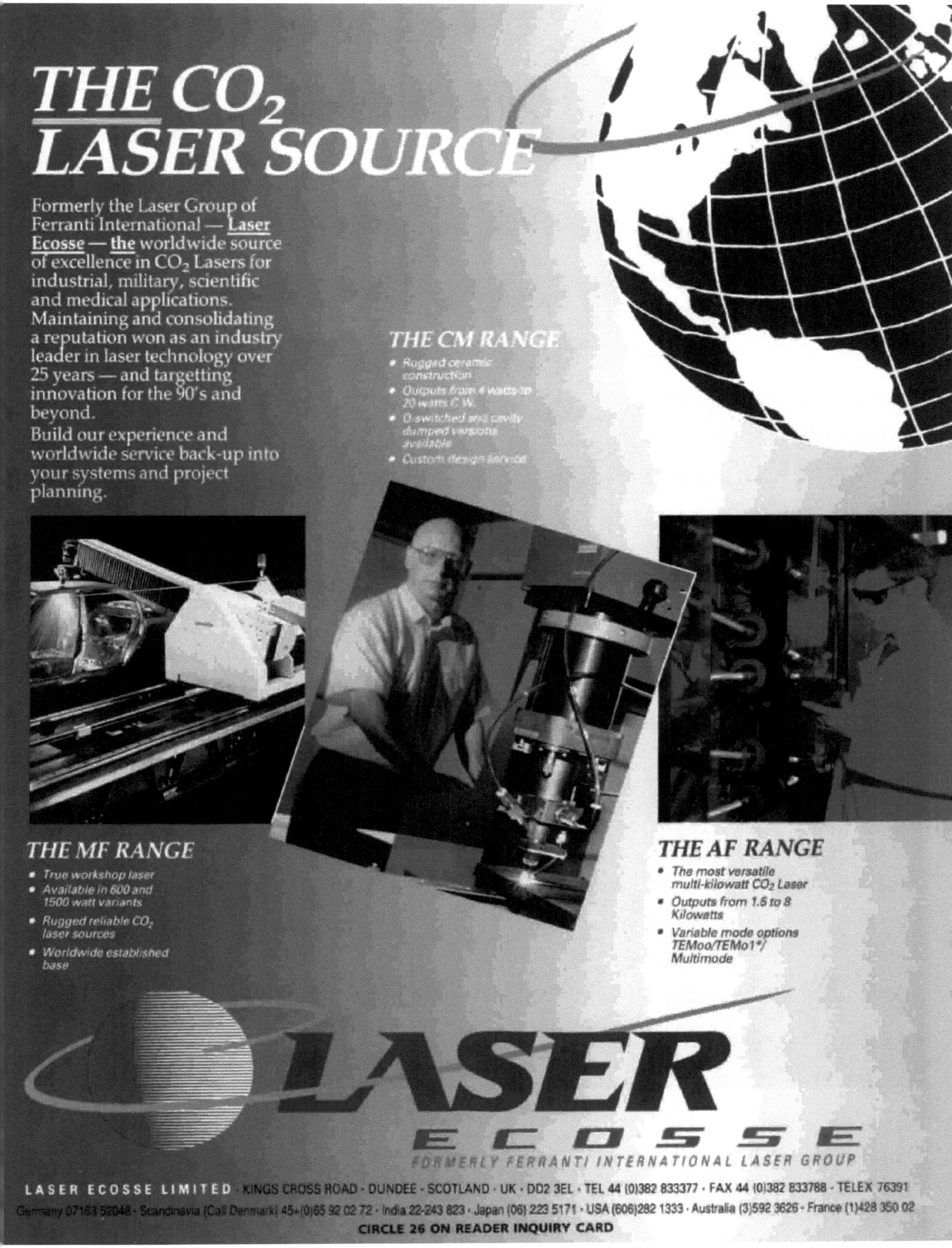
THE CO_2 LASER SOURCE
Formerly the Laser Group of Ferranti International — Laser Ecosse — the worldwide source of excellence in CO_2 Lasers for industrial, military, scientific and medical applications. Maintaining and consolidating a reputation won as an industry leader in laser technology over 25 years — and targetting innovation for the 90's and beyond.
Build our experience and worldwide service back-up into your systems and project planning.
THE CM RANGE
• Rugged ceramic construction
• Outputs from 4 watts to 20 watts C.W.
• Q-switched and cavity dumped versions available
• Custom design service
THE MF RANGE
• True workshop laser
• Available in 600 and 1500 watt variants
• Rugged reliable CO_2 laser sources
• Worldwide established base
THE AF RANGE
• The most versatile multi-kilowatt CO_2 Laser
• Outputs from 1.5 to 8 Kilowatts
• Variable mode options TEMoo/TEMo1*/ Multimode
LASER ECOSSE
FORMERLY FERRANTI INTERNATIONAL LASER GROUP
LASER ECOSSE LIMITED · KINGS CROSS ROAD · DUNDEE · SCOTLAND · UK · DD2 3EL · TEL 44 (0)382 833377 · FAX 44 (0)382 833788 · TELEX 76391
Germany 07183 52048 · Scandinavia (Call Denmark) 45+(0)65 92 02 72 · India 22-243 823 · Japan (06) 223 5171 · USA (606)282 1333 · Australia (3)592 3626 · France (1)428 350 02
CIRCLE 26 ON READER INQUIRY CARD

Yes,
we are
ready
please contact

INTERNATIONAL JOB SHOPS (CHART PAGE 5)

	Lasers			Applications									Services								Motions	
	Solid-State	CO2	Excimer	Welding	Cutting (Metal)	Cutting (nonmetal)	Drilling (Metal)	Drilling (nonmetal)	Scribing	Heat-Treating	Other Surface Treat.	Marking	Engraving	Part Design	Prototype	Pilot Plant	Production	Spec. Sys. Supp.	Metallurgy & QC	Processing R&D	MIL Spec Qual	2-Axes	Multi-Axes
Dubi & Company Herzogenbuchsee Switzerland 63-60-12-12 *		•			•										•		•						
Dyno AG Aefligen Switzerland 34-45-23-33*		•			•										•		•						
Ecole Centrale Chatenay Malabry France 16.1.46.83.63.15		•		•	•							•			•					•			•
Eitrac St. Martin D'Heres France 76 62 98 52		•			•																		
Elcede GmbH Kirchheim Germany *		•				•									•		•						
Eldim BV Lomm Netherlands 04703-1919	•			•	•		•								•	•	•			•			
Elettronica MI AR Salerno Italy		•			•																		
Ellickson Laser Cutting Ltd Waterford Ireland 353 51-32316		•			•	•	•	•				•			•		•						
Emaillierwerk Hannover Haselbacher GmbH & Co Wedemark/Mellendorf 1 Germany 51-(30) 40066*		•			•										•		•						
ERABA Livingston UK 0506-31234		•		•	•	•	•	•							•		•					•	
ETRA Nova Milanese (Milano) Italy		•			•																		
Eurofar Ornago (Milano) Italy 039/6010137		•			•																		
Eurofer Saneguernines France 87 98 59 59					•																		
Eurosteel Izano (Cremona) Italy		•		•	•																		
Exico Sevres France (1) 46 26 77 87		•			•	•																	
Fancareggi Bologna Italy		•			•																		
Fedele 82 Roma Italy		•				•																	
Feinwerktechnik Hago GmbH Kadelburg Germany 77-(41) 3551*		•			•										•		•						
Fila Orgiano (Vicenza) Italy 044/874191		•			•																		

INTERNATIONAL JOB SHOPS
(CHART PAGE 6)

	Lasers			Applications										Services								Motions	
	Solid-State	CO2	Excimer	Welding	Cutting (Metal)	Cutting (nonmetal)	Drilling (Metal)	Drilling (nonmetal)	Scribing	Heat-Treating	Other Surface Treat.	Marking	Engraving	Part Design	Prototype	Pilot Plant	Production	Spec. Sys. Supp.	Metallurgy & QC	Processing R&D	MIL Spec Qual	2-Axes	Multi-Axes
Formatype Vernauillet France 37 43 89 52		●			●																		
Weihbrecht GmbH Schwabisch-Hall Germany (0791) 51002		●			●	●									●		●					●	●
FPC ZA Montelier France 75 59 66 34		●			●																		
Fratelli Giusti Novellara(Reggio Emila Italy 0522/654219		●			●																		
Fratelli Rossi S. Lazzare Di Salena Italy		●				●											●						
Fraunhofer-Institute for Produktion Technology Aachen Germany 0241/8904-0		●				●				●				●	●	●	●		●	●			
Fraunhofer-Institut fur Lasertechnik Aachen Germany 0241/8906-0	●	●	●	●	●	●	●	●	●	●	●	●	●		●	●			●	●		●	●
Friedrich Alexander Universitat Erlangen Germany 0049/9131/85 75 20				●	●	●					●												
Fritz Born AG Langenthal Switzerland 63 22 14 07*		●			●										●		●						
Fustellifico Aioifi Carauaggio (Bergano) Italy						●		●							●								
Fustellifico Veronese Verona Italy		●				●									●								
Fustezbox Marconi (BO) Italy		●			●												●						
Gasser Ravussin SA Lucens Switzerland 021 906 8159						●		●															
GEAT Gesellschaft fur angewandte Technologie mbH Nurnberg 70 Germany 911/947 2080		●		●	●					●					●		●			●			
General Kamet Agaate Brianzh (MO) Italy		●			●																		
Gerhard Werhlbrecht Schwabisch Hall Germany 79 15 10 57*		●			●										●		●						
Gregis Alessandro Alzano Lombardo Italy 035/512001		●			●																		
Gunther Frey GmbH & Co KG Berlin 47 Germany 030/6257078								●															

INTERNATIONAL JOB SHOPS (CHART PAGE 7)

	Lasers			Applications										Services							Motions	
	Solid-State	CO2	Excimer	Welding	Cutting (Metal)	Cutting (nonmetal)	Drilling (Metal)	Drilling (nonmetal)	Scribing	Heat-Treating	Other Surface Treat.	Marking	Engraving	Part Design	Prototype	Pilot Plant	Production	Spec. Sys. Supp.	Metallurgy & QC	Processing R&D	MIL Spec Qual	2-Axes	Multi-Axes
Harlow Sheet Metal PLC Harlow UK 0279-414475		•			•	•	•	•	•			•	•	•	•	•	•						
Harr GmbH & Company Sindelfingen Germany 70-31-694-0		•			•										•								
Hebert Novellara Italy		•			•																		
G. Heise Laser Bandstahlschnitte GmbH Kirchheim Germany 70 21 5 41 11		•			•																		
Heurchrome Laserama St. Ouety L'Aumone France (1) 34 40 36 36	•			•	•			•															
Hexagone Cholet France 41 58 81 57		•			•																		
Hoesch Industrielaser GmbH Kleinostheim Germany 6027/6061*		•		•	•	•				•	•				•	•	•	•		•			
Holloway Sheet Metal Works Ltd London UK 071-607 4296		•		•	•		•																
Humber Sheet Metal Toronto, Ontario Canada		•		•	•	•	•	•						•	•	•	•				•	•	•
Hydee Bussolfno (TO) Italy		•				•																	
Hywel-Connection b.v. Doetinchem Netherlands 0031-8340 26266		•			•	•	•	•	•		•	•	•	•	•	•	•			•		•	
ICOS Zola Predosh (BO) Italy 051/752325-406		•			•																		
ICT Sorgues France 90 39 67 42		•			•																		
Inductoheat (Tewkesbury) Ltd Gloucestershire UK		•		•	•	•				•	•				•	•	•	•	•	•			
Industria Le Plessis Trevise France (1) 45 76 53 78	•			•	•					•													
Industriele Laser Toepassingen (ILT) B.V. Enschede Netherlands 053-355677	•	•		•	•	•	•	•	•	•		•	•		•	•	•			•		•	
Institut De Soudure Paris France (33) 1 42 03 94 05*		•		•																			
Institut Universitaire De Technologie Bethune France 21 57 60 80/2156 8460	•	•		•	•	•	•	•												•			

INTERNATIONAL JOB SHOPS
(CHART PAGE 8)

	Lasers			Applications										Services								Motions	
	Solid-State	CO2	Excimer	Welding	Cutting (Metal)	Cutting (nonmetal)	Drilling (Metal)	Drilling (nonmetal)	Scribing	Heat-Treating	Other Surface Treat.	Marking	Engraving	Part Design	Prototype	Pilot Plant	Production	Spec. Sys. Supp.	Metallurgy & QC	Processing R&D	MIL Spec Qual	2-Axes	Multi-Axes
Institut fur Werkstoffkunde Clausthal-Zellerfeld Germany (05323) 722120*		•		•	•																		
Interatom GmbH Bergisch Gladbach 1 Germany (0) 2204/84-2300		•		•	•																		
International R&D Ltd Newcastle-on-Tyne UK 091-2650451	•			•											•			•	•	•	•		•
Irepa Illkirch France (00.33) 88.67.93.00	•	•		•	•	•	•	•		•	•				•				•	•			
IRTM Vico Canavese (TO) Italy 0125/74362	•	•	•	•	•	•	•	•		•	•				•								
Isollima Bentivoglio (BO) Italy		•			•																		
Italdesign Moncalieri (TO) Italy 011/6470219		•			•										•								
ITCA Grugliasco (TO) Italy 011/70941		•			•										•								
Izumi Sogyo Co Ltd Tokyo Japan 3-902-3491*		•			•										•		•						
HP Kaysser Stuttgart Germany 07195/88-0		•			•	•	•	•						•	•	•	•					•	•
KHI Neckarsulm Germany *		•			•	•									•		•						
Kitamura Kikai Co Ltd Toyama Japan 766-63 1100*		•			•	•									•		•						
Label SA Boussu Belgium 065-78.18.08*	•		•	•			•	•		•	•			•	•			•		•			
Laboratorie National D'Essais Trappes France 16 (1) 34 61 22 22*	•	•													•								
Lacura Laser Cut W. Schoenenbon GmbH Radevomwald Germany 21 95 70 19*		•			•	•									•		•						
LA Deertip Vernouillet France 37 42 74 72		•				•																	
Lamier LaLoggia (TO) Italy 011/9629485		•			•										•								
LAMP Srl Villanova D'Asti Italy 0141/916614		•			•										•								

INTERNATIONAL JOB SHOPS (CHART PAGE 9)

	Lasers			Applications									Services								Mo-tions	
	Solid-State	CO2	Excimer	Welding	Cutting (Metal)	Cutting (nonmetal)	Drilling (Metal)	Drilling (nonmetal)	Scribing	Heat-Treating	Other Surface Treat.	Marking	Engraving	Part Design	Prototype	Pilot Plant	Production	Spec. Sys. Supp.	Metallurgy & QC	Processing R&D	MIL Spec Qual	2-Axes	Multi-Axes
LAP Illkirsh Graffenstaden France 88 67 33 45		•			•																		
Lappeenranta University of Technology Lappeenranta Finland +358 53 5711		•		•	•	•	•	•		•	•								•	•			
Laser & Allied Cutting Service Pty Ltd Welshpoo 1 Western Australia (09) 4701751		•			•	•									•		•					•	
Laser Alsace Production Ilkirch Graffenstaden France (33) 88 67 33 45		•		•	•	•	•	•	•	•	•				•		•		•				
Laser Application & Information Center Amsterdam WS Amsterdam Netherlands 20 5224014*	•	•		•	•															•			
Laser-Automation Gekatronic SA La Chaux-de-Fonds Switzerland 39 25.21.75	•			•	•	•	•	•	•			•	•	•	•	•	•	•		•		•	•
Laser Bullard Magland France 50 34 74 10		•			•																		
Lasercut Products Ltd Herts UK 0279-600521		•			•	•									•	•	•			•			
Laser Cut W A Pty Ltd Welshpool Perth, W.A. Australia 619-362-5255		•			•	•	•	•	•			•	•		•		•					•	
Laser Decoup Sarcelles France *		•			•	•									•		•						
Laser Expertise Ltd Nottingham UK 0602 587452		•		•	•	•	•	•		•										•			
Laser 53 Bazaugers France 43 02 82 20		•			•																		
Laser Industrie St.Marie Aux Chenes France 87 61 80 39	•	•		•	•	•	•	•	•			•			•		•						
Laser Industrie Atlantic Perigwy France 46 44 60 85		•				•																	
Laser Laboratorium Gottingen e.V Gottingen Germany 551-50350			•								•									•			
Laserlife Australia Clayton, Victoria Australia 61-3-5440666		•											•					•				•	
Laser Line Inc New Market, Ontario Canada *		•										•	•										
Laser Machining Center Inc Toronto, Ontario Canada 416-752-7372	•	•		•	•	•	•	•	•	•		•	•	•	•	•	•		•	•	•		•

INTERNATIONAL JOB SHOPS
(CHART PAGE 10)

	Lasers			Applications										Services							Motions	
	Solid-State	CO2	Excimer	Welding	Cutting (Metal)	Cutting (nonmetal)	Drilling (Metal)	Drilling (nonmetal)	Scribing	Heat-Treating	Other Surface Treat.	Marking	Engraving	Part Design	Prototype	Pilot Plant	Production	Spec. Sys. Supp.	Metallurgy & QC	Processing R&D	MIL Spec Qual	2-Axes	Multi-Axes
Laser Materialbearbeitung GmbH Worth Germany 09372/5008		•		•	•	•					•												•
Laser Melis SA Aretxabaleta Spain 43 79 75 11*				•	•					•	•				•			•	•	•			
Laser Optronic Srl Milano Italy 02/27000435	•																						
Laser Plastic Cutting Service Pty Ltd Huntingdale Australia (03)5438433		•			•	•	•	•	•				•	•	•		•			•			
Laser Precision Engineering Ltd Manchester UK 061 876 7273		•			•	•	•	•				•			•		•					•	
Laser Process Ltd Cannock, Staffs UK 0543 466676		•			•	•									•	•	•			•		•	
LaserProdukt GmbH Alfeld Germany 0049-5181-26347	•	•	•	•	•	•	•	•	•	•				•	•		•		•			•	•
Laser Profiles Kingsgrove, NSW Australia *		•			•	•									•	•	•						
Laser Profiles Eastern Ltd Basildon, Essex UK 0268 729292*		•			•	•									•	•	•						
Laser Profiles Southern Ltd Wimborne, Dorset UK 0202 861438*		•			•	•									•	•	•						
Laser Rhone-Alpes Fontanil Cornil France 76 56 07 57	•			•			•																
Laser Roma Anzio (Roma) Italy 06/9875990		•			•												•						
Laser Service France St Laurent Sur Gorre France 55 00 00 30		•			•																		
Laser Services Cesena C Forli Italy		•				•																	
Laser Services LM Suresnes France 33 1 45066854*		•				•		•	•			•	•		•		•						
Laserstream Cutting Ltd Auckland New Zealand 09 884 093		•			•	•									•		•	•					
Laser Techniques St. Laurent Sur Gorre France 55 00 04 11		•			•																		
Laser Top Villetaneuse France (1) 48 27 59 30		•			•																		

INTERNATIONAL JOB SHOPS
(CHART PAGE 11)

	Lasers			Applications										Services								Motions	
	Solid-State	CO2	Excimer	Welding	Cutting (Metal)	Cutting (nonmetal)	Drilling (Metal)	Drilling (nonmetal)	Scribing	Heat-Treating	Other Surface Treat.	Marking	Engraving	Part Design	Prototype	Pilot Plant	Production	Spec. Sys. Supp.	Metallurgy & QC	Processing R&D	MIL Spec Qual	2-Axes	Multi-Axes
Lasertron Pty Ltd Nerang Australia (75) 963711		•			•	•									•	•	•						
Laser 2000 Aixe Sur Vienne France 55 70 31 31		•			•																		
Laser 2005D Gonesse France (1) 39 87 25 66		•			•																		
LaserWest Ltd Plympton, Devon UK (752) 348840	•	•		•	•	•	•								•		•	•					
Lase Tech Ltd Daventry, Northants UK 0327 79666		•		•	•	•								•			•			•			
Lasindustria Technologia Laser S.A. Oeiras Portugal 351 1 4430557		•			•	•	•	•					•	•			•						
Lastec AG Brugg/Biel Switzerland *		•			•	•																	
LA Tolerie Fine Illkirch Graffenstaden France 88 67 30 20		•			•																		
Laws Laser Nundah Australia 266 3888		•			•	•									•	•	•						
LBC Laser Bearbeitung Ceresa AG Flamatt Switzerland 0041 31 741 2021		•			•	•								•	•							•	•
LEM Laser Basaldella Formido Italy		•			•	•																	
LEM Laser Salt Di Povoletto Italy		•			•	•																	
Levigne et Fils Pont du Chateau France 73 83 00 52	•			•				•															
LIDIS Cherisy France		•			•																		
LIVA Fiume Veneto (PN) Italy 0434/959102		•			•																		
LM Laser Metalworking srl Cambiago (Milano) Italy 39-2-95308132		•		•	•					•	•					•	•			•			
Lowara Montecchio Maggiore Italy *		•			•	•									•		•						
LTTA Sartrouville France (1) 39 57 50 02		•		•	•										•								•

INTERNATIONAL JOB SHOPS
(CHART PAGE 12)

	Lasers			Applications									Services								Motions	
	Solid-State	CO2	Excimer	Welding	Cutting (Metal)	Cutting (nonmetal)	Drilling (Metal)	Drilling (nonmetal)	Scribing	Heat-Treating	Other Surface Treat.	Marking	Engraving	Part Design	Prototype	Pilot Plant	Production	Spec. Sys. Supp.	Metallurgy & QC	Processing R&D	MIL Spec Qual	2-Axes	Multi-Axes
Lulea University of Technology Lulea Sweden +46 920 91000		●		●	●															●			
Lutral Luttange France 82 83 95 11		●			●																		
Malaval Le Coteau France 77 70 44 65		●			●																		
Manfred Toller GmbH Hagenburg Germany 05033-7024		●				●																●	
MAN Technologie AG Munchen, 50 Germany	●	●		●	●	●				●	●				●	●	●		●	●			●
Masson Bobigny France (1) 18 45 57 04		●			●																		
Matsuda Co Ltd Houfu, Yamaguchi Japan 835-24-2111 *		●			●										●		●						
MECC 2000 SNC Villalunga Casalgrande Italy 0522/849893		●			●																		
Meccanica DS Sanlazzaro D Savena Italy 051/463090		●			●																		
Meccaniche Marcon Villaraspa (Vicenza) Italy		●			●																		
MeKo Sarstedt/Hannover Germany +49 5066 4035	●			●	●	●	●	●		●	●						●	●					●
Metafot GmbH Wuppertal Germany 0202/6090		●			●	●	●	●															
Metal Mobil Mas D'Agenay France 53 89 50 34		●			●																		
Metalplasma Srl Padova Italy 049/5968146		●			●																		
Michiper Robassomero (TO) Italy		●			●																		
MIRAP AG Jona Switzerland 0041-55-28 17 28		●			●							●		●	●		●					●	
Model Master Moncalieri (TO) Italy 011/6811958		●			●										●								
Modensider Modena (MO) Italy 059/360236		●			●																		
Montes Albaredo D'Adige Italy		●			●																		

INTERNATIONAL JOB SHOPS
(CHART PAGE 13)

	Lasers			Applications										Services								Motions	
	Solid-State	CO2	Excimer	Welding	Cutting (Metal)	Cutting (nonmetal)	Drilling (Metal)	Drilling (nonmetal)	Scribing	Heat-Treating	Other Surface Treat.	Marking	Engraving	Part Design	Prototype	Pilot Plant	Production	Spec. Sys. Supp.	Metallurgy & QC	Processing R&D	MIL Spec Qual	2-Axes	Multi-Axes
Mundinus & Buss Lasertechnik Luneburg Germany 41-31-89030	•																						
Nailam Rovereto (Trento) Italy		•			•	•																	
NC Laser Cutting Services Ltd Cumbria UK *		•			•	•																	
Nimbus Laser Services W. Sussex UK 0444 87 03 86 *		•		•	•	•								•	•		•						
Nissan Shatai Co Ltd Kanagawa Japan 463-55-4800 *		•			•										•		•						
Nobels BV AV Katwijk Netherlands *		•			•	•									•	•	•						
Nollet Lys Les Lannoy France		•			•										•		•						
Numerlaser La Chapelle Basse Mer France 40 06 33 01		•			•																		
Nuova Cantro Torino Italy 011/7395640		•			•																		
Nuova CMB Ozzegna Canadese (TO) Italy		•			•																		
N.U. - Tech GmbH Neumunster Germany 04321 3 06 20 *	•	•		•	•	•	•	•	•						•								
OBM Di Baccigaluppi & C Srl Mesfro (MI) Italy 02/9786182		•			•																		
Octal Jassans Riothier France 74 60 72 29		•			•																		
Officine De Zan Rozzano (MI) Italy		•			•																		
OMA AG Aarau Switzerland (64)24-4924 *		•			•										•		•						
OMEP Srl Pistola (PT) Italy		•			•																		
Omes SpA Santorso, IV Italy *		•			•	•									•	•	•						
OMP Proserpio (Como) Italy 031/621434		•			•																		

INTERNATIONAL JOB SHOPS
(CHART PAGE 14)

	Lasers			Applications										Services								Mo-tions	
	Solid-State	CO2	Excimer	Welding	Cutting (Metal)	Cutting (nonmetal)	Drilling (Metal)	Drilling (nonmetal)	Scribing	Heat-Treating	Other Surface Treat.	Marking	Engraving	Part Design	Prototype	Pilot Plant	Production	Spec. Sys. Supp.	Metallurgy & QC	Processing R&D	MIL Spec Qual	2-Axes	Multi-Axes
OMR SpA Remeldello Sotto (BS) Italy 030/957212		•			•																		
Ontario Hydro Research Division Toronto, Ontario Canada 416-231-4111		•		•	•	•	•	•		•	•									•		•	
Ontario Laser & Lightwave Research Center Toronto, Ontario Canada 416-978-2939	•	•	•			•		•	•			•	•						•	•			
OPN Noisy Le Grand France (1) 45 92 18 18		•			•																		
Orskovs Maskinfabrik A/S Frederikshavn Denmark 8-47-9033 *		•			•										•		•						
OTS Malestroit France 97 75 12 99		•			•																		
Ouag Italia Refrontolo (Treviso) Italy		•				•																	
Ouest Tolerie Fougeres France 99 99 71 06		•			•																		
Oxymetal Bordeaux Cedex France 56 49 37 46		•			•																		
Palbam Metal Works En-Harod Ihud Israel 06-531703		•			•	•								•	•		•				•	•	
Parimor Castlemaggiore (BU) Italy 051/326784		•			•										•								
Perform Metal Industries Pty Ltd Brisbane, Queensland Australia 0-61-7-395-5499		•		•	•	•	•	•					•	•	•	•	•					•	
PermaNova Lasersystem AB Ostersund Sweden +46 63 132455	•	•	•	•	•		•	•	•			•	•				•	•					
Permascand AB Ljungaverk Sweden +46-691-329 40	•	•	•	•	•	•	•	•		•	•	•	•		•		•	•	•	•		•	•
Photonics Systems Laboratory Strasbourg France 88 36 90 01			•	•	•	•	•	•		•				•	•	•		•		•			
Pilosio SpA Feletto Umberto (Udine Italy 0432/570983		•			•																		
Piroux St. Etienne Du Bois France 74 51 32 10		•			•																		
Plasti Center Napoli Italy		•				•																	

INTERNATIONAL JOB SHOPS
(CHART PAGE 15)

	Lasers			Applications										Services								Motions	
	Solid-State	CO2	Excimer	Welding	Cutting (Metal)	Cutting (nonmetal)	Drilling (Metal)	Drilling (nonmetal)	Scribing	Heat-Treating	Other Surface Treat.	Marking	Engraving	Part Design	Prototype	Pilot Plant	Production	Spec. Sys. Supp.	Metallurgy & QC	Processing R&D	MIL Spec Qual	2-Axes	Multi-Axes
Plastiques De Champagne Pont Sainte Marie France 25 81 14 98		•			•																		
Podetti Marcello Camin (PD) Italy 049/760073		•			•																		
Power Beam Technologies Risley/Warrington UK 0925-25 2845	•	•		•	•	•	•			•	•				•		•		•	•	•	•	•
PPP Mestre (Venezia) Italy		•				•																	
Preci Spark Ltd Leicestershire UK *	•																						
Prestoforme Jagney S/Bois France (01) 34 71 13 30		•			•	•																	
Profils Chaponost France 78 56 53 38		•			•																		
PS Laseranwendungstechnik GmbH Thedinghausen Germany 04204-7604		•		•	•	•	•	•		•	•			•	•	•	•	•	•	•			
PUM Reims France 26 87 96 96		•			•																		
Puricelli Gallarate (VA) Italy 0331/795732		•			•																		
Quality Plastics Supplies Pty Ltd Sefton, NSW Australia 743-8722		•				•		•							•		•				•		
Raimkeh CIMT Marly France 27 45 92 20		•			•																		
Rapid Lamiera Sala Bolognese Italy 051/954165		•			•																		
R Audemars SA Cadempino-Lugano Switzerland 091-58 26 62	•			•				•						•			•	•					•
RBD St. Pierre D'Albigwy France 79 28 58 40	•		•					•															
Redditch Laser Cutting Ltd Redditch UK 0527-510474		•			•	•																	
Renoult Pontault France (1) 60 28 04 67		•			•																		
Reydes SA Gondecourt France 20 90 92 50		•			•																		

INTERNATIONAL JOB SHOPS
(CHART PAGE 16)

	Lasers			Applications										Services								Motions	
	Solid-State	CO2	Excimer	Welding	Cutting (Metal)	Cutting (nonmetal)	Drilling (Metal)	Drilling (nonmetal)	Scribing	Heat-Treating	Other Surface Treat.	Marking	Engraving	Part Design	Prototype	Pilot Plant	Production	Spec. Sys. Supp.	Metallurgy & QC	Processing R&D	MIL Spec Qual	2-Axes	Multi-Axes
Riva Officine Caluso D'Lurea (TO) Italy 011/9831143		•			•																		
Rjukan Metall A/S Rjukan Norway 47-36-94316		•		•													•						
Romi Montecchio Maggiore Italy		•			•																		
Rossi Luigi Basso Marino Italy 0736/402016		•			•																		
Rossi Officine Carmagnola (TO) Italy 011/9770244		•			•										•								
RYG Argenteuil France (1) 39 61 31 38		•			•																		
SACTI Venissieux France 72 50 32 04		•			•																		
Sadela SA Serrieres/Felines France 75348631		•		•	•																		
Sallig Nicheline (TO) Italy 011/624480		•			•										•								
Sarzi Lamiere Mantova Italy		•			•																		
SBA Interior Mustio Finland 358 12 48201		•			•																		
SCGI Le Breuil France 85 56 09 81		•			•																		
Schurholz KG Plettenberg Germany 23-91-8-10-30 *		•			•	•									•		•						
Schurholz/VS Kirchhundem Germany 2723 12033	•	•		•	•		•																
Schweiko Indust GmbH Huckeswagen Germany 21-92-4175 *		•			•																		
Secatol Saint Benois France 19 57 25 20		•			•																		
Seibu Taiyo Laser Kitakyushu, Fukouka Japan 93-861-2664 *		•			•										•		•						
Semtp-Zahra Bubres Sur Yvette France (1) 69 07 57 81		•			•																		

INTERNATIONAL JOB SHOPS
(CHART PAGE 17)

	Lasers			Applications										Services								Motions	
	Solid-State	CO2	Excimer	Welding	Cutting (Metal)	Cutting (nonmetal)	Drilling (Metal)	Drilling (nonmetal)	Scribing	Heat-Treating	Other Surface Treat.	Marking	Engraving	Part Design	Prototype	Pilot Plant	Production	Spec. Sys. Supp.	Metallurgy & QC	Processing R&D	MIL Spec Qual	2-Axes	Multi-Axes
SER Sanint Ouen France (1) 40 11 14 57		•			•																		
Sercas Poluerigi (Anluna) Italy 071/908146		•			•																		
Sermeto Cussey France 70 31 76 88		•			•																		
Serrieres Changs Sur Marne France 64 35 92 53		•			•																		
Shanghai Institute of Laser Technology Shanghai PRC 4700458	•	•		•	•	•	•	•	•	•	•	•	•	•	•	•	•	•		•		•	
Shin Meiwa Kogyo Co Ltd Takarazuke, Hyogo Japan *		•			•										•		•						
SIAC Pontiroli Nuoua (BG) Italy 0363/88437		•			•										•								
SIAPE Nerviano (MI) Italy 0331/585101		•			•																		
Sider Cesenate Diegaro (FO) Italy		•			•																		
Signer Ag Riedt B. Erlen Switzerland 72-48-11-11 *		•			•										•								
Silap Srl Vimercate (MI) Italy 039/6085966-7		•			•																		
Smada Elettromeccanica Avellino Italy		•			•																		
Socamcaser Toulouse France 61 47 38 02		•			•																		
Socomec La Souterraine France 55 89 49 49		•			•																		
Sodery SA Charleville, Mezieres France 24.33.91.47 *		•			•	•									•		•	•					
Sorba Precisieplaatwerk BV Winterswijk Netherlands 05430-14666		•			•										•		•						
Sorenco Roedovre Denmark 1-412244 *		•			•										•		•						
SOTIAG Amplepuis France 74 89 45 93		•			•																		

INTERNATIONAL JOB SHOPS
(CHART PAGE 18)

	Lasers			Applications									Services								Motions	
	Solid-State	CO2	Excimer	Welding	Cutting (Metal)	Cutting (nonmetal)	Drilling (Metal)	Drilling (nonmetal)	Scribing	Heat-Treating	Other Surface Treat.	Marking	Engraving	Part Design	Prototype	Pilot Plant	Production	Spec. Sys. Supp.	Metallurgy & QC	Processing R&D	MIL Spec Qual	2-Axes	Multi-Axes
South Yorkshire Laser Cutting Ltd Dinnington, Sheffield UK (0909) 568682		•		•	•	•						•			•		•						
Specitec Paris France 33 (1) 43 70 50 50 *	•			•	•	•	•	•						•	•		•	•	•	•			
STECYR Chateaudun France 37 45 40 50		•			•																		
Steec Brindas France 78-45-14-63	•			•	•	•	•	•				•					•		•				
STILL Sartrouville France (1) 39 14 29 92		•			•																		
Strathclyde Fabricators Ltd Strathclyde UK 0698-283452		•		•	•		•				•	•			•	•	•						
Subcon Laser Cutting Ltd Nuneaton, Wark UK 0203 642221		•		•	•	•									•		•					•	•
Sulzer Bros Ltd Winterthur Switzerland 41 52 2625155		•		•						•	•				•	•	•		•	•			
Swiss Federal Institute of Technology Lausanne Switzerland 021/6934914		•		•						•	•									•			•
Symtec Laser Processing Inc Guelph, Ontario Canada 519-763-2420	•	•		•	•	•						•	•	•	•		•			•		•	
Systrel Lesulis France (1) 69 97 80 39	•	•			•																		
Taiyo Laser Co Ltd Midor-ku, Nagoya Aichi Japan 52-622-8553 *		•			•										•		•						
Taniyama Co Ltd Osaka-shi, Osaka Japan 6-931-7802 *		•			•										•		•						
TCIN Bourguebus France 31 23 13 61		•			•																		
Techmeta Annecy France (33) 50 27 20 90	•			•																			
Techno Filtres Bueil France 32 336 58 26	•			•																			
Technoplex Di Capmodarsego (P) Italy		•			•																		
Tecnica Marchent (BS) Italy 030/861635		•			•																		

INTERNATIONAL JOB SHOPS
(CHART PAGE 19)

	Lasers			Applications										Services								Motions	
	Solid-State	CO2	Excimer	Welding	Cutting (Metal)	Cutting (nonmetal)	Drilling (Metal)	Drilling (nonmetal)	Scribing	Heat-Treating	Other Surface Treat.	Marking	Engraving	Part Design	Prototype	Pilot Plant	Production	Spec. Sys. Supp.	Metallurgy & QC	Processing R&D	MIL Spec Qual	2-Axes	Multi-Axes
Tecnisud SpA Pescara (PE) Italy 085/576242		●			●																		
Tecnolaser Marsango Di Campo Italy		●			●																		
Tepro-Prazionstechnik GmbH Monchweiler Germany 77 21 7 30 91		●			●	●																●	●
The Laser Institute Edmonton, Alberta Canada 403-436-9750		●		●	●	●				●	●				●		●		●	●		●	●
Torbrex Engineering Ltd Dundee UK 0382-825253		●			●	●											●						
Triefus Industries/Australian Pty Ltd Mona Vale, NSW Australia 997-7033		●					●	●				●		●			●						
TWI Abington, Cambridge UK 223 891162	●	●	●	●	●	●	●	●		●	●			●	●	●		●	●	●		●	●
University of Liverpool Liverpool UK 051.794.4892		●	●			●	●	●		●	●												
University of Stuttgart Stuttgart 80 Germany (711) 685-6841		●		●	●					●										●			
Usines Chausson Asnieres Cedex France (1) 47 91 76 42		●			●																		
Vantage Laser Cutting Ltd Cambridge, Ontario Canada 519-653-1588		●			●										●		●					●	
Vermec Beinasco (TO) Italy 011/3497385		●			●																		
Verona Lamiere S. Maria Zerbio (VR) Italy		●			●																		
VGM Rovagnasco Di Segrate Italy		●			●																		
W&B Machining Service Ltd Redditch UK *		●			●										●		●						
WBM Staalservice Centrum BV Stramproy Netherlands 4956 1781		●			●										●		●						
Wirth & Company AG Buchrain Switzerland 41-89-16-31		●			●										●		●						
Yagi's Job Shop Remy Renaud Hille 1 Germany 57 03 32 70 *	●						●																

INTERNATIONAL JOB SHOPS
(CHART PAGE 20)

	Lasers			Applications										Services								Motions		
	Solid-State	CO2	Excimer	Welding	Cutting (Metal)	Cutting (nonmetal)	Drilling (Metal)	Drilling (nonmetal)	Scribing	Heat-Treating	Other Surface Treat.	Marking	Engraving	Part Design	Prototype	Pilot Plant	Production	Spec. Sys. Supp.	Metallurgy & QC	Processing R&D	MIL Spec Qual	2-Axes	Multi-Axes	
Yamato Kogyo Co Ltd Yamoto, Kanagawa Japan *		•			•										•		•							
Zucchelli Furniture Verona Italy 0451501433		•			•																			
Zumstein AG Batterkinden Switzerland 65-45-35-31 *		•			•										•		•							

INTERNATIONAL CONTRACT LASER PROCESSING SERVICES DIRECTORY

ATS, Via L. Sampietro 109, Saronno (Varese), 21047; 02/9609465, FAX 02/96700497

Australia

Action Laser Pty Ltd, 105 Delhi Rd, North Ryde, NSW2113, Australia; 61-2-8878290, FAX 61-2-8878291
mng dir, Dr. Ken Crane; 1988
Laser materials processing, in particular perforation of metals to produce fine holes or slots; production of very fine filters, sieves, screens, and aerators, for food processing, minerals-processing, and sewerage treatment. Unique capabilities for drilling stainless steel sheets and other metals by laser techniques.

Alpha Laser Pty Ltd, 1 Summer Lane Ringwood, Victoria, 3134, Australia; +613-870-2345*, FAX +613-879-3676
emp 30, 1984
Contract laser cutting of steel, and plastics. Equipment Strippit Laser center. Agent for Laser Power Optics.

Aunger Plastics, Womma Rd, Elizabeth West SA, 5113, Australia; 08 252-1344, FAX 08 255-1169
mng dir, Chris Tormay; sls mgr, Tony Evans; mfg mgr, Andrew Martin; 1967
Manufactures automotive accessories. Processes performed include plastic vacuum forming, acrylic sheet forming, metal stamping and contract laser cutting of steels and plastics in 2, 3, and 5 axis. Four state-of-the-art laser cutting machines are used 24 hours per day.

Brooks-Koochew Pty Ltd, Laser Plus, 4 Aileen Ave, Heidelberg West, Victoria, 3081, Australia; 458-2611, FAX 458-1358
emp 80, 1972
Sheet-metal shop with laser cutting capability.

Cobra Signs, Spectrum Photographic Color Services, 20 Brennan St, Slacks Creek, Brisbane, Queensland, 4127, Australia; 208 0722, FAX 208 4361
1981
Major job shop contractors in laser cutting and manufacturers to the sign, architectural and building industries.

Cols Cutting Formes P/L, 78 Queen Victoria St, Bexley, NSW, 2207, Australia; 588 5999*, FAX 587 7081
emp 280, 1967
Cutting form manufacturer.

CSIRO Industrial Laser Centre, PO Box 218, Lindfield, NSW, 2070, Australia; (02) 413-7194, FAX (02) 413 7637
mgr, Dr. M. Brandt; 1967
Offers research and development of new products and processes based on laser processing. Particular emphasis is placed on existing Australian industry to fully evaluate the widespread capabilities of laser processing and improve their competitive positions.

Laser & Allied Cutting Service Pty Ltd, 1 Unit 17 Kew St, Welshpoo 1 Western, 6106, Australia; (09) 4701751, FAX (09) 4702807
gen mgr, Neil Johnson; 1987
Contract cutting of mild steel, stainless steel, and acrylics.

Laser Cut W A Pty Ltd, 83 President St, Welshpool Perth, W.A., 6106, Australia; 619-362-5255, FAX 619-362-5456
gen mgr, Ed Hauswirth; mgr, Brendan Cullinan; 1986
Provides laser profile cutting of steels, stainless steel, plastics, wood, and other non-ferous materials.

Laserlife Australia, 46-50 Bendix Dr, Clayton, Victoria, 3168, Australia; 61-3-5440666
1986
Offers laser engraved ceramic coated anilox rolls, engraved rubber sleeves and rolls.

Laser Plastic Cutting Service Pty Ltd, 3/86 Fenton St., Huntingdale, 3167, Australia; (03)5438433, FAX (03)5436257
mng dir, Terry Leonard; 1986
Offers laser cutting of most steels and plastics, prototype and production runs.

Laser Profiles, 17 Commercial Rd, Kingsgrove, NSW, 2208, Australia
1986

Lasertron Pty Ltd, PO Box 518, Nerang, QLD4211, Australia; (75) 963711, FAX (075) 962195
emp 70, 1987
Laser cutting shop and plastic fabrication factory.

Laws Laser, 61 Robinson Rd, Nundah, QLD4012, Australia; 266 3888
pres, Ken Laws; mfg mgr, Geoff Laws; emp 100, 1985
Provides full cutting service to most industries.

Perform Metal Industries Pty Ltd, 38 Love St, Bulimba,, Brisbane, Queensland, 4171, Australia; 0-61-7-395-5499, FAX 61-7-399-6132
emp 400, 1981
Offers precision shearing computer numerical controlled accuracy in punching, laser cutting, louvering and bending, robot welding, spot welding, croping, stamping and full sheet-metal workshop with qualified tradesperson flexability.

Quality Plastics Supplies Pty Ltd, 5 Marjorie St, Sefton, NSW, 2162, Australia; 743-8722, FAX 743-8725
ch exec, R.W. Townsend; eng mgr, S. Reid; 1948
Specializes in supplying laser-cut blanks of acrylic (Plexiglas), supply blanks for export.

Triefus Industries/Australian Pty Ltd, 81 Bassett St, Mona Vale, NSW, 2103, Australia; 997-7033, FAX 997-8313, telex AA 122314
dir & gen mgr, S.W. Matthews; dir & ops mgr, D.M. Vaughan; dir & secty, A. Mandalidis; 1935
Manufactures industrial diamond products, including natural and synthetic wire drawing dies.

Belgium

Actif Industries SA, Rue Des Fraisiers 89, Herstal, B-4040, Belgium; 032 41 40 0299, FAX 0032 41 40 06 86
1935

Label SA, Grand Hornu,rue Sainte Louise, Boussu, B-7320, Belgium; 065-78.18.08*, FAX 065-78.20.35, telex 57727 (Label)
emp 120, 1987
Feasibility studies, development studies, prototype production. A laser processing service to industry for trial or small production runs. Development of application-specific laser processing equipment to meet customer specific-requirements. Consultancy and training courses in laser processing techniques. Specialists in laser chemical vapour deposition and laser chemical surface processing.

Canada

Die X, 176 Strmach Crescent, London, Ontario, NVS3A1, Canada; 519-451-9254
1987

Humber Sheet Metal, 700 Kipling Ave, Toronto, Ontario, M8Z 5G3, Canada
pres, Irving Weiss; vp, Joseph Vital; plant mgr, Dave Anderson; 1962
Components manufacturer for: public transportation equipment (buses, subways, etc.), locomotives, trucks, and military equipment. Design and development of new components and assembly capability.

IKS

IKS, 19689 Tlegraph Trail, Langley, BC, Z34 4P8, Canada
1962

Laser Line Inc, 1224 Twinney Cresent, New Market, Ontario, L3Y 4W1, Canada
1962

Laser Machining Center Inc, 64 Carnforth Rd, Toronto, Ontario, M4A 2K7, Canada; 416-752-7372
pres/gen mgr, D.R. Bellerby; eng mgr, Steve Hicks; vp tech serv, Hugh Bissett; prod mgr, Iain Ferguson; 1986
Offers laser cutting, drilling, and welding. Localized heat treating and marking of metals and most nonmetal products. Prototype to full production runs. 100% job shop facility.

Ontario Hydro Research Division, Industrial Laser Laboratory, 800 Kipling Ave KR128, Toronto, Ontario, M8Z 5S4, Canada; 416-231-4111, FAX 416-231-5799, telex 06-984706
sr physicist, Dr. L. Mannik; tech, S.K. Brown; 1987
Provides assistance to Ontario manufacturers in assessing the feasibility of laser materials processing applications.

Ontario Laser & Lightwave Research Center, Central Facility, 60 St. George St Rm 331, c/o Univ. of Toronto, Toronto, Ontario, M55 1A7, Canada; 416-978-2939, FAX 416-978-3936
sls mgr, Dr. M. Hubert; staff scis, Dr. X. Gu, Dr. A. Othonos, V. Isbrucker; emp 70, 1988
Lasers and laser based instrumentation available for use in-house or on loan to industry for feasibility experiments, development of new manufacturing, quality control, and process control techniques, and calibration of light sources and detectors. Consulting and contract services in laser and optical technology, and educational and training courses.

Symtec Laser Processing Inc, 79 Regal Rd, Unit 5, Guelph, Ontario, N1H 6L8, Canada; 519-763-2420, FAX 519-763-2420
pres, Keith Drinkwater; vp, Jan Drinkwater; 1985

The Laser Institute, 9924 45 Ave, Edmonton, Alberta, T6E 5J1, Canada; 403-436-9750, FAX 403-437-1240
ch exec, D.C.D. McKen; mktg mgr, Brenda Pullen; prog dir r&m, D.F. Allsopp; prog dir sensors, M.R. Cervenan; prog dir mp, V.E. Merchant; bus dev mgr, Allan F. Peirce; jobshop mgr, Dietmar Seibt; 1985
Job shop work, process development, and research and development in materials processing. Design of specialized fiberoptic and visual sensors for process control and inspection of manufactured components.

Vantage Laser Cutting Ltd, 1600 Industrial Rd, Unit 10, P.O. Box 153, Cambridge, Ontario, N1R 5S9, Canada; 519-653-1588, FAX 519-650-2261
pres, B. Snider; 1989
Laser cutting job shop utilizing Amada LC667 II and Cincinatti CL-5 cutting systems. Fully CAD/CAM. Prototypes and samples run for a nominal minimum charge.

Denmark

IB Andresen Industri A/S, Industrivei 12-14, Langeskov, 5550, Denmark; +45 65 38 12 34, FAX +45 65 38 31 95, telex 59834
mng dir, Ib Andresen; sls mgr, Villi Skov; vp/mktg mgr, Ib Rasmussen; 1967
Subcontractor operating in all fields of sheet metal work. Coil departments specializing in slitting, blanking, and cut-to-length work. Sheet metal department offers cutting-in-size, bending, automatic punching, laser cutting, and automatic bending. Roll forming department can shape exactly to profile.

Covi Gravure, APS, Kastagervej 3, Herlev, 2730, Denmark; 42915161, FAX 42915169
1963
Industrial engraving in all kind of materials including TESA 6930 TAPE.

Denmarks Tekniske Hijskole, Procesteknisk Institut, Lyngby, DK-2800, Denmark
1963

Force Institutes, Park Alle 345, Broendby, 2605, Denmark; 45 4296 8800, FAX 45 4296 2636, telex 33388 svc dk
mng dir, Knud Rimmer; 1941

Orskovs Maskinfabrik A/S, Vangen, Frederikshavn, DK-9900, Denmark; 8-47-9033 *, telex 67178 ocborg dk
1941

Sorenco, Roedovreves 222, Roedovre, 2610, Denmark; 1-412244 *, telex 19382SOREMOK
emp 300, 1964
Provides custom-tailored laser applications.*

Finland

Aspa Oy, Asponkotu 4, Jarvenpaa, S-04400, Finland; (90) 280 122*, telex 122894 ASPO SF
1964

Jet Laser OY, Tehdaskylankatu 5, Riihimaki, 11710, Finland
1964

Lappeenranta University of Technology, Laser Processing Laboratory Div, PO Box 20, Lappeenranta, SF53851, Finland; +358 53 5711, FAX +358 53 5712499
hd of lab, Prof. Tapani Moisio; 1985
Laser processing laboratory conducts scientific research and provides training for students and engineers from industry on laser processing. It acts as a national laser center representing Finland in international long-range research programs and offers contract R&D for the industry.

SBA Interior, Mustio, 10360, Finland; 358 12 48201, FAX 358 12 48487, telex 1429 SBA
pres, Thor Johansson; mktg mgr, Ralf Hansson; prod mgr, Per-Olof Flythstrom; matl, Bjorn Siggberg; 1985
Offers sheet-metal products and laser cutting.

France

Alpha Laser, Z.A. Les Primeveres, Colombey Les Deuxeglis, F-52330, France; 25 01 51 94, FAX 25 01 53 10
1985

ARES, 1 rue Charles Cros, St. Lev La Foret, 95320, France; 030 40 88 32, FAX 30 40 89 76
pres, M. Mathieu; 1985

Armor, BP 51, Ploemeur, F-56270, France; 97 83 24 61, FAX 97 87 99 01
1985

Ateca, ZI Verlhaguet, Montauban, F-82000, France; (33) 6363 7897, FAX (33) 63 2007 96
1981
Study and realise elements and systems with high technology assembly: vacuum heat treatment, brasing under vacuum, induction brasing, laser soldering, cutting, drilling, vacuum, pressure, UV, x-ray controls. Military agreement RAQ2.

Ateliers Maitre, La Foret du Temple, Lourdoueix Saint Pierr, F-23360, France

Atim, 101 Rue Van Wyngene, Courtry, F-77181, France; (1) 60 08 22 54, FAX (1) 60 20 58 87
1981

Atom, 24 Rue Camille Didier, Charleville Mezieres, F-08000, France; 24 37 66 77, FAX 24 87 77 88
1981

ATS, 3 Allee Sarments, Cruissy, F-77183, France; (1) 60 17 80 23, FAX (1) 64 80 59 85
1981
Laser cutting capability.

Bauchet, 12 Blvd de la Zeme Division, Relhel, F-08300, France; 24 38 51 43
1981

Bellion, Z.I. du Cormier, Chollet, F-49300, France; 41 71 90 90, FAX 41 71 91 00
1981

Berard et fils, 12 Chemin des Basses Vallieres, Brignais, F-69530, France; 78 05 30 22, FAX 78 05 42 25
1981

Cacir Nicco, 72 Rue Jacquard, Lagny Sur Marne, F-77480, France; (1) 64 30 43 64
1981

Calfetmat, INSA, Batiment 403- INSA, Villeurbanne, F-69621, France; 72-43-83-04, FAX 72-43-85-28
J.M. Pellitier, A.B. Vannes, C. Vialle; 1985
Research center with goals as follows: study of various material processing using lasers (welding, cutting, heat treatments). Study of laser-material interactions. Three different activities are developed: fundamental and applied research, initial and continuous information for engineers, and technological transfer.

CBS Industrie SA, Rue Moines, Nemours, F-77140, France; (1) 64 28 02 34, FAX (1) 64 28 99 30
1985

CERCA, 9-11 Rue Georges Enesco, Creteil Cedex, F-94008, France; 43.77.12.63*, FAX 43.99.91.40, telex 213051 F
emp 220, 1957
Nuclear core components, plate type fuel elements, instrumentation, fuel handling, in-pool inspection and fuel survices equipment, and depleted uranium parts.

Cermip, 13 Rue de L'Industrie, Lagnieu, F-01150, France; 74 85 75 42, FAX 74 85 75 01
1957

CGM, Z.I. D'Igny, Igny, F-91430, France; (1) 69 41 13 01
1957

Cheval Freres SA, PO Box 3004, Besancon Cedex, F-25045, France; 81.53.75.33*, FAX 81.53.75.33, telex 361162 F
1848
Manufactures machining centers for welding, cutting, drilling, and marking following specifications of industrial customers. Laser powers are in the range of 100 to 600 W. Nd-YAG laser sources of 100 to 300-W Multifiber systems. Laser subcontracting department. A hard materials machining department (silicium, ruby, sapphire, ceramic, carbide).

Claux, Z.I. Beauregard, Brive La Gaillarde, F-19100, France; 55 86 90 07
1848

Collot Technologies, BP 1085, Laxou Cedex, F-54523, France; 83 97 20 10, FAX 83 98 70 75
1848

Comela, Z.I. St. Bel, L'Arbresle, F-69210, France; 74 26 41 55, FAX 74 26 44 49
1848

CTM, Rue Lavosier, Igny, F-91430, France; (1) 69 41 13 01
1848

Decoupe Laser, 11 Grande Rue, Granciere En Drouai, F-28500, France; 37 43 62 02, FAX 37 43 64 11
1848

Delcros, Echery, St. Marie Aux Mines, F-68160, France; 89 58 75 49, FAX 89 58 67 19
1848

Deville Sodery, 79 Rue Forest, Charleville Mezieres, F-08102, France
1848

Ecole Centrale, Grde Voie des Vignes, Chatenay Malabry, F-92295, France; 16.1.46.83.63.15, FAX (1) 46.83.64.37
prof, J.P. Longuemard; mgr, A. Tarrats; eng, F.X. De Contencin; 1986
Provides laser assisted machining, laser machining without cutting, control laser beam, waves material interaction, detection and generation of ultrasound by lasers, and process control.

Eitrac, 16 Rue Beal, St. Martin D'Heres, F-38000, France; 76 62 98 52, FAX 76 25 65 64
1986

Etablissement Technique Central de L'armement, Centre De Recherches Et D'Etude D'Arcueil (sub of Department of Defense), 16 Bis Ave, Prieur De La Cote D'or, Arcueil-Cedex, F-94114, France; (33) 1 42 31 95 71, FAX (33) 1 42 31 99 92
ch exec, Eng. D. Gerbet; resch unit, Dr. R. Fabbro, Dr. D. Kechemair; 1980
Research and development in the field of high energy lasers. Industrial applications including beam diagnostic and fundamental studies about laser-matter-interaction with CW CO_2 lasers with a power level of up to 25 kW and extended in 1990 to pulsed 1.06 um laser (2 x 80J; 25 ns; 0.1Hz).

Eurofer, BP211, Saneguernines, F-57202, France; 87 98 59 59
1980

Exico, 30 Rue Troyon, Sevres, F-92310, France; (1) 46 26 77 87
1980

Formatype, Z.I. les Forts Cherisy, Vernauillet, F-28500, France; 37 43 89 52
1980

FPC ZA, Qurat Petits Champs, Montelier, F-26120, France; 75 59 66 34, FAX 75 59 60 84
1980
Sheet metal shop with laser cutting capability.

Game Ingeniere, Laser Techniques, 135 Nue Pierre Pamond Caupian, Saine Medard en Jallos, F-33160, France; 56 05 86 60*, FAX 56 51 82 26, telex 580 201 F
emp 160, 1980

Heurchrome Laserama, Z.I. du Vert Galant Ave, Des Gros Cheveaux, St. Ouety L'Aumone, F-95310, France; (1) 34 40 36 36, FAX (1) 34 21 97 37
1980

Hexagone

Hexagone, Boulevard Cormier, Cholet, F-49300, France; 41 58 81 57
1980

ICT, Z.I. de Lavelle Sud, Sorgues, F-84700, France; 90 39 67 42, FAX 90 39 42 77
1980

Industria, 28 Avenue Clara, Le Plessis Trevise, F-94420, France; 45 76 53 78, FAX 45 76 47 44, telex 264849
mng dir, Patrick Dubois; comm dir, Alain Dehais; fin mgr, Daniel Michel; 1980

Industria, 24 Au Clara, Le Plessis Trevise, F-94420, France; (1) 45 76 53 78, FAX (1) 45 76 47 44
1980

Institut De Soudure, 32 Boulevard de la Chapelle, Paris, F-75000, France; (33) 1 42 03 94 05*, telex 210 335 F
emp 510, 1930

Institut Universitaire De Technologie, Calfa, Rue Du Moulin A Tabac, Bethune, F-162408, France; 21 57 60 80/ 2156 8460, FAX 21 68 49 57
1930
Research and development, studies of feasibility of cutting, drilling and heat treatment by continuous and pulsed high-power lasers.

Irepa, Critt Laser, 4 rue Hoelzel, Illkirch, F-67400, France; (00.33) 88.67.93.00, FAX 88.67.93.52
dir, Freneaux Olivier; 1982
Laser center for application and technology transfer. Feasibility studies on cutting, drilling, welding, and heat treatment by lasers. Technical counsel and aid to industry. High-level practical training of engineers, technicians, and researchers. R&D studies on lasers CO_2 50 W, 1.5 kW, 7.5 kW, and lasers YAG 400 W, and 1200 W.

Laboratorie National D'Essais, Metrologie et Caracterisation des Lasers, Z.A. de Trappes-Elancourt, 5 rue Enrico Fermi, Trappes, F-78190, France; 16 (1) 34 61 22 22*, FAX 34 61 22 22, telex 602 319 F
emp 450, 1982
Develops specific means for laser beams characterization, in particular for laser power and energy measurements; for CO_2 lasers performing high power measurements up to 50 kW.

LA Deertip, 20 Route Chateauneuf, Vernouillet, F-28500, France; 37 42 74 72
1982

LAP, 33 Route de Lyon, Illkirsh Graffenstaden, F-67400, France; 88 67 33 45, FAX 88 25 61 00
1982

Laser Alsace Production, Z.A. *Rue Girlenhirsch, Ilkirch Graffenstaden, F-67400, France; (33) 88 67 33 45, FAX (33) 88 67 48 25
1988
Laser job shop specializing in cutting (laser Bystronic 1500 W), welding, and heat treatment (laser Rofin Sinar 7.5 kW). Works in collaboration with a research group for new laser development.

Laser Bullard, 64 Route Gravin, Magland, F-74300, France; 50 34 74 10, FAX 50 90 21 79
1988

Laser Decoup, Rue D'Ableval Parc Industriel, Sarcelles, F-95200, France
1988

Laser 53, 31 Rue Soulae, Bazaugers, F-53170, France; 43 02 82 20, FAX 43 02 35 02
1988

Laser Industrie, Zone Industrielle B P 11, St.Marie Aux Chenes, F-57118, France; 87 61 80 39, FAX 87 61 96 73, telex 961 239
Robert Mauchoffe; emp 250, 1982
Laser job shop cutting, welding, drilling.

Laser Industrie Atlantic, 44 Rue Frederil Sauvage, Perigwy, F-17180, France; 46 44 60 85, FAX 46 45 26 11
1982

Laser Rhone-Alpes, 44 bis Rue Pre' St Didier, ZI Le Fontanil, Fontanil Cornil, F-38120, France; 76 56 07 57, FAX 76 56 10 38
1982

Laser Service France, Gorre, St Laurent Sur Gorre, F-87310, France; 55 00 00 30, FAX 55 00 09 99
1982

Laser Services LM, 17 Rue Edouard Nieuport, Suresnes, F-92150, France; 33 1 45066854*, FAX 33 1 47284548
emp 60, 1987
Job shop cutting plastics with CO_2 equipment, engraving woods & plastics.

Laser Techniques, St. Cyr, St. Laurent Sur Gorre, F-87310, France; 55 00 04 11, FAX 55 48 11 15
1987

Laser Top, 21 Rue Raymond Brosse, Villetaneuse, F-93430, France; (1) 48 27 59 30, FAX (1) 48 22 09 14
1987

Laser 2000, 48 Route de Bordeaux, Aixe Sur Vienne, F-87700, France; 55 70 31 31, FAX 55 70 37 39
1987

Laser 2005D, 17 Rue Gay Lussac, Gonesse, F-95550, France; (1) 39 87 25 66, FAX (1)39 85 20 08
1987

LA Tolerie Fine, Rue Papin, Illkirch Graffenstaden, F-67000, France; 88 67 30 20, FAX 88 67 44 77
1987

Levigne et Fils, 52 Ave de Lyon, Pont du Chateau, F-63430, France; 73 83 00 52, FAX 73 83 15 28
1987

LIDIS, Z.A. Les Forts, Cherisy, F-28500, France
1987

LTTA, 172 Rue Leon Jouhaux, BP 112, Sartrouville, F-78500, France; (1) 39 57 50 02, FAX (1) 39 57 07 70
mgr, Guy Aubertel; 1986

Lutral, Chemin Mancy, Luttange, F-57144, France; 82 83 95 11
1986

Malaval, Z.I. Rue Des Guerins, Le Coteau, F-42120, France; 77 70 44 65, FAX 77 71 23 75
1986

Masson, 11 Rue De L'Industrie, Bobigny, F-93000, France; (1) 18 45 57 04, FAX (1) 48 45 99 55
1986

Mecadex, 70 Rue St. Hippolyte BP 121, Scionzier Cedex, F-74954, France; 50 98 80 71, FAX 50 98 85 94
1986

Metal Mobil, Impasse Venteuil, Mas D'Agenay, F-47730, France; 53 89 50 34, FAX 53 89 59 41

Nollet, Z.I. Robaix Est, Lys Les Lannoy, F-59452, France
1986
Job shop with laser cutting of sheet metal.

Numerlaser, Z.I. Saint Clement, La Chapelle Basse Mer, F-44450, France; 40 06 33 01, FAX 40 06 31 54
1986

Octal, BP 29, Jassans Riothier, F-02480, France; 74 60 72 29
1986

OPN, 7 Allee Du Closeau, Noisy Le Grand, F-93160, France; (1) 45 92 18 18
1986

OTS, Z.I. Tirpety, Malestroit, F-56140, France; 97 75 12 99, FAX 97 75 20 91
1986

Ouest Tolerie, ZAC Aumaillerie, Selle en Luitre, Fougeres, F-35133, France; 99 99 71 06, FAX 99 99 26 69
1986

Oxymetal, BP 53, Bordeaux Cedex, F-33038, France; 56 49 37 46
1986

Photonics Systems Laboratory, National School of Physics Strasbourg (sub of Louis Pasteur Univ-Strasbourg), 7 rue de l'universite, Strasbourg, F-70000, France; 88 36 90 01, FAX 88 60 75 50
1986
Working in research under contract for photonics applications in laser, optical fiber, optical memory, optical metrology imaging with lasers and computers, optical communication, optical processors and computing, laser processing and laser processing systems, diffractive optics, head up displays, optical fibers networks. The laboratory creates new products or assists companies or any organization in developing new technologies in photonics. Activities are developed in the context of the French-German Institute for Application of Research.

Piroux, Z.A. Treffort, St. Etienne Du Bois, F-01370, France; 74 51 32 10, FAX 74 51 32 10
1986

Plastiques De Champagne, 19 Rue Marc Verdier, Pont Sainte Marie, F-10150, France; 25 81 14 98, FAX 25 80 39 71
1986

Prestoforme, 13 Rue Foflot, Jagney S/Bois, F-95850, France; (01) 34 71 13 30, FAX (1) 34 71 13 83
1986

Profils, Z.I. La Combe, Chaponost, F-69630, France; 78 56 53 38
1986

PUM, 110 Rue Courcelle, Reims, F-51100, France; 26 87 96 96, FAX 26 87 58 01
1986

Raimkeh CIMT, 40 Rue Jean Jaures, Marly, F-59000, France; 27 45 92 20
1986

RBD, Z.I., St. Pierre D'Albigwy, F-3250, France; 79 28 58 40, FAX 79 28 58 82
1986

Renoult, 29 Au Chardons, Pontault, F-77000, France; (1) 60 28 04 67, FAX (1) 60 28 05 50
1986

Reydes SA, 37 Reu Jen Baptiste Marquant, Gondecourt, F-59147, France; 20 90 92 50
1986

RYG, 5 Av 5 Mars, Argenteuil, F-95100, France; (1) 39 61 31 38, FAX (1) 39 61 63 03
1986

SACTI, 18 Rue Eugene Henaff BP 566, Venissieux, F-69636, France; 72 50 32 04, FAX 72 50 76 86
1986

Sadela SA, Route Nationale 'Le Flacher', Serrieres/Felines, F-07340, France; 75348631, FAX 75348590
1981

SCGI, 109 Route Couche, Le Breuil, F-71670, France; 85 56 09 81, FAX 85 56 29 63
1981

Secatol, Route Liguge, Saint Benois, F-86280, France; 19 57 25 20, FAX 49 55 42 28
1981

Semtp-Zahra, 27 Rue du Dr. Robert Colle, BP 01, Bubres Sur Yvette, F-91440, France; (1) 69 07 57 81, FAX (1) 69 28 77 87
1981

SER, 36 Quai Seine, Sanint Ouen, F-93000, France; (1) 40 11 14 57, FAX (1) 40 10 09 23
1981

Sermeto, BP 35, Cussey, F-03301, France; 70 31 76 88, FAX 70 31 83 28
1981

Serrieres, Les Ambruises, Changs Sur Marne, F-77660, France; 64 35 92 53, FAX 64 35 70 81
1981

SIATA, 12 bis rue du Dr. Roux, Choisey le roi, F-94600, France; (1) 46 82 21 56, FAX (1) 45 73 02 85
1981

Socamcaser, 40 Chemin de Chantelle, Toulouse, F-31200, France; 61 47 38 02
1981

Socomec, Z.I. Route de Versillat, La Souterraine, F-23300, France; 55 89 49 49, FAX 55 89 49 89
1981

Sodery SA, 19 Ave de Montcy Notre Dame, Charleville, Mezieres, F-08000, France; 24.33.91.47 *, telex 84001GF CHAMCO
emp 600, 1981
Laser cutting of steel up to 10 mm, hardened steel up to 8 mm, stainless steel up to 8 mm, aluminum up to 8 mm, copper up to 1.5 mm, and brass up to 1.5 mm.

SOTIAG, Z.I. La Gaite, Amplepuis, F-69550, France; 74 89 45 93
1981

Specitec, Laser Machining & Welding Div, 3 & 5, rue de Nice, Paris, F-75011, France; 33 (1) 43 70 50 50 *, FAX 33 (1) 43-704944, telex 218 925 F
emp 350, 1965
Offers laser welding, cutting, boring, and development of new products or processes and feasibility studies and tests. Manufactures prototypes in small and medium size-series.

STECYR

STECYR, Z.I. de Vilsain, Route d'Orleans, Chateaudun, F-28200, France; 37 45 40 50
1965

Steec, ZA Les Andres, BP4, Brindas, 69126, France; 78-45-14-63, FAX 78-45-40-77
1980
Offers subcontracting for micro drilling, cutting, and welding of metals and nonmetals.

STILL, 97 Route De Comeilles, Sartrouville, F-78500, France; (1) 39 14 29 92, FAX (1) 39 57 06 62
1980

STIMY, 13 Rue de la Marne, Paulse, F-44270, France; 40 26 02 36
1980

Systrel, 9 Av du Hoggar, Z.A. De Courtasoeuf, Lesulis, F-91940, France; (1) 69 97 80 39
1980

TCIN, Z.I., Bourguebus, F-14540, France; 31 23 13 61, FAX 31 23 79 94
1980

Techmeta, Mets Tessy, Annecy, F-74370, France; (33) 50 27 20 90, FAX (33) 50 27 33 18
1980
Job shop in laser (YAG) and electron beam applications: welding, heat treatment, drilling, and cutting (laser). Electron beam welding machine manufacturer.

Techno Filtres, 8 Grande Rue, Bueil, F-27730, France; 32 336 58 26, FAX 32 36 54 27
1980

Universite De Bretagne Occidentale, L.E.L.T., Guidel-Plages, Guidel, F-56520, France; 97 05 92 78
dir, H.P. LeBodo, PhD; 1981
Main fields of research include determination of laser emission characteristics; design and development of accurate calorimeters for medical, industrial, and military applications; control of radiation characteristics of lasers used in medical and industrial centers; determination of the safety level of laser plants; determination of thermal properties of materials by the laser flash technique; research on welding, machining, and thermal processing by laser; studies developed under governmental contracts; and technical collaboration with manufacturers.

Usines Chausson, BP 236, Asnieres Cedex, F-92601, France; (1) 47 91 76 42, FAX (1) 47 83 52 66
1981

Germany

Aclas Laseranwenddungen GmbH, Metzgerstrasse 2, Aachen, D-5100, Fed Rep Germany; 0241-161480, FAX 0241-167120
1981

Alfa Lasersystem GmbH, Peutestrasse 51, 2000 Hamburg 26, Germany; 49 (0)40 789 83 87, FAX 49 (0)40 789 25 33
Bernd Riemann; 1990
Offers standard cutting installations from 500x500 mm up to 2000x3000 mm for cutting acrylics, wood, metals, glass, etc. Laser installations according to client' s specifications. Also, application research and contract work in acrylics, wood, glass, etc.

Alfa Lasertechnik GmbH, Pfingstrasse 12, Braunschweig, D-3300, Germany; 49 (0)531-89 51 67, FAX 49 (0)531 89 39 38
K.H. Heine, S. Schubert; 1989
Offers laser welding installations according to client' s specifications. Application research and contract work, feasibility studies for welding and cutting of metals with max 12kW output power. 3 dimensional workpiece treatment for max size 3000x3000x2000 mm.

Alfred Gnida GmbH & Co, Postfach 1163, Bad Waldsee, D7967, Germany; 07524 40070, FAX 07524-400727
1989

Alfred Hermann GmbH & Co KG, Carl Zeiss Strasse 43-45, Schorndorf-Weiler, D-7060, Germany; FAX 07181 4 37 49
1989

Alfred Rexroth GmbH & Co, Motzener Strasse 29, Berlin 48, D-1000, Germany; 030 721 20 97, FAX 030 721 20 90
1989

Alfred Rexroth Maschinenbau GmbH, Bahnofstrasse 11, Rhinow, D-188331, Germany
1989

ALL Applikationslabor fur Lasertechnik, Westendstrasse 123, Munchen 2, D-8000, Germany; 089-716033*, telex 5215 546 all d
emp 140, 1977
Drilling and scribing of ceramic substrates for thick- and thin-film technology; laser cutting, welding, and drilling of metals and nonmetals; construction of laser accessories and computer-aided laser beam diagnostic systems, and laser penetration monitor.

Anderssen GmbH & Co KG, Untere Neckarstrasse 1, Neckarsulm, D-7107, Germany; 07132 327-0, FAX 0732 328-49
1977

Applikationslabor Fur Lasertechnik, Westendstrasse 123, Munich 2, D-8000, Germany; (49) (89) 507080, FAX (49) (89) 507010
ch exec, Dr. Peter Arnold; sls mgr, Rainer Weiss; mktg mgr, Michael Zimmermann; eng mgr, Harold Stegbauer; 1977
Drilling and scribing of ceramic substrates for thick- and thin-film technology; laser cutting, welding, and drilling of metals and nonmetals; construction of laser accessories and computer-aided laser beam diagnostic systems, and laser penetration monitor.

Arnold GmbH & Co, Industriestr 6, Friedrichsdorf/Taunus, D-6382, Germany; 06172 50 26*, FAX 06172 7 97 80
1977

Autz & Herrmann, Carl Benz Strasse 10-12, Heidelberg 1, D-6900, Germany; 62-21-50-60*
1977

Auwarter Schider Stempel, Helgdlandstrasse 34, Stuttgart 40, D-7000, Germany; 0711 82 40 98
1977

B&B Metall und Kunststoffverarbertung GmbH, Bochumer Str 12, Wuppertal 2, D-6000, Germany; 0202 1662988*, FAX 02021 663724
1977
Laser processing shop.

Beiersdorf AG, Unnastrasse 48, Hamburg 20, D-2000, Germany; 040-5690, FAX 040-569 34 34
1977

Bernhard Lasertechnik, Ludwig-Thoma-Str.5, Bad Aibling, D-8202, Germany; 08061/6064*, telex 051933521 dmbox g/box
emp 50, 1986
Laser job shop.

BIAS Forschungs und Entwicklungs Labor, fur Angewandte Strabbtechnik, Klagenfurter Strasse 2, Bremen 33, D-2800, Germany; (49)421/21801, FAX 0421/2185063
dir, Prof. Dr. Sepold; dir, Prof. Dr. Juptner; 1977
CO_2, YAG, and excimer-laser applications and process development laboratory. Holographic interferometry, vision inspection, and quality inspection techniques.

Buchwald GmbH, Mausegatt 44, Bochum-Wattenscheid, D-4630, Germany; 02327 8 98 89, FAX 92327 616 14
1977

Carl Baasel Lasertechnik, Petersbrunnerstr 1B, Starnberg, D-8130, Germany
1977

CEFRA Interanational Handels, Ungererstrasse 40, Munchen 40, D-8000, Germany; 089 38 10 09-0, FAX 089 38 1009-44
1977

Contura Halbzeuge Laserschneiden GmbH, Nikolaus-Otto Strasse 1, Hilden/Rhld, D-4010, Germany; (02103) 5982, FAX (02103) 52208
1977

DFVLR, Institut Fur Technische Physik, Pfaffenwaldring 38-40, Stuttgart 80, D-8000, Germany; 0711-6862-770, FAX 0711/6862-788, telex 7255689 (dfvsd)
Dr. rer. nat. H. Opower; 1977
R&D of high-power gas lasers. Material processing with lasers (cutting, drilling, welding, surface treatment). Process development for laser material processing.

Dienes Werk GmbH & Co KG, Kolner StraBe, Postfach 13 20, Overath Vilkerath, D-5063, Germany; 022206 605-0, FAX 02206 605-111
1977

Doll, Postfach 27, Sasbach, D-7591, Germany
1977

Dorries Scharmann GmbH, Laser Technique Div, Dorriesstrasse 2, Mechernich, D-5353, Germany; 02484 12 123, FAX 02484 41 40, telex 833619 DOEWE D
Rudiger Rothe, Ralf Louis, Adolf Starkens, Joseph Lehle; emp 1100, 1977
Machine tool manufacturer utilizing lasers integrated with products and services: laser systems for material processing (PROLAS), complete industrial systems, job shop in laser material processing (welding, surface treatment, cutting), engineering R&D laser material processing.

EGO, Rohr-Heizkorper GmbH, Flehinger Strabe, Oberderdinggen, D-7519, Germany
1977

Elcede GmbH, Otto Hahn Strasse 7, Kirchheim, D-7312, Germany
1971
Dieboard cutting.

Emaillierwerk Hannover Haselbacher GmbH & Co, Postfach 100252, Andreas-Haselbacherstr 47-49, Wedemark/Mellendorf 1, D-3002, Germany; 51-(30) 40066*
1977

EPP Electronic Production Partners GmbH, Marschnerstr 93, Munchen 60, D-8000, Germany; 089 834 50 24, FAX 089 834 98 10
pres, Henry Stenger; 1987
Laser trimmers and markers, laser lamps and spares, fiber test systems, fiber splicers, laser diodes.

ESAB Held GmbH, Industriestrasse 26, Heusenstamm, D-6056, Germany; 06104 6431, FAX 06104 62221
1987

Eugen Kotter Metaly Oberflachen GmbH, Carl Zeiss Strass 26, Mainz-Hechtsheim, D-6500, Germany; 06131 959 95 7, FAX 06131 959 95 77
1987

Faust Sonderbearbeitungen GmbH, Karlstruherstrasse 7, Leinfelden-Echterdingen, D-7022, Germany; 0711 79 60 21*
1987

Feinwerktechnik Hago GmbH, Industriegebeil, Kassaberg I, Kadelburg, D-7897, Germany; 77-(41) 3551*
1987

Fink Kunststoffverarbeitung, Reidweg 57, Ulm-Soflingen, D-7900, Germany; 0731-381501, FAX 0731 38 15 06
1987

Weihbrecht GmbH, Daimlerstr. 40, Schwabisch-Hall, D-7170, Germany; (0791) 51002, FAX 0791 51058
1987
Laser cutting job shop.

Fraunhofer-Institute for Produktion Technology, Steinbachstrasse 17, Aachen, D-5100, Germany; 0241/8904-0, FAX 0241/8904-198, telex 8329411 IPT
prof di-ing, Dr. h.c.W. Konig; 1980
Main fields of research in laser material processing are: surface treatment (transformation hardening, remelting, alloying, cladding, dispersing); cutting of fiber-reinforced plastics; process control and monitoring; layout and assessment of laser machine tools; and planning and organization of laser machining processes in manufacturing sequences.

Fraunhofer-Institut fur Lasertechnik, Steinbachstrasse 15, Aachen, D-5100, Germany; 0241/8906-0, FAX 0241/8906-121
dir, Prof.Dr.-Ing. G.Herziger; 1985
Offers research & development in material processing by laser beams; flexible manufacturing; development and optimization of laser sources, laser components, beam shaping and guiding systems; development, installation, test and control of laser processing devices; laser diagnostics; laser measurement and test techniques; laser analysis; pulsed plasma devices; training; laser safety and security.

Friedrich Alexander Universitat, Forschungsverbund Lasertechnologie Erlangen, Martensstrasse 5, Erlangen, D-8520, Germany; 0049/9131/85 75 20, FAX 0049/9131/36403
spksmn, Prof. Dr.-Ing. M. Geiger; mgr appls lab, Dr. A. Tinschmann; 1987
FLE is a combination of 4 institutes at the Univ. of Erlangen working in the field of laser material processing and laser development. FLE Applications Lab offers a wide variety of laser systems and diagnostic tools for development in the field of laser material processing.

GEAT Gesellschaft fur angewandte Technologie mbH, Bahnhofsplatz 6, Nurnberg 70, D-8500, Germany; 911/947 2080, FAX 911/223456
gen mgr, Dipl.-Ing H. Burckhardt; prog mgr, Dr. F. Ernst; sls/app mgr, Dr. R. Nuss; 1985
Supplies UTIL CO_2 lasers within a power range of 6 kW to 45 kW. Main production uses are welding, surface finishing (hardening), surface melting in automotive and other metal working industries. Offers complete turnkey solutions and provides process development in local application laboratory.

Gebr. Staiger GmbH, Postfach 1432, Georsen-Schwarzwald, D-7742, Germany; 7724/ 60 44
1985

Gerhard Werhlbrecht

Gerhard Werhlbrecht, Daimlerstrasse 40, Schwabisch Hall, D-7170, Germany; 79 15 10 57*, telex 74 828 LASMA-D
1985

Gotz & Sohn GmbH, Schwarzbachstrasse 11, Aglasterhausen, D-6955, Germany; 06262 470, FAX 06262 4769
1985

Grosse Lasertechnic, Gutersloher Str. 44, Steinhagen Brockhagen, D-4803, Germany; 05204 20 54
1985

Gunter Konig, Pilgerspfad 8, Wittighausen, D-6978, Germany; 09347 12 87, FAX 09347 7 17
1985

Gunther Frey GmbH & Co KG, Walkenrieder Strasse 19, Berlin 47, D-1000, Germany; 030/6257078, FAX 030/6257077
1925
Pinholes for laser application systems.

Harr GmbH & Company, Mahdentalstrasse 84, Sindelfingen, D-7032, Germany; 70-31-694-0, FAX 07031-69440, telex 7-265-857
1925

Hassink GmbH Maschinew, Am Uffelnor Moor 7, Ibbenburen 4, D-4530, Germany; 05459 1212, FAX 05459 1217
1925

Heinz Dreeskornfeld KG, Erpestrasse 53, Bielefeld 14, D-4800, Germany; (0521) 48486, FAX (0521) 489172
1925

G. Heise Laser Bandstahlschnitte GmbH, Gausstrasse 26, Kirchheim, D-7312, Germany; 70 21 5 41 11
1925

HEKO Electronik GmbH, Gutenbergstr. 55, Postfach 1228, Lilienthal/Bremen, D-2804, Germany; 04298/1069, FAX 04298/5569
gen mgr, W. Koch; 1978

Heribert Lehner, Assbrook 4-6, Wiemersdorf, D-2351, Germany; (041-92) 50070, FAX (04192) 500711, telex 2180146
1978
Manufactures and distributes diode lasers; micro positioning components; optical components; fiberoptic components; CCD-line and array cameras; acousto-optic devices; LCD-displays; special semiconductor components.

Hoedtke & Boes GmbH, Industriestrasse 2-6, Pinneberg, D-2080, Germany; 04101 7983
1978

Hoesch Industrielaser GmbH, Reinhard-Heraeus-Ring 21, Kleinostheim, D-8752, Germany; 6027/6061*, FAX 6027/8544
1988
Job shop for welding, cutting, and surface treatment applications. Operations research, consulting and planning.

Holzrichter GmbH, Schonebecker Platz 11, Wuppertal 2, D-5600, Germany; 0202 562 0, FAX 0202 562-278
1988

Huber Lasertechnik, Belchenweg 5, Ehnigen, D-7044, Germany
1988

Innovat Laseranwendungs GmbH, Viktoria Strasse 28, Minden, D-4950, Germany; 0571 36607, FAX 0571 36678
1988

Institut fur Werkstoffkunde, Agricolastrasse 2, Clausthal-Zellerfeld, D-3392, Germany; (05323) 722120*
emp 400, 1978
Development of use of lasers in materials working- cutting, welding, and in particular, heat treatment. *

Interatom GmbH, Instruments Div, Friedrich-Ebert-Strasse, Bergisch Gladbach 1, D-5060, Germany; (0) 2204/84-2300, FAX (0) 2204/84-3045
1978
Provides accelerator and magnet technology, synchrotron radiation sources, linear accelerators, and free electron lasers.

Johann Schurhulz KG, Albaumerstrasse 32, Kirchhunden, D-5942, Germany; 2723/20 33, FAX 2723/20 34 18
1978

Joseph Rabb GmbH, Postfach 2261, Neuweid 1, D-5450, Germany; (02631) 509-0
1978

Karl Keppler Werkzeugbau, Wittkullerstrasse 175-177, Solingen, D-5650, Germany; 0212 214377, FAX 0212 31 48 85
1978

Karl Schutz Stahl & Maschinenbau GmbH, Hofarten 17, Schlittberg, D-8896, Germany; 08259 1027, FAX 08259 1029
1978

Kaygasse Nr O Laserschneidtechnik, Kaygasse Nr O, Koln 1, D 5000, Germany; 0221 23 45 18
1978

HP Kaysser, Industriestr 4, Leutenbach 3, Stuttgart, D-7057, Germany; 07195/88-0, FAX 07105/18830
Thomas Kaysses; 1947
Constructions in sheet-metals (aluminum, stainless, normal, and nonmetals) for industry.

KHI, Klauenfusse 3, Neckarsulm, D-7101, Germany
1947

Lacura Laser Cut W. Schoenenbon GmbH, Leimhholer Muhle 1-5, Radevomwald, D-5608, Germany; 21 95 70 19*
1947

Lascript GmbH, Sebald Kopp Strasse 5, Foreheim, D-8550, Germany; 09191 28 49
1947

Lascript GmbH, Oppenheimer Strasse 20, Gross-Geram, D-6080, Germany; 06152 7727
1947

Laser Laboratorium Gottingen e.V, Im Hassel 21, Gottingen, D-3400, Germany; 551-50350, FAX 551-503599
pres, Dr. H. Gerhardt; vp r&d, Dr. K. Mann; 1987
Provides R&D service in the field of excimer and dye lasers, e.g. damage testing facility, material processing, micro machining, combustion diagnostics. Manufacturers femtosecond excimer lasers, laser beam profilers esp. UV, fiberoptic beam-delivery system for UV.

Laser Materialbearbeitung GmbH, Gartenstrasse 22, Worth, D-8767, Germany; 09372/5008, FAX 09372/72808
1989

LaserProdukt GmbH, Brunker Stieg 8, Alfeld, D-3220, Germany; 0049-5181-26347, FAX 0049-5181-25020
Dipl.-Ing. W. Hinrichs, Dr.-Ing. J. Balbach; 1985
Job shop services. Consulting in the use of CO_2 and Nd:YAG lasers in manufacturing. Software production: NC programming systems for CNC laser machines.

Laserschneidtechnik GmbH, OttoHahn Strasse 12, Houelfof, D4799, Germany; 05257/08580, FAX 05257/6362
1985

Laser Walzen Center, Essener Starsse 259, Oberhausen, D4200, Germany; 0208/889293
1985

Laser Zentrum Hannover e.V., Hollath-Allee 8, Hannover 21, D-3000, Germany; +49-511-3563-290, FAX +49-511-3563-299
H.K. Tonshoff, H. Welling, H. Haferkamp; 1986
Offers R&D activities and consulting covering material processing technology with lasers (Excimer, Nd:YAG, and CO_2), controlling, laser measurement, planning and organization of laser processes, investment planning, CAD/CAM, laser safety, laser source development and coatings for optical devices; training and education of experts.

LPKF CAD/CAM Systeme GmbH, Scheffelstr. 17, Hannover, D-3000, Germany; 511 708 390, FAX 511 717277
dipl ing Jorg Kickelhain
Multifunction laser system for prototype production of printed circuit boards and hybrid circuits in fine line technology and machining of micromaterials. High speed coordinate tables.

Lusebrink & Teuber, Ziegelstrasse 46, Plettenberg, D5970, Germany; 02391/164, FAX 02391/10708

Manfred Toller GmbH, Wilhelm Suhr Strasse 7, Hagenburg, D-3055, Germany; 05033-7024, FAX 05033-7311
mgmt, M. Toller; 1946
Laser job shop for cutting with CO_2 lasers from 50 - 1900 W. Tables up to 2.1 x 3.1 mm (82 x 122 in.). Specialist in laser or waterjet cutting of letters and signs from acrylic or other materials.

MAN Technologie AG, Bauschinger str. 20, Postfach 500426, Munchen, 50, D-8000, Germany; FAX 89-1-507168, telex 523211-21 ma d
dirs, Dr.-Ing W. Amende; emp 1280, 1980
Application development with high power lasers. Job shop for laser applications. Engineering and consulting.

Martin Fink KG, Riedweg 57, Ulm, D7900, Germany; 0731/381501, FAX 0781/381506
1980

MeKo, Am Teinkamp 1, Sarstedt/Hannover, D-3203, Germany; +49 5066 4035, FAX +49 5066 4036
Clemens Meyer-Kobbe; 1991
Laser job shop for drilling, cutting, welding and surface treatment of metals, hard metals and ceramics. Speciality: precision material processing. Sales and service of VORTEK high-power arc lamps. Surface treatment with lamps: surface layer hardening and remelting.

Metafot GmbH, Karl Bamler Strasse 40, POB 22 01 65, Wuppertal, D-5600, Germany; 0202/6090, FAX 0202/6090080, telex 8592293 metya d
Hr. Dr. R. Florian; 1969
Provides R&D based on wide knowledge. Offers comprehensive in-house know-how in fields of etching, electroforming, lasercutting, and watercutting techniques. Experience means that every customer can be sure of obtaining the very highest precision even with the most complex metal shaped parts.

M&L GmbH, Zettachring 8A, Stuttgart 80, D-7000, Germany; (0700) 72-87440, FAX (0711) 72-87441
1969

Mundinus & Buss Lasertechnik, Willelm Fressel Str 6, Luneburg, D-2120, Germany; 41-31-89030, FAX 4131-8903-33
1985
Laser job shop.

N.U. - Tech GmbH, Ilsahl 5, Neumunster, D-2350, Germany; 04321 3 06 20 *, FAX 04321 3 06 31
emp 200, 1985
R&D institute in the fields of laser technology, optics, and surface technology. Offers contract R&D feasibility studies, and job shop work. Specialist in precise and fine-mechanics and optics. Develops optoelectronic-mechanical prototypes.

PS Laseranwendungstechnik GmbH, Bahnhofstr 56, Thedinghausen, D-2819, Germany; 04204-7604, FAX 04204-1627
1988
Job shop for cutting all materials, and welding. Offers planning and engineering of laser systems. Representative for NVL Balliu Lasersystems.

Roggenbuck KG, Dortmund, D4600, Germany; 0231/170026, FAX 0231/179709
1988

RWTH Aachen ISF, Pontstasse 49, Aachen, D-5100, Germany; 241-803871/3872, FAX 241-806268
1988

Schilder-Bischiff Lichtwerbung GmbH, Roermonder Str 145, Aachen, D5100, Germany; 0241/8 40 84
1988

Schurholz KG, Industriestrasse 9, Plettenberg, D-5970, Germany; 23-91-8-10-30 *
1988

Schurholz/VS, Albaumer Strasse 35, Kirchhundem, D-5942, Germany; 2723 12033, FAX 2723 3222
H. W. Reichling; dir, Hans Otto Schurholz; tech mgr, Heinz Otto Lenchter; 1928

Schweiko Indust GmbH, Stahlschmidtsbrucke, Huckeswagen, D-5609, Germany; 21-92-4175 *
1928

Spektrum Laser Entwicklungs und Vertriebs GmbH, Mehringdamm 33, Berlin, D-W1000, Germany; (49-30) 69 00 88-0, FAX (49-30) 69 00 88-10
pres, Dr. Frank Massmann; pres, Dr. Manfred Voss; 1981
Manufactures solid-state laser systems for industry, medicine and research; mainly Nd:YAG (CW, Q-switched, mode-locked), Er:YAG and Alexandrit. Distributor of CCD cameras.

Technische Universitat Clausthal, Institut fur Werkstoffkunde und Weskstofftechnik, Adolf Romer Strasse 2 A, Agricolastr. 6, Clausthal-Zellerfeld, D-3392, Germany; 5323-722120, FAX 05323-723148
Prof. B.L. Mordike, Dr.-Ing. D. Burchards; 1978
Offers laser treatment of metals and ceramics with CO_2 and Nd:YAG lasers. Also process development, application and material research.

Tepro-Prazionstechnik GmbH, Obere Muhlenstrasse 66, Monchweiler, D-7733, Germany; 77 21 7 30 91, FAX 07721/70867
1978

Trenntechnik Ing Beyermann GmbH, Muhlstrasse 18, Michelfeld, D-7171, Germany
1978

University of Stuttgart

University of Stuttgart, Institut fur Strahlwerkzeuge, Pfaffenwaldring 43, Stuttgart 80, D-7000, Germany; (711) 685-6841, FAX (711) 685-6842
dir, Prof.Dr.-Ing. H. Hugel; head staff, Dr.rer.nat. F. Dausinger, Dr.rer.nat. A. Giesen, Dipl.Ing. P. Berger; emp 390, 1986
Develops high-power lasers, and designs optical systems. Qualification of optical elements, and R&D laser materials processing applications.

Wagner Lasertechnik GmbH, Postfach 1058, Dauchingen, D-7735, Germany
1986

Wolfgang Deutscher, Rigaerstrasse 2115, Munchen 50, D8000, Germany; 089 1402637, FAX 0891414367
1986

Yagi's Job Shop Remy Renaud, Inden Eichen 7, Hille 1, D-4955, Germany; 57 03 32 70 *
1986

Ireland

Ellickson Laser Cutting Ltd, Kilmurry, Waterford, Ireland; 353 51-32316, FAX 353 51 32907
chmn, D.J. Ellickson; joint mng dirs, John Drewnan, Michael C. Finnegan; 1990
Laser cutting, welding, marking, and fabrication shop. Laser scribing and drilling.

Israel

Palbam Metal Works, En-Harod Ihud, 18960, Israel; 06-531703, FAX 972-65-31904
gen mgr, N. Hen; prd mgr, I. Nur; chf eng, C. Geyari; 1955
Offers laser cutting of stainless steel, carbon steel, armor plate, composite materials, plastics and plywood. Uses fully computerized CNC system including CAS/CIM and quality control to MIL-I45208 A and ISO-9002.

Italy

MECC 2000 SNC, Via Canele 96, Villalunga Casalgrande, I42013, Italy; 0522/849893, FAX 0522/819891
1955

ADB, Via G. DiVittorio 8, Mazzano (Brescia), I25080, Italy; 030/2591831, FAX 030/2591221
1955

Berto Lamet, Strada Del Portone 18, Torino, 10137, Italy; 011/30021
1955

Bonomi, Via Sempione 5, Vergiate (Varese), I21029, Italy; 0331/946 251, FAX 0331/947 545
1955

Bordogna, via Europea 37, Palazzolo Sull'Oglio, I25036, Italy; 030/7300261, FAX 030/7301760
1955

Bruscoli Marcello, Via Delle Tenaglie 360, Serrungarina (Pesaro), I61030, Italy
1955

Bussetti Officine, Strada Del Molino 25, Moncalieri (TO), I10024, Italy
1955

Car Mec, Via Di Vittorio, Malnate (Varese), I21046, Italy
1955

Cassetto, Via Provinciale I, Lugnacco (TO), I10080, Italy; 0125/789088, FAX 0125/789200
1955

CBI Italia SpA, Via Della Taccona 77, Monza, I-20052, Italy; 039 741741, FAX 039-737125, telex 331645 CBIIT I
pres, Corrado Maveri; eng mgr, Giuseppe Lenardon; emp 100, 1962
Sheet-metal shop with laser cutting capability of surfaces up to 1500 x 3000 mm with 1000-W and 1500-W equipment.

CEM, Via Carruaccio 5, Magnago (Milano), I20040, Italy
1962

Centro Laser-Soc Cons a r.l., Str. Prov. per Casamassima Km3, Valenzano (Bari), I-70010, Italy; 080-8774314*, FAX 080-8774457
emp 130, 1979
Laser material processing, R&D laser industrial application, laser photolytography, feasibility studies, laser photobiology, laser and processing consultants, laser safety consultant, laser material processing short courses, and surface testing laboratory services (thermometry, colorimetry, reflectrometry, profilometry).

CIM, Via Toscana 29/31, Vignate (Milano), I20060, Italy; 02/9566266, FAX 02/9560160
1979

Cipiemme, Via Cardorna 30, Carygo (Como), I22060, Italy; 031/749455, FAX 031/745162
1979

CISE, Materials & Technologies Div, Casella Postale 12081, Milano, I-20134, Italy; (02) 2167.1, FAX (02) 2167.2620, telex 311643 CISE I
dir, A. Rota; 1946
A research company developing innovation technologies and transferring them to industry. Performs technological services and produces equipment and instrumentation. About 40 people are involved at present in the development of Nd:YAG laser source prototypes, CO_2 laser source prototypes, material processing, and laser systems.

CMB Srl, Via Candia 26, Barone Canauese (TO), I10010, Italy; 011/9898095-62, FAX 011/989806
1946

CMS SpA, Via Strada Nuova Consortile, Fisciano (Salerno), I84084, Italy; 089/879233, FAX 089/890992
1946

Compel, Via Donizzetti, Cornate D'Adda, I20040, Italy; 039/6926555, FAX 039/6927060
1946

Cosmetal, Zona Ind. E. Mattei, Recanati (Marcerta), I62019, Italy; 071/7570426, FAX 071/7570083
1946

CREAS, Via Marsala 6, Nichelino (TO), I10042, Italy; 011/621959, FAX 011/6272104
1946

Derby, Via Kennedy, Carauaggio (Bergano), I24034, Italy; 0363/52003
1946

Elettronica MI AR, Via Acquasanta 15, Salerno, I84100, Italy
1946

ETRA, Via Fermi 11, Nova Milanese (Milano), I200251, Italy
1946

Eurofar, Via A Ciucani 17, Ornago (Milano), I202060, Italy; 039/6010137
1946

Eurosteel, Via Crema 9, Izano (Cremona), I26010, Italy
1946

Fancareggi, Via B. da Verignana 5, Bologna, I40100, Italy
1946

Fedele 82, Via Del Bruzi 14, Roma, I00100, Italy
1946

Fila, Via Tenochio 48, Orgiano (Vicenza), I36040, Italy; 044/874191
1946

Fratelli Giusti, Provinciale Nord 32, Novellara(Reggio Emila, I42017, Italy; 0522/654219
1946

Fratelli Rossi, Via Della Technica 54, S. Lazzare Di Salena, I40068, Italy
1946

Fustellifico Aiolfi, Via Guzzasete 77, Carauaggio (Bergano), I24043, Italy
1946

Fustellifico Veronese, Via Porto S. Michele 15C, Verona, I37133, Italy
1946

Fustezbox, Via I Maggio 7/11, Pontecchio Di Sasso, Marconi (BO), Italy
1946

General Kamet, Via Del Capitanei 14/16, Agaate Brianzh (MO), I20041, Italy
1946

Gregis Alessandro, Via Giustinelli 9, Alzano Lombardo, I24020, Italy; 035/512001, FAX 035/510105
1946

Hebert, Via Provinciale Nord 10, Novellara, I42017, Italy
1946

Hydee, Corso Susa 18, Bussolfno (TO), I10053, Italy
1946

ICOS, Via Curiel 4, Zola Predosh (BO), I40069, Italy; 051/752325-406
1946

IRTM, Regione Lime, Vico Canavese (TO), I-10080, Italy; 0125/74362, FAX 0125/72255
1946

Isollima, Via Romagnoli 8, Bentivoglio (BO), I40010, Italy
1946

Italdesign, Via Grandi II, Moncalieri (TO), I10024, Italy; 011/6470219
1946

ITCA, Via D. Vittorio 22, Grugliasco (TO), I10095, Italy; 011/70941, FAX 011/7094693
1946

Lamier, Via Baracca 4, LaLoggia (TO), I10040, Italy; 011/9629485, FAX 011/9627259
1946

LAMP Srl, Stroda Valminiar 17, Villanova D'Asti, I14019, Italy; 0141/916614, FAX 0141/946645
1946

Laser Metalworking Srl, Via Tiziano 20, Cambiago(Milano), I14019, Italy; 02/95308132, FAX 02/95067398
1946

Laser Optronic Srl, Via Iglesias 30A, Milano, I20128, Italy; 02/27000435, FAX 02/2570171
1946

Laser Roma, Via Nettunese 91, Anzio (Roma), I00042, Italy; 06/9875990, FAX 06/9875991
1946

Laser Services, Via Cadorna 78, Cesena C Forli, I47023, Italy
1946

LEM Laser, Via Adriatica 170, Basaldella Formido, I33040, Italy
1946

LEM Laser, Via Cadorna 44, Salt Di Povoletto, I33040, Italy
1946

LIVA, Via A. Vespucci, Fiume Veneto (PN), I33080, Italy; 0434/959102, FAX 0434/957712
1946

LM Laser Metalworking srl, Via Tiziano 20, Cambiago (Milano), I-20040, Italy; 39-2-95308132, FAX 39-2-95067398
gen mgr, Albino O. Fiorini; 1990
Laser processing job shop offering CO_2 laser welding in the range of 1-4 mm thick joints. Offers job-shop service, feasibility studies on laser processing and pilot production.

Lowara, Via Lowara, Montecchio Maggiore, I-30675, Italy
1990

Meccanica DS, Via Dell Artigiano 8, Sanlazzaro D Savena, I40068, Italy; 051/463090
1984

Meccaniche Marcon, Via De Gasperi 41, Villaraspa (Vicenza), I36060, Italy
1990

Metalplasma Srl, Via Degli Alberi 19, Galliera Veneta, Padova, I35015, Italy; 049/5968146, FAX 049/5968880
1990

Michiper, Via Einaudi 3, Robassomero (TO), I10020, Italy
1990

Model Master, Via Uittime di Piazza, Fontana 38, Moncalieri (TO), I10024, Italy; 011/6811958, FAX 011/6171267
1990

Modensider, Via Dei Tornitori 30, Modena (MO), I41100, Italy; 059/360236, FAX 059/280361
1990

Montes, Via Delle Industrle 18, Albaredo D'Adige, I37041, Italy
1990

Nailam, Vai Delle Artigiano 51, Rovereto (Trento), I38068, Italy
1990

Nuova Cantro, Via Cuniberti 45, Torino, I10151, Italy; 011/7395640, FAX 011/7395640
1990

Nuova CMB, Regione Praolino, Ozzegna Canadese (TO), I10080, Italy
1990

OBM Di Baccigaluppi & C Srl, Via E. Matteri 6, Mesfro (MI), I20010, Italy; 02/9786182, FAX 01/97289288
1990

Officine Bussetti S.P.A., Str Molino Del Pascolo 25, Moncalieri (TO), I-10024, Italy; 0039-11/6 470879, FAX 0039-11/647 0365
1990

Officine De Zan, Via Ariosto 63, Rozzano (MI), I20089, Italy
1990

OMEP Srl, Via Ombrone Vecchio 9/II, Pistola (PT), I51100, Italy
1990

Omes SpA, via Marconi 15, Santorso, IV, I-36014, Italy
1990

OMP, Via Filli Rizzi 20, Proserpio (Como), I22030, Italy; 031/621434, FAX 031/622471
1990

OMR SpA, Via Carauaggio 3, Remeldello Sotto (BS), I25010, Italy; 030/957212, FAX 030/957644
1990

Ouag Italia, Via Crevada 59, Refrontolo (Treviso), Italy
1990

Parimor, Via Massarenti 2, Castlemaggiore (BU), I40013, Italy; 051/326784, FAX 051/326784
1990

Pilosio SpA, Via Fermi 45, Feletto Umberto (Udine, I33010, Italy; 0432/570983, FAX 0432/5704
1990

Plasti Center, Via Nuoua Del Campo 21, Napoli, I80141, Italy
1990

Podetti Marcello, Via Dell Artigiamio 18, Camin (PD), I35020, Italy; 049/760073, FAX 049/8700583
1990

PPP, Via Torino 184, Mestre (Venezia), I30172, Italy
1990

Puricelli, Via Da Vinci 48, Gallarate (VA), I21013, Italy; 0331/795732, FAX 0331/795084
1990

Rapid Lamiera, Via Bizzeri 23, Sala Bolognese, I40010, Italy; 051/954165
1990

Riva Officine, Via Nuova Circonvallazione 53, Caluso D'Lurea (TO), I10014, Italy; 011/9831143
1990

Romi, Via Dell Artigianto 6, Montecchio Maggiore, I36075, Italy
1990

Rossi Luigi, Zona Industriale, Basso Marino, I63100, Italy; 0736/402016, FAX 0736/402016
1990

Rossi Officine, Via Racconigi 269, Carmagnola (TO), I10022, Italy; 011/9770244, FAX 011/9770244
1990

Sallig, Via Calatafimi 27, Nicheline (TO), I10042, Italy; 011/624480
1990

Sarzi Lamiere, Via Di Giunti 27, Mantova, I46018, Italy
1990

Sercas, Via S. Egidio, Poluerigi (Anluna), I60020, Italy; 071/908146, FAX 071/907118
1990

SIAC, Via Bergamo, Pontiroli Nuoua (BG), I24040, Italy; 0363/88437
1990

SIAPE, Via Dei Boschi 39, Nerviano (MI), I20014, Italy; 0331/585101, FA. 0331/586866
1990

Sider Cesenate, Via Brighi 12, Diegaro (FO), I47020, Italy
1990

Silap Srl, Via PO 5, Vimercate (MI), I20059, Italy; 039/6085966-7
1990

Smada Elettromeccanica, Via Bisogno 5, Avellino, I83100, Italy
1990

Technoplex, Via Otte 12/B, Reschigliano, Di Capmodarsego (P), Italy
1990

Tecnica, Via Gitti, Marchent (BS), I25060, Italy; 030/861635
1990

Tecnisud SpA, Via Raiace 305/307, Pescara (PE), I65100, Italy; 085/576242, FAX 085/57025
1990

Tecnolaser, Via Venezia 23, Marsango Di Campo, I35010, Italy
1990

Vermec, Strada Dei Boschi, Beinasco (TO), I10092, Italy; 011/3497385
1990

Verona Lamiere, Via Pascoli, S. Maria Zerbio (VR), I37100, Italy
1990

VGM, Via Galilei I, Rovagnasco Di Segrate, I20098, Italy
1990

Zucchelli Furniture, Via Edison 22, Verona, I37100, Italy; 0451501433
1990

Japan

Azuma Co Ltd, 5-5-11 Soto-kanda, Chiyoda-ku, Tokyo, Japan; 3-834-4541*
1990
Sheet-metal shop.

Chugoku Giko Co Ltd, 2-14-26 Yoshijima-nishi, Naka-ku Hiroshima-shi, Hiroshima, Japan; 82-249-4450*
1990

Izumi Sogyo Co Ltd, 2-47-17 Kamiya Kita-ku, Tokyo, Japan; 3-902-3491*
1990
Sheet-metal shop.

Kitamura Kikai Co Ltd, 1870 Toide-machi Takaoka, Toyama, 939-11, Japan; 766-63 1100*
1990
Sheet-metal shop.*

Matsuda Co Ltd, Aza-ohhama, Ohama-hamakata, Houfu, Yamaguchi, Japan; 835-24-2111 *
1990

Nissan Shatai Co Ltd, 2909 Onkami Hiratsuka, Kanagawa, Japan; 463-55-4800 *
1990

Osaka Unv Dept Welding, 11-1 Mibogaoku-Ybarake, Osaka, 567, Japan
1990

Seibu Taiyo Laser, 3-19-12 Senbou, Tobata-ku, Kitakyushu, Fukouka, Japan; 93-861-2664 *
1990

Shin Meiwa Kogyo Co Ltd, 1-1 Shin-Melwa-cho, Takarazuke, Hyogo, Japan
1990

Taiyo Laser Co Ltd, 58 Aza-kawazoe, Ohtakacho, Midor-ku, Nagoya Aichi, Japan; 52-622-8553 *
1990

Taniyama Co Ltd, 2-3-17 Chuoa Joto-ku, Osaka-shi, Osaka, Japan; 6-931-7802 *
1990

Yamato Kogyo Co Ltd, 3825 Shimo-tsuruma, Yamoto, Kanagawa, Japan
1990

Netherlands

De Schelde Apparatenbauu, Postbus 16, Vlissingah, 4300AA, Netherlands; 01184-82545, FAX 01184-82914, telex 37815
1965
Offers sheet, plate, tube, pipe and forging processing for equipment for nuclear and chemical processes. Also shearing, forming, welding and brazing. Engineering facilities. Special knowledge and experience in non-common materials like high alloyed stainless steel, nickel alloys, zirconium, titanium, tantalum, etc.

DRU Industrial Products & Services, DRU Business Development Div, Frank Daamenstraat 4, PO Box 151, Ulft, 7071 AC, Netherlands; 08356-89911, FAX 08356-32040
1965
Manufactures dashboards and other components for the automotive industry; frames and assemblies for the copier and computer industry; the manufacture of heating and kitchen appliances; internal transport; and stainless steel single and double walled tanks for the manufacture of drinks, beer, dairy products, chemicals and pharmaceuticals.

Eldim BV, Spiktweg 24, Lomm, NL5943AD, Netherlands; 04703-1919, FAX 04703-2485
mgr tech devlmt, Martien H.H. van Dijk; gen mgr, Ger de Vlieger; emp 200, 1983
Laser job shop. Manufactures turbine components. Provides production and prototype parts in addition to conventional machining facilities for EDM, ECM, ECD and laser processing. Offers development of laser materials processing for turbine and non-turbine applications.

Hoogovens Groep BV, Vondellaan 10, Ismuiden, NL-1970, Netherlands; 0031 2514 99911, FAX 0031 2514 70057
1983

Hywel-Connection b.v., Thero Electronics Div, Voltastraat 61-65, Doetinchem, 7006 RT, Netherlands; 0031-8340 26266, FAX 0031-8340 45450
Dr. R. Cornelissen; 1991
Offers hybrid heating elements, hybrid electronics, hybrid sensors, hot plates and laser services.

Industriele Laser Toepassingen (ILT) B.V., Hengelosestraat 705, Postbus 545, Enschede, NL7500AM, Netherlands; 053-355677, FAX 053-36317
ch exec, P. Bant; 1985
Cutting, welding, drilling, soldering of fine mechanical parts with CO_2 and Nd:YAG lasers. Production of parts from metal, plastic, wood, paper and more. Process development and application-research. Laser engraving of numbers, logos, etc. Scanning of logos for reproduction purposes.

Laser Application & Information Center Amsterdam, Nieuwe achtergracht 127, WS Amsterdam, 1018, Netherlands; 20 5224014*, telex 16460 facwn nl
emp 2061, 1982

Nobels BV, Nijverheidstraat 2, AV Katwijk, NL-2222, Netherlands
1982
Sheet-metal shop with laser cutting capability.*

Sorba Precisieplaatwerk BV, Industrieweg 20, PO Box 54, Winterswijk, 7100 AB, Netherlands; 05430-14666, FAX 05430-12461
pres, R.J. Goedhart; eng mgr, H. Thomassen; mfg mgr, H. Asbreuk; sls/mktg mgr, H. Vuyk; 1946
Specialized contractors and/or subcontractors in precision sheet-metal working. Specialty industrial use of CNC laser cutting combined with CNC turret pressing, CAD/CAM facilities: Medusa 3-D solid modeling.

Staalservice Centrum, Industrieweg 2, Stamproy, 6039-AP, Netherlands; 04956-1781
1946

Staalservice Weeat, Postbus 380, Weeat, 6000-AJ, Netherlands; 04850-40838
1946

Suplacon bv, Assamblagweg 11, Emmeloord, 8301 BB, Netherlands; 05270-98006
1946

T.N.O., Institute of Production and Logistic Research Center for Materialprocessing with Lasers, Laan Van Westenenk 501, Postbox 541, Apeldoorn, 7300 AM, Netherlands; +31 55 49 34 93, TWX +31 55 49 33 45
A.H. van Krieken, W. Husslage, J. Snoeij, H.B. Zeedijk; 1931
Offers consultancy and feasibility studies; prototype development; systems engineering; contract research for national and international industries. Research projects, with emphasis on laser micro-machining, laser surface treatments and laser (micro) welding and joining.

WBM Staalservice Centrum BV, (sub of WBM Staalhold BV), Industrieweg 2, PO Box 3108, Stramproy, 6039 ZG, Netherlands; 4956 1781, FAX 4956 1981
ch exec, P.A. Govaert; sls mgr, J. Esmeyer; 1968
Subsuppliers of nonwelded, semimanufactured sheet products in small and midranged series. Shearing, flame-cutting, bending, CNC punching, and CO_2 lasercutting according to customer' s specifications.

Zantech, Lekstraat 26, Oss, 5347-KY, Netherlands
1968

New Zealand

C&R Equipment Ltd, 17 Wickham St, Christchurch, New Zealand; 0011 64 3 3843154, FAX 0011 64 3 3843152
dir, K. Gunter; dir, R.G. Malcolmson; 1954
Offers laser cutting of most materials including plastic, timber, steel, bakelite, leather, rubber, cork, slate, ceramics, etc. to the most intricate pattern. Ideal for long runs.

Laserstream Cutting Ltd, 439 Rosebank Rd, PO Box 19528, Avondale, Auckland, New Zealand; 09 884 093, FAX 09 881 132
gen mgr, Glyn Dickson; emp 120, 1989
Laser and high-pressure water-jet contract cutting service. Computer controlled cutting of all materials from stone to titanium. Agents for Waterjet and Laser systems.

Norway

Rjukan Metall A/S, Rjukan, N-3660, Norway; 47-36-94316, FAX 47-36-9397
emp 250, 1970
Welding of aluminium tubes, anodising, and machinery.

P.r.c.

Laser Processing Centre, Beijing Mach & Elec Institute, Sun Li Tun Chaoyang, Beijing, P.R.C.
1970

Portugal

Instituto de Soldadura e Qualidade, Rua Thomas de Figueiredo 16-A, Lisbon, 1500, Portugal; 351-1-4429599, FAX 351-1-4429799
mech eng, J.F. Oliveira Santos,PhD; metallurgical eng, R.M. Miranda; physical eng, A.P. Martins; 1965
Offers metallic and non-metallic materials processing development with CO_2, Nd:YAG and excimer lasers. Also applications and feasibility studies, applied R&D, technology transfer and training.

Lasindustria Technologia Laser S.A., Rua Francisco Antonio da Silva, Oeiras, 2780, Portugal; 351 1 4430557, FAX 351 1 4430567
1988
Laser cutting and engraving on metal and nonmetal parts. The maximum thickness is 8 mm for steel, 5 mm for stainless steel and mm for Al. Precision of the systems is 0.05 mm.

Prc

Shanghai Institute of Laser Technology, Yi Shan Rd 770, Shanghai, 200233, PRC; 4700458, FAX 4700037, telex 33422
dir, Nie Bao Cheng; sls & mktg mgr, Wang Chunyao; mfg mgr, Fu Hequn; eng mgr, Hun Zhendong; 1970
Laser materials processing division is one of SILT' s major units. Engaged in R&D in application of laser materials processing. Designs, manufactures, and sells laser systems. Also offers technology services (prototype design) and job shop services.

Spain

Arbucias Industrial SA, Hermana Assumpta,, 46, Bajos, ISDA, Arbucies (Girona), Spain
1970
Laser cutting capability.

CETENASA, Laser Div, Poligono de Elorz s/n, Noain (Navarra), 31110, Spain; (48) 238258, FAX (48) 238354
div dir, Pablo Garriz; tech dir, Kazie Jasnowski; 1986
Carries out R&D projects involving laser materials processing (cutting, welding, surface treatment, etc.) Processes small batches of parts or prototypes. Provides technical and economical feasibility studies, for the installation of industrial lasers.

Laser Melis SA, Poligono Basabe, Pab D 1 2, Aretxabaleta, 20 550, Spain; 43 79 75 11*, FAX 43 79 14 76
emp 80, 1986
Tailored laser processing systems for cutting, welding, and heat treating. Laser processing job shop services and applications development studies.

Sweden

Control Laser PTI, Box 203, Ludvika, S-77101, Sweden; +46 240 13660*, FAX 46 240 10601, telex 74276
emp 100, 1986
Laser systems for industrial applications and laser job work-cutting, welding, heat treatment, and marking. *

Lulea University of Technology, Materials Processing, Lulea, S-95187, Sweden; +46 920 91000, FAX +46 920 97288
Prof. Claes Magnusson; 1972
Leading R&D center for laser materials processing in Sweden. The laser laboratory is equipped with CO_2 lasers in the power range from 1.7 to 6 kW and both 3D and 2D work stations. A Nd:YAG laser of 260 W average output power is used for high-precision processing.

PermaNova Lasersystem AB, (sub of Permascand AB), Fagerbacken 28, Ostersund, S-83146, Sweden; +46 63 132455, FAX +46 63 106128
gen mgr, Nils Ohrberg; 1985
Manufactures laser systems for material processing service. Maintenance and spare parts. Job shop.

Permascand AB, (sub of EKA Nobel), Box 42, Ljungaverk, S-84010, Sweden; +46-691-329 40, FAX 46-691-330 04
vp, Nils Ohrberg; emp 230, 1971
Subcontract work (CO_2 laser cutting, welding, heat treatment). Distributor for companies such as Rofin-Sinar, Prima Industrie, etc. Special CO_2 laser systems and R&D application studies.

Switzerland

Alex Neher AG, Felsensteinstrasse 8, Ebnat-Kappel, CH-9642, Switzerland; 74 314 14*, FAX 74 3 3690, telex 88 41 33
emp 700, 1937
Laser cut, copy punched parts from 0.5 to 6 mm steel and laser cut parts from 0.3 to 12 mm thickness.

Bal Laser AG, IM Feld 3, Schinznach Dorf, 5107, Switzerland; 56 43 28 21*, telex 56 340 CONST
1937
Laser job shop.

Buchi AG Metallwarenfabrik, Hubstrasse 11, Wil, CH-9500, Switzerland
1937

Duap Engineering AG, N-eue Winterthurstrasse 30, Dietlikon, CH-8305, Switzerland; 0041 1 833 1500, FAX 0041 1 833 5035
1937

Dubi & Company, Hofmattstrasse 12, Herzogenbuchsee, 3360, Switzerland; 63-60-12-12 *
1937
Sheet-metal shop with laser capability.

Dyno AG, Schalunenstrasse 54, Aefligen, 3426, Switzerland; 34-45-23-33*
1937
Sheet-metal shop with laser capability.

Fritz Born AG, Keltenweg 2, Langenthal, 4900, Switzerland; 63 22 14 07*
1937
Sheet-metal shop with laser cutting capability.

Gasser Ravussin SA, Postfach 56, Lucens, CH 1522, Switzerland; 021 906 8159, FAX 021 906 9472
1925
Offers watch jewels drilling with laser, industrial jewels drilling with laser, and laser cutting for sapphire, ruby and ceramics.

Laser-Automation Gekatronic SA, Rue Louis Chevrolet, La Chaux-de-Fonds, 2300, Switzerland; 39 25.21.75, FAX 39 26.03.88
pres/gen mgr, J.C. Kullman; eng/mfg mgr, F. Obrist; mktg mgr, Ted Hagen, J. Chr. Kullman; 1968
Job shop application in laser micro-, spot, and seam welding, drilling and precision cutting, feasibility studies, etc. Manufactures fully automated Nd:YAG laser systems to customer specifications.

Lastec AG, Mattenstrasse 6, Brugg/Biel, CH-2555, Switzerland
1968

LBC Laser Bearbeitung Ceresa AG, Chrummatt 42, Flamatt, CH-3175, Switzerland; 0041 31 741 2021, FAX 0041 31 741 0949
pres, A. Ceresa; gen mgr, A. Luthi; 1989

MIRAP AG, Buechstrasse 5, Jona, CH-8645, Switzerland; 0041-55-28 17 28, FAX 0041 55 28 1776
ceo, E. Wille; prod mgr, H. Zuger; sls mgr, E. Rudisuhli; 1982
Laser and plasma cutting. Sheet metal job shop. Sheet metal from .2 - 12 mm and 1500 x 3000 mm. Steel, alloy, and alloy high quality support. Water abrasive cutting, 30 mm steel and 60 mm alloy. Subcontractor to more than 200 industrial customers in Switzerland and Germany.

Nivarox-Far SA, Rue Dr. Schwab 32, Saint-Imier, 2610, Switzerland; 039 414646, FAX 039 318364, telex 952 340
sls/mktg mgr, Berger Maurice; 1982
Offers welding of miniature objects.

OMA AG, Industriestrasse 44, Aarau, CH-5000, Switzerland; (64)24-4924 *
1982

PB Laser Systems AG, Kupfgasse 10A, Lengnau, CH-2543, Switzerland; 41-65/52 96 94, FAX 41-65/53 07 81
1982

Projectina AG, Dammstrasse, Heerbaugg, CH-9435, Switzerland; 41-71/72 20 44
1982

R Audemars SA, Via Ponteggia, Cadempino-Lugano, CH-6814, Switzerland; 091-58 26 62, FAX 091-56 69 52, telex 843013
r&d, Ing. R. Zafferri; sls mgr, Ing. M. Maffioli; 1899
Manufactures and assembles very small stepping motors. Manufactures micro-permanent magnets for industrial use, industrial micro-coils, and fiberoptic connections. Operates micro-laser for soldering and drilling. Also manufactures very hard materials, corundum and ceramic. Development and manufacture of micro-electronic coil windings.

Signer Ag, Ennetaach, Riedt B. Erlen, CH-8586, Switzerland; 72-48-11-11 *, telex 719-232-SIAG
1899

SULZERINNOTEC
LASER SURFACE ENGINEERING

Laser Surface Engineering (sub of Sulzer Innotec), PO Box 65, Winterthur, CH-8404, Switzerland; 41 52 2625155, FAX 41 52 220703, telex 896 060 SZCH
proj mgr, Dr. Roger Dekumbis; 1988
Offers laser surface treatment R&D; jobbing services such as: local hardfacing of iron-, titanium-, cobalt- and nickel based alloys, local repair of worn parts (missing material added), and local hardening.

Swiss Federal Institute of Technology, C.T.M.L., CH-Ecublens, Lausanne, 1015, Switzerland; 021/6934914, FAX 021/6934916
pres, Prof. Wilfried Kurz; sls mgr, Dr. Charles Marsden; 1985
Applications R&D consultant. Applications research in surface treatment of metals including transformation hardening, remelting, alloying, cladding, and welding.

Wirth & Company AG, Schachen, Buchrain, CH-6033, Switzerland; 41-89-16-31, FAX 41-89-10-53
1962

Zumstein AG, Industriestrasse 6, Batterkinden, CH-3315, Switzerland; 65-45-35-31 *
1962

United Kingdom

W.R. Anderton & Co Ltd, 189 Drake St, Castleton, Rochdale, Lancaster, UK
1962

Applied Cutting Technology, Unit 6, Block 2, Lawford Dale Ind Est, Lawford, Manningtree, Essex, CO111US, UK; 0206-395858, FAX 0206-396545
1962
Subcontract laser cutting, welding, engraving (2,3,4 and 5-axis applications); precision engineering; CNC machining and turning centers; conventional milling, drilling, turning and grinding; sheetmetal work; plastics cut and fabricated, all to BS5150.

BG Industrial Laser Group Plc, Stukeley Meadows Ind Estate, Huntingdon, Camb, PE186EL, UK; 0480-455441, FAX 0480 411930, telex 32865 LSSMIC G
ceo, B.G. Green; works dir, S. Adamson; tech dir, D. Crawford; prod dir, B.C. Sandford; 1977
Subcontract laser engineering using CO_2 lasers for profiling flat and formed sheet materials. Specialize in large panels and difficult materials. Laser welding, engraving, scribing/profiling of ceramics and design/building of CO_2 laser systems.

Capital Lasers, 36 Bear Lane, London, SE1 OUH, UK; 071-928-6235, FAX 071 261 0994
1977

Carlton Laser Service Ltd

Carlton Laser Service Ltd, 470 Thurmaston Blvd, Troon Ind Area, Leicester, LE4 7LN, UK; 533 761 177*
1977

ERABA, Grange Road, Livingston, EH54 5DE, UK; 0506-31234, FAX 0506-32664
mng dir, J. Jamieson; 1981
Sub-contract company supplying auto, computer industries with metal parts from laser cutting, precision sheet metal, CNC turning and milling, press work, tool making, and assembly work.

Hamilton Fabrications Ltd, Crab Tree Lane Atherton, Manchester, M29 0AG, UK
1981

Harlow Sheet Metal PLC, Allen House Edinburgh Way, Harlow, UK; 0279-414475, FAX 0279 635192
mng dir, L.L. Simmons; sls dir, Kevin White; 1975
Provides laser profiling, prototype work, CNC fabrication, metal pressings, painting, silk-screening and assembly work. A complete precision sheet metal facility. Approvals held: BS5750 part 2; 1509002.

Holloway Sheet Metal Works Ltd, 3034 Eden Grove Holloway, London, N7 8EL, UK; 071-607 4296, FAX 071-7001214
T. Bigland, M. Reeves; 1928
Offers subcontract work on sheet metal items, laser cutting service and CNC punching.

Inductoheat (Tewkesbury) Ltd, Industrial Estate, Northway Ln, Tewkesbury, Gloucestershire, UK; FAX 6084 850442, telex 43382 Tewsaw Attn Inducto
mng dir (lsr), I.C. Hawkes; mng dir (induction), N.A. Hawkes; 1962
Laser hardening of complex press tooling (eg. automotive die sets). Also laser welding, profile cutting, induction hardening and brazing.

Intec Laser Services, 16 Dunlop Rd Hunt End Ind Est, Redditch Worc, B97 5XP, UK
1962

International R&D Ltd, Fossway, Newcastle-on-Tyne, NE6 2YD, UK; 091-2650451, FAX 091-2760177
ch exec, P.R. Whitehouse; sls mgr, Dr. W. MacFarlane; eng mgr, D. Tuck; 1960
Offers contract R&D, problem solving, applications development, design & construction of special or prototype systems, limited production of laser, optical, electronic, and laser, energy measuring systems, IR & UV radiometers, and manufacture of IR filters.

Lasercraft (Design) Ltd, Stonehill Rd, Stukeley Meadows Ind Est, Huntingdon, Cambs, PE18 6LR, UK; (0480) 432 433, FAX 0480 432 567
mng dir, Colin Ward; 1983
Laser engraving of wood, acrylic & many other materials. Offers a standard range of ready made corporate gift items and have a popular range of award/presentation plaques that are customized to the client' s needs. A subcontract service is available to engrave customer' s materials.

Lasercut Products Ltd, Sun Street, Sawbridgeworth, Herts, CM21, UK; 0279-600521, FAX 0279-600225
mng dir, John Bishop; gen mgr, Kevin Willet; works mgr, Mike Willson; 1978
Laser cutting job shop.

Laser Expertise Ltd, Units 1-3 Trent South Indus Pk, Little Tennis St, Nottingham, NG2 4EQ, UK; 0602 587452, FAX 0602 411 620
mng dir, A.J. Schwarz; tech dir, Dr. J. Powell; prod dir, C.D. Young; 1983
Laser applications job shop and consultancy-CO_2 laser cutting specialists.

Laser Precision Engineering Ltd, 21 Westbrook Rd, Trafford Pk, Manchester, M17 1AY, UK; 061 876 7273, FAX 061 876 7261
mrktg dir, Keith Hillard; 1983
Offers laser profile cutting service, cutting of metals & non-metals. Quantities from 1 off development pieces to 10,000 off production runs. Small guage precision pieces and clean edge stainless cutting performed as well as laser identification marking.

Laser Process Ltd, Hemlock Way, Hawks Green, Cannock, Staffs, WS112GB, UK; 0543 466676, FAX 0543 466679
mng dir, D. Lindsey; sls mgr, W. Whorton; contracts mgr, J. Lindsey; 1990
Laser cutting and profiling sub-contractors. Full CAD/CAM facilities for rapid turn-around of prototype components.

Laser Profiles Eastern Ltd, 1 Ryder Way, Burnt Mills Industrial Estate, Basildon, Essex, SS155TE, UK; 0268 729292*
1990

Laser Profiles Southern Ltd, 71 Haviland Rd, Ferndown, Industrial Estate, Wimborne, Dorset, BH217PY, UK; 0202 861438*
1990

LaserWest Ltd, Unit 1A, Garden Close, Langoye Industrial Estate, Plympton, Devon, PL7 5AW, UK; (752) 348840, FAX (752) 348860
ch exec, Andrew Greenslade; 1984
Laser trade shop offering 2-axis CO_2 lasers-powers up to 3kW and table sizes to 4 meters X 2.25 meters. 7-axis YAG using a Laserdyne 780 beam director with a 702 Lumonics laser. This combination of machines provides a unique capability to process the smallest of components to thick plate.

Lase Tech Ltd, Stephenson Close, Drayton Field Bus Park, Daventry, Northants, NN115RF, UK; 0327 79666, FAX 0327 300241
dirs, R.W. Turner, T.J. Thomas; 1986
Offers sub-contract laser cutting and welding, sub contract R&D cutting and welding, and laser spares and service.

NC Laser Cutting Services Ltd, (sub of WR Anderton & Co Ltd), Sta Approach, Cark in Cartmel, Cumbria, UK; FAX 0706 358201
emp 360, 1977
Profile laser cutting intricate and simple shapes. Thirteen machines available cutting a wide range of materials, for example: mild steel 12 mm, stainless steel 8 mm, gauge plate 12 mm, wood 50 mm, plastics 100 mm. Table capacity for components 1650 mm by 1150 mm. All NC tapes are produced in-house by computer system with either digitization or dimensional input.

Nimbus Laser Services, Albert Drive, Burgess Hill, W. Sussex, RH159TN, UK; 0444 87 03 86 *, FAX 0444 87 02 97
1977
CO_2 laser cutting and welding. 2 m x 1 m cutting table, *xyz* rotary workstation, *xz* rotary (flat) workstation. Additional expertise with laser-welded diamond saws and core drills for construction industry. Full CAD/CAM programming aids.

Power Beam Technologies, Building 5A11, AEA Technology, Risley/Warrington, WA3 6AT, UK; 0925-25 2845, FAX 0925 25 2206
1977
Contract production facility offering: cutting (LP and VHP processes), welding (autogenous and wire feed), surface modifcation (transformation and powder cladding). Uses six laser systems including YAG CO_with powers 400W-10kW and includes three multi- axis/multi-process work cells. Specializes in cutting Titanium and superalloys to aerospace standards. BS 5750 approved and Rolls Royce approved contracter.

Preci Spark Ltd, Railway Terrace, Loughborough, Leicestershire, UK
1977

Redditch Laser Cutting Ltd, 80A Arthur Street, Lakeside, Redditch, B98 8YP, UK; 0527-510474, FAX 0527-510432
dir, W. Bewick; 1986
Provides subcontract lasercutting of simple/complex shapes in mild steel, stainless steel, wood, plastics, special materials, etc. Full in-house CAD/CAM system.

South Yorkshire Laser Cutting Ltd, (sub of Coventry Lasers Ltd), Unit 22, Bookers Way,, Todwick Rd, Industrial Estate, Dinnington, Sheffield, S31 7SE, UK; (0909) 568682, FAX (0909) 565648
pres, A.J. Dadley; sls mgr, I. Circuit; eng mgr, K. Hambleton; 1988
Laser metal cutting and marking shop. The group operates 8 systems utilizing 2 and 3 axis some with 4th axis rotary capability. Plus its latest acquisition, a totally programmable 5 axis laser cutting system for 3 dimensional trimming. The company is registered and approved to quality standards BS 5750 Part II and ISO 9002.

Strathclyde Fabricators Ltd, Bothwell Rd, Hamilton, Strathclyde, ML3 0SF, UK; 0698-283452, FAX 0698-283478
mng dir, J. Gunn; sls/mktg dir, D. Lemon; qualty dir, D. Nolan; manf dir, G. Gilmour; prdn dirs, D. Brown, G. Dickenson; 1974
Manufactures precision sheet metal and electronic assemblies for the computer and electronic industries worldwide. Uses conventional presswork for 15T to 400T capacity, alternatively, laser or CNC press manufacture, CNC forming, finishing-powder or wet paint, auto turned parts, silk screening. Mechanical or electrical assembly approved to BS 5750, ISO 9000 part 2.

Subcon Laser Cutting Ltd, Unit 7, Tident Bus Pk, Park St, Nuneaton, Wark, CV11-4NS, UK; 0203 642221, FAX 0203 342180
mng dir, W.J. Brown; 1988
Subcontract laser cutting services to all industries. Prototype, small batch and large batch production undertaken.

Torbrex Engineering Ltd, Rutherford Rd, Kingsway West, Dundee, DD2 3XH, UK; 0382-825253, FAX 0382-814508
1974
High quality metal components, competitively priced with low tooling costs and short delivery time. Services offered include CNC punching, CNC laser cutting, CNC forming and bending, and robotic welding. Finishing facilities include stove enameling and powder coating.

TWI, Laser Centre, Abington Hall, Abington, Cambridge, CB1 6AL, UK; 223 891162, FAX 223 892588
ch exec, A.B.M. Braithewaite; hd of mktg, J.G. Wylde; lsr ctr mng, I.J. Spalding; emp 420, 1923
A R&D organization concerned with all aspects of joining and related technologies. The Laser Centre is a collaboration between TWI and AEA Technology, covering the range of industrial material processing using lasers.

University of Liverpool, Laser Group, Mechanical Eng Dept, Liverpool, L69 3BX, UK; 051.794.4892, FAX 051.794.4848, telex 627095 UNILPL G
Prof. W.M. Steen, Dr. W. O'Niell, Dr. G. McCartney, Dr. P. Modern, Dr. K. Watkins; emp 5000, 1892
Laser material processing R&D. Training to doctorate level in laser cutting, welding, surface treatment, modelling, automation, in-process sensing, process development process, and laser physics.

W&B Machining Service Ltd, Unit 48,, Pipers Rd Park Farm Ind Estate, Redditch, B98 0HU, UK
1892

Product specifications

The following tables provide product specifications for CW gas lasers, pulsed gas lasers, CW solid-state lasers, and pulsed solid-state lasers. The data is presented in tabular form for easy accessibility.

Wave-lengths (μm)	Output TEM₀₀	(W) Multi-mode	*Beam dia (mm)	*Beam diverge (mrad)	Special Features	Manufacturer	Model no.
Carbon dioxide							
	100		7	2	Invar stabilized	CRILASER SA	A-100
	200		8	2	Invar stabilized	CRILASER SA	A-200
5.5	5		3	4		Synrad Inc	48-4
9-11	8		4.5	3.5	Sld,grtng tnd,ultra stbl	Line Lite Laser Corp	950
10.2-10.6	3		4	4	Sld,PZT-tnd,ultra stbl	Line Lite Laser Corp	941S
10.2-10.6	5		4	4	Sld,pwr supply in head	Line Lite Laser Corp	945
10.2-10.6	10		4.5	3.5	Sealed, ultra stable	Line Lite Laser Corp	948
10.2-10.6	20		4.5	3.5	Sealed, ultra stable	Line Lite Laser Corp	952
10.2-10.6	60W		6	2.2	Sld,cw/pulsed,18mo warrn	Line Lite Laser Corp	990-6
10.2-10.6	100 W		6	2	Sld,cw/pulsed,18mo warrn	Line Lite Laser Corp	990
10.6					40x43x81 (135)sealed off	Lasercut SA	B25S
10.6		up to 60 W		<7	Sealed low cost laser	Edinburgh Instruments Ltd	SL 4-60
10.6			7	1.2	40x43x81 (135)cm 51kg	Lasercut SA	B25
10.6			7	1.2	40x43x81 (135)cm 51kg	Lasercut SA	B40
10.6		100	10	2	Single enclosure	Electrox Ltd	M80
10.6		160	10	2	Single enclosure	Electrox Ltd	M140
10.6		700		2	RF, 10 kHz ultra compacy	MLI Lasers Ltd	UCL-700
10.6		800 W	16	1.5	Integrated turbine laser	Lasers Industriels SA	Modulas 2 M
10.6		1000 W	16	1.5	Integrated turbine laser	Lasers Industriels SA	Modulas 3 M
10.6		10000	40	3	Silent dischrge excitatn	Mitsubishi International Corp	100R
10.6		10000	45	1.7	Fast transverse flow	Laser Ecosse Ltd	CL10
10.6		10000	50	2	Fast axial flow, RF exc	Rofin-Sinar Inc	RS10000RF
10.6		10000	50	2	Fast axial flow, RF exc	Rofin-Sinar Laser GmbH	RS10000RF
10.6		1150	15	2	Single enclosure	Electrox Ltd	Pegasus
10.6		1500 W	20	1.5	Integrated turbine laser	Lasers Industriels SA	Modulas 4 M
10.6		2000	13	<2	TURBO FLOW™/gas econ	PRC Corp	PRC 2001
10.6		2000	14	<1	RF exec pulsed/CW	TRUMPF Inc	TLF 2000
10.6		2000	15	<2.2	+/-.5% power stability	OPL (Oerlikon- PRC Laser SA)	OPL2001
10.6		300	10	19	Shot to shot to 1KHz	Quantel SA	IQ 10
10.6		600	10	19	Shot to shot to 1KHZ	Quantel SA	IQL 20
10.6		900	10	19	Shot to shot to 1KHz	Quantel SA	IQL 30
10.6		1200	10	22	Shot to shot to 1Khz	Quantel SA	IQL 40
10.6		2000	10	22	Shot to shot to 1KHz	Quantel SA	IQL 60

Wave-lengths (μm)	Output TEM$_{00}$	(W) Multi-mode	*Beam dia (mm)	*Beam diverge (mrad)	Special Features	Manufacturer	Model no.
Carbon dioxide							
10.6		2500		1.5	Fast axial,pulsing,RFexc	Rofin-Sinar Inc	RS2500RF
10.6		2500		1.5	Fast axial,pulsing,RFexc	Rofin-Sinar Laser GmbH	RS2500RF
10.6		2500W	20	<1.5	RF exc, pulsed/CW	TRUMPF GmbH & Co	TLF 2500
10.6		2500	20	<3	RF exc pulsed/CW	TRUMPF Inc	TLF 2500
10.6		2500	23	1		Messer Griesheim GmbH	Eurolas 3000
10.6		3000	16	2.3	+/-.5% power stability	OPL (Oerlikon- PRC Laser SA)	OPL3000
10.6		3000	16	<2.5	TURBO FLOW™/gas econ	PRC Corp	PRC3000
10.6		3000W	22	<1	RD exc, pulsed/CW	TRUMPF GmbH & Co	TLF2000 turbo
10.6		3000	30	1.5	Fast axial,pulsing,RFexc	Rofin-Sinar Inc	RS3000RF
10.6		3000	30	1.5	Fast axial,pulsing,RFexc	Rofin-Sinar Laser GmbH	RS3000RF
10.6		3000	41	3	Trnsvrse,microproc-ctrl	Rofin-Sinar Laser GmbH	RS825
10.6		3000	41	3	Trnsvrse,microproc-ctrl	Rofin-Sinar Inc	RS825
10.6		4000	20	<1	RF exec pulsed/CW	TRUMPF Inc	TLF 4000
10.6		4000	23	1		Messer Griesheim GmbH	Eurolas 4000
10.6		4000	41	3	Trnvrse,microproc-ctrl	Rofin-Sinar Inc	RS840
10.6		4000	41	3	Trnvrse,microproc-ctrl	Rofin-Sinar Laser GmbH	RS840
10.6		5000W	22	<1	RF exc, pulsed/CW	TRUMPF GmbH & Co	TLF 5000 turbo
10.6		5000	22	<3	RF exc pulsed/CW trbine	TRUMPF Inc	TLF 5000T
10.6		5000	41	3	Trnsvrse, microproc-ctrl	Rofin-Sinar Laser GmbH	RS850
10.6		5000	41	3	Trnsvrse, microproc-ctrl	Rofin-Sinar Inc	RS850
10.6		5000	45	1.7	Fast transverse flow	Laser Ecosse Ltd	CL5
10.6		5000	70	1	High freq discharge	MLI Lasers Ltd	ML-5000
10.6		6000W	24	<1.5	RF exc, pulsed/CW	TRUMPF GmbH & Co	TLF6000
10.6		6000	24	<3	RF exec pulsed/CW	TRUMPF Inc	TLF 6000
10.6		6000	35	1.5	Fast axial,pulsing,RFexc	Rofin-Sinar Inc	RS6000RF
10.6		6000	35	1.5	Fast axial,pulsing,RFexc	Rofin-Sinar Laser GmbH	RS6000RF
10.6		9000	70	1	High freq discharge	MLI Lasers Ltd	ML 108
10.6	2x250		8	<2	Dual beam indiv control	OPL (Oerlikon- PRC Laser SA)	OPL600/2
10.6	2x250		8	<2	TURBO FLOW™/dual beam	PRC Corp	PRC600/2
10.6	2x700		10	<2	Dual beam indiv control	OPL (Oerlikon- PRC Laser SA)	OPL1500/2
10.6	2x750		10	<2	TURBO FLOW™/dual beam	PRC Corp	PRC1500/2
10.6	2x1250		11	<2	Dual beam indiv control	OPL (Oerlikon- PRC Laser SA)	OPL3000/2
10.6	2x1250		12	<2	TURBO FLOW™/dual beam	PRC Corp	PRC3000/2

Wave-lengths (μm)	Output (W) TEM_{00}	Multi-mode	*Beam dia (mm)	*Beam diverge (mrad)	Special Features	Manufacturer	Model no.
Carbon dioxide							
10.6	3		1.4	10	Compact (9in)	Hughes Aircraft Co	4908 H
10.6	3		1.4	10	Stabilized output	Hughes Aircraft Co	4908-H-P
10.6	4		1.3	10	RF exc, EH_{11} mode	Laser Ecosse Ltd	CM3044
10.6	4		1.3	10	Waveguide hardsealed AB	Edinburgh Instruments Ltd	LM-4
10.6	4x700		10	<2	TURBO FLOW™/quad beam	PRC Corp	PRC 3001/4
10.6	5-50		12	<3	Slow flow, enh pulse	Avimo Ltd	PLT 50
10.6	5-100		15	<3	Slow flow, enh pulse	Avimo Ltd	PLT 100
10.6	5-200		15	<3	Slow flow, enh pulse	Avimo Ltd	PLT 200
10.6	8		1.3	10	RF exc, EH_{11} mode	Laser Ecosse Ltd	CM3054
10.6	8		1.3	10	Waveguide, RF shielded	Edinburgh Instruments Ltd	LM-8
10.6	10			<1.5	Super pulse to 5 kW	NTC/Marubeni America Corp	RS1200-SM
10.6	10			<1.5	Super pulse to 10 kW	NTC/Marubeni America Corp	RS-1700-SM
10.6	10		1.7	8.5	Stabilized output	Hughes Aircraft Co	3915 H-P
10.6	10		1.7	8.5	RF pulsable	Hughes Aircraft Co	3915 H
10.6	10		3	5		Synrad Inc	48-1
10.6	10		3.5	4	Invar stabilized	Synrad Inc	48I-1
10.6	10		3.5	4	Grating tuned	Synrad Inc	48G-1
10.6	20		1.3	10	Waveguide	Edinburgh Instruments Ltd	LM-20
10.6	20		1.3	10	RF exc, EH_{11} mode	Laser Ecosse Ltd	CM2000
10.6	20		3	5		Synrad Inc	48-2
10.6	20W		4	3.5	Sealed/RF excited	Melles Griot	05CRF220
10.6	20		6	2	Invar stabilized	CRILASER SA	A-20
10.6	25		1.7	8.5	Stabilized output	Hughes Aircraft Co	3926 H-P
10.6	25		1.7	8.5	RF pulsable	Hughes Aircraft Co	3926H
10.6	30		6	3	Slow flow	Laser Electronics Pty Ltd	LE-30C
10.6	35		6	3	RF excited,sealed	Coherent Hull Ltd	WG35
10.6	40		4	5.7	RF exc, sealed off	Laser Ecosse Ltd	CT-40
10.6	40W		4.6	3	Sealed/RF excited	Melles Griot	05CRF440
10.6	50		1.7	7	Folded waveguide	Edinburgh Instruments Ltd	LM-50
10.6	50		3	5		Synrad Inc	48-5
10.6	50		8	2.	Slow flow	Laser Electronics Pty Ltd	LE-50C
10.6	50	75	6	4	Microprcsor cntrl & opr	Adron Sources S.A.	ALS 50
10.6	50-500		15	<2	Fast flow, enh pulse	Avimo Ltd	PLT 500

Wave-lengths (μm)	Output (W) TEM_{00}	Multi-mode	*Beam dia (mm)	*Beam diverge (mrad)	Special Features	Manufacturer	Model no.
Carbon dioxide							
10.6	50-1000		17	<2	Fast flow, enh pulse	Avimo Ltd	PLT 1000
10.6	50-2000		23	<2	Fast flow, pulse opt	Avimo Ltd	PLT 2000
10.6	60		5	2	Slow flow	A-B Lasers Inc	BLS60C
10.6	60		5	5		Baasel Lasertech	BLS 60C
10.6	60		6	2.2	Inc gas recycler	Laser Dynamics Ltd	LX-60
10.6	60		7.5	2	Stable, low cost	Edinburgh Instruments Ltd	PL5
10.6	70		6	≥3.	Sealed-off, refillable	MPB Technologies Inc	IN-70
10.6	100		4	3.5		Synrad Inc	57-1
10.6	100		11.5	6.5	Compact unit	Laser Ecosse Ltd	LE100
10.6	100		12	2.	Slow flow	Laser Electronics Pty Ltd	LE-501C
10.6	100 - 1200	1000 - 25000	6 - 100	.1 - 5 m	Dif beam sources	ESAB Group	
10.6	120W		5.5	3	Sealed/RF excited	Melles Griot	05CRF1200
10.6	120		7	2	TEM 01	Baasel Lasertech	BLS 170 C
10.6	120		8	1	Slow flow	Laser Valfivre Sorgenti e Sist	L-120
10.6	120		8	2	Slow flow	A-B Lasers Inc	BLS120C
10.6	140		7	≥2.	Sealed-off, refillable	MPB Technologies Inc	IN-100
10.6	140		13	3	Enhanced pulsing	Lumonics Corp	V150
10.6	150		8	1.5	Slow axial flow, pulsing	Rofin-Sinar Inc	RS150
10.6	150		8	1.5	Slow axial flow, pulsing	Rofin-Sinar Laser GmbH	RS 150
10.6	150		8	1.5	Slow axial flow	Rofin-Sinar Laser GmbH	RS150
10.6	150		8	2	Inc gas recycler	Laser Dynamics Ltd	LX-150
10.6	150	180	10	2.6	Microprcsor cntrl & opr	Adron Sources S.A.	ALS 150
10.6	200		4	3.5		Synrad Inc	57-2
10.6	200		11	2	Folded laser	Edinburgh Instruments Ltd	PL6
10.6	200		14	2	Slow flow	Laser Electronics Pty Ltd	LE-502C
10.6	220		8		Slow flow	A-B Lasers Inc	BLS220C
10.6	240W		8	2	Sealed/RF excited	Melles Griot	05CRF2400
10.6	250		10	1	Fast axial flow	Laser Valfivre Sorgenti e Sist	L 250
10.6	250		10	1.7	Inc gas recycler	Laser Dynamics Ltd	LX-250
10.6	270		10	1.5	TEM 01	Baasel Lasertech	BLS 220 C
10.6	275		12.25		Compact	Coherent General Inc	Everlase E3500
10.6	350		11	2	Fast axial flow	Laser Electronics Pty Ltd	LE-503F
10.6	350	400	7.5	1.8	Rugged,bltin diagnostics	GSR Technologies Ltd	Modulase-400

Wave-lengths (μm)	Output TEM∞	(W) Multi-mode	*Beam dia (mm)	*Beam diverge (mrad)	Special Features	Manufacturer	Model no.
Carbon dioxide							
10.6	375		8	1 1	Fine cutting	Shibuya Kogyo Co Ltd	M46
10.6	400		10		Slow flow	A-B Lasers Inc	BLS400C
10.6	400		18	1.5		Tungsram Laser Technology	TLS 400
10.6	450		14	2	TEM 01	Baasel Lasertech	BLS 450 C
10.6	475		15	2	Enhanced pulsing	Lumonics Corp	V505
10.6	475		15	2	Cut/weld mode	Lumonics Corp	V500
10.6	500		12	1.5	Fast axial flow	Laser Electronics Pty Ltd	LE-505F
10.6	500		16	1.4	Inc gas recycler	Laser Dynamics Ltd	LX-500
10.6	500		18	1		Messer Griesheim GmbH	Eurolas 500
10.6	500		18	1.5	Fast axial flow	Matsushita Ind Equip Co Ltd	YB0506LA4
10.6	500	50-500	18	<1.5	Compact, pulsable	Panasonic Factory Automation C	YB0506LA-5
10.6	500	600	7	<2	Enhanced pulsing avail	Raytheon Co, Laser Products	GS-600R
10.6	500	1000	16	2	External mode switching	Mitsubishi International Corp	10C
10.6	550		7	1.7	Ultra compact, plsg aval	Laser Ecosse Ltd	MF600
10.6	550		10	1.5	Fine cutting	Shibuya Kogyo Co Ltd	S47
10.6	550		20	<1	Compact	Coherent General Inc	Everlase E7000
10.6	600		9	<2	Dual 600W beams	Raytheon Co, Laser Products	GS-1200-2
10.6	600		10	<2	+/-.5% pwr stab,plsing F	OPL (Oerlikon- PRC Laser SA)	OPL600
10.6	600		10	<2	TURBO FLOW™/gas econ	PRC Corp	PRC600
10.6	600		12	1	Fast axial flow	Laser Valfivre Sorgenti e Sist	LFA-700
10.6	600		18	1.5	Fast axial flow type	Toshiba Corp	C006-PSSB
10.6	700		17	2	Excellent compact, stabl	Daihen Corp	EL-751
10.6	700		18	1.5	Fast axial flow, pulsed	Rofin-Sinar Inc	RS700SM
10.6	700		18	1.5	Fast axial flow, pulsed	Rofin-Sinar Laser GmbH	RS700SM
10.6	700	800	7.5	1.8	Dual 400W beams	GSR Technologies Ltd	Modulase-800
10.6	750				2 beams @ 750 W ea	Coherent General Inc	Everlase S51-2
10.6	750		10	<1	RF plsd/100 kHz/CW/trbin	TRUMPF Inc	TLF 750T
10.6	750W		10	<1	RF exc, pulsed/CW	TRUMPF GmbH & Co	TLF 750 turbo
10.6	750		12	≥2	Enhanced pulse av. 750W	Mitsubishi Electric Corp	ML15SRP
10.6	750		15	1.5	Fast axial flow	Laser Electronics Pty Ltd	LE-508F
10.6	750		19	2	+/- 1% stability	Mitsubishi International Corp	15 SRP
10.6	750		28		Gas recycles, temp stab	Coherent General Inc	Everlase S48
10.6	750	800	15	1.7	Microprcsor cntrl & opr	Adron Sources S.A.	ALS 800

Wave-lengths (μm)	Output (W) TEM₀₀	Multi-mode	*Beam dia (mm)	*Beam diverge (mrad)	Special Features	Manufacturer	Model no.
Carbon dioxide							
10.6	750	1200	15	2	Compact and portable	Optomic Lasers Ltd	1CCL-750
10.6	750	1200	15	2	Compact and portable	Optomic Lasers Ltd	1CCL-750
10.6	775		16	1.5	Fine cutting	Shibuya Kogyo Co Ltd	S48
10.6	800	1000	12	<2	TURBO FLOW™/gas econ	PRC Corp	PRC1001
10.6	800	1000	15	1.8	Microprcsor cntrl & opr	Adron Sources S.A.	ALS 1000
10.6	900		12	<2	+/-.5% pwr stab,pls ftrs	OPL (Oerlikon- PRC Laser SA)	OPL1000
10.6	900		19			Coherent General Inc	S48 Plus
10.6	1000		12	≥2	Enhanced pulse av. 1000W	Mitsubishi Electric Corp	ML25SRP
10.6	1000W (00/01>)		12	<1	RF exc, pulsed/CW	TRUMPF GmbH & Co	TLF 1000 turbo
10.6	1000		13	<1	RF exc pulsed/CW/trbine	TRUMPF Inc	TLF 1000T
10.6	1000		15	≥1	Fast axial flow	Bystronic Inc	BL 1000
10.6	1000		16	<1	1-4000 Hz pulse/superpul	Bystronic Laser Ltd	BL 1000
10.6	1000		16	1.4	Inc gas recycler	Laser Dynamics Ltd	LX-1000
10.6	1000		16	1.5	Fast axial flow	Laser Electronics Pty Ltd	LE-510F
10.6	1000W		<17	<2	RF excited fast axial fl	GMFanuc Robotics Corp	C-1000
10.6	1000		17	<2	RF excitation, fst axial	Fanuc Ltd	C1000
10.6	1000		18	1.5	Fast axial flow	Tungsram Laser Technology	TLS 1000
10.6	1000		19	1.5	Fast axial flow	Matsushita Ind Equip Co Ltd	YB1006LA4
10.6	1000		19	2	Compact, easy use	Daihen Corp	EL-1101
10.6	1000		21	2	+/- 1% stability	Mitsubishi International Corp	25 SRP
10.6	1000	200-1000	20	<1.5	Compact, pulsable	Panasonic Factory Automation C	YB1006LA-5
10.6	1100		15.5		Slow axial flow	Coherent General Inc	Everlase Arrow
10.6	1200		9	<1.5	Fast axial flow	Raytheon Co, Laser Products	GS-1200
10.6	1200		18		Fine cutting	Shibuya Kogyo Co Ltd	S66
10.6	1200		18	1.5	Super pulse	Toyama America (NTC)	RS-1200SM
10.6	1200		18	1.5	Fast axial flow, pulsing	Rofin-Sinar Inc	RS1200SM
10.6	1200		18	1.5	Fast axial flow, pulsing	Rofin-Sinar Laser GmbH	RS1200SM
10.6	1200		20	1.5	Fast axial flow type	Toshiba Corp	C012-PSSB
10.6	1400		18	1.5	Fine cutting	Shibuya Kogyo Co Ltd	P15 (SS1500)
10.6	14000		50	1-2	Unstable resonator	United Technologies Corp	SM21-14
10.6	1500			<2	RF excited fast axial fl	GMFanuc Robotics Corp	C-1500
10.6	1500		9	<1.5	Fast axial flow	Raytheon Co, Laser Products	GS-1200 upgrade
10.6	1500		11	1.7	Moveable, pulsing avail	Laser Ecosse Ltd	MF1500

Wave-lengths (μm)	Output (W) TEM00	Multi-mode	*Beam dia (mm)	*Beam diverge (mrad)	Special Features	Manufacturer	Model no.
Carbon dioxide							
10.6	1500		13	<1	RF pulsed/CW	TRUMPF Inc	TLF 1500
10.6	1500W		13	<1	RF exc, pulsed/CW	TRUMPF GmbH & Co	TLF 1500
10.6	1500		13	<2	+/-.5% pwr stab,pls ftrs	OPL (Oerlikon- PRC Laser SA)	OPL1501
10.6	1500		13	<2	TURBO FLOW™/gas econ	PRC Corp	PRC1501
10.6	1500		15	2	Single enclosure	Electrox Ltd	M1200
10.6	1500		16		Enhanced fast axial flow	Coherent General Inc	Everlase S51
10.6	1500		16	1.5	Fine cutting	Shibuya Kogyo Co Ltd	SF52
10.6	1500		17	≥1	Fast axial flow	Bystronic Inc	BL 1500
10.6	1500		17	<1	1-4000 Hz pulse/superpul	Bystronic Laser Ltd	BL 1500
10.6	1500		19	2	Excellent compact, stabl	Daihen Corp	EL-1501
10.6	1500		20	<2	RF excitation, fst axial	Fanuc Ltd	C1500
10.6	1500		21	1.5	Fast axial flow	Matsushita Ind Equip Co Ltd	YB1506LA4
10.6	1500	300-1500	20	<1.5	Compact, pulsable	Panasonic Factory Automation C	YB1506LA-5
10.6	1500	2000	19	1.4	Trnsvrse, microproc-ctrl	Rofin-Sinar Inc	RS820
10.6	1500	2000	19	1.4	Trnsvrse, microproc-ctrl	Rofin-Sinar Laser GmbH	RS820
10.6	1500	2500	21/31	3	External mode switching	Mitsubishi International Corp	25C
10.6	1500	5000	14	≥3	Cutting wlding heat trtm	Mitsubishi Electric Corp	ML50C
10.6	1500	5000	20/40	3	External mode switching	Mitsubishi International Corp	50C
10.6	1600	1800	13	<2	TURBO FLOW™/gas econ	PRC Corp	PRC 2000
10.6	1600	2500	20	Adj	Pulsing avail	Laser Ecosse Ltd	AF2L
10.6	1700		17	1.5	Fast axial,pulsing RFexc	Rofin-Sinar Laser GmbH	RS1700RF
10.6	1700		17	1.5	Fast axial,pulsing RFexc	Rofin-Sinar Inc	RS1700RF
10.6	1700		20	1.5	Super pulse	Toyama America (NTC)	RS-1700SM
10.6	1700		20	1.5	Fast axial flow, pulsing	Rofin-Sinar Inc	RS1700SM
10.6	1700		20	1.5	Fast axial flow, pulsing	Rofin-Sinar Laser GmbH	RS1700SM
10.6	1750W (00/01>)		13	<1	RF exc, pulsed/CW	TRUMPF GmbH & Co	TLF1950
10.6	1750		18	1	Fast axial flow	Laser Valfivre Sorgenti e Sist	LFA-1750
10.6	1800		14	<2.1	+/-.5% pwr stab,pls ftrs	OPL (Oerlikon- PRC Laser SA)	OPL2000
10.6	1800		15	2	Single enclosure	Electrox Ltd	M1500
10.6	2000W (00/01>)		14	<1	RF exc, pulsed/CW	TRUMPF GmbH & Co	TLF2000
10.6	2000		<19	<2	RF excited fast axial fl	GMFanuc Robotics Corp	C-2000
10.6	2000		20	<2	RF excitation, fst axial	Fanuc Ltd	C2000
10.6	2000		22	2	Powerful, compact	Daihen Corp	EL-2001

Wave-lengths (μm)	Output (W) TEM₀₀	Multi-mode	*Beam dia (mm)	*Beam diverge (mrad)	Special Features	Manufacturer	Model no.
Carbon dioxide							
10.6	2000 TEM_{01}		25	2	Fast axial flow type	Toshiba Corp	C020-PLSB
10.6	2000	400-2000	22	<l.5	Compact, pulsable	Panasonic Factory Automation C	YB2006LA-5
10.6	2300					Bystronic Laser Ltd	BL 2300
10.6	2300		15	2	Single enclosure	Electrox Ltd	M2000
10.6	2300		20	≥1	RF excited	Bystronic Inc	BL 2300
10.6	2400	3600	20	Adj	Pulsing avail	Laser Ecosse Ltd	AF3L
10.6	25000		50	1-2	Unstable resonator	United Technologies Corp	SM41-25
10.6	3000		18		Enhanced axial flow	Coherent General Inc	Everlase Vulcan
10.6	3000		22	<3	RF exc pulsed/CW trbine	TRUMPF Inc	TLF 3000T
10.6	3000		<23	<2	RF excited fast axial fl	GMFanuc Robotics Corp	C-3000
10.6	3000		23	<2	RF excitation, fst axial	Fanuc Ltd	C3000
10.6	3000 TEM_{11}		27	3	Fast axial flow type	Toshiba Corp	C030-PMSB
10.6	3200	4800	20	Adj	Pulsing avail	Laser Ecosse Ltd	AF4L
10.6	4000	800	20	Adj	Pulsing avail	Laser Ecosse Ltd	AF8
10.6	4000	6000	20	Adj	Pulsing avail	Laser Ecosse Ltd	AF5L
10.6	45000		50	2	Unstable resonator	United Technologies Corp	SM61-45
10.6	5000 TEM_{11}	34	4		Fast axial flow type	Toshiba Corp	PMSB
10.6	5500		28x20	3.5x1.5	Rec bm shp wldg&ht trtmt	Laser Valfivre GmbH	C76
10.6	6000		50	1-2	Unstable resonator	United Technologies Corp	SM11-6
10.6	4000		34	3	Cut, wldg, htt trtmt	Laser Valfivre GmbH	C84
10.6	8000		32x20	3.5x1.5	Rec bm shp,wldg, httrtmt	Laser Valfivre GmbH	C88

Wave-lengths (μm)	Output TEM∞	(J) Multi-mode	Rep rate (Hz)	Pulse-length (μs)	*Beam dia (mm)	*Beam diverge (mrad)	Adv.Avg. Power, W	Special Features	Manufacturer	Model no.
.193-.351		.1-.25	1-200	.025-.014	23.5x7.5	<4x2		mat proc, mrkg	Lambda Physik GmbH	LAMBDA 1000 Series
.248 & .308		.075-.16	1-400	.01	23.5x7.5	<4x2		mat proc, mrkg	Lambda Physik GmbH	LAMBDA 2000 Series
.248 & .308		.5	1-300	.026-.03	28x8	<4x2		mat proc, mrkg	Lambda Physik GmbH	LAMBDA 3000 Series
9-11	100mJ		6mm	<3				Grating tuned TEA	Edinburgh Instruments Ltd	MTI-GT1
10.6	100 mJ	150 mJ	6mm	<3				Mini Tea	Edinburgh Instruments Ltd	MTL-1
Argon fluoride										
.193		.10	300	.010	6-8x25	2x3		Industrial	Lumonics Inc	INDEX® 210
193		.5	1-300	ns				Fully indust sys turnkey	Resonetics	REX193
Carbon dioxide										
9-11	.075		10	65	20x20	Multimode 5-7		100 lines inc R2, P2	Pulse Systems Inc	LP-30G
9-11	2.4		8-(25)	35	40x40	Multimode 5-7		Grating-tuned	Pulse Systems Inc	LP-140G
10.6			≥5000	≥150	8	<2		Dual beam,G,S,h-pulse	OPL (Oerlikon- PRC Laser SA)	OPL600/2
10.6			≥5000	≥150	10	<2		Dual beam,G,S,h-pulse	OPL (Oerlikon- PRC Laser SA)	OPL1500/2
10.6			≥5000	≥150	10	<2		Gated,super,hyperpulse	OPL (Oerlikon- PRC Laser SA)	OPL600
10.6			≥5000	≥150	11	<2		Dual beam,G,S,h-pulse	OPL (Oerlikon- PRC Laser SA)	OPL3000/2
10.6			0-10	250-50	5	5			Baasel Lasertech	BLS 60 C
10.6			0-10	250-50	7	2			Baasel Lasertech	BLS 170 C
10.6			0-2.5	250-50	10	1.5			Baasel Lasertech	BLS 270 C
10.6			10	1	23x16	10		Easy service	Shibuya Kogyo Co Ltd	SQ 2000
10.6			20	1	27x24	10		Constant energy output	Shibuya Kogyo Co Ltd	SQ 1000
10.6			10000	$100\text{-}10^7$	16	1.5		Integrated turbine laser	Lasers Industriels SA	Modulas 3 M
10.6			10000	$100\text{-}10^7$	20	1.5		Integrated turbine laser	Lasers Industriels SA	Modulas 4 M
10.6			10000	$100\text{-}10^7$	16	1.5		Integrated turbine laser	Lasers Industriels SA	Modulas 2 M
10.6			2500-10000	250-50-25	14	2			Baasel Lasertech	BLS 450 C
10.6		.035	20	65	20x20	Multimode 8		Complete system	Pulse Systems Inc	LP-15
10.6		.15	10	65	20x20	Multimode 8		Complete system	Pulse Systems Inc	LP-30
10.6		.25	10	20	20x20	Multimode 8		Built-in HeNe/IR detectr	Pulse Systems Inc	LP 60
10.6		.5	100	.001	25x25				Lumonics Inc	LaserMark® 948HS
10.6		.6	100	2	11x12	3		Low operating cost	Alltec GmbH & Co KG	Allmark 855
10.6		1	60	.001	25x25				Lumonics Inc	LaserMark® 948
10.6		1.2	50	1.5				TEA	A-B Lasers Inc	864
10.6		1.3	50	1.5	.8x.8	2.5		TEA	A-B Lasers Inc	854
10.6		1.3	50	2.5	18x16	3		Low operating cost	Alltec GmbH & Co KG	Allmark 854

Wave-lengths (μm)	Output (J) TEM$_{00}$	Multi-mode	Rep rate (Hz)	Pulse-length (μs)	*Beam dia (mm)	*Beam diverge (mrad)	Adv.Avg. Power, W	Special Features	Manufacturer	Model no.
Carbon dioxide										
10.6		1.5	50	2.5	13x26	3		Low operating cost	Alltec GmbH & Co KG	Allmark 864
10.6		1.6	25	.001	25x25				Lumonics Inc	LaserMark® 936
10.6		2	5	2	18x18	4			Coherent Hull Ltd	XM230
10.6		2	18	.001	25x25				Lumonics Inc	LaserMark® 934
10.6		2	30	.001	25x25				Lumonics Inc	LaserMark® 946
10.6		2.4	25	1.5	.75x1.34			TEA	A-B Lasers Inc	863
10.6		2.5	25	3	32x18	3		Low operating cost	Alltec GmbH & Co KG	Allmark 863
10.6		2.5	25	3	20x20	3		Low operating cost	Alltec GmbH & Co KG	Allmark 853
10.6		2.7	25	2	30x30	4		Thyratron	Coherent Hull Ltd	XL315
10.6		3	7	.001	25x25	L5		Microproc. Ctrl	Lumonics Inc	LaserMark® 930
10.6		3	8-(25)	35	40x40	Multimode 8		Complete systems	Pulse Systems Inc	LP-140
10.6		3	13	.001	25x25				Lumonics Inc	LaserMark® 932
10.6		3	20	.001	25x25				Lumonics Inc	LaserMark® 944
10.6		3	25	1.5	.8x.8	2.5		TEA	A-B Lasers Inc	853
10.6		3-5	12-20	2	29x22	7x5		Low cost consumables	Lasertechnics Inc	Blazer 6000
10.6		3.7	20	2	30x30	4		Thyratron	Coherent Hull Ltd	XL412
10.6		4	15	.001	25x25				Lumonics Inc	LaserMark® 942
10.6		4	15	1.5	.75x1.34			TEA	A-B Lasers Inc	862
10.6		4	15	1.5	.8x.8	2.5		TEA	A-B Lasers Inc	852
10.6		4	15	3	20x20	3		Low operating cost	Alltec GmbH & Co KG	Allmark 852
10.6		4	15	3	32x18	3		Low operating cost	Alltec GmbH & Co KG	Allmark 862
10.6		5	5	1.5	.8x.8	2.5		TEA	A-B Lasers Inc	851
10.6		5	5	3	21x21	3		Low operating cost	Alltec GmbH & Co KG	Allmark 851
10.6		5	12	.001	25x25				Lumonics Inc	LaserMark® 940
10.6		5	12	2	30x30			Thyratron	Coherent Hull Ltd	XL572
10.6		5.5	5	1.5	.75x1.34			TEA	A-B Lasers Inc	861
10.6		6	5	3	32x18	3		Low operating cost	Alltec GmbH & Co KG	Allmark 861
10.6		7	2	2	30x30	4		Thyratron	Coherent Hull Ltd	XL712
10.6		25-30	1-100		6			Price	Laser Nucleonics Inc	C-25
10.6		2000W	to 10000	30-CW	13	<2		GTD,super,Hyperpulse®	PRC Corp	PRC2001
10.6		2500 W	100-10000	10-CW	20	<1 half angle		RF exc pulsed/CW	TRUMPF Inc	TLF 2500
10.6		2500 W	100-100000	10-CW	20	<1.5		RF excitation	TRUMPF GmbH & Co	TLF 2500
10.6		3000W	to 10000	30-CW	16	<2.5		GTD,super,Hyperpulse®	PRC Corp	PRC3000

Wave-lengths (μm)	Output TEM∞	(J) Multi-mode	Rep rate (Hz)	Pulse-length (μs)	*Beam dia (mm)	*Beam diverge (mrad)	Adv.Avg. Power, W	Special Features	Manufacturer	Model no.
Carbon dioxide										
10.6		3000	100-100000	10-CW	22	<1		RF excitation	TRUMPF GmbH & Co	TLF3000 turbo
10.6		3500 W	100-10000	10-CW	24	<1 half angle		RF exc pulsed/CW	TRUMPF Inc	TLF 3500
10.6		4000 W	100-10000	10-CW	24	<1 half angle		RF plsd/100 kHz optn	TRUMPF Inc	TLF 4000
10.6		5000 W	100-10000	10-CW	22	<1 half angle		RF pulsed/CW/turbo	TRUMPF Inc	TLF 5000
10.6		5000 W	100-100000	10-CW	22	<1		RF excitation	TRUMPF GmbH & Co	TLF 5000 turbo
10.6		5000	1000	100-CW	23	1			Messer Griesheim GmbH	Eurolas 5000
10.6		6000W	100-100000	10-CW	24	<1.5		RF excitation	TRUMPF GmbH & Co	TLF6000
10.6		6000 W	100-10000	10-CW	24	<1 half angle		RF plsed/100 kHz optn	TRUMPF Inc	TLF 6000
10.6	1.3		≥5000	≥150	13	<2		Gated,super,hyperpulse	OPL (Oerlikon- PRC Laser SA)	OPL1501
10.6	1.7		≥5000	≥150	12	<2		Gated,super,hyperpulse	OPL (Oerlikon- PRC Laser SA)	OPL1000
10.6	2x1250 W		to 10000	30-CW	12	<2		GTD,super,Hyperpulse®	PRC Corp	PRC3000/2
10.6	2x250 W		to 10000	30-CW	8	<2		GTD,super,Hyperpulse®	PRC Corp	PRC600/2
10.6	2x750 W		to 10000	30-CW	10	<2		GTD,super,Hyperpulse®	PRC Corp	PRC1500/2
10.6	2.2		≥5000	≥150	14	2.1		Gated,super,hyperpulse	OPL (Oerlikon- PRC Laser SA)	OPL2000
10.6	2.5		≥5000	≥150	15	2.2		Gated,super,hyperpulse	OPL (Oerlikon- PRC Laser SA)	OPL2001
10.6	2.7		≥5000	≥150	16	2.3		Gated,super,hyperpulse	OPL (Oerlikon- PRC Laser SA)	OPL3000
10.6	4x700		to 10000	30-CW	10	<2		GTD,super,Hyperpulse®	PRC Corp	PRC3001/4
10.6	10		0-1000	200 - cw		<1.5		Super pulse to 10 kW	NTC/Marubeni America Corp	RS-1700-SM
10.6	10		0-1000	200 - cw		<1.5		Super pulse to 5 kW	NTC/Marubeni America Corp	RS-1200-SM
10.6	100		3300	100ns	11.5	6.5		Compact unit	Laser Ecosse Ltd	LE100
10.6	100 - 1200	1000 - 25000	6 - 100	.l - 5 m				Dif beam sources	ESAB Group	
10.6	140		3300	Up to CW	13	3		Enhanced pulsing	Lumonics Corp	V150
10.6	150				8	1.5		Slow axial flow	Rofin-Sinar Inc	RS150
10.6	250 W		50 kHz	00 nS	1.3	10		Q-switch waveguide	Laser Ecosse Ltd	CM2500
10.6	275 CW		2500	100	12.5	<1		Modular design	Coherent General Inc	Everlase E3500
10.6	475		2500	Up to CW	15	2		Cut/weld mode	Lumonics Corp	V500
10.6	475		2500	Up to CW	15	2		Enhanced pulsing	Lumonics Corp	V505
10.6	500 W		0-2500	100-CW	18	1.5		Fast axial	Matsushita Ind Equip Co Ltd	YB0506LA4
10.6	550 CW		2500	100	12.5	<1		Modular design	Coherent General Inc	Everlase E7000
10.6	550 W		3300	.01	7	1.7		CW available	Laser Ecosse Ltd	MF600P
10.6	600 W		to 10000	30-CW	10	<2		GTD,super,Hyperpulse®	PRC Corp	PRC600
10.6	750 W		100-10000	10-CW	10	<1 half angle		RF plsd/100 kHz/CW/turbo	TRUMPF Inc	TLF 750
10.6	750 W		100-100000	10-CW	10	<1		RF excitation	TRUMPF GmbH & Co	TLF 750 turbo

Wavelengths (μm)	Output TEM00	(J) Multimode	Rep rate (Hz)	Pulse-length (μs)	*Beam dia (mm)	*Beam diverge (mrad)	Adv.Avg. Power, W	Special Features	Manufacturer	Model no.
Carbon dioxide										
10.6	750 CW		2500	100	28			Temp-stabilizer	Coherent General Inc	Everlase S48
10.6	800	1000X	to 10000	30-CW	12	<2		GTD,super,Hyperpulse®	PRC Corp	PRC1001
10.6	900			100-CW	19				Coherent General Inc	Everlase S48 Plus
10.6	1000		0-2kHz	0-100	<17	<2		RF excited fast axial fl	GMFanuc Robotics Corp	
10.6	10000 (TEM_{20})		0-25000	.02-CW	50	2			Rofin-Sinar Laser GmbH	
10.6	10000 (TEM_{20})		0-25000	.02-CW	50	2			Rofin-Sinar Inc	
10.6	1000		4 kHz	10-150	15	≥1		Enhanced pulsing	Bystronic Inc	BL 1000
10.6	1000 W		50 kHz	100 nS	2.5	7		Q-switch waveguide	Laser Ecosse Ltd	CM3500QS
10.6	1000 W (00/01>)		100-100000	10-CW	12	<1		RF excitation	TRUMPF GmbH & Co	TLF 1000 turbo
10.6	1000W		100-10000	10-CW	12	<1 half angle		RF plsd/100 kHz/CW/turbo	TRUMPF Inc	TLF 1000
10.6	1000		10000	50	50-CW	1		RF-exited	Messer Griesheim GmbH	Eurolas 1000 HF
10.6	1100		CW-3.3 kHz	100	10	1.3		CW/pulsed circular polar	International Lasersmiths	Ferranti MFKP
10.6	1100 CW		1000		15.5				Coherent General Inc	Everlase Arrow
10.6	1200W		0-2500	.1-CW	18	1.5		Gated	Rofin-Sinar Laser GmbH	RS1200SM
10.6	1200W		0-2500	.1-CW	18	1.5		Gated	Rofin-Sinar Inc	RS1200SM
10.6	1200		2500	100-CW	18	1.5		Fast axial flow	Rofin-Sinar Inc	RS1200SM
10.6	1200		2500	100-CW	18	1.5		Fast Axial flow	Rofin-Sinar Laser GmbH	RS1200SM
10.6	1500		to 10000	30-CW	13	<2		GTD,super,Hyperpulse®	PRC Corp	PRC1501
10.6	1500		0-2kHz	0-100		<2		RF excited fast axial fl	GMFanuc Robotics Corp	C-1500
10.6	1500-W		0-2500	100-CW	19	1.5		Fast axial	Matsushita Ind Equip Co Ltd	YB1006LA4
10.6	1500-W		0-2500	100-CW	21	1.5		Fast axial	Matsushita Ind Equip Co Ltd	YB1506LA4
10.6	1500		4 kHz	10-150	17	≥1		Enhanced pulsing	Bystronic Inc	BL 1500
10.6	1500W (CW)		5-2000	5	14	2		RF	Niigata Engineering Co Ltd	C1500
10.6	1500		20-3000	Variable	19	2		Rectangular pulse	Mitsubishi International Corp	15 SRP
10.6	1500 W		100-100000	10-CW	13	<1		RF excitation	TRUMPF GmbH & Co	TLF 1500
10.6	1500 W		100-10000	10-CW	13	<1 half angle		RF pulsed/CW	TRUMPF Inc	TLF 1500
10.6	1500 CW		1000	250	16			Fast axial flow	Coherent General Inc	S51
10.6	1500 W		3300	.01	11	1.7		CW available	Laser Ecosse Ltd	MF1500P
10.6	1600W	1800W	to 10000	30-CW	13	<2		GTD,super,Hyperpulse®	PRC Corp	PRC2000
10.6	1600 W	2500 W	2500	.01	20	Adj		CW available	Laser Ecosse Ltd	AF2L
10.6	1700				12	1.5		Fast axial flow, RF-exc	Rofin-Sinar Inc	RS1700RF
10.6	1700				12	1.5		Fast axial flow, RF-exc	Rofin-Sinar Laser GmbH	RS1700RF
10.6	1700W (TEM_{10})		0-25000	.02-CW	17	1.5		RF-exc, pulsable	Rofin-Sinar Inc	RA1700RF

Wave-lengths (μm)	Output (J) TEM$_{00}$	Multi-mode	Rep rate (Hz)	Pulse-length (μs)	*Beam dia (mm)	*Beam diverge (mrad)	Adv.Avg. Power, W	Special Features	Manufacturer	Model no.
Carbon dioxide										
10.6	1700W (TEM$_{10}$)		0-25000	.02-CW	17	1.5		ARF-exc, pulsable	Rofin-Sinar Laser GmbH	RA1700RF
10.6	1700		0-2500	100-CW	20	1.5		Fast axial flow	Rofin-Sinar Inc	RS1700SM
10.6	1700		0-2500	100-CW	20	1.5		Fast axial flow	Rofin-Sinar Laser GmbH	RS1700SM
10.6	1750W (00/01>)		100-100000	10-CW	13	<1		RF excitation	TRUMPF GmbH & Co	TLF1750
10.6	2000		0-2kHz	0-100	<19	<2		RF excited fast axial fl	GMFanuc Robotics Corp	C-2000
10.6	2000W (00/01>)		100-100000	10-CW	14	<1		RF excitation	TRUMPF GmbH & Co	TLF2000
10.6	2000 W		100-10000	10-CW	14	<1 half angle		RF plsd/100 kHz optn	TRUMPF Inc	TLF 2000
10.6	2300		4 kHz	10-150	20	≥1		RF	Bystronic Inc	BL 2300
10.6	2400 W	3600 W	2500	.01	20	Adj		CW available	Laser Ecosse Ltd	AF3L
10.6	2500W (TEM$_{10}$)		0-2500	.02-CW	24	1.5		RD-exc, pulsable	Rofin-Sinar Laser GmbH	RS2500RF
10.6	2500W (TEM$_{10}$)		0-2500	.02-CW	24	1.5		RF-exc, pulsable	Rofin-Sinar Inc	RS2500RF
10.6	2500		0-25000	20-CW	22	1.5		Fast axial flow, RF-exc	Rofin-Sinar Inc	RS2500RF
10.6	2500		0-25000	20-CW	22	1.5		Fast axial flow, RF-exc	Rofin-Sinar Laser GmbH	RS2500RF
10.6	2500W (CW)		5-2000	5	20	2		RF	Niigata Engineering Co Ltd	C2000
10.6	2500		20-3000	Variable	19	2		Rectangular pulse	Mitsubishi International Corp	25 SRP
10.6	3000		0-2kHz	0-100	<23	<2		RF excited fast axial fl	GMFanuc Robotics Corp	C-3000
10.6	3000W (TEM$_{10}$)		0-25000	.02-CW	24	1.5		RFexc,pulsable,superpls	Rofin-Sinar Laser GmbH	RS3000RF
10.6	3000W (TEM$_{10}$)		0-25000	.02-CW	24	1.5		RFexc,pulsable,superpls	Rofin-Sinar Inc	RS3000RF
10.6	3000		0-25000	20-CW	30	1.5		Fast axial flow, RF-exc	Rofin-Sinar Laser GmbH	RS3000RF
10.6	3000		0-25000	20-CW	30	1.5		Fast axial flow, RF-exc	Rofin-Sinar Inc	RS3000RF
10.6	3000 CW		2 kHz		18				Coherent General Inc	Everlase Vulcan
10.6	3000W (CW)		5-2000	5	22	2		RF	Niigata Engineering Co Ltd	C3000
10.6	3200 W	4800 W	2500	.01	20	Adj		CW available	Laser Ecosse Ltd	AF4L
10.6	4000 W	6000 W	2500	.01	20	Adj		CW available	Laser Ecosse Ltd	AF5L
10.6	4000W	8000W	3300	100ns	20	Adj		CW available	Laser Ecosse Ltd	AF8
10.6	5000W		0-2500	.1-.4	18	1.5		FAF enhanced	Rofin-Sinar Inc	RS1200SM
10.6	5000W		0-2500	.1-.4	18	1.5		FAF enchanced	Rofin-Sinar Laser GmbH	RS1200SM
10.6	5000		50kHz	10ns	1.8	8		Cavity Dumped w/g	Laser Ecosse Ltd	CM3500CD
10.6	6000W (TEM$_{20}$)		0-25000	.02-CW	32	1.5		Remotely switchable	Rofin-Sinar Laser GmbH	RS6000RF
10.6	6000W (TEM$_{20}$)		0-25000	.02-CW	32	1.5		Remotely switchable	Rofin-Sinar Inc	RS6000RF
10.6	6000		0-25000	20-CW	35	1.5		Fast axial flow, RF exc	Rofin-Sinar Laser GmbH	RS6000RF
10.6	6000		0-25000	20-CW	35	1.5		Fast axial flow, RF exc	Rofin-Sinar Inc	RS6000RF
10.6	6000 W		2.5 kHz	100	18	1.5		Super pulse	Toyama America (NTC)	RS-1200SM

Wave-lengths (μm)	Output (J) TEM$_{00}$	Multi-mode	Rep rate (Hz)	Pulse-length (μs)	*Beam dia (mm)	*Beam diverge (mrad)	Adv.Avg. Power, W	Special Features	Manufacturer	Model no.
Carbon dioxide										
10.6	7000 W		2.5 kHz	100	20	1.5		Super pulse	Toyama America (NTC)	RS-1700SM
10.6	8000W		0-25000	.02-.4	17	1.5		Superpulsed	Rofin-Sinar Inc	RS1700RF
10.6	8000W		0-25000	.02-.4	17	1.5		Superpulsed	Rofin-Sinar Laser GmbH	RS1700RF
Krypton fluoride										
.248		.25	200	.016	8x25	2x3		Industrial	Lumonics Inc	INDEX® 200-K
248		1	1-500	ns				Fully indust sys turnkey	Resonetics	REX248
Neodymium:YAG										
1.06	25		0-25000					Q-switched, CW	Rofin-Sinar Inc	RSY25Q
1.06	25		0-25000					Q-switched, CW	Rofin-Sinar Laser GmbH	RSY25Q
1.06	30		0-25000					Q-switched	Rofin-Sinar Laser GmbH	RSY90Q
1.06	30		0-25000					Q-switched	Rofin-Sinar Inc	RSY90Q
1.06	40		0-25000					Q-switched	Rofin-Sinar Inc	RSY120Q
1.06	40		0-25000					Q-switched	Rofin-Sinar Laser GmbH	RSY120Q
1.06	50		0-25000					Q-switched	Rofin-Sinar Inc	RSY150Q
1.06	50		0-25000					Q-switched	Rofin-Sinar Laser GmbH	RSY150Q
1.06	90 (low order)							CW, OEM package avail	Rofin-Sinar Laser GmbH	RSY90CW
1.06	90 (low order)							CW, OEM package avail	Rofin-Sinar Inc	RSY90CW
1.06	120 (low order)							CW, OEM package avail	Rofin-Sinar Laser GmbH	RSY120CW
1.06	120 (low order)							CW, OEM package avail	Rofin-Sinar Inc	RSY120CW
1.06	150 (low order)							CW, OEM package avail	Rofin-Sinar Inc	RSY150CW
1.06	150 (low order)							CW, OEM package avail	Rofin-Sinar Laser GmbH	RSY150CW
1.06	500	80	0-500	100-20000				Active pulse shaping	Rofin-Sinar Inc	RSY500P
1.06	500	80	0-500	100-20000				Active pulse shaping	Rofin-Sinar Laser GmbH	RSY500P
1.06	1000	120	0-1000	100-20000				Active pulse shaping	Rofin-Sinar Laser GmbH	RSY1000P
1.06	1000	120	0-1000	100-20000				Active pulse shaping	Rofin-Sinar Inc	RSY1000P
Xenon										
480-520	.0005		2	1	4.8	1		Failure analysis	Florod Corp	LFA
480-520	.0007		30	1	4.8	1		Semiconductor uses	Florod Corp	MEL-30

Wave-lengths (μm)	Output TEM$_{00}$	(J) Multi-mode	Rep rate (Hz)	Pulse-length (μs)	*Beam dia (mm)	*Beam diverge (mrad)	Adv.Avg. Power, W	Special Features	Manufacturer	Model no.
Xenon chloride										
.308		.15	200	.012	8x25	2x3		Industrial	Lumonics Inc	INDEX® 200-X
.308		.3 long term	500	.025	29x19	5 full angle		Industrial 150W avg pwr	XMR Inc	5100
.308		.5 long term	300	.040	33x19	5 full angle		Industrial 150W avg pwr	XMR Inc	5110
.308		1	1-2 kHz	ns				Fully indust sys turnkey	Resonetics	REX308
Xenon fluoride										
.351		.5		ns				Fully indust sys turnkey	Resonetics	REX351

Wave-lengths (μm)	Output TEM$_{00}$	(W) Multi-mode	*Beam dia (mm)	*Beam diverge (mrad)	Special Features	Manufacturer	Model no.
Neodymium:YAG							
1.06		250	6	10	Efficient, single lamp	Lee Laser Inc	7250M
	12	40	4-6			Laser Material Processing Ltd	
.266	.25		1	d.l.	Q-sw, KTP, BBO	US Laser Corp	403TQU
.266	.6		.2	2.5	External BBO/Q-sw	Quantronix Corp	354-266
.355	.5		1.2	.6	External BBO/Q-sw	Quantronix Corp	354-355
.53	.002		1.4	1.5	Diode pumped	A-B Lasers Inc	DPY105
.53	.005		1.4	1	Diode pumped	A-B Lasers Inc	DPY115
.53	.010		1.4	1.5	Diode pumped	A-B Lasers Inc	DPY205
.53	.020		.1	7.5	Diode pumped	A-B Lasers Inc	DPY215
.53	.040		.1	7.5	Diode pumped	A-B Lasers Inc	DPY305
.53	.080		.1	7.5	Diode pumped	A-B Lasers Inc	OPY315
.53	1					Baasel Lasertech	BLS 642
.53	1					Baasel Lasertech	BLS 652
.53	1		.5	1.4	LiJ03 crystal	Baasel Lasertech	BLS 620
.53	1		1.2	2	Q-switched,freq dbld	Lee Laser Inc	708TQG
.53	2		.5	1.4	KTP crystal	Baasel Lasertech	BLS 621
.53	3.5-4.5		1	d.l.	Q-sw, KTP double	US Laser Corp	403TQG
.532		8	2.5	5	Intracavity crystal/Q-sw	Quantronix Corp	532R-M/QS-8.
.532	.002		.7	1.5	Diode laser pumped	Adlas GmbH & Co KG	DPY 105
.532	.005		.7	1.5	Diode laser pumped	Adlas GmbH & Co KG	DPY 115
.532	.01		.35	6	Frequency doubled	Amoco Laser Co	ALC532-10
.532	.010		.7	1.5	Diode laser pumped	Adlas GmbH & Co KG	DPY 205
.532	.015		.35	6	Frequency doubled	Amoco Laser Co	ALC532-15
.532	.020		.7	1.5	Diode laser pumped	Adlas GmbH & Co KG	DPY 215
.532	.040		.15	7	Diode laser pumped	Adlas GmbH & Co KG	DPU 305
.532	.080		.15	7	Diode laser pumped	Adlas GmbH & Co KG	DPY 315
.532	.150		.15	7	Diode laser pumped	Adlas GmbH & Co KG	DPY 2305
.532	.3		.5	2	40 ns Q-switch pulse	Quantronix Corp	532F-O/QS-.3
.532	1.5		.4	3	Intracavity crystal/Q-sw	Quantronix Corp	532F-O/QS-1.5
.532	2		1.5 max	2 max	Q-sw, up to 50 kHz	Controllaser	612QTG
.532	4		.5	2	Intracavity crystal/Q-sw	Quantronix Corp	532R-O/QS-4.
1	12		1	1.5	Diode laser pumped	Adlas GmbH & Co KG	DPY 201
1.06		6	2.7	6	Modular, hi power stabl	A-B Lasers Inc	BLS600

Wavelengths (μm)	Output TEM$_{00}$	(W) Multimode	*Beam dia (mm)	*Beam diverge (mrad)	Special Features	Manufacturer	Model no.
Neodymium:YAG							
1.06		25	4	8	Q-switched, compact	Lee Laser Inc	725MQ
1.06		25	4	8	CW, OEM special	Lee Laser Inc	725M
1.06		50	4		Options available	NEC Electronics Inc	YL114K
1.06		50	4		Q-switched/marking	Miyachi Technos Corp	ML-4220A
1.06		50	4	10	Q-switched, OEM special	Lee Laser Inc	750MQ
1.06		50	4	10	CW, fiberoptic option	Lee Laser Inc	750M
1.06		70	5	10	Q-swch, 125 kW peak pwr	Lee Laser Inc	775MQ
1.06		70-85	5	4-8	Low divergence, Q-sw	US Laser Corp	403H
1.06		75	5	10	CW, fiberoptic option	Lee Laser Inc	775M
1.06		100	6		Options available	NEC Electronics Inc	YL115K
1.06		100	6	10	CW, fiberoptic option	Lee Laser Inc	7100M
1.06		100	6	10	Q-swch, 150 ns @ 1 kHz	Lee Laser Inc	7100MQ
1.06		100	6.3		Q-switched/marking	Miyachi Technos Corp	ML-4210A
1.06		100-200	5	4-8	Low divergence, Q-sw	US Laser Corp	404L
1.06		120	2	3	Dry laser cavity	mls munich laser systems gmbh	MLS 50
1.06		120	3	7.5	Modular, hi power stabl	A-B Lasers Inc	BLS650
1.06		125-150	6.3	4-10	Low divergence, Q-sw	US Laser Corp	405L
1.06		150	6	10	Q-swch, 200 kW peak pwr	Lee Laser Inc	7150MQ
1.06		150	6	10	CW,	Lee Laser Inc	7150M
1.06		200			Laser cutter	NEC Electronics Inc	M702A
1.06		250	6.3	12-16	Lsr only, or turnkey	US Laser Corp	405-1
1.06		250	8	15	Modulation, 500 W peak	Miyachi Technos Corp	ML-1330C
1.06		300	8		Options available	NEC Electronics Inc	YL117C
1.06		500	6x25	5x15	"Slab" crystal	mls munich laser systems gmbh	MLS P500
1.06		500	6	12	Rugged industrial	Lee Laser Inc	7250MX2
1.06		500	6.3	12-18	Lsr only, or turnkey	US Laser Corp	405-2
1.06		600	8		Options available	NEC Electronics Inc	YL117-2C
1.06		750	6	12	OEM configurations	Lee Laser Inc	7250MX3
1.06		900	8		Options available	NEC Electronics Inc	YL117-3C
1.06		1000	6x25	5x15	'Slab' crystal	mls munich laser systems gmbh	MLS P1000
1.06		1000	6	12	220,380/415,440 VAC	Lee Laser Inc	7250MX4
1.06		1000	6.3	14-20	Lsr only, or turnkey	US Laser Corp	405-4
1.06		1200	8		Options available	NEC Electronics Inc	YL117-4C

Wave-lengths (μm)	Output TEM$_{00}$	(W) Multi-mode	*Beam dia (mm)	*Beam diverge (mrad)	Special Features	Manufacturer	Model no.
Neodymium:YAG							
1.06		1800	8		Options available	NEC Electronics Inc	YI117-6C
1.06	.05		.3	1.5	Diode pumped, pulsable	A-B Lasers Inc	DPY101
1.06	.12		1	1.5	Diode pumped, pulsable	A-B Lasers Inc	DPY201
1.06	.25	1	1	<2	Low cost, gen purpose	Line Lite Laser Corp	605E
1.06	.35		1	1.8	Diode pumped	A-B Lasers Inc	DPY301
1.06	1	7	1	<2	Gen purpose, stable	Line Lite Laser Corp	607C
1.06	2		1.2	2	Stability <1%	Lee Laser Inc	703ST
1.06	3	20	1.2	2	Q-switched, compact	Lee Laser Inc	703TQ
1.06	3	20	1.2	2	CW, stable, durable	Lee Laser Inc	703T
1.06	6		1.2	2	Stability <1%	Lee Laser Inc	708ST
1.06	6	50	4	10		JEC Lasers Inc	2500
1.06	6	60	6	5	Extended lamp life	Electrox Inc	Y60C
1.06	7				Options available	NEC Electronics Inc	YL114L
1.06	7	70	4	12		JEC Lasers Inc	3000
1.06	7-9		1	d.l.	Tunable, Q-sw	US Laser Corp	403C
1.06	8	40	1.2	2	CW,durable,hi stabl optn	Lee Laser Inc	708T
1.06	8	40	1.2	2	Q-swch, 15 kW peak power	Lee Laser Inc	708TQ
1.06	10		1.2	2	Stability <1%	Lee Laser Inc	712ST
1.06	10	75	2	.8	Dry laser cavity	mls munich laser systems gmbh	MLS 35
1.06	10	120	6	5	Extended lamp life	Electrox Inc	Y102C
1.06	12				Options available	NEC Electronics Inc	YL115L
1.06	12		.7	2	Modular, hi power stabl	A-B Lasers Inc	BLS615
1.06	12	45	1.2	2	CW,durable,hi stabl optn	Lee Laser Inc	712T
1.06	12	45	1.2	2	Q-swch, <100 ns @ I kHz	Lee Laser Inc	712TQ
1.06	12 W	60 W	5	1.2	Gen purpose, Kr pumped	Line Lite Laser Corp	611
1.06	12-14		1	d.l.	Tunable, Q-sw	US Laser Corp	403T
1.06	15		.7	2	Modular, hi power stabl	A-B Lasers Inc	BLS610
1.06	15		1.2	2	Stability <1%	Lee Laser Inc	718ST
1.06	15	50	1.2	2	Rugged industrial	Lee Laser Inc	715T
1.06	15	50	1.2	2	Q-swch, 37 kW peak power	Lee Laser Inc	715TQ
1.06	15	60			Marking system, to 120 W	A-B Lasers Inc	LBI6000
1.06	16				Options available	NEC Electronics Inc	YL115J
1.06	18	60	1.2	2	CW-high stability option	Lee Laser Inc	718T

Wavelengths (μm)	Output (W) TEM∞	Multi-mode	*Beam dia (mm)	*Beam diverge (mrad)	Special Features	Manufacturer	Model no.
Neodymium:YAG							
1.06	18	60	1.2	2	Q-swch, 45 kW peak power	Lee Laser Inc	718TQ
1.06	18-20		1	d.l.	High power, tunable	US Laser Corp	404T
1.06	50	60	5	5	Q-switched	Synrad Inc	Y50-1
1.064		25	4	4.7	Self cooled, light cont	Elettronica Valseriana	EV 25 MED
1.064		25	4	10	OEM	Elettronica Valseriana	AMG 25
1.064		35	3	7		Quantronix Corp	114F-M/CW
1.064		40	2.3	5.5	High brightness	Quantronix Corp	117F-L/CW
1.064		50	3	7		Quantronix Corp	116F-M/CW
1.064		50	4	10 max		Controllaser	612
1.064		50	4	10 max	Q-sw, up to 50 kHz	Controllaser	612Q
1.064		60	2.7	6		Baasel Lasertech	BLS 600
1.064		60	4	5.5	Self cooled, light cont	Elettronica Valseriana	EV 60 AS
1.064		60	5	10	OEM	Elettronica Valseriana	AMG 60
1.064		75	6	10 max	Q-sw, up to 50 kHz	Controllaser	630Q
1.064		80	3	6		Baasel Lasertech	BLS 601
1.064		80	5	7	Rugged and compact	Quanta System Srl	QY-100
1.064		100	3	5.5	High brightness	Quantronix Corp	118F-L/CW
1.064		100	5	4.9	Self cooled, light cont	Elettronica Valseriana	EV 132 AS
1.064		100	5	4.9	Light controlled	Elettronica Valseriana	EV 100 AS
1.064		100	5	10	OEM	Elettronica Valseriana	AMG 100
1.064		100	6	10 max		Controllaser	630
1.064		100	6.	11.		Quantronix Corp	117F/M/CW
1.064		120	4	9		Baasel Lasertech	BLS 602
1.064		325	7	20		Quantronix Corp	118F-M/CW
1.064	.05		.4	7		Amoco Laser Co	ALC1064-50P
1.064	.05		1	1.5	Diode laser pumped	Adlas GmbH & Co KG	DPY 101
1.064	.15		.4	5		Amoco Laser Co	ALC1064-150P
1.064	.35		.5	4		Amoco Laser Co	ALC1064-350P
1.064	.35		1	1.5	Diode laser pumped	Adlas GmbH & Co KG	DPY 301
1.064	.75		1	1.5	Diode laser pumped	Adlas GmbH & Co KG	DPY 2301
1.064	6		.8	2.3	High stability,Q-swtched	Quantronix Corp	114F-O/CW
1.064	6		1.5 max	2 max	Q-sw, up to 50 kHz	Controllaser	612QT
1.064	8		1.5 max	2 max		Controllaser	612T

Wave-lengths (μm)	Output TEM₀₀	(W) Multi-mode	*Beam dia (mm)	*Beam diverge (mrad)	Special Features	Manufacturer	Model no.
Neodymium:YAG							
1.064	8	60	4	1	Small size laser	Chromatron Laser Systems	Induwrite
1.064	8	60	4	2	Modular, serviceable	Chromatron Laser Systems	Chromat 1000
1.064	9		.7	2		Baasel Lasertech	BLS 613
1.064	10		.7	2		Baasel Lasertech	BLS 650
1.064	10		.7	2		Baasel Lasertech	BLS 640
1.064	10 W	100 W			ILM marking system	International Lasersmiths	ILM #0120
1.064	12		.7	2		Baasel Lasertech	BLS 610
1.064	12		.8	2.5	Available for all TEMoo	Quantronix Corp	116R-O/CW
1.064	15		.7	2		Baasel Lasertech	BLS 611
1.064	15		1.5 max	2 max		Controllaser	620T
1.064	15		1.5 max	2 max	Q-sw, up to 50 kHz	Controllaser	620QT
1.064	16		.7	2.5	Models & most MM models	Quantronix Corp	117R-O/CW
1.064	20		.5	3	High power temoo	Quantronix Corp	126R-O/CW-20
1.064	20		.7	2		Baasel Lasertech	BLS 612
1.064	50		1.5	2	High beam quality Q-SW	Quanta System Srl	QY-50 SM
1.064	75	730				CONTEC GmbH	
1.064	75	770				CONTEC GmbH	
1.3	3		.7	2		Baasel Lasertech	BLS 630
1.319	.02		1	2	Diode laser pumped	Adlas GmbH & Co KG	DPY 103
1.319	.04		1	2	Diode laser pumped	Adlas GmbH & Co KG	DPY 203
1.319	.12		1	2	Diode laser pumped	Adlas GmbH & Co KG	DPY 303
1.319	1.		.9	2.7	Q-sw available	Quantronix Corp	114SRW-O/CW-2L
1.319	2		1.1	2	Q-sw available	Quantronix Corp	116SRW/O/CW-2L
1.319	4		.8	2.8	Q-sw available	Quantronix Corp	117SRW-O/CWQ-2L
1.32	up to 3	up to 30	1.2-6	2-10	CW & Q-switched	Lee Laser Inc	Series 700
1.32		25	5	4.9	Self cooled, light cont	Elettronica Valseriana	EV 132 AS
1.32		75-100	6.3	6-12	High pwr spcl wvlgth	US Laser Corp	400 Series
1.32	.02		.9	1.5	Diode pumped, pulsable	A-B Lasers Inc	DPY103
1.32	.04		1	1.5	Diode pumped, pulsable	A-B Lasers Inc	DPY203
1.32	.12		1.5	1.6	Diode pumped	A-B Lasers Inc	DPY303
1064		50	4	10	Q-switched	Quantronix Corp	612 Q
1064		70	6	10	Q-switched	Quantronix Corp	630 Q
1064		100	6	10	CW	Quantronix Corp	630

Wave-lengths (μm)	Output (W) TEM$_{00}$	Multi-mode	*Beam dia (mm)	*Beam diverge (mrad)	Special Features	Manufacturer	Model no.
Neodymium:YAG							
1064	6		1	2.5	Q-switched	Quantronix Corp	612 QT
1064	8		1	2.5	CW	Quantronix Corp	612 T
1064	15		1.5	2	Q-switched	Quantronix Corp	620 QT

Wave-lengths (μm)	Output TEM∞	(J) Multi-mode	Rep rate (Hz)	Pulse-length (μs)	*Beam dia (mm)	*Beam diverge (mrad)	Special Features	Manufacturer	Model no.
Neodymium									
1.06		50	5-300	300-200000	10	.3	Lumonics	Madrid Laser	JK 704
Neodymium:glass									
1.06		50	3	10000	7	8		Haas Laser GmbH	AG 10
1.06		60	3	10000	7	10		Haas Laser GmbH	AG 15
1.06	15		3	65,100,150	6		High rep rates	Coherent General Inc	M9A
1.06	40		1	.4,.6,.95,1.2	6			Coherent General Inc	M11E
1.06	40		1	.5,1,3	3x9		Rectangular beam	Coherent General Inc	M14D
1.06	40		1	.6,1,3	6			Coherent General Inc	M11F
1.06	40		1	1,3,5	3x9		Rectangular beam	Coherent General Inc	M14C
1.06	40		1	1,3,5	6			Coherent General Inc	M11D
1.06	40		1	1,3,5,8	6			Coherent General Inc	M11G
1.06	40		1	3	6			Coherent General Inc	M11A
1.06	40		1	3,4,8	3x9		Rectangular beam	Coherent General Inc	M14E
1.06	40		1	3,5	6			Coherent General Inc	M11B
1.06	40		1	3,5	3x9		Rectangular beam	Coherent General Inc	M14B
1.06	40		1	3	3x9		Rectangular beam	Coherent General Inc	M14A
1.06	40	1	1,3	6				Coherent General Inc	M11C
1.06	75		1	.75,.95	8		Drilling holes	Coherent General Inc	M18
Neodymium:YAG									
.266	$2x10^{7}$		5000	.025	1	.5	Diode laser pumped	Adlas GmbH & Co KG	DPY 201 Q quadrupled
.266	$5x10^{8}$		5000	.035	1	.5	Diode laser pumped	Adlas GmbH & Co KG	DPY 101 Q quadrupled
.266	10^{6}		5000	.015	1	.5	Diode laser pumped	Adlas GmbH & Co KG	DPY 301 Q quadrupled
.53	.2		l-50 K	.l	.5	l.4	Q-switched	A-B Lasers Inc	BLS611
.53	.4		l-50 K	.1	.5	l.5	Q-switched	A-B Lasers Inc	BLS612
.53	1.210^{-5}		76/84 M	$7x10^{-5}$			Mode locked	A-B Lasers Inc	BLS635
.53	10^{-5}		100 M	$7x10^{-5}$			Mode locked	A-B Lasers Inc	BLS625
.532	$3x10^{6}$		5000	.025	1	.8	Diode laser pumped	Adlas GmbH & Co KG	DPY 201 Q doubled
.532	10^{5}		5000	.015	1	.8	Diode laser pumped	Adlas GmbH & Co KG	DPY 301 Q doubled
.532	10^{6}		5000	.035	1	.8	Diode laser pumped	Adlas GmbH & Co KG	DPY 101 Q doubled
1.06			40,40,20	30,40,50J				Coherent General Inc	Everpulse Gamma

Wave-lengths (μm)	Output TEM$_{00}$	(J) Multi-mode	Rep rate (Hz)	Pulse-length (μs)	*Beam dia (mm)	*Beam diverge (mrad)	Special Features	Manufacturer	Model no.
Neodymium:YAG									
1.06		.1	.5-5	.008-.012	6	5	Single-shot, burst mode	Line Lite Laser Corp	622
1.06		.1-26	.1-50	.1-20 ms	6	2-10	Multi-spot beam scanning	Lasag Corp	KLS 026
1.06		.1-35	.1-300	.1-20 ms	6	2-10	Multi-spot beam scanning	Lasag Corp	KLS 112
1.06		.1-70	.1-300	.1-20 ms	6	2-10	Fiberoptic beam delvy	Lasag Corp	KLS 322
1.06		.1-100	.1-300	.1-20 ms	9	2-12	Fiberoptic beam delvy	Lasag Corp	KLS 522
1.06	NA	1	100	.1 ms		2	Hole U, kerf widths<10um	Lumonics Ltd	MS35LD
1.06		2.5	0-100	200-5000	5	2-6	Lnr lamp, low divergence	US Laser Corp	303
1.06		6	0-100	200-5000	6.3	2-8	Lnr lamp, low divergence	US Laser Corp	304
1.06		10	0-50	500-10000	6.3	4-10	Linr lamp,low divergence	US Laser Corp	305
1.06		10	1-50 K	.15	2.7	6	Q-switched	A-B Lasers Inc	BLS600
1.06		12.5	1	600-10000	6.5	10		Cheval Freres SA	CF 11-10 P
1.06		15	0-500	50-20000	6.5	15		Cheval Freres SA	CF 11-100 P
1.06	NA	15	500	.5 ms-20000		2	Pulse-shaping	Lumonics Ltd	JK703
1.06		18	0-500	50-20000	6.5	13		Cheval Freres SA	CF 11-150 P
1.06		18	1-50 K	.1	3	7.5	Q-switched	A-B Lasers Inc	BLS650
1.06		20	0-200	100-9900	6.3	6	Welding	Miyachi Technos Corp	ML-2100A
1.06		20	2	8	6		Welding	Raytheon Co, Laser Products	SS384
1.06		25	200	.35-10	8		Enhanced pulse	Raytheon Co, Laser Products	SS525
1.06		30	0-5	300-9900	8	8	Desktop/spot welding	Miyachi Technos Corp	ML-2220A
1.06		30	0-10	300-20000	6.3	6	Compuct/spot welding	Miyachi Technos Corp	ML-2230B
1.06		30	0-200	100-9900	8	8	Welding	Miyachi Technos Corp	ML-2200A
1.06		30	200	10000	9	4		Haas Laser GmbH	AY 50
1.06		35 W	1-100	.1			Micro-driller/cutter	Lumonics Corp	MS35LD
1.06	NA	35	500	.5 ms-cw	10-20	<2-7	Microprocessor control	Lumonics Ltd	JK702
1.06		40	200	10000	7	9		Haas Laser GmbH	AY 100
1.06		40	200	10000	8	8		Haas Laser GmbH	AY 75
1.06		45	0-500	50-20000	7.5	22		Cheval Freres SA	CF 12-200 P
1.06		50	0-10	300-20000	8	8	Compuct/spot welding	Miyachi Technos Corp	ML-2330B
1.06		50	0-200	100-9900	10	10	Welding	Miyachi Technos Corp	ML-2300A
1.06		50	0-200	100-9900	10	15	High repitition/400 W	Miyachi Technos Corp	ML-2600B
1.06		50	0-500	50-20000	7.5	22		Cheval Freres SA	CF 12-300 P
1.06		50	10	8	6		Welding	Raytheon Co, Laser Products	SS484
1.06		50	200	.35-10	8	8-20	Enhanced pulse	Raytheon Co, Laser Products	SS550

Wave-lengths (μm)	Output TEM$_{00}$	(J) Multi-mode	Rep rate (Hz)	Pulse-length (μs)	*Beam dia (mm)	*Beam diverge (mrad)	Special Features	Manufacturer	Model no.
Neodymium:YAG									
1.06	NA	50	300	.3-5 ms	15-30	1-4	Peak power	Lumonics Ltd	JK704
1.06		50	500	10000	10	10		Haas Laser GmbH	AY 300
1.06	NA	55	500	.5 ms-cw	15-30	2-10	Fixed plug-in optics	Lumonics Ltd	JK701
1.06	NA	55	500	3-20 ms	15-30	2-10	Drills holes	Lumonics Ltd	JK704
1.06		75	0-500	50-20000	9.5	22		Cheval Freres SA	CF 12-450 P
1.06		80	0-200	100-9900	10	10	Welding/250W	Miyachi Technos Corp	ML-2500A
1.06		95	0-500	50-20000	7.5	20		Cheval Freres SA	CF 24-600 P
1.06		100	0-50	100-20000	10	10	Pulse shape control	Miyachi Technos Corp	ML-2360A
1.06		100	1-200	.5-20	6	5	Extended lamp life	Electrox Inc	Y100P
1.06	NA	100	500	300-5000	10-30	2-6		Lumonics Ltd	JK708
1.06	NA	100	500	500-20000	10-30	2-10		Lumonics Ltd	JK706
1.06	NA	100	500	500-20000	10-30	2-6	High speed thick cutting	Lumonics Ltd	JK707
1.06		120 W	1-500	.5-20			Spot welding	Lumonics Corp	JK703
1.06		150	0-500	50-20000	9.5	24		Cheval Freres SA	CF 24-900 P
1.06		200	1-200	.5-20	9	5	Extended lamp life	Electrox Inc	Y200P
1.06		250 W	1-500	.5-20			Industrial duty	Lumonics Corp	JK702
1.06		400 W	1-500	.3-20			Deep-hole drilling	Lumonics Corp	JK704
1.06		400 W	1-500	.5-20			Industrial duty	Lumonics Corp	JK701
1.06		400	1-500	.5-20	9	5	Extended lamp life	Electrox Inc	Y400-HP
1.06		600 W	1-500	.5-20			Deep cutting	Lumonics Corp	JK707
1.06		800	1-500	.2-20	10	7	Extended lamp life	Electrox Inc	Y800P
1.06		800 W	1-500	.3-20			High speed drilling	Lumonics Corp	JK708
1.06		1000 W	1-500	.5-20			High power	Lumonics Corp	JK706
1.06		1000 W	1-1500	.5 to CW			Fiberoptic beam dlvy	Lumonics Corp	Multilase
1.06		1600	1-1000	.2-20	10	9	Extended lamp life	Electrox Inc	Y1600P
1.06	.008-.030		.5-5	.01-.015	1	<2	Short pulsed, air cooled	Line Lite Laser Corp	620
1.06	.030		5	.01-.015	1	<2	Short pulsed, air cooled	Line Lite Laser Corp	621
1.06	.1-28	.1-300	.1-20ms	6	2-10		Cruise ctrl,sgl flslamp	Lasag Corp	KLS312
1.06	.1-130	.1-300	.1-20ms	9	2-12		60 kW peak power	Lasag Corp	KLS532
1.06	.5		1-50 K	.18	.7	2	Q-switched	A-B Lasers Inc	BLS615
1.06	1.210^{-4}		76/84 M	10^{-4}	.7	2	Mode locked	A-B Lasers Inc	BLS630
1.06	3		1-50 K	.1	.7	2	Q-switched	A-B Lasers Inc	BLS610
1.06	10^{-4}		100 M	10^{-4}	.7	2	Mode locked	A-B Lasers Inc	BLS620

Wave-lengths (μm)	Output TEM∞	(J) Multi-mode	Rep rate (Hz)	Pulse-length (μs)	*Beam dia (mm)	*Beam diverge (mrad)	Special Features	Manufacturer	Model no.
Neodymium:YAG									
1.06	15	0-300	.1-.5ms	5x5	.2-1.5		Full In beam delv attach	Lasag Corp	Lasag Slab Laser
1.06	15m5	25m5	1	$14x10^3$	4	1.5	Compact	Coteglade Photonics Ltd	GM105
1.06	50		40,40,20	30,40,50J	10		Deep hole drilling	Coherent General Inc	Everpulse Omega
1.064	$\geq$60		015	<20 ns	$\geq$.4	$\geq$5	Collimator available	Amoco Laser Co	ALC1064-1000Q
1.064	$\geq$25		015	<23 ns		$\geq$.3 $\geq$6	Collimator available	Amoco Laser Co	ALC1064-500Q
1.064		1	96	.65 ms	18	5		Controllaser	480-16
1.064		1.5	30	500-10000	5	4	OEM	Elettronica Valseriana	ILEVI 5
1.064		3	20		6	3	Air-cooled	Chromatron Laser Systems	K
1.064		3	56	1.5 ms	18	5		Controllaser	480-16
1.064		4	48	.65 ms	18	5		Controllaser	440-8
1.064		5	20	1000-6000	6	4	OEM	Elettronica Valseriana	ILEV 5
1.064		5	32	4 ms	18	5		Controllaser	480-16
1.064		8	30	100-5000	6	6	OEM	Elettronica Valseriana	ND 3000
1.064		15	1-300	100-20000	6	6		Baasel Lasertech	BLS 700
1.064		16	13	.65 ms	18	5		Controllaser	440-8
1.064		20	0-300	500-20000	7	7	Compact design	Quanta System Srl	QS100W
1.064		20	300	500-25000	6	6	OEM	Elettronica Valseriana	ND 5000
1.064		25	1-300	100-20000	6	7		Baasel Lasertech	BLS 710
1.064		32	6	1.5 ms	18	5		Controllaser	440-8
1.064		32	12	.65 ms	18	5		Controllaser	480-16
1.064		35	0-300	500-20000	7	10	Field proven tech	Quanta System Srl	QS250W
1.064		35	1-600	100-20000	6	8		Baasel Lasertech	BLS 720
1.064		50	l-600	300-20000	6	12		Baasel Lasertech	BLS 730
1.064		50	0-300	500-20000	8	15	Field proven tech	Quanta System Srl	QS400W
1.064		60	4	4 ms	18	5		Controllaser	440-8
1.064		60	6	1.5 ms	18	5		Controllaser	480-16
1.064		75	1-500	0-20	.375	15	800 W average power	International Lasersmiths	ILM 1800
1.064		120	4	4 ms	18	5		Controllaser	480-16
1.064		150	1-500	0-20	.375	20	1600 W average power	International Lasersmiths	ILM
1.064	.2		9.9	200			115Vac self contained	Quantronix Corp	294YO
1.064	1.		9.9				115Vac	Quantronix Corp	294YM
1.064	$5x10^6$		5000	.035	1	1.5	Diode laser pumped	Adlas GmbH & Co KG	DPY 101 Q
1.064	5	30	100-10000					CONTEC GmbH	

Wavelengths (μm)	Output TEM$_{00}$	(J) Multimode	Rep rate (Hz)	Pulse-length (μs)	*Beam dia (mm)	*Beam diverge (mrad)	Special Features	Manufacturer	Model no.
Neodymium:YAG									
1.064	8	10	170			.5	Compact	Chromatron Laser Systems	CLP 2000
1.064	$15x10^6$		5000	.025	1	1.5	Diode laser pumped	Adlas GmbH & Co KG	DPY 201 Q
1.064	$40x10^6$		5000	.015	1	1.5	Diode laser pumped	Adlas GmbH & Co KG	DPY 301 Q
1.32				gated	.7	2	Q-switched	A-B Lasers Inc	BLS613
2.1		.8					115Vac	Quantronix Corp	
2.936		1	9.9	200	1.1	3	115Vac cooling	Quantronix Corp	294EO
2.936		1	9.9	200	3.5	9	115Vac	Quantronix Corp	294EM
1064			6000	60 ns	3	4	Spec modfd to saw dimnds	Nanavati Sales Pvt Ltd	986
1064 nm	10 mJ		1	70	2 um		Microscope mount	Alessi Inc	LYI
Neodymium:YLF									
.262	$2x10^6$		2500	.015	1	.5	Diode laser pumped	Adlas GmbH & Co KG	DPF 301 Q quadrupled
.262	$4x10^7$		2500	.025	1	.5	Diode laser pumped	Adlas GmbH & Co KG	DPF 201 Q quadrupled
.262	10^7		2500	.035	1	.5	diode laser pumped	Adlas GmbH & Co KG	DPF 101 Q quadrupled
.523	$2x10^5$		2500	.015	1	.8	Diode laser pumped	Adlas GmbH & Co KG	DPF 301 Q doubled
.523	$2x10^6$		2500	.035	1	.8	Diode laser pumped	Adlas GmbH & Co KG	DPF 101 Q doubled
.523	$3x10^6$		2500	.025	1	.8	Diode laser pumped	Adlas GmbH & Co KG	DPF 201 Q doubled
1.047	$3x10^5$		2500	.025	1	1.5	Diode laser pumped	Adlas GmbH & Co KG	DPF 201 Q
1.047	$8x10^5$		2500	.015	1	1.5	Diode laser pumped	Adlas GmbH & Co KG	DPF 301 Q
1.047	10^5		2500	.035	1	1.5	Diode laser pumped	Adlas GmbH & Co KG	DPF 101 Q
Ruby									
.69		10	3	1.5	.250		Computer operated	Advanced Laser Systems Inc	404 (Microwelder)
.69		10	5	Var	.200		Computer operated	Advanced Laser Systems Inc	604 (Driller)
6.943		25-30	1-100		6		Price	Laser Nucleonics Inc	C-25

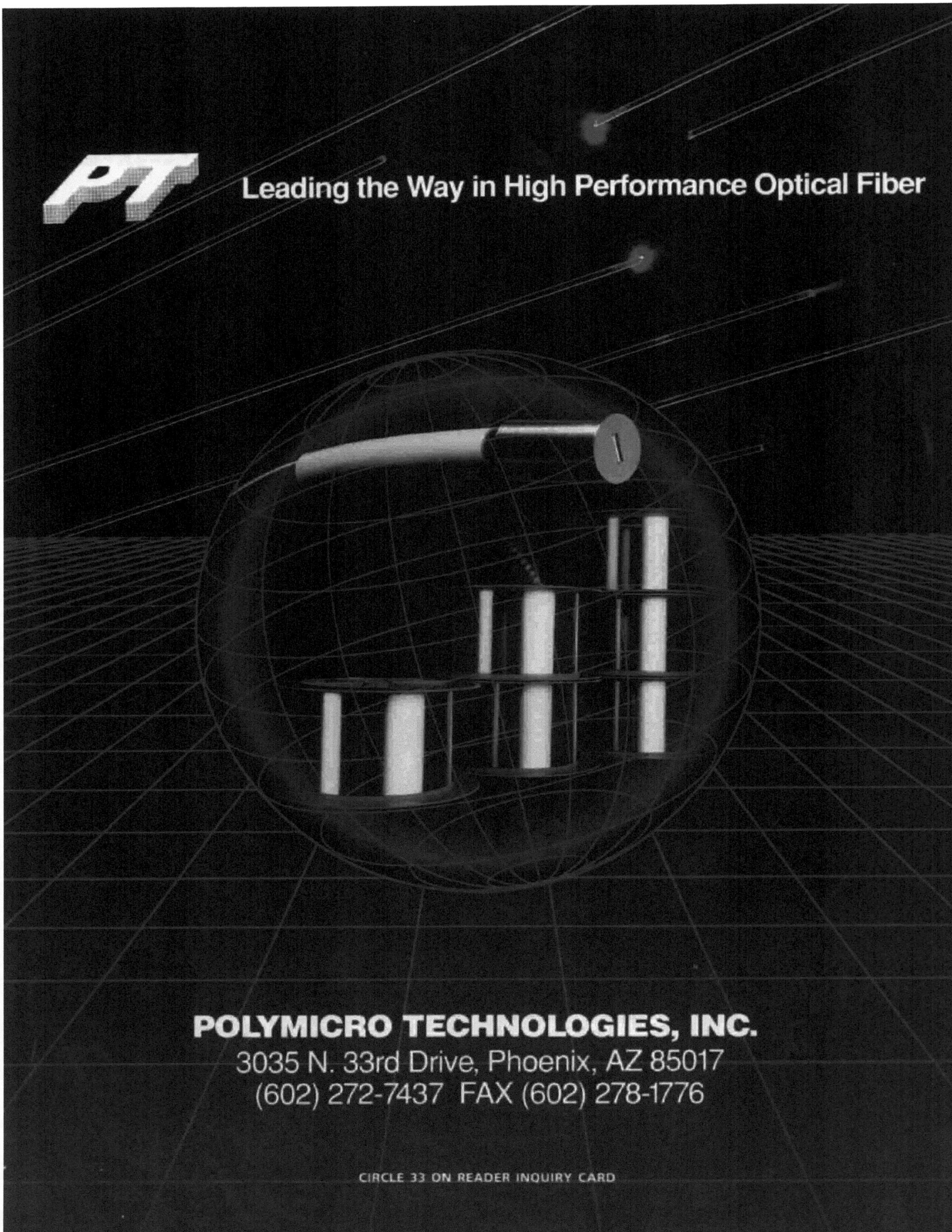
Leading the Way in High Performance Optical Fiber
POLYMICRO TECHNOLOGIES, INC.
3035 N. 33rd Drive, Phoenix, AZ 85017
(602) 272-7437 FAX (602) 278-1776
CIRCLE 33 ON READER INQUIRY CARD

Industrial Laser IQL

Nd : YAG 1200 W

With the IQL modular lasers series, QUANTEL is the partner in the development of new industrial applications. The different units already installed on production lines and the studies carried out in industrial process research centres have shown the high abilities of a single beam delivering more than 1 kilowatt at a wavelength of 1.06 µm.

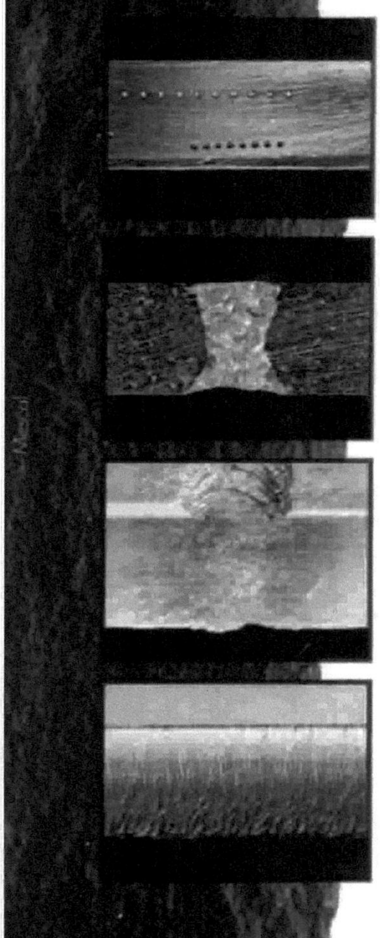

Drilling in cobalt (diameter 0,3 mm, thickness 2 mm)

Welding in titanium (thickness 3 mm).

Welding in titanium (thickness 6 mm).

Cutting in aluminium (thickness 3 mm)

quantel

17, avenue de l'Atlantique - ZA de Courtabœuf - BP 23
91941 Les Ulis Cedex - France
Tél : (1) 69 29 17 00 - Télex : 601329 F - Fax : (1) 69 29 17 29

CIRCLE 34 ON READER INQUIRY CARD

SECTION 4

Related products & services

Suppliers of replacement parts & materials

Beam-delivery equipment (nozzles and accessories)

Edinburgh Instruments Ltd., Riccarton, Currie, Edinburgh UK EH14 4AP; Tel: 031-449-5844; FAX: 031-449-5848

Laser Machining Inc., 500 Laser Dr, Somerset, WI 54025; Tel: 715-247-3285; FAX: 715-247-5650

Laser Mechanisms Inc., PO Box 2064 Southfield, MI 48037; Tel: 313-474-9480; FAX: 313-474-9277

MPB Technologies Inc., 1725 TransCanada Hwy, Dorval, Quebec Canada H9P 1J1; Tel: 514-683-1490; FAX: 514-683-1727

V&S Scientific Ltd., Brosanhouse, Darkes Lane, Pottersbar, Hertfordshire, UK EN61BW; Tel: 0707-45293

Weidmuller Sensor Systems, 5400 West Elm St., McHenry, IL 60050; Tel: 815-344-4141; FAX: 815-344-4152

Beam-delivery fibers

CeramOptec Inc., 188 Moody St., Enfield, CT 06082; Tel: 203-763-4855

Cheval Freres SA, PO Box 3004, Besancon Cedex, 25045 France; Tel: 81.53.75.33; FAX: 81.53.72.39

Ensign Bickford Optics Co, 150 Fisher Dr, Avon, CT 06001; Tel: 203-678-0371; FAX: 203-674-8818

Fiberguide Industries, 1 Bay St., Stirling, NJ 07980; Tel: 201-647-6601; FAX: 201-647-8467

General Fiber Optics Inc., 98 Commerce Rd, Cedar Grove, NJ 07009; Tel: 201-239-3400; FAX: 201-239-4258

Lasag Corp., 702 W. Algonquin Rd, Arlington Heights, IL 60005; Tel: 708-593-3021; FAX: 708-593-5062

Laser Power Optics, 12777 High Bluff Dr, San Diego, CA 92130; Tel: 619-755-0700; FAX: 619-259-9093

Mitsubishi Cable America Inc., 520 Madison Ave, New York, NY 10022; Tel: 212-888-2270

Polymicro Technologies Inc., 3035 N. 33rd Dr, Phoenix, AZ 85017; Tel: 602-272-7437; FAX: 602-278-1776

3M Specialty Optical Fibers, 420 Frontage Rd, W. Haven, CT 06516-41900.; Tel: 203-934-7961; FAX: 203-932-3883

Beam-focusing equipment

Laser Mechanisms Inc., PO Box 2064, Southfield, MI 48037; Tel: 313-474-9480; FAX: 313-474-9277

Weidmuller Sensor Systems, 5400 West Elm St., McHenry, IL 60050; Tel: 815-344-4141; FAX: 815-344-4152

Beam-profiling instrumentation

Big Sky Software Corp., PO Box 3220, Bozeman, MT 59772-3220; Tel: 406-586-0456; FAX: 406-586-5462

Photon Inc., 970 University Ave, Los Gatos, CA 95030; Tel: 408-354-5600; FAX: 408-354-7138

Spiricon Inc., 2600 N. Main, Logan, UT 84321; Tel: 801-753-3729; FAX: 801-753-5231

Cooling equipment

Advantage Engineering Inc., 525 E. Stop 18 Rd, Greenwood, IN 46142; Tel: 317-887-0729

Application Engineering, 801 AEC Dr, Wood Dale, IL 60106; Tel: 312-595-6641

Astro Div, Harsco Corp., 3225 Lincoln Way WeSt., Wooster, OH 44691; Tel: 216-264-8639

Barnstead Co, 160 Wells Ave, Newton, MA 02159; Tel: 617-969-8400

Filtrine Mfg Co, 10 Main St., Harrisville, NH 03450; Tel: 603-827-3321

FTS Systems Inc., PO Box 158, Stone Ridge, NY 12484; Tel: 914-687-0071; FAX: 914-687-7481

Koolant Koolers, 2625 Emerald Dr, Kalamazoo, MI 49001-4542; Tel: 616-349-6800; FAX: 616-349-8951

Lytron Inc., 55 Dragon Ct, Woburn, MA 01801-1039; Tel: 617-933-7300

Neslab Instruments Inc., PO Box 1178, Portsmouth, NH 03802-1178; Tel: 603-436-9444; FAX: 603-436-8411

Rapp Machinery, 2805 Mitchell, Bldg 500, Greenville, TX 75401; Tel: 214-455-2311

Remcor, 500 Regency Dr, Glendale Heights, IL 60139,; Tel: 312-671-7140

Spirec, 141 Lanza Ave, Garfield, NJ 07026; Tel: 201-478-1701; FAX 201-478-0972

Flashlamps for solid-state lasers

AIMS Laser Products, 455 Aldo Ave, Santa Clara, CA 95054; Tel: 408-980-0801; FAX: 408-980-9859

Candela Laser Corp., 530 Boston Post Rd, Wayland, MA 01778; Tel: 508-358-7637; FAX: 508-358-5602

Cascade Laser Corp., 7720 SW Nimbus Ave, Beaverton, OR 97005; Tel: 800-443-5561; FAX: 503-641-4363

Directed Light, 1270 Lawrence Station Rd, Suite A, Sunnyvale, CA 94089; Tel: 408-745-7300; FAX: 408-745-7726

Edinburgh Instruments Ltd., Riccarton, Currie, Edinburgh, UK EH14 4AP; Tel: 031-449-5844; FAX: 031-449-5848

EG&G Inc., Electro-Optics Div, 35 Congress St., Salem, MA 01970; Tel: 508-745-3200; FAX: 508-745-0894

Electro Scientific Industries Inc., 13900 NW Science Park Dr, Portland, WA 97229; Tel: 503-641-4141; FAX: 503-643-4873

Heimann GmbH, Weher Koeppel 6, PO Box 3007, Wiesbaden, FRG 6200; Tel: 49-611-4920; FAX: 49-611-492260

Heraeus-Noblelight Ltd., 161 Science Park, Milton Rd, Cambridge, UK CB4 4GQ; Tel: 44 0223-423324; FAX: 44 0223 423999

ILC Technology Inc., 399 Java Dr, Sunnyvale, CA 94089; Tel: 800-347-2474; FAX: 408-744-0829

Kentek, 4 Depot St., Pittsfield, NH 03263; Tel: 603-435-7201; FAX: 603-435-7441

Q-Arc Ltd., 1 Trafalgar Way, Bar Hill, Cambridge, UK CB3 8SQ; Tel: 0954-82266; FAX: 0954-782993

The Specialty Bulb Co, 345A Central Ave, PO Box 231, Bohemia, NY 11716-0231; Tel: 516-589-3393; FAX: 516-563-3089

TJS Inc., 9 Mt Pleasant Trnpk, PO Box 95, Denville, NJ 07834; Tel: 201-575-3527

VBI Technologies, 11850 Kemper Rd, Auburn, CA 95603; Tel: 916-823-9999; FAX: 916-823-9094

Verre et Quartz, 30 Route d'Aulnay, Bondy, France F-93147; Tel: 1 48 49 7421; FAX: 1 48 48 4422

Xenon Corp., 20 Commerce Way, Woburn, MA 01801; Tel: 617-938-3594; FAX: 617-933-8804

Gases

L'Air Liquide Headquarters, 75 Quai D'Orsay, Paris, Cedex, 07, France, F-75321; Tel: 40 62 55 55

Air Products & Chemicals Inc., 7201 Hamilton Blvd, Allentown, PA 18195; Tel: 215-481-8257; FAX: 215-481-5036

Airco Gases, 575 Mountain Ave, Murray Hill, NJ 07974; Tel: 908-771-1882; FAX: 908-771-1460

Als Dansk Ilt-& Brintfabrik, 29 Scandiagade, Copenhagen, Denmark, SV 2450; Tel: 453121 8840; FAX: 453121 8899

BOC Ltd., The Priestly Ctr, Surrey Research Pk, Guildford, Surrey, UK GU2 5XY; Tel: 048-579857

Carboxyque Francaise, 91 rue du Faubourg, St. Honore, Paris, France 75362; Tel: 161 266 92 30

Cryogenic Rare Gas Labs Inc., 913 Commerce Circle, Hanahan, SC 29406; Tel: 803-747-0956; FAX: 803-747-0958

Icon Services Inc., 19 Ox Bow Lane, Summit, NJ 07901; Tel: 201-273-0449; FAX: 201-273-0449

Isotec Inc., 3858 Benner Rd, Miamisburg, OH 45342; Tel: 513-859-1808

Linde Div, Union Carbide Corp., 200 Cottontail Lane, PO Box 6744, Somerset, NJ 08873; Tel: 800-982-0030

Line Lite Laser Corp., 430 Ferguson Dr Bldg 4, Mountain View, CA 94042; Tel: 415-969-4900; FAX: 415-969-5480

Liquid Air Corp., 2121 N California Blvd, Suite 350, Walnut Creek, CA 94596; Tel: 415-977-6504; FAX: 415-977-6261

Liquid Carbonic Corp., 135 S. LaSalle St., Chicago, IL 60603; Tel: 312-855-2500

Matheson Gas Products, 30 Seaview Dr, Secaucus, NJ 07094; Tel: 201-867-4100

MG Industries, 2460 Blvd of the Generals, Valley Forge, PA 19482; Tel: 215-630-5492

Scientific Gas Products, 2330 Hamilton Blvd, South Plainfield, NJ 07080; Tel: 201-754-7700

Scott Environmental Tech, Rte 611 North, Plumsteadville, PA 18949; Tel: 215-766-8861

Southland Cryogenics Inc., PO Box 110669, Carrollton, TX 75011; Tel: 214-243-1311; FAX: 214-243-1370

Spectra Gases Inc., 277 Coit St., Irvington, NJ 07111; Tel: 201-372-2060; FAX: 201-372-8551

Industrial laser optics

Acton Research, PO Box 2215, Acton, MA 01720; Tel: 508-263-3584; FAX: 508-263-5086

Applied Light Inc., 3640 Main St., Springfield, MA 01107; Tel: 413-731-8818; FAX: 413-737-1218

Boston Electronics, 72 Kent St., Brookline, MA 02146; Tel: 617-566-3821

Coherent Components Group, 2301 Lindbergh St., Auburn, CA 95603; Tel: 916-823-9550

CVI Laser Corp., PO Box 11308, 200 Dorado Place SE, Albuquerque, NM 87192; Tel: 505-296-9541; FAX: 505-298-9908

Davin Optical Ltd., 9A Chester Rd, Borehamwood, UK WD6 1LD; Tel: 081 9051414; FAX: 081-2076581

Hoya Optics Inc., 3400 Edison Way, Fremont, CA 94538; Tel: 415-490-1880; FAX: 415-490-1988

Janos Technology Inc., HCR#33, Box 25, Townshend, VT 05353-7702; Tel: 802-365-7714; FAX: 802-365-4596

Laser Optics Inc., PO Box 127, Danbury, CT 06813; Tel: 203-744-4160; FAX: 203-798-7941

Laser Power Optics, 12777 High Bluff Dr, San Diego, CA 92130; Tel: 619-755-0700; FAX: 619-259-9093

Lincoln Laser Co, 234 E. Mohave St., Phoenix, AZ 85004; Tel: 602-257-0407

Melles Griot/Optics, 55 Science Parkway, Rochester, NY 14620; Tel. 716-244-7220; FAX 716-244-6292

Ophir Optronics Inc., 21 Main St., No Reading, MA 01864; Tel: 508-664-1176; FAX: 508-664-0309

Optics for Research, PO Box 82, Caldwell, NJ 07006; Tel: 201-228-4480; FAX: 201-228-0915

Optovac, E. Brookfield Rd, North Brookfield, MA 01535; Tel: 508-867-6444; FAX: 508-867-8349

Spawr Optical Research Inc., PO Box 1899 Corona, CA 91718-1899; Tel: 714-735-0433; FAX: 714-735-4123

Spectra-Physics Optics Corp., 1250 W Middlefield Rd, Mountain View, CA 94039-7013; Tel: 415-961-2550; FAX: 415-969-7939

Spindler & Hoyer, 459 Fortune Blvd, Milford, MA 01757; Tel: 508-478-6200; FAX: 508-478-5980

II-VI Inc., 375 Saxonburg Blvd, Saxonburg, PA 16056; Tel: 412-352-1504; FAX: 412-352-4980

V&S Scientific Ltd., Brosanhouse, Darkes Lane, Pottersbar, Hertfordshire, UK EN6 1BW; Tel: 0707 45293

Laser rods for solid-state lasers

AIMS Laser Products, 455 Aldo Ave, Santa Clara, CA 95054; Tel: 408-980-0801; FAX: 408-980-9859

Cascade Laser Corp., 772 SW Nimbus Ave, Beaverton, OR 97005; Tel: 800-443-5561; FAX: 503-641-4363

Directed Light Inc., 1270 Lawrence Station Rd, Suite A, Sunnyvale, CA 94089; Tel: 408-745-7300; FAX: 408-745-7726

Kentek Corp., 4 Depot St., Pittsfield, NH 03263; Tel: 603-435-7201; FAX: 603-435-7441

Kigre Inc., 100 Marshland Rd, Hilton Head, SC 29926; Tel: 803-681-5800; FAX: 803-681-4559

Laser Crystal Corp., 154 Edison Rd, Lake Hopatcong, NJ 07849; Tel: 201-663-1322; FAX: 201-663-0721

Lasermetrics Inc., 196 Coolidge Ave, Englewood, NJ 07631; Tel: 201-894-0550; FAX: 201-894-8829

Laser Photonics Inc., 12351 Research Pkwy, Orlando, FL 32826; Tel: 407-281-4103; FAX: 407-281-4114

Laser SOS Limited, 4B Bartholomew's Walk, Cambridgeshire Business Park, Angel Drove, Ely Cambridge, UK CB7 4EAG; Tel: 0353-666334; FAX: 0353-666375

Lasertech Industries, 3173 Texas Ave, Simi Valley, CA 93063; Tel: 805-583-3406; FAX: 818-889-5605

Lee Laser Inc., 3718 Vineland Rd, Orlando, FL 32811; Tel: 407-422-2476; FAX: 407-839-0294

Litton Airtron, 1201 Continental Blvd, PO Box 410168, Charlotte, NC 288241-0168; Tel: 704-588-2340; FAX: 704-588-2516

Photox Optical Systems, PO Box 274 Headington, Oxford, UK OX3 0BJ; Tel: 0 865 64563; FAX: 0-865-741778

Quantronix Corp., 49 Wireless Blvd, PO Box 9014, Smithtown, NY 11787; Tel: 516-273-6900; FAX: 516-273-6958

Raytheon Laser Products, 465 Center St., Quincy, MA 02169; Tel: 617-479-5300; FAX: 617-472-5084

Roditi International Corp., Carrington House, 130 Regent St., London, UK W1R 6BR; Tel: 071-439-4390; FAX: 071-434-0896

TJS Inc., 9 Mt. Pleasant Tpk, PO Box 95, Denville, NJ 07834; Tel: 201-575-3527

Positioning equipment

Aerotech, 101 Zeta Dr, Pittsburgh, PA 15238; Tel: 412-963-7470; FAX: 412-963-7459

Ambrit Inc., 231 Andover St., PO Box 538, Wilmington, MA 01887; Tel: 508-658-2291

Anorad Corp., 110 Oser Ave, Hauppauge, NY 11788; Tel: 516-231-1995; FAX: 516-435-1612

Daedal, PO Box 500, Sandy Hill Rd, Harrison City, PA 15636; Tel: 800-245-6903; FAX: 412-744-7626

Dover Instrument Corp., 200 Flanders Rd, PO Box 200, Westborough, MA 01581; Tel: 508-366-1456; FAX: 508-366-9774

Klinger Scientific, 999 Stewart Ave, Garden City, NY 11530; Tel: 516-745-6800; FAX: 516-745-6812

Oriel Corp., 250 Long Beach Blvd, Stratford, CT 06497; Tel: 203-377-8282; FAX: 203-378-2457

Schneeberger Inc., 7 DeAngelo Dr, Bedford, MA 01730; Tel: 617-271-0140; FAX: 617-275-4794

Power/energy meters

Coherent Components, 2301 Lindbergh St., Auburn, CA 95603; Tel: 916-823-9550

Gentec, 2627 Dalton St., Ste. Foy, Quebec, Canada G1P 3S9; Tel: 418-651-8000; FAX: 418-651-6695

International Light, 17 Graf Rd, Newburyport, MA 01950; Tel: 508-465-5923; FAX: 508-462-0759

Laser Precision Corp., 1231 Hart St., Utica, NY 13502; Tel: 315-797-4492; FAX: 315-797-0696

Molectron Detector Inc., 7470 SW Bridgeport Rd, Portland, OR 97224; Tel: 503-620-9069; FAX: 503-620-8964

nm Laser Products, 140 San Lazaro Ave., Sunnyvale, CA 94086; Tel.: 408-733-1520; FAX: 408-736-1152

Ophir Optronics, 21 Main St., North Reading, MA 01864; Tel: 508-664-1176; FAX: 508-664-0309

Spectra Physics Inc., 3333 North First St., San Jose, CA 95134-1955; Tel: 408-432-3333; FAX: 408-432-0203

Xenon Corp., 20 Commerce Way, Woburn, MA 01801; Tel: 617-938-3594; FAX: 617-933-9804

Repairs/service

Laser Tech Industries, 3173 Texas Ave, Simi Valley, CA 93063; Tel: 805-583-3406; FAX: 818-889-5605

Safety equipment

American Allsafe Co, 99 Wells Ave, Tonawanda, NY 14150-5104; Tel: 716-695-8300; FAX: 716-695-6905

Echo Engineering, PO Box 1595, Grants, NM 87020; Tel.: 302-774-1000; FAX: 302-773-5876

Fred Reed Optical Inc., PO Box 27010, SE B7106 Albuquerque, NM, 87125-7010; Tel: 505-265-3531; FAX: 505-266-8507

General Scientific Safety Equip Co, 1821 JF Kennedy Blvd, Philadelphia, PA 19103; Tel: 215-504-6366

Glendale Protective Tech Inc., 130 Crossways Park Dr, Woodbury, NY 11797; Tel: 516-921-5800; FAX: 516-364-3212

Laser Peripherals Inc., 10395 W 70th St., Eden Prairie, MN 55344; Tel: 800-966-5273

Lase-R-Shield Inc., PO Box 91957, Albuquerque, NM 87199; Tel: 800-288-1164; FAX: 505-888-4627

Rockwell Laser Industries, 7754 Camargo Rd, PO Box 43010 Cincinnati, OH 45243; Tel: 513-271-1568; FAX: 513-271-1598

Uvex Safety Inc., 10 Thurber Blvd, Smithfield, RI 02917; Tel: 401-232-1200; FAX: 401-232-1830

Wilson Industries, 2112 Santa Anita, South El Monte, CA 91733; Tel: 818-444-7781

Used laser sales

Excitek Inc., 277 Coit St., Irvington, NJ 07111; Tel.: 201-372-1669, 800-932-0624; FAX: 201-372-8551

Holo Spectra, 7742 B Gloria Ave, Van Nuys, CA 91406; Tel: 818-994-9577; FAX: 818-994-4709

Laser Automation Inc., 10357 Kinsman Rd, PO Box 373, Newbury, OH 44065; Tel: 216-543-9291

Laser Resale Inc., 54 Balcom Rd, Sudbury, MA 01776; Tel: 508-443-8484; FAX: 508-443-7620

LPT Laser Physiktecnik, Leobersdorfer str 26, Industrieareal, Obj 85, Berndorf Vienna, Austria 2560; Tel: 02672 34880; FAX: 02672-5164

MWK Industries, 1269 Pomona Rd #110, Corona, CA 91720; Tel: 714-278-0563; FAX: 714-278-4887

Subject index

U

V

W

X

Y

Z

Index to corporate sponsors

Advertisements from leading suppliers of industrial laser equipment can be found in four special color sections with pages denoted as A1–A8, B1–B8, C1–C8, and D1–D12.

Further information on these companies and industrial products can be obtained by circling the corresponding numbers on the Reader Service Cards.

MIX
Papier aus verantwortungsvollen Quellen
Paper from responsible sources
FSC® C105338

If you have any concerns about our products,
you can contact us on
ProductSafety@springernature.com

In case Publisher is established outside the EU,
the EU authorized representative is:
Springer Nature Customer Service Center GmbH
Europaplatz 3, 69115 Heidelberg, Germany

Printed by Libri Plureos GmbH
in Hamburg, Germany